LONG-TERM ENERGY RESOURCES

LONG-TERM ENERGY RESOURCES

Volume I

An International Conference

Sponsored by

The United Nations Institute for
Training and Research (UNITAR)

and

Petro-Canada

in cooperation with

United Nations Development Programme
Canadian International Development Agency
Petroleos de Venezuela
Petroleos Mexicanos
Hydro-Quebec
Compagnie Francaise des Petroles

November 26 - December 7, 1979
Montreal, Quebec, Canada

Joseph Barnea
Scientific Secretary

R. F. MEYER
Editor

J. C. OLSON
Associate Editor

Pitman

Boston · London · Melbourne · Toronto

PITMAN PUBLISHING INC.
1020 Plain Street
Marshfield, Massachusetts 02050

PITMAN BOOKS LIMITED
39 Parker Street
London WC2B 5PB

Associated Companies:
Pitman Publishing Pty Ltd., Melbourne
Pitman Publishing New Zealand Ltd., Wellington
Copp Clark Pitman, Toronto

Library of Congress Cataloging in Publication Data
Main entry under title:

Long-term energy resources.

Includes index.
1. Energy industries--Congresses. 2. Energy development--Congresses. I. Meyer, R. F. (Richard F.), 1921- . II. Olson, Jerry Chipman, 1917- III. United Nations Institute for Training and Research. IV. Petro-Canada.
HD9502.A2L66 333.79 81-11902
ISBN 0-273-08534-4 (set) AACR2

Manufactured in the United States of America

10 9 8 7 6 5 4 3 2 1

LIST OF PARTICIPANTS

AFGHANISTAN

Mr. Ahmad Arsalan
Engineer
Mineral Exploitation Department
Ministry of Mines and Industries
Government of Afghanistan
Kabul, Afghanistan

ALGERIA

M. Mohamad Larbi Kateb
Juriste - SONATRACH
46, Boulevard Mohamed V
Alger, Algeria

AUSTRALIA

Mr. J. B. Carter
Principal Advisor
National Energy Office
Canberra, A.C.T. Australia

Dr. Ron J. Knight
Counsellor (Atomic Energy)
Embassy of Australia
1601 Massachusetts Avenue, N.W.
Washington, D.C. 20036 USA

AUSTRIA

Dr. Rudolf Partl
Alseggerstrasse 18
A-1180
Wien, Austria

BAHAMAS

Mr. James Smith
Deputy Permanent Secretary
Ministry of Economic Affairs
P.O. Box N-4596
Nassau, Bahamas

BANGLADESH

Mr. A. K. Rashid Ahmed
Deputy Secretary
Government of Bangladesh
Ministry of Petroleum and Mineral
Resources
35, Road No. 12 A
Dhanmondi RA,
Dacca 5, Bangladesh

Mr. Mozammel Haque
Director of Planning Research and Gas
Petrobangla
Motisheel Commercial Area
Dacca, Bangladesh

BARBADOS

Mr. W. A. Cox
Deputy Research Director
Central Bank of Barbados
P.O. Box 1016
Treasury Building
Bridgetown, Barbados

Mr. Frank McConney
Managing Director
Barbados Light and Power Company
P.O. Box 142
Bridgetown, Barbados

BELGIUM

Prof. R. Van Den Damme
Directeur de la Société Intercom
1, Place du Trône
1000 Bruxelles, Belgique

BENIN

M. Janvier Elégrédé
Ingénieur
SONACOP
B.P. 975
Cotonou, Benin

M. Emile Louis Paraiso
Directeur Général de la Société
Beninoise d'Electricité et d'Eau
B.P. 73
Cotonou, Benin

BRAZIL

Mr. Amaldo V. de Carvalho Jr.
Centro de Tecnologia Promon
Praia do Flamengo 154
22210 Rio de Janeiro, Brazil

Mr. Marcio de Paula Fernandes
Coordinator of Technological
Development
Secretary of Industrial Technology
Government of Brazil
SAS, Quadra 2
Brasilia, Brazil

Mr. Thelmo Dutra de Rezende
Vice-President, Petrobras
Avenida Republica do Chile
65-23 Andar
Rio de Janeiro, Brazil

Mr. Sergio C. Trindade
Executive Director
Centro de Tecnologia Promon
Praia do Flamengo, 154 - 12° andar
20.000 Rio de Janeiro, Brazil

Mr. Jose Roberto Moreira
Instituto de Fisica
Universidade de Sao Paulo
C.P. 20516
Sao Paulo, S. P. Brazil

BURMA

Lt. Colonel Maung Maung Aye
Chief Engineer
Electric Power Corporation
197-199 Lower Kemmen-dine Road
Rangoon, Burma

Mr. U Than Nyunt
Director, Petroleum Operations
Myanma Oil Corporation
604 Merchant Street
P.O. Box 650
Rangoon, Burma

Dr. Mehm Thet San
Director General
Central Research Organization
Ministry of Industry
Kaba Aye Pagoda Road
Rangoon, Burma

BURUNDI

M. Louis Kagisye
Division de l'Energie et des Mines
B.P. 224
Bujumbura, Burundi

M. Albert Mukubano
Division de l'Energie et des Mines
B.P. 745
Bujumbura, Burundi

M. Evariste Simbarakiye
Directeur Technique de la REGIDESO
B.P. 660
Bujumbura, Burundi

CANADA

Mr. Pierre Albert
Total Eastcan Exploration Ltd.
P.O. Box 9720
1600, 202 - 6th Avenue, S.W.
Calgary, Alberta T2P 2W6 Canada

Mr. P. J. Arnopoulos
Prof. of Political Science
Concordia University
1455 de Maisonneuve Boulevard West
Montreal, P.Q. Canada

Mr. Fred Bain
Vice-President of Technology
Suncor Incorporated
20 Eglinton Avenue West
Toronto, Ontario Canada

Mr. Paul Beaulieu
Policy Analyst
Ministry of State for Science and
Technology
270 Albert Street
Ottawa, Ontario Canada

Mr. John Bechtold
Manager, Crude Oil & Supply
Petro-Canada
P.O. Box 2844
Calgary, Alberta Canada

Mr. Joel Bell
Senior Vice-President
Finance and Planning
Petro-Canada
P.O. Box 2844
Calgary, Alberta Canada

Dr. Roger A. Blais
Directeur de la Recherche
Ecole Polytechnique de Montreal
Case Postale 6079, Succursale "A"
Montreal, P.Q. Canada

M. André Bolduc
Hydro-Québec
Revue Forces
75, boulevard Dorchester ouest
Montreal, P.Q. Canada

M. Jacques M. Bonneville
Directeur, Mécanique et Thermo-Dynamique
Institut de Recherche d'Hydro-Québec (IREQ)
C.P. 1000
Varennes, P.Q. J0L 2P0
Canada

M. Gilles A. Bouchard
Ingenieur A.C.D.I. Canada
200 Promenade Portage
Hull, P.Q. Canada

M. Michel P. Boudriau
Vice-President
Developpement des Marchés
Hydro-Québec International
75, boulevard Dorchester ouest
Montreal, P.Q. H2Z 1A4
Canada

Mr. Douglas Bowie
Director, Environmental and Social Affairs
Petro-Canada
P.O. Box 2844
Calgary, Alberta Canada

M. Robert Boyd
President Directeur General
Hydro-Québec
75, boulevard Dorchester ouest
Montreal, P.Q. Canada

Mr. Leslie Breiner
Senior Project Officer
Canadian International Development Agency (CIDA)
200 Promenade du Portage
Hull, P.Q. Canada

Mr. Tom Brett
Director
Conservation and Renewable Energy Policy
580 Booth Street
Ottawa, Ontario Canada

Mr. D. James Cameron
Manager, Renewable Energy
Imperial Oil Limited
111 St. Clair Avenue West
Toronto, Ontario Canada

Ms. Marilyn J. Cayer
Sr. Policy Advisor
Canadian International Development Agency (CIDA)
200 Principal Street
Hull, P.Q. Canada

M. André Chamberland
Institut de Recherche
Hydro-Québec
C.P. 1000
Varennes, P.Q. J0L 2P0
Canada

Mr. Marc Chappel
Task Co-ordinator
Renewable Energy Research & Development
National Research Council of Canada
Building M-50
Montreal Road
Ottawa, Ontario Canada

Mr. R. H. Clark
Senior Engineer Advisor
Inland Waters Directorate
Environment Canada
Ottawa, Ontario K1A 0E7
Canada

M. Bernard Cloutier
President
SOQUIP
3340 de la Perade
Ste-Foy, P.Q. G1X 2L7
Canada

Dr. J. H. Dagher
General Manager, Corporate Planning
B.P. Canada
1245 Sherbrooke Street West
Montreal, P.Q. Canada

Mr. A. de Hoog
General Energy Policy Division
External Affairs Department
125 Sussex Drive
Ottawa, Ontario Canada

M. Gilles Desautels
Gouvernement du Canada
Environnement Canada
100 rue Alexis Nihon, salle 300
St-Laurent, P.Q. Canada

M. Claude Desjarlais
Ministère de l'Energie du Québec
1050, rue St-Augustin
Quebec, P.Q. Canada

Mr. R. C. Dobbs
Canadian Forestry Service
19th Floor
Place Vincent Massey
Ottawa, Ontario Canada

M. Gilles Dupont
Centre de la Recherche Industrielle
Problème sur l'Energie
333, rue Franquet
C.P. 9038
Ste Foy, P.Q. Canada

Mr. Jim Eckford
Petro-Canada
P.O. Box 2844
Calgary, Alberta T2P 2M7
Canada

Dr. Gordon Edwards
Chairman
CCNR
2010 Mackay Street
Montreal, P.Q. Canada

Mr. John Ferbey
Department of Renewable Resources
Yukon Territory, Canada

Mr. J. Paul Filion
Petro-Canada
External Relations
P.O. Box 2844
Calgary, Alberta T2P 2M7
Canada

Mr. Joel F. Finlay
Special Assistant to Ontario Minister
of Energy
Government of Ontario
56 Wellesley Street West, 12th Floor
Toronto, Ontario Canada

Dr. S. T. Fisher
President
F. T. Fishers' Sons Limited
P.O. Box 396
Montreal, P.Q. H3P 3C6
Canada

Mr. John Foster
Senior Economic Advisor
Petro-Canada
350 Sparks Street, Suite 306
Ottawa, Ontario K1R 7S8
Canada

Mr. Giovanni Gaggi
Canada Integrated Technical
Information
P.O. Box 852
La Cité
Montreal, P.Q. H2W 2P5
Canada

M. Robert Gagnon
Economiste
Petro-Canada
350, rue Sparks, Suite 306
Ottawa, Ontario K1R 7S8
Canada

Mr. Bernard H. Garfunkel
Toronto Stock Exchange
Andras, Hatch & Hetherington Ltd.
4 King Street West
Suite 1100
Toronto, Ontario Canada

M. Marcel Gervais
André Marsan & Associés
1130 ouest, rue Sherbrooke
Montreal, P.Q. H3A 2R5
Canada

Mr. Ralph J. Giguere
Manager, New Energy Resources
Amoco Canada Petroleum Company
Calgary, Alberta Canada

M. T. Gilsig
Directeur, Projects de Centrales
Hydro-Québec
855 est, rue Ste. Catherine
Montreal, P.Q. H2L 4P5
Canada

Mr. John Godfrey
International Energy Development Corporation
2208 Paliswood Place, S.W.
Calgary, Alberta T2V 3R2
Canada

Mr. Richard Gold
Deputy Director
Canadian International Development Agency (CIDA)
7 Saguenay Drive
Aylmer, P.Q. J9J 1A3
Canada

Mr. William Griggs
Senior Resource Economist
18th Floor, P.U.M.
Hull, P.Q. Canada

Mr. P. J. Haines
Director General
Canadian International Development Agency (CIDA)
200 Promenade du Portage
Hull, P.Q. Canada

Mr. R. W. D. Hanbidge
President
B.P. Canada
1245 Sherbrooke Street West
Montreal, P.Q. Canada

Dr. Peter Harrison
Department of Geography
University of Ottawa
78 Laurier Street East
Ottawa, Ontario K1N 6N5
Canada

Dr. James P. Hea
Senior Advisor, Petroleum Resources
Department of Energy & Resources
580 Booth Street
Ottawa, Ontario Canada

Dr. David Henry
Senior Programme Officer
Centre de Recherche Internationale
P.O. Box 8500
Ottawa, Ontario Canada

The Honorable R. J. Hnatyshyn
Minister for Energy, Mines and Resources
580 Booth Street
Ottawa, Ontario K1A 0E4
Canada

Mr. W. H. Hopper
Chairman and Chief Executive Officer
Petro-Canada
P.O. Box 2844
Calgary, Alberta T2P 2M7
Canada

Mr. A. W. Hyndman
General Manager, Upgrading
Syncrude Canada Limited
10030 - 107th Street
Edmonton, Alberta Canada

Mr. Andrew Janisch
President and Chief Operating Officer
Petro-Canada
P.O. Box 2844
Calgary, Alberta T2P 2M7
Canada

Mr. Michael Jarvis
Head, International Energy Group
Energy Power Sector
Department of Energy, Mines and
  Resources
580 Booth Street
Ottawa, Ontario Canada

Mr. Alan M. Jessop
Earth Physics Branch
Department of Energy, Mines and
  Resources
580 Booth Street
Ottawa, Ontario Canada

Mr. Michael Kaire
Polyenergy
P.O. Box 639, Station Outremont
Montreal, P.Q. Canada

Mr. Keshav Kant
Department of Chemical Engineering
McGill University
Sherbrooke & University Street
Montreal, P.Q. Canada

Mr. Joseph Karam
Manager, Corporate Planning Group
Petro-Canada
P.O. Box 2844
Calgary, Alberta Canada

M. Adam Karpiuski
Hydro-Québec
855 est, rue Ste. Catherine
Montreal, P.Q. Canada

Mr. Albert Khazoom
Senior Director, External Relations
Petro-Canada
P.O. Box 2844
Calgary, Alberta T2P 2M7
Canada

Mr. Roger Khazoom
1411 Fort, Suite 2506
Montreal, P.Q. H3H 2N7
Canada

Mr. Paul Krukowski
Director, Bilateral Programs Advisory
  Group
Canadian International Development
  Agency (CIDA)
200 Principale Street
Hull, P.Q. Canada

Ms. Hélène Lajambe
Economiste
1599, rue Bourdon
St. Bruno, P.Q. Canada

M. André Latulipe
Ingénieur Conseil
3375, rue Toupin
St-Laurent, P.Q. Canada

M. V. Bernard Lavigueur
President
Ecole Polytechnique de Montreal
C.P. 6079
Montreal, P.Q. Canada

Mr. Thomas A. Lawand
Director
Brace Research Institute
MacDonald College of McGill
  University
1 Stewart Park
Ste Anne de Bellevue, P.Q. Canada

M. Roger Lepage
Directeur de la Production
Hydro-Quebéc
75 ouest, boulevard Dorchester
Montreal, P.Q. Canada

Mr. Gerry Lewis-Steer
General Manager, Special Projects
Petro-Canada
P.O. Box 2844
Calgary, Alberta Canada

Ms. Diane Loucks
Special Assistant, Environmental and Social Affairs
Petro-Canada
P.O. Box 2844
Calgary, Alberta Canada

Mr. John Lucus
Domtar Incorporated
Senneville, P.Q. Canada

Mr. Michael E. Lynch
Market Analyst
275 Slater Street
Ottawa, Ontario K1A 0S4
Canada

Mr. Joseph Martinelli
Group Vice President, Production
Petro-Canada
P.O. Box 2844
Calgary, Alberta Canada

Mr. George Matheson
Friends of the Earth
P.O. Box 569, Station B
Ottawa, Ontario K1A 5PC
Canada

Mr. John McMillan
Manager, Transportation Department
Petro-Canada
P.O. Box 2844
Calgary, Alberta Canada

Mr. H. B. Merlin
Senior Advisor, Uranium and Nuclear Energy
Department of Energy, Mines and Resources
580 Booth Street
Ottawa, Ontario Canada

Dr. Morris Miller
Senior Economic Advisor
Department of Energy, Mines and Resources
580 Booth Street, Room 2160
Ottawa, Ontario Canada

Mr. John Neate
John Neate & Associates
Energy Planning Consultants
366 Sunnyside Avenue
Toronto, Ontario Canada

Mr. A. Niitenberg
Vice President
Ontario Hydro
700 University Avenue
Toronto, Ontario Canada

Mr. Bob Niven
Vice President, Planning
Petro-Canada
P.O. Box 2844
Calgary, Alberta T2P 2M7
Canada

Mr. Henry Oliver
Petro-Canada
P.O. Box 2844
Calgary, Alberta T2P 2M7
Canada

Dr. Ralph Overend
National Research Council of Canada
Montreal Road, Building M-50
Ottawa, Ontario K1A 0R6
Canada

Mr. K. J. Parbhakar
I.N.R.S. - Energie
1650 Montée Ste Julie
Varennes, P.Q. Canada

Mr. Truman Phillips
Research Associate
Prom Technology Center
School of Agricultural Economics
University of Guelph
Guelph, Ontario N1G 2W1
Canada

Mr. J. Ploeg
Hydraulics Laboratory
National Research Council of Canada
Montreal Road, Building M-32
Ottawa, Ontario K1A 0R6
Canada

M. Gaby Polisois
Ministère de l'Energie et Resources
1305, chemin Ste Foy
Quebec, P.Q. Canada

Mr. Adrian M. Poplawski
Senior Policy Advisory
Bilateral Programs Branch
Canadian International Development Agency (CIDA)
200 Principale Street
Hull, P.Q. Canada

Dr. Andrija Puharich
Planetary Association for Clean Energy
100 Bronson Avenue, Suite 1001
Ottawa, Ontario K1R 6G8
Canada

Mr. Peter Quinn
Operations Manager
Petro-Canada
P.O. Box 2844
Calgary, Alberta Canada

Mr. Don Quinsey
Policy Advisor
Ministry of State for Science and Technology
270 Albert Street
Ottawa, Ontario Canada

Mr. David L. Robinson
Environment Canada
Environmental Management Service
Ottawa, Ontario K1A 0H3
Canada

Prof. John Runnalls
Energy Studies
University of Toronto
Toronto, Ontario M5S 1A1
Canada

Mr. James Scurr
Vice-President, Human Resources
Petro-Canada
P.O. Box 2844
Calgary, Alberta Canada

Mr. Walter Seaton
Marketing Department
Stone & Webster Canada Limited
Yonge-Eglinton Centre
2300 Yonge Street
Toronto, Ontario Canada

Mr. Adel Shelaby
Petroleum Economist
Department of Energy, Mines and Resources
580 Booth Street
Ottawa, Ontario Canada

Mr. Donald Sinclair
Department of External Affairs
125 Sussex Drive
Ottawa, Ontario Canada

Mr. J. Graham Smith
A.E.C.L.
275 Slater Street
Ottawa, Ontario Canada

Mr. James Stanford
General Manager, Conventional Production
Petro-Canada
P.O. Box 2844
Calgary, Alberta Canada

Mr. E. R. Stoian
Science Advisor
Science Council of Canada
100 Metcalf Street
Ottawa, Ontario Canada

Mr. Maurice F. Strong
Chairman
International Energy Development Corporation
7 Charles Street
London WLX 7HB England

Ms. Barbara Summers
Bioenergy Program
National Research Council
Room 206, Building M-50
Montreal Road
Ottawa, Ontario Canada

Mr. J. F. Tanguay
Director, United Nations Political and Institutional Affairs Division
External Affairs Department
125 Sussex Drive
Ottawa, Ontario Canada

Mr. R. J. Templin
Head, Low Speed Aerodynamics
National Research Council
Montreal Road, Laboratory M-2
Ottawa, Ontario Canada

Mr. Don Towson
Manager, Technical Services
Petro-Canada
P.O. Box 2844
Calgary, Alberta Canada

Mr. David Tregunno
Market Analyst
275 Alater Street
Ottawa, Ontario K1A 0S4
Canada

M. Yvon Turcotte
Ingenieur
Departement de l'Energie
820, rue Eymard
Quebec, P.Q. Canada

Mr. Ian Watson
Minister of Parliament
House of Commons
Ottawa, Ontario Canada

Dr. T. S. Wauchop
Manager, Energy Studies
Union Gas Limited
50 Keil Drive North
Chatham, Ontario Canada

Ms. Mardi Weltman
Policy Advisor
Canadian International Development Agency (CIDA)
Bilaterial Programs Branch
200 Principale Street
Hull, P.Q. Canada

Mr. Roger Wilson
Advisor, Policy Branch
Canadian International Development Agency (CIDA)
200 rue Principale
Hull, P.Q. Canada

CHAD

M. Mahamet Saleh Alhabbo
Ingenieur des Mines
Direction des Mines et de la Geologie
B.P. 816
Ndjamena, Chad

M. Brahim Ngarmane
Ingenieur Petro-Chimiste Technologue
Direction des Mines et de la Geologie
B.P. 816
Ndjamena, Chad

CHILE

Mr. Patricio De Solminihac Tampier
Comision Nactional de Energia
Ingeniero Jefe Projecto
Teatinos 120, Piso 14
Santiago, Chile

CHINA, PEOPLE'S REPUBLIC OF

Mr. Chen Xi
Leading Member of the Fifth Bureau
Energy and Natural Resources
Chinese Academy of Sciences
Beijing, China

Dr. Zhu Xia
Chief Geologist
Department of Petroleum Geology
Ministry of Geology
Beijing, China

Prof. Yang Qi
Post-Graduate Department of Wuhan College of Geology
Xue Yuan Road
Beijing, China

Mr. Bao Han-Chen
Associate Professor
Deputy Director, The Institute
of Coal Chemistry
Chinese Academy of Sciences
Ying-ze Street
Taiyuan, Shanxi, China

Mr. Chen Guan-Qian
Research Associate
M.A. Agricultural Economy
Deputy Chief of Information Division
Chengdu Institute of Biology
Chinese Academy of Sciences
Hua-xi Campus
Chengdu, China

Mr. Li Nian-Guo
Research Associate
Head of Information Division
Guangzhou Institute of Energy Sources
Chinese Academy of Sciences
81 Central Martyrs' Road
Guangzhou, China

COLOMBIA

Mr. Fernando Lecaros
Instituto SER de Investigacion
Apartado Aereo 55089
Bogota, Colombia

Mr. Hugo Salazar Marciales
Assistant Professor
Chemical Engineer
Dpto. Procesos Quimicos
Universidad Del Valle
Apartado Aereo 2188
Cali, Colombia

CONGO

M. Paul-Robert Mangouta
Attaché économique à la Présidence
de la République, Pétrole et Mines
B.P. 1256
Brazzaville, Congo

COSTA RICA

Ing. Agustin Rodriguez Matamoros
Instituto Costarricense de
Electricidad
Apartado 10032
San Jose, Costa Rica

Mr. Frank Schnabel
Consulate of Costa Rica
1170 Peel Street, Suite 1172
Montreal, P.Q. H3B 2T4
Canada

CYPRUS

Mr. George Christodoulides
Director, Higher Technical Institute
Government of Cyprus
P.O. Box 2423
Nicosia, Cyprus

DEMOCRATIC YEMEN

Mr. Abubaker Ahmed Ayash
Head, Energy Section
Ministry of Planning
Government of the People's Democratic
Republic of Yemen
Aden, Yemen

Mr. Mohamed Salem Bamatraf
Chief Geophysicist
Petroleum and Minerals Board
Government of the People's Democratic
Republic of Yemen
Aden, Yemen

DENMARK

Mr. Bent Sørensen
Associate Professor
Niels Bohr Institute
Blegdamsvej 17, DK-2100
Copenhagen, Denmark

ECUADOR

Mr. Eduardo Moran Fierro
Advisor, Instituto Nacional
de Energia
Sta Prisca #223
Quito, Ecuador

FRANCE

M. Claude Bienvenu
Directeur Adjoint des Etudes et
Recherches d'Electricité de France
2, rue Louis Murat
75008 Paris, France

M. M. Catta
Total Coopération Industrielle
Compagnie Française des Pétroles
5, rue Michel-Ange
75781 Paris Cedex 16
France

M. Carlos Cramez
Total C.F.P.
39-43 Quai A. Citroen
75739 Paris Cedex 15
France

Ms. Patricia M. Jurisic
Corporate Services Manager
Total Eastcan Exploration Limited
1600, 202 - 6th Avenue, S.W.
Calgary, Alberta Canada

M. Jean d'Herbès
President
Total Coopération Industrielle
Compagnie Française des Pétroles
5, rue Michel-Ange
75781 Paris Cedex 16
France

M. Pierre Maréchal
Vice-President
Finance et Administration
Total Coopération Industrielle
Compagnie Française des Pétroles
5, rue Michel-Ange
75781 Paris Cedex 16
France

M. H. Montaige
Total Coopération Industrielle
Compagnie Française des Pétroles
5, rue Michel-Ange
75781 Paris Cedex 16
France

M. Philippe Saint Andrew
Total Coopération Industrielle
Compagnie Française des Pétroles
5, rue Michel-Ange
75781 Paris Cedex 16
France

THE GAMBIA

Mr. Amadou Taal
Ministry of Economic Planning and
Industrial Development
Government of the Republic of
the Gambia
Banjul, Gambia

GERMANY, FEDERAL REPUBLIC OF

Mr. E. O. Kantelberg
Veba A.G.
Energy Staff
4000 Dusseldorf 30
Karl-Arnold-Platz 3
Germany

Mr. Hans Weise
Vice President
Rheinbraun-Consulting U.S.A., Inc.
7501 East Marin Drive
Englewood, Colorado 80111 USA

GHANA

Mr. Godfried Opong Kesse
Director
Geological Survey Department
P.O. Box M 80
Accra, Ghana

GUINEA

M. Boubacar Diallo
Ingénieur Electrotechnicien au
Bureau d'Etudes
Ministère de l'Energie et du Konkouré
Conakry, Guinea

HUNGARY

Dr. T. Boldizsar
Department of Petroleum
Technical University of
Heavy Industry
3515 Miskolc
Egyetemvaros, Hungary

INDIA

Mr. A. D. N. Reddy
Convener
Indian Institute of Science
Bangalore, Astra 560012
India

Mr. H. R. Srinivasan
Director, Gobar Gas Scheme
Khadi & V.I. Commission
3, Irla Road, Vile Pare
West Bombay
Bombay, 400056
India

INDONESIA

Mr. Harsono Hadipoetro
Pertamina
Jalan Perwira No. 2-4-6
P.O. Box 12
Djakarta, Indonesia

ISRAEL

Dr. Joel Schechter
Director of Applied Research
Institute
Ben-Gurion University of Negev
P.O. Box 1025
Beer-Sheva 84100
Israel

Mr. Harry Tabor
The Scientific Research Foundation
P.O. Box 3745
Jerusalem, Israel

Mr. Avraham Talmor
Consul, Economic Affairs
Israel Trade Commission
550 Sherbrooke Street West, Suite 1675
Montreal, P.Q. H3Z 1B9
Canada

ITALY

Mr. Calamita Ercole
Energy Planning Department
E.N.I.
Rome, Italy

M. Gian Carlo Ferrara
Geochemist
ENEL (National Electricity Agency)
Study and Research Center
Piazza Bartolo Da Sasso Ferrato, 14
Pisa, Italy

Mr. Mario Grosso
Instituto de Technologia
dell' Ambiente Costruito
Viale Mattioli 39
Torino 10100
Italy

Dr. Aldo Peta
Vice-Secretary General
Fondazione Adriano Olivetti
Viale Mazzini 88
Roma 00195
Italy

Mr. Vincenzo Spaziante
Head, Energy Research Program
Fondazione Adriano Olivetti
Viale Mazzini 88
Roma 00195
Italy

JAMAICA

Dr. Henry Lowe
Director of Energy, Division
of Energy
Ministry of Mining and Natural
Resources
2 St-Lucia Avenue
Kingston 5
Jamaica

JAPAN

Mr. Shigeaki Tsukagoshi
Mitsui Mining and Smelting Co. Ltd.
c/o P.O. Box 53, Mitsui and Co.
Royal Bank Plaza
Toronto, Ontario Canada

KENYA

Mr. T. H. Hutchinson
Managing Director
Tunnel Company Limited
Tunnel Estate
Fort Ternan, Kenya

KUWAIT

Mr. Salem Hamzah Al-Jumaie
Petroleum Engineer
Technical Affairs
Ministry of Oil
P.O. Box 22795
Safat, Kuwait

Dr. Jassim Al Kummer
Senior Research Economist
Ministry of Oil
P.O. Box 5077
Safat, Kuwait

Mr. Sadiqu J. Abul
Research Economist
Ministry of Oil
P.O. Box 5077
Safat, Kuwait

LIBERIA

Dr. Miedi-Himie Neufville
Director, Bureau of Hydrocarbon
Ministry of Lands and Mines
Government of Liberia
Monrovia, Liberia

MADAGASCAR

M. Wilson Rakotoarivelo
Chef du Service des Projets du Ministère de l'Economie et du Commerce
Président de la Société des Energies Nouvelles Hery Vao
B.P. 527
Antananarivo, Madagascar

MALAWI

Mr. P. A. B. Kalirangwe
Principal Economist
Economic Planning Division O.P.C.
P.O. Box 30136
Lilongwe 3
Malawi

Mr. M. M. Z. Njolwa
Industrial Economist
Ministry of Trade, Industry and Tourism
P.O. Box 30366
Lilongwe 3
Malawi

Mr. R. S. Nkaonja
Senior Conservator of Forests
Forestry Research Institute of Malawi
Ministry of Agriculture & Natural Resources
P.O. Box 270
Zomba, Malawi

Mr. Nigel Derek Hunter
Senior Physical Planner and Environmental Officer
Ministry of Agriculture & Natural Resources
P.O. Box 30134
Lilongwe 3
Malawi

MALAYSIA

Mr. Wau Zahir Bin Sheikh Abdul Rahman
Deputy Secretary-General
Ministry of Energy, Telecommunications & Posts
Wisma Damansara
Kuala Lumpur, Malaysia

Miss Yap Siew Ying
Assistant Director, Petroleum Unit
Government of Malaysia
Prime Minister's Department
Wisma Damansara
Kuala Lumpur, Malaysia

MALI

M. Bouba Camara
Ingénieur
Direction Nationale de l'Hydraulique et de l'Energie
B.P. 66
Bamako, Mali

MAURITANIA

M. Mohamed Lemine Benahi
Chef du Service des Etudes et Programmation
Ministère de l'Industrie et des Mines
Gouvernement de la République Islamique
Nouakchott, Mauritania

MAURITIUS

Mr. F. Georges Carver
Senior Lecturer
Sugar Technology Division
School of Industrial Technology
University of Mauritius
Reduit, Mauritius

MEXICO

Mr. Hector Alonso Espinosa
Coordinator Ejecutivo de Geotermica
Comision Federal de Electricidad - Mexico
P.O. Box 248
Calexico, California 92231 USA

Mr. Miguel Castaneda Perez
Ingeniero de Minas y Metalurgista
Pisco #584
Mexico City, DF 14
Mexico

Mr. Arturo Del Castillo
Deputy Director for Economic Studies and Industrial Planning
Mexican Institute for Petroleum
Av. de las Cien Metros No. 152
Mexico City, DF 14
Mexico

Mr. Fernando Hiriart
Under-Secretary, Mines and Energy
Secretariat for Patrimony & Industrial Development
Insurgentes Sur 552
Mexico City, DF 7
Mexico

Mr. Abundio Juarez Mendez
Gerente de Planeacion de Exploracion Petroleos Mexicanos
Marina Nacional 329, Col. Anahuac
Mexico City, M DF
Mexico

Mr. Jerzayn Leon-Alvarez
Assistant Manager to the Resources
Management of Exploration
Pemex Exploration
Marina Nacional 329
Mexico City, Mexico

Mr. Javier Matus Pacheco
Director General for Multi-Lateral Economic Relations
Secretariat of Foreign Affairs
Flores Magon 1
Mexico City, Mexico

Mr. Fernando Schutz Estrada
Researcher
Electrical Power Research
Leibritz 14-3
Mexico City, Mexico

NEPAL

Dr. G. R. Bhatta
Senior Engineer
Ministry of Water Power and
Irrigation
Kathmandu, Nepal

Mr. Achyut P. Pradhan
Chief Electrical Inspector
Ministry of Water Power and
Irrigation
HMS Nepal Babar Mehal
Kathmandu, Nepal

NETHERLANDS

Mr. Edward Bakhuis
Fundashon Antiyano Pa Energia
Admiraalsweg 26A
Curaçao
Netherlands Antilles

Dr. A. L. N. Bongaarts
Ministry of Economic Affairs
Bezuidenhoutseweg 30
The Hague, Netherlands

Mr. Th. Evers
Attaché for Health and Environmental Protection
Netherlands Embassy
4200 Linnean Avenue, N.W.
Washington, D.C. 20008 USA

Mr. J. A. Suurland
Head Division Economic Affairs and
Policy Department
Ministry of Public Health and
Environmental Protection
Government of Netherlands
The Hague, Netherlands

NIGERIA

Mr. James Babatunde Owokalu
Deputy Manager (Conservation)
Nigerian National Petroleum
Corporation
7 Kojo Abayomi Street
PMB 12701
Victoria Island
Lagos, Nigeria

NORWAY

Mr. Knut Am
Manager of Research and Development
Department
Statoil
Postboks 300
N-4000 Stavanger, Norway

Mr. Paul Hofseth
Head of Division
Ministry of Environment
Oslo Dep.
Oslo 1, Norway

PANAMA

Mr. Virgilio Luque
Director Projecto Geotermico
Apartado 5312
Panama 5, Panama

PAPUA NEW GUINEA

Mr. Lose Kose
Department of Minerals and Energy
P.O. Box 2352
Konedobu, Papua New Guinea

Dr. Kenneth Newcombe
Department of Minerals and Energy
P.O. Box 2352
Konedobu, Papua New Guinea

Mr. Larry J. Weick
Department of Minerals and Energy
P.O. Box 2352
Konedobu, Papua New Guinea

PERU

Ms. Teresa Chipoco Malmborg
Director de Acondicionamiento de
Territorio
Instituto Nacional de Planificacion
Republica de Chile 262
Lima 11, Peru

Mr. Luis A. Haro Zavaleta
Oficina Sectorial De Planificacion
Ministerio de Energia & Minas
Au Las Artes S/N, Surquillo
Lima 34, Peru

POLAND

Dr. Zygmunt Falecki
Scientific and Development Centre
for Energy Problems
UP Owocowa 6a
(OBR-GPE)
Katowice, Poland

Mr. Wieslaw Filipczak
Principal Advisor on Energy
Ministry for Science and Technology
6/8/10 Miodowa Str
Warsaw, Poland

PORTUGAL

José Isidoro d'Oliveira Garvalho Netto
General Director of Energy
Director for Economical Studies and
Planning
Rua da Beneficência, 241
Lisbon, Portugal

QATAR

Mr. Rashid Awaida Al-Thani
Deputy Managing Director
Qatar General Petroleum Corporation
P.O. Box 3212
Arabian Gulf
Doha, Qatar

Dr. Zuhair Nafi
Head of Projects and Research
Division
Economics Department
Qatar General Petroleum Corporation
P.O. Box 3212
Arabian Gulf
Doha, Qatar

ROMANIA

Dr. Dan Vamanu
Scientific Secretary
Program on Alternative Energy Sources
National Council for Science and
Technology
32-34 Roma Street
Bucharest 1, Romania

RWANDA

M. Jean-Bosco Balinda
Chef de Service
"Etudes Techniques"
Bureau National d'Etudes de Projets
(BUNEP)
BP 1337
Kigali, Rwanda

SENEGAL

M. Mamadou Ba
Ingénieur
Ministère du Développement Industriel
et de l'Artisanat
Direction de l'Energie
B.P. 4037
Dakar, Senegal

SPAIN

Mr. Jose Pozancos
Commercial Office
Embassy of Spain
151 Slater Street, Suite 201
Ottawa, Ontario K1P 5H3
Canada

SRI LANKA

Mr. James Henry Lanerolle
Secretary to the Ministry of Power
and Highways
Government of Sri Lanka
50, Sir Chittampalam A. Gardiner
Mawatha
P.O. Box 540
Colombo 2, Sri Lanka

Prof. K. K. Y. W. Perera
Chairman
Ceylon Electricity Board
50, Sir Chittampalam A. Gardiner Mawatha
P.O. Box 540
Colombo 2, Sri Lanka

SUDAN

Dr. Hashim El Thuami
Director, Energy Department
Ministry of Energy and Mines
c/o Mr. Michael Hyland
Resident Representative UNDP
Khartoum, Sudan

SWEDEN

Mr. Harald Haegermark
Secretary of Swedish Commission for Oil Substitution
Ministry of Industry
Fack
S-10310 Stockholm, Sweden

Prof. Francois Hsieh
Vice-President, Directeur
Centre for the Study of International Relations
Dobelnsgatan 81
Box 19112
S-10432 Stockholm, Sweden

SWITZERLAND

Prof. Felice J. Jaffe
Department of Minerology
University of Geneva
13, rue des Maraichers
1211 Geneva, Switzerland

Mr. Riccardo E. Mueller
Energy Counselor
Embassy of Switzerland
2900 Catherdral Avenue, N.W.
Washington, D.C. 20008 USA

SYRIAN ARAB REPUBLIC

Prof. Adham Al Samman
Faculty of Sciences
Department of Physics
University of Damascus
Damascus, Syria

M. Jamil Hariri
Ministry of Petroleum
Damascus, Syria

Mr. Burham Khiami
Director of Training
Ministry of Electricity
P.B. 4900
Damascus, Syria

THAILAND

Mr. Sompongse Chantavorapap
Chief, Design & Energy Research Branch & Alternative Energy Study & Development Project
National Energy Administration
Ministry of Science, Technology & Energy
Pibultham Villa, Yodse
Bangkok, Thailand

Dr. Sermpol Ratasuk
Director, Environmental & Resources Research Division
Thailand Institute of Scientific & Technological Research
Ministry of Science, Technology & Energy
196 Phonyothin Road
Bangkhen
Bangkok, Thailand

Mr. Phol Songpongs
Director, Investigation & Planning Division
National Energy Administration
Ministry of Science, Technology & Energy
Pibultham Villa, Yodse
Bangkok, Thailand

TURKEY

Dr. Tolga Yarman
Associate Professor of Nuclear Engineering
Institute for Nuclear Energy
Technical University of Istanbul
Emirgan, Istanbul
Turkey

UNION OF SOVIET SOCIALIST REPUBLICS

Dr. Alexander A. Arbatov
U.S.S.R. State Committee for Science and Technology
11 Gorky Street
Moscow, U.S.S.R.

Mr. Valery M. Dobrynin
Professor and Vice-Rector
Moscow Gubkin Institute of Petrochemical and Gas Industry
Lenin Avenue, 65
117296, Moscow, U.S.S.R.

Dr. Mikhail V. Tolkachev
U.S.S.R. State Committee for Science and Technology
11 Gorky Street
Moscow, U.S.S.R.

UNITED ARAB EMIRATES

Mr. Saleh A. Akasha
Director
Petroleum Department
Government of the United Arab Emirates
Abu Dhabi, United Arab Emirates

UNITED KINGDOM

Dr. F. J. P. Clarke
Research Director of Energy Department
Aere Harwell
Building 329
Harwell, Didcot
Oxfordshire, England

Dr. Leo Hepner
L. Hepner & Associates Limited
Tavistock House North
Tavistock Square
London, England

Mr. John James Syrett
Head of Strategic Studies Section
Central Electricity Generating Board
Courtenay House
18 Warwick Lane
London EC4, England

UNITED REPUBLIC OF TANZANIA

Mr. P. V. Irengo
Principal Research Development and Planning Officer
Tanzania Petroleum Development Corporation
P.O. Box 2774
Dar es Salaam
Tanzania

Dr. S. R. Nkonoki
University of Dar es Salaam
P.O. Box 35169
Dar es Salaam
Tanzania

UNITED STATES

Mr. Martin Aaron
Materials Consultant
212 East Broadway
New York, New York 10002 USA

Mr. Jack Allentuck
Acting Head, Economic Analysis
Brookhaven National Laboratory
Building 4-5
Upton, New York 11973 USA

Dr. Carl Aspliden
U.S. Department of Energy
Wind Section
600 E Street, N.W.
Washington, D.C. 20585 USA

Mr. William Bifano
NASA Manager, Photovoltaic Development and Support Project Office
Lewis Research Center
21000 Brookpark Road
Cleveland, Ohio 44135 USA

Mr. Lawrence C. Bonham
Senior Research Associate
Chevron Oil Field Research Company
P.O. Box 446
La Habra, California 90631 USA

Mr. Edward L. Burwell
Program Manager for Underground Coal Gasification
Department of Energy
Germantown, Maryland 20545 USA

Mr. Thomas R. Carr
Production Superintendant
Burlington Electrical Department
585 Pine Street
Burlington, Vermont 05401 USA

Ms. Elizabeth Cecelski
Resources for the Future
1755 Massachusetts Avenue, N.W.
Washington, D.C. 20036 USA

Dr. Frank Collins
Consultant
Oil, Chemical and Atomic Workers International Union
1126-16th Street, N.W.
Washington, D.C. 20036 USA

Dr. Barry Commoner
Director
Center for the Biology of Natural Systems
Washington University, Campus Box 1126
St. Louis, Missouri 63130 USA

Mr. Wallace de Witt, Jr.
U.S. Geological Survey
Mail Stop 955
U.S. Geological Survey
National Center Building
Reston, Virginia 22092 USA

Dr. Louis Divone
Department of Energy
Chief of Wind Systems
600 E Street, N.W.
Washington, D.C. 20585 USA

Mr. Michael Hoyt du Pont
President
DuPont Energy Control Corporation
2040 Avenue of the Stars
Los Angeles, California 90067 USA

Mrs. Peggy Vaughan du Pont
Treasurer
DuPont Energy Control Corporation
2040 Avenue of the Stars
Los Angeles, California 90067 USA

Mr. K. O. Emery
Woods Hole Oceanographic Institution
74 Ransom Road
Falmouth, Massachusetts 02540 USA

Mr. Giancarlo Facca
Geoexplor International
81 Buckeye Avenue
Oakland, California 94618 USA

Dr. W. R. Finnerty
Chairman
Department of Microbiology
University of Georgia
Athens, Georgia 30602 USA

Mr. Malcolm D. Fraser
Director, Alternate Energy Engineering
Inter Technology/Solar Corporation
100 Main Street
P.O. Box 340
Warrenton, Virginia USA

Mr. Medard Gabel
Director, World Resources Inventory Incorporated
3500 Market Street
Philadelphia, Pennsylvania 19104 USA

Dr. Amal K. Ghosh
Exxon Research and Engineering Company
Linden, New Jersey 07036 USA

Dr. Peter E. Glaser
Vice-President
Arthur D. Little Inc.
20 Acorn Park
Cambridge, Massachusetts 02140 USA

Dr. Augusta Goldin
The Staten Island Advance
590 Bard Avenue
Staten Island
New York, New York 10310 USA

Mr. Lincoln Gordon
3069 University Terrace, N.W.
Washington, D.C. 20016 USA

Mr. William P. Harrison, Jr.
Virginia Polytechnical Institute &
State University
P.O. Box 364
Blacksburg, Virginia 24060 USA

Dr. Kenneth K. Hekimian
President
Hekimian & Associates
16162 Beach Boulevard, Suite 308
Huntington Beach, California USA

Mr. Robert H. Hendron
Los Alamos Scientific Laboratory
(MS 980)
Los Alamos, New Mexico 87545 USA

Mr. Claude Hocott
Senior Research Consultant
Gulf University Consortium
4538 Ivanhoe Street
Houston, Texas 77027 USA

Dr. John M. Hunt
Woods Hole Oceanographic Institute
Woods Hole, Massachusetts 02543 USA

Dr. William Illuminate
Manager, Plant Energy Supply
Operations
General Motors Corporation
Room 204, Boulevard Center Building
3044 West Grand Boulevard
Detroit, Michigan 48202 USA

Dr. Oleg D. Jefimenko
Department of Physics
West Virginia University
Morgantown, West Virginia 26506 USA

Mr. Howard R. Johnson
P.O. Box 199
314 North Main Street
Blacksburg, Virginia 24060 USA

Ms. Chris Kelly
United Nations International School
24-50 East River Drive
New York, New York 10010 USA

Mr. James A. Knight
Principal Research Scientist
Engineering Experiment Station
Georgia Institute of Technology
Atlanta, Georgia 30332 USA

Dr. Joseph Knox
Division Leader
Atmospheric Science
P.O. Box 808
Livermore, California USA

Prof. Abrahim Lavi
Carnegie-Mellon University
Pittsburgh, Pennsylvania 15217 USA

Dr. Sidney Loeb
Inter Technology/Solar Corporation
100 Main Street
Warrenton, Virginia 22186 USA

Mr. Robert E. Lumpkin
Group Head, Oil Shale
Occidental Research Corporation
P.O. Box 19601
Irvine, California 92713 USA

Dr. Peter Markovitch
Executive Vice-President
DuPont Energy Control Corporation
2040 Avenue of the Stars
Suite 400
Los Angeles, California 90067 USA

Dr. Sullivan S. Marsden
Department of Petroleum Engineering
Stanford University
Stanford, California 94305 USA

Dr. Richard D. McIver
President
GeoChem Research Incorporated
16920 Park Row
Houston, Texas 77084 USA

Mr. Gurmukh D. Mehta
Manager
Advanced Thermal Systems
Inter Technology Solar Corporation
100 Main Street
Warrenton, Virginia 22186 USA

Dr. H. Tsvi Meidav
Consultant
Suite 600
3 Embarcadero Center
San Francisco, California 94111 USA

Dr. Richard F. Meyer
Chief
Office of Resource Analysis
U.S. Geological Survey
920 National Center
Reston, Virginia 22092 USA

Mr. Vinod Mubayi
Associate Physicist
Brookhaven National Laboratory
Upton, New York 11973 USA

Mr. Charles W. Muller
President
Murden and Company
39 E. 51st Street
New York, New York 10022 USA

Mr. Richard Nehring
The Rand Corporation
1700 Main Street
Santa Monica, California 90406 USA

Prof. Wheeler J. North
Environmental Science Department
California Institute of Technology
Kerckhoff Marine Laboratory
101 Dahlia Street
Corona Del Mar, California 92625 USA

Dr. Ernest J. Oppenheimer
40 Central Park South, Apt. 9-D
New York, New York 10019 USA

Mr. Stanford R. Ovshinsky
President
Energy Conversion Devices Inc.
1675 West Maple Road
Troy, Michigan 48084 USA

Mr. Harry Perry
Resources for the Future
1755 Massachusetts Avenue, N.W.
Washington, D.C. 20036 USA

Prof. John T. Pfeffer
Department of Sanitary Engineering
University of Illinois
Urbana, Illinois 61801 USA

Mr. Kenneth Pipes
Manager, Exploration Division
Exxon Corporation
1251 Avenue of the Americas
New York, New York 10020 USA

Dr. Elizabeth A. Rauscher
DuPont Energy Control Corporation
2040 Avenue of the Stars
Los Angeles, California 90067 USA

Mr. Charles Richardson
McGraw-Hill Inc.
1221 Avenue of the Americas
New York, New York 10020 USA

Mr. Milton Russel
Director, Center for Energy Policy Research
Resources for the Future
1755 Massachusetts Avenue, N.W.
Washington, D.C. 20036 USA

Dr. E. J. Salmon
National Academy of Sciences
2101 Constitution Avenue, N.W.
Washington, D.C. 20418 USA

Ms. Lorna Salzman
Friends of the Earth
72 Jane Street
New York, New York 10014 USA

Mr. Frederick N. Seebach
President
Aquarian Agency, Ltd.
1030 North State Parkway 41J
Chicago, Illinois 60610 USA

Mr. Robert B. Swandby
Office of International Affairs
U.S. Department of Energy
1000 Independence Avenue, Room 5H 039
Washington, D.C. 20585 USA

Mr. Blair Swezey
Energy Study Centre
Electric Power Research Institute
3412 Hillview Avenue
Palo Alto, California 94304 USA

Prof. Walter E. Tengelsen
Cooper Union & Transystems Inc.
P.O. Box 81
Murray Hill Station
New York, New York 10016 USA

Mr. Samuel Weimersheimer
President
U.S. Magnet & Alloy Corporation
85 North Main Street
Yardley, Pennsylvania 19067 USA

Prof. Alvin M. Weinberg
Director
Institute for Energy Analysis
Oak Ridge Associated Universities
P.O. Box 117
Oak Ridge, Tennessee 37830 USA

Prof. Robert L. Whiting
Department of Petroleum Engineering
Texas A & M University
College Station, Texas 77843 USA

Prof. Teh Fu Yen
Department of Chemical Engineering
University of Southern California
University Park, PCE 201
Los Angeles, California 90007 USA

URUGUAY

Mr. Jose L. Bozzo
Sub Directeur Energie
Rincon 723
Montevideo, Uruguay

VENEZUELA

General Rafael Alfonzo-Ravard
President
Petroleos de Venezuela S.A.
Apartado 169
Caracas, Venezuela

Dr. Jose Rafael Dominguez
Senior Advisor
Petroleos de Venezuela S.A.
Apartado 169
Caracas, Venezuela

Dr. Hugo Finol
Director
Petroleos de Venezuela S.A.
Apartado 169
Caracas, Venezuela

Mr. M. Gurfinkel
Petroleos de Venezuela S.A.
Apartado 169
Caracas, Venezuela

Dr. F. J. Gutierrez
School of Petroleum Engineering
Universidad de Venezuela
Caracas, Venezuela

Mr. Richard Irving
Planning Coordinator
Petroleos de Venezuela S.A.
Apartado 169
Caracas, Venezuela

Mr. Carlos Kulka
Exploration and Production Planning Manager
Maraven S.A.
Apartado 829
Caracas, Venezuela

Mr. Jose Ramon Solano
Executive Assistant
Minister of Energy and Mines
of Venezuela
Caracas, Venezuela

Mr. Boris Tarantaev
Planning Department
Lagoven S.A.
Ave. Leonardo da Vinci
Los Chaguaramos
Caracas, Venezuela

Mr. Arnold Volkenborn
General Manager, Oil Trading
and Marine
Apartado 829
Caracas, Venezuela

ZAIRE

M. Vangu Wa Kinanga
Service Présidentiel d'Etudes
B.P. 9797
Kinshasa I, Zaire

M. Yimbu-Nzau Manzukula
Expert-Chimiste
Service Présidentiel d'Etudes
B.P. 9797
Kinshasa I, Zaire

M. Ngoy-Muana Mutanda
Directeur de l'Equipement
Société Nationale d'Electricité
Snel B.P. 500
Kinshasa I, Zaire

ZAMBIA

Mr. P. J. Chisanga
Executive Director
Transport and Energy
Zambia Industrial Development
Corporation
Zimco, Zambia

Mr. Roy G. Miti
General Manager
Kariba North Bank Company
Lusaka, Zambia

Dr. E. H. B. Mwanang'onze
Lecturer
Department of Geology
School of Mines
University of Zambia
Lusaka, Zambia

UNITED NATIONS AND OTHER INTERNATIONAL ORGANIZATIONS

Dr. Joseph Barnea
Special Fellow
UNITAR
801 U.N. Plaza
New York, New York 10017 USA

Mr. Michael Bloome
Research Associate
UNITAR
801 U.N. Plaza
New York, New York 10017 USA

Mr. Bernard Boubli
International Air Transportation
Association
1000 Sherbrooke Street West
Montreal, P.Q. H3A 2R2
Canada

Mr. Klaus J. Brendow
Deputy Director, Energy Division
Economic Commission for Europe (ECE)
Palais des Nations
CH-1211 Geneva 10
Switzerland

Mr. Daniel Camp
Inter-American Development Bank
801 17th Street, N.W.
Washington, D.C. 20577 USA

Mr. John James Carney
Deputy Director
Audio-Visual Information
Vision Habitat
Centre on Human Settlements
2075 Westbrook Mall
Vancouver, B.C. V6T 1W5
Canada

Mr. Jorge Cavallotti
Senior Power Engineer
World Bank
1818 H Street, N.W.
Washington, D.C. 20433 USA

Mr. Bertrand Chatel
Office for Science and Technology
United Nations
One United Nations Plaza
New York, New York 10017 USA

Mr. Maurice Del Carril
Senior Technical Advisor
United Nations Development Programme
(UNDP)
New York, New York 10017 USA

Mr. William Epstein
Special Fellow
UNITAR
801 U.N. Plaza
New York, New York 10017 USA

Mr. Russell Ferguson
Industry and Mining Specialist
Inter-American Development Bank
801 17th Street, N.W.
Washington, D.C. 20577 USA

Mr. J.M. Harrison
U.N. University
4 Kippewa Drive
Ottawa, Ontario Canada

Mr. Roderick Heitmeyer
Chief, Economics Section
International Civil Aviation
Organization
1000 Sherbrooke Street West
Montreal, P.Q. H3A 2R2
Canada

Mr. Kalmen Kaplansky
Director, Canada Branch
International Labour Organization
75 Albert Street, Suite 202
Ottawa, Ontario Canada

Mr. M.L. Pijit Kambhu
Chief, Energy Resources Section
Natural Resources Division
United Nations Economic and Social
Commission for Asia and the Pacific
(ESCAP)
United Nations Building
Rajadamnern Avenue
Bangkok 2, Thailand

Mr. George Komissarov
Industrial Development Officer
United Nations Industrial Development
Organization
(UNIDO)
Vienna International Centre
P.O. Box 300
Vienna 1400 Austria

Mr. Derek Lovejoy
Department of Technical Co-operation for
Development
United Nations
New York, New York 10017 USA

Ms. Claudia Machaver
Programme Officer
UNITAR
801 U.N. Plaza
New York, New York 10017 USA

Mr. M.B. Marin-Curtoud
United Nations Conference on Trade
and Development
Bureau 10032
Palais des Nations
CH-1211 Geneva 10 Switzerland

Mr. James R. McNitt
Technical Advisor
Department of Technical Co-operation
for Development
Room DC-892
United Nations
New York, New York 10017 USA

Mr. Joseph Mullen
Economic Affairs Officer
Economic Commission for Latin America
(ECLA)
Edificio Naciones Unidas
Avenida Dag Hammarskhold
Casilla 179-D
Santiago Chile

Mr. Jean Nauwelaers
First Economic Affairs Officer
United Nations Economic Commission for Western Asia (ECWA)
P.O. Box 4656
Beirut, Lebanon

Dr. Davidson Nicol
Executive Director
UNITAR
801 U.N. Plaza
New York, New York 10017 USA

Dr. Valeriy Ovcharenko
Energy Section
Department of Technical Cooperation for Development
United Nations, Room DC-888
New York, New York 10017 USA

Mr. Emmanuel Pederico
Senior Officer, Organization of Petroleum Exporting Countries (OPEC)
Obere Donaustrasse 93
1020 Vienna II, Austria

Mr. James Reddington
International Energy Agency
Organization for Economic Cooperation and Development (OECD)
2 rue André-Pascal, Cedex 16
Paris 72775, France

Mr. G.V. Subba Rao
Senior Officer for Energy
Department of International Economic and Social Affairs
United Nations
New York, New York 10017 USA

Mr. Thomas Szekely
Training Advisor
International Civil Aviation Organization
P.O. Box 400
1000 Sherbrooke Street West
Montreal, P.Q. H3A 2R2 Canada

CONFERENCE ORGANIZING COMMITTEE

Chairman	Mr. Albert Khazoom	Petro-Canada
	Mr. Philippe de Seynes	Director, Project on the Future UNITAR
	Dr. Joseph Barnea	UNITAR
	Mr. Michael Bloome	UNITAR
	Ms. Claudia Machaver	UNITAR
	Mr. Donald Sinclair	Department of External Affairs Government of Canada
	Mr. J. Paul Filion	Petro-Canada
	Mr. Jim Eckford	Petro-Canada
	Mr. Charles Muller	President, Murden and Co. New York
	Mr. Roger Khazoom	Chief, Liaison Office, Montreal

BUREAU OF THE CONFERENCE

Co-Chairmen:	General Rafael Alfonzo-Ravard President Petroleos de Venezuela
	Mr. Maurice F. Strong Chairman International Energy Development Corporation
Vice-Chairmen:	Dr. Jassim Al Kummer (Kuwait)
	Mr. Robert Boyd (Canada)
	Dr. Zygmunt Falecki (Poland)
	Mr. Harsono Hadipoetro (Indonesia)
	Mr. Jean d'Herbes (France)
	Mr. Abundia Juarez Mendez (Mexico)
	Mr. Louis Kagisye (Burundi)
	Dr. Godfried Opong Kesse (Ghana)
	Mr. Adrian Poplawski (Canada)
Secretary-General:	Mr. Albert Khazoom Senior Director for External Relations Petro-Canada
Scientific Secretary:	Dr. Joseph Barnea Special Fellow UNITAR
Rapporteur:	Mr. John Foster Senior Economic Advisor Petro-Canada
Assistant Secretary-General:	Mr. Michael Bloome Research Associate UNITAR

TABLE OF CONTENTS

TABLE OF CONTENTS (continued)

TABLE OF CONTENTS (continued)

TABLE OF CONTENTS (continued)

TABLE OF CONTENTS (continued)

TABLE OF CONTENTS (continued)

TABLE OF CONTENTS (continued)

TABLE OF CONTENTS (continued)

TABLE OF CONTENTS (continued)

INTRODUCTION

The papers included in this volume are those which were presented to the Conference on Long-Term Energy Resources held in Montreal, Quebec, Canada, 26 November to 7 December 1979. The Conference was sponsored jointly by the United Nations Institute for Training and Research (UNITAR) and Petro-Canada, on behalf of the Government of Canada. Cooperating organizations included the United Nations Development Programme, the Canadian International Development Agency, Petroleos de Venezuela, Petroleos Mexicanos, Hydro-Quebec, and the Compagnie Francaise des Petroles.

In the foreword to the Summary Report[1] of the Conference, Mr. Albert Khazoom, Secretary-General of the Conference, wrote:

> Several months after the UNITAR Conference on Long-Term Energy Resources, interest in the event continues unabated. Because of the quality of the documentation, the high level of the discussions, and the broad geographical representation at the Conference, requests for copies of the documentation and the Conference report are reaching UNITAR at an ever-increasing rate. The present report is an initial answer to these demands.
>
> A book containing the documentation and the list of participants in the conference is scheduled to be published later this year.
>
> With my appreciation to the Rapporteur, John Foster, and to Philippe de Seynes' Project for the Future for editing an enlightening summary of the proceedings of the conference, I wish to reiterate my thanks to the host country, Canada, and to the International Civil Aviation Organization (ICAO) for their hospitality. I am also grateful to Petro-Canada for their financial assistance and the services and facilities they provided, as well as to the other co-sponsors of the event, whose generosity and support contributed in no small measure to the success of the Montreal meeting.

This introduction is derived from the excellent Summary Report of the Conference, prepared by the Rapporteur, Mr. John Foster (Petro-Canada), with the assistance of Ms. Claudia Machaver (UNITAR). It is intended to provide a nontechnical guide to the content of the reports, which are grouped together in related sections. The large number of chapters prevents the abstracting of each one individually.

---

[1]UNITAR, 1980: Important for the Future, v. 5, no. 1, p. 3-22.

Over 400 delegates from approximately 90 countries were in attendance. They heard and discussed 120 technical presentations during the 2 weeks of intensive deliberations. Indeed, one of the outstanding features of the Conference was the constructive spirit of cooperation and enthusiasm shown by experts from a diversity of social and economic systems, while considering a vast amount of material and arriving at a common point of view.

The purpose of the Conference was to evaluate the variety of energy sources from which production is obtained today or which are at various stages of research and development. Papers were presented on some 45 different energy sources and on energy policies and issues of a number of countries. However, the Conference evaluated neither the outlook for energy demand nor the capital, trained manpower, and infrastructure needed for energy production and consumption.

The Conference was designed to be of specific significance to the developing countries, whose economic development is being greatly affected by higher energy prices and the prospect of rapid expansion in energy consumption. During the Conference, it became apparent that many countries have the potential to become self-sufficient.

Some of the most pressing and urgent questions which face mankind today involve the future availability and cost of energy supplies. Are the available resources sufficient to meet the growing demand? Will individual countries be able to cope with the price of energy? How will the development process of poor countries be affected and will people, even in developed countries, experience a drastic decline in their standard of living?

The fact that great importance has been placed on the necessity to assess the energy situation carefully and solve related problems is not surprising. The use of energy plays a very prominent role in the growth process of any economy, whether it is developing or developed. In either case, energy is needed for industry, agriculture, transport, domestic use, and many other activities which create employment, produce goods and services, and generally affect the day-to-day lives and the standard of living of the people. Therefore, the ability of any country to provide the basic needs of its people is closely tied to whether it has access to adequate energy supplies and whether it can afford the cost of those supplies.

The concern being expressed today stems largely from the fact that the world has come to depend predominantly on conventional or light oil, and petroleum products in general, to supply its energy needs, and the price of these commodities has risen dramatically in recent years. This has had a debilitating effect on those developing countries which derive more than 75 percent of their energy supplies from imported oil. These countries have had to spend a large proportion of their limited foreign exchange to purchase oil, a state of affairs that has induced them into heavy borrowing and rendered more precarious their attempts to pursue successfully their development programs.

The dependence on conventional oil is a cause for great concern not only because of its expense, but also because of its availability. Many experts have warned that oil supplies are being rapidly depleted and should the process not be arrested, the resource may be exhausted by the end of this century or the early part of the 21st century. However, as the price of conventional oil continues to rise, increasing quantities of "unconventional" oil and gas will enter the market. In the long term, therefore, the resource potential and cost of "unconventional" oil will increasingly determine the availability of future oil and other energy supplies.

Thus, the uncertainties associated with "unconventional" energy sources have highlighted the need for diversification and prompted increased research all over the world into sources which are potential alternatives. A diversity of energy sources is needed because: (1) there is a diversity of energy needs, and (2) there are technological and economic uncertainties surrounding many individual energy sources.

The four chapters in Section I of this volume are in the nature of an overview of the world energy situation and deal with important sources of future energy supply. Chapter 1 was prepared by the Scientific Secretary to the Conference and sets the stage for the succeeding papers. Particular emphasis is given to what are presently considered to be unconventional energy sources; many of these will be the staple sources of the future, the mix depending mainly on the balance of costs and technology.

If it were possible to predict future energy needs with precision, then filling them at least cost would require a diversity of sources. This is because the sources differ in at least three ways: their thermodynamic nature (high or low temperature), place of occurrence, and the timing of demand, whether with respect to developed or developing countries there is a necessity to develop diverse sources and herefore for immediate research and development on requisite technologies.

A transition to solar energy can be compatible with conventional energy systems. A most effective system involves cogeneration of space heat, cooling, and electricity through inputs of natural gas and solar, that is, biomass, methane. The key to this arrangement is that a cogenerator fueled by natural gas releases to the environment only carbon dioxide and water (nitrogen oxides being controlled with available technology) and the natural gas can be substituted by biomass gas with no change in equipment. Combined with conservation, such approaches to energy supply imply the efficiency with which economic output is used to produce energy and the efficiency with which energy is converted to work and economic output.

Numerous sources and potential sources of oil and natural gas exist which are presently underdeveloped but which may prove to be of great significance in the future. Already, data show a trend to the discovery and development of small oil and gas fields, a result of both declining field sizes and improved prices. The occurrence of tar sand and oil shale deposits in the world is fairly well known but not the quantity of their recoverable resources. Knowledge of heavy oil occurrence is considerable in such areas as Alberta, Canada, Venezuela, and parts of the USA; much is still to be learned about its recovery and the discovery methods applicable to it. Understanding of the occurrence and resources of natural gas in other than conventional reservoirs can best be described as primitive. A major difficulty, even in speculating on such resources, is the lack of information on the nature of the world's geologic basins or even the total number of them. The estimates given for the world of 1.4 trillion barrels of recoverable oil in small fields and occurring as heavy oil, tar sands, and shale oil, and 22,600 trillion cubic feet of natural gas dissolved in water, associated with coal or shale, found in tight sands, or present in hydrated form, must indeed be considered as speculative but serve as useful indicators of future possibilities.

Thirteen chapters make up Section II, which considers the vital area of energy policy and planning. Four of the chapters deal with the efforts of individual countries to deal with rational energy balances to suit their national needs. Such efforts, for the most part, utilize up-to-date modeling techniques. Other chapters give attention to industrial cooperation as a means to achieve energy efficiency and to the ever-present spectre of environmental hazards which generally accompany energy production and utilization.

Choice by government among the many energy alternatives becomes highly complex and often highly subjective. It is difficult to generalize because of the many different sets of circumstances facing planners: largely urban versus largely agrarian populations, climate, presence of indigenous resources. Perhaps most difficult is the definition and measurement of external costs: possible mining fatalities, acres of disturbed land, life-style changes. In developing countries, attention must be given to obtaining shadow prices for labor and imported technologies and integrating them into the cost analysis. Furthermore, the energy costs imposed by national, political, or social policies and goals, such as "rural electrification" or "energy independence" should be taken explicitly into account.

One aspect of energy supply receives close attention, centralization as compared with decentralization. In most developed countries, energy has become centralized as a matter of cost-saving and efficiency, and moves toward decentralization of sources generally will be made through existing infrastructure. Most developing countries lack the infrastructure--electricity grids, for example--and are thus inherently decentralized. In many of them, future industrialization will entail power centralization simply because the large nonhydrocarbon energy sources, such as hydropower or geothermal, are point sources. Such dispersed sources as solar power, wind power, or biomass are insufficient to yield large-scale, concentrated production with existing technology.

Section III is comprised of five chapters dealing with coal. The world's presently known reserves of solid fuels (coal, brown coal, lignite, and peat) are huge, much larger than those of oil and gas. However, they are unevenly distributed geographically, being concentrated largely in three countries: the Soviet Union, China, and the United States. The rest is in a few other countries: Australia, Czechoslovakia, DR Germany, FR Germany, Poland, South Africa, and some developing countries, such as Botswana, Colombia, India, Mozambique, Republic of Korea, Vietnam, and Zimbabwe. Both developed and developing countries seem ready to move strongly toward greater coal production and utilization.

Developing countries have new opportunities to exploit small coal fields which still may be large enough to contribute to their domestic energy needs. For these countries, the key to resource development lies with government policies framed to encourage such exploitation, which UN agencies should be increasingly ready to aid.

The geographical concentration of known coal reserves reflects to a large extent differences in intensity of exploration effort. The world's ultimately recoverable resources are still poorly known, but may be greatly underestimated. Worldwide, a large production base exists which could be expanded, if markets were developed.

The oil price increases since 1973 have made coal competitive for many end uses and have created new opportunities for its expanded use. While entirely new technologies do not need to be developed, as coal was at one time the dominant commercial fuel worldwide, improved technology is desirable to reduce costs and adverse environmental impacts. The single largest end use for coal in recent years has been for electric power generation, where the technology for coal burning has significantly improved, including the introduction of fluidized bed combustion. In power generation, the use of coal in place of oil or nuclear power is likely to expand and new opportunities for coal usage, particularly in developed countries, could include the development of plants to produce synthetic oil and gas.

A critical issue for the expansion of coal demand is its acceptability to end users. Particularly in developed countries, this includes such things as air pollution and supply from an industry which has not enjoyed the best of labor-management relations. In addition, there are serious constraints on the expansion of coal supplies which need governments' attention, such as: the environmental impact of strip-mining, recruitment of miners and engineers, the lead-times and considerable investments needed to establish and expand mines, transportation (pipelines and trains) and storage, and handling and port facilities. Despite these constraints, there seems likely to be a substantial increase in coal demand and output.

At present, most exports are from the USA, Australia, South Africa, FR Germany, and Poland; Canada is a growing net exporter. Major importers include France, Italy, and Japan. While most trade is now in coking-coal for steel-making, some steam coal is also traded, and it is thought that there will be a very sharp increase in the international trade of coal, particularly steam coal. The USA and Australia are likely to expand their coal exports dramatically during the next two decades, and Canada and South Africa could do the same. Although Poland will continue to export, it may not be able to expand its shipments significantly. Of the developing countries, Botswana, Colombia, Venezuela, India, and Mozambique have a good potential to become coal exporters in the future.

Often neglected in earlier years, solid fuels of low calorific value have more recently seen an upsurge in development. In descending order of heating value, these include sub-bituminous coal, brown coal, lignite, and peat.

World reserves of brown coal and lignite are large, some 10 percent by heating value of all solid fuels. Production in 1978 amounted to more than 900 million tons of hard coal equivalent and represented about 14 percent of all solid fuel production. Some countries have exploited brown coal, lignite, and peat for many decades, above all the Soviet Union, Czechoslovakia, DR Germany, FR Germany, and, to a lesser extent, some other European countries, Australia, and some Asian countries. Recently there have been large-scale plans to expand the use of lignite in the Gulf states of the United States.

The chapter on brown coal indicates that it can be delivered economically in the form of briquettes by barge or railcar as far as 400 miles. Processes to produce synthetic oil or gas from brown coal have experienced greater financial and environmental setbacks than expected, and there are no production plants yet in existence. Other uses for brown coal involve the direct reduction of iron ore and production of special coke. The environmental impact of strip-mining brown coal and lignite can be great if suitable precautions are not taken; however, FR Germany has enforced well-designed laws which have resulted in a compatible confrontation of modern technology with nature.

Solid fuels of low heating value are sensitive in their economic feasibility to transport costs. Consequently, it typically makes economic sense to burn these fuels close to the mine. For example, they can be used as fuel in mine-mouth electric power stations. The technology of power generation from large peat-fuel boilers has reached an advanced stage of development in several countries, including the Soviet Union, Finland, and Ireland.

The nine chapters in Section IV deal with conventional oil and natural gas with respect to resources, exploration, and transportation. There have been many estimates of ultimate recoverable resources. In a sense they are no more than academic, because the amount of recoverable reserves is unknown, but they do give some indication of the constraint on the rate at which additions may be made to presently known reserves. Estimates are typically defined to exclude unconventional oil, such as deposits of heavy oil, tar sands, and oil shales, as well as oil in polar regions, in small fields or in very deep water. They usually assume some arbitrary increase in the rate of recovery from oil reservoirs.

One chapter suggests that the world's conventional oil in place might amount to 3,200 billion barrels (BB). Of that amount, 370 BB had been produced at the end of 1977 and 740 BB might in the future be recovered by primary and secondary recovery methods. Hence the combined total recoverable oil might be some 1,110 BB or 35 percent of the original oil in place. The application of enhanced recovery processes might recover another 200 BB, resulting in total recovery of 1,310 BB or more than 40 percent of the oil in place.

Another, more conservative estimate presents an assessment of world cumulative production of crude oil of approximately 420 BB through 1978, of known remaining reserves at about 680 BB, and of potential resources at another 500-900 BB, a rounded total of 1,600-2,000 BB. While recognizing the uncertainty of resource estimates, the author speculates that a major breakthrough in technology might add as much as 500 BB to his estimate as an absolute upper limit.

The world's known conventional oil reserves remaining to be recovered at the end of 1978 are concentrated in a relatively few countries, the 10 largest containing 80 percent of the total: in order, Saudi Arabia, the Soviet Union, Kuwait, Iran, USA, Iraq, United Arab Emirates, Mexico, Libya, and China. The world's reserves are also concentrated in a relatively few sedimentary basins. Of the approximately 600 basins worldwide, there are producible hydrocarbons in nearly 240, and 96 percent of the world's known recoverable oil is located in 30 of these, only 2 of which (Arabian-Iranian and West Siberian) contain over half. There is a similar concentration of world reserves in a small number of large fields; of about 30,000 fields worldwide, fewer than 1,900 (or 6 percent) each contain the equivalent of 50 million barrels or more of hydrocarbons (oil and gas).

In 1978, 19 percent of the world's oil came from the continental shelves. Prospective oil provinces in deeper waters could push the proportion of petroleum produced from under the ocean far higher. The continental slopes and rises contain much larger volumes of compacted sediments than the total of sediments on land and the continental shelves. However, knowledge of the petroleum potential of deep water basins has been obtained by seismic testing, geophysical studies and rock dredgings, and supported most importantly by a relatively few drill-hole samples obtained by the Deep Sea Drilling Project ("Glomar

Challenger" and "Glomar Explorer"). The total information about the ocean's internal structure and underlying crust is consequently quite limited, and this situation will only be remedied by a major drilling program, deep seismic exploration by multi-ship arrays, and other new approaches. Additionally there is the legal issue regarding the common heritage of resources beyond the sovereignty of coastal states, which is yet unresolved.

Today's higher oil prices have also improved the operation of small fields. Some 29,000 fields (95 percent of the world total of fields so far discovered) each hold 50 million barrels or less of recoverable oil, or in aggregate the equivalent of just 75 BB of hydrocarbons, or only 7 percent of the world total. More than 26,000 of them are in the USA, while outside North America there are only 1,500 small fields which contribute 15 BB or 3 percent of the total oil found there. Because the USA is the most intensely drilled area of the world, it has a greater share of its reserves in nongiant fields. An intensive drilling effort in other places could have similar results. In many countries small fields will continue to be found for many years to come. Their cumulative impact could be great, and their contribution to the economies of the oil-importing developing countries would be especially significant.

Exploration techniques have improved significantly in recent years; nevertheless, evaluation of geological provinces is largely dependent on the degree to which they are tested by drilling.

In 1978, exploration worldwide reached record levels. By far the greatest amount took place in North America, with the lion's share of the rest in those countries which already produce oil. A relatively small amount of exploration took place in oil-importing countries, no more than 130 seismic party-months (1.5 percent of the total for developed and developing countries) and 63 wells (0.4 percent of the total).

Additions to world oil reserves have depended heavily on the discovery of giant oil fields. Nevertheless, a substantial portion of the future increases in world oil reserves will come from enhanced recovery from existing fields rather than from discovery of new fields. Supply prospects are heightened by second-phase discoveries in established basins, encouraged by improved economic returns of smaller fields at today's higher oil prices, and enhanced by better recovery technology and deeper drilling capabilities.

Natural gas is likely to make an increasingly important contribution to world energy supplies. Though gas expansion in the past has been held back by economic constraints, these have recently been greatly reduced. As gas is relatively expensive to transport over long distances compared with oil or coal, its production has basically been developed in countries close to large energy markets, particularly the developed and socialist countries, and has been relatively little exploited in developing countries. However, specific problems associated with natural gas, such as storage and transportation, have been eased considerably by improvements in the cryogenic process and technologies involving conversion of gas into liquids.

Gas resources and usage have been inadequately pursued in the developing countries, but a substantial part of the growth of conventional natural gas production during the next two decades is expected to occur in these countries. Expansion of local gas usage should be considered a major policy aim. Where the domestic market is limited, exports by pipeline or liquefied natural gas (LNG) tanker may be advantageous, but where the export market is very distant, it may be more feasible to convert the gas into methanol prior to shipment. In terms of the contribution that natural gas is likely to make to future world energy supplies, one chapter concludes that "over the period through 1985, natural gas will be the fastest growing energy source in the developing countries as well as in the world."

LNG technology is highly advanced and readily available. It probably represents no higher level of risk to the public and the environment than conventional fuels. The main markets are Japan, Western Europe, and the United States, and the main LNG sources are OPEC countries (Algeria, Indonesia, Libya, and UAE), Brunei, and Alaska (USA). Other LNG projects are under construction or in the planning stage.

In recent years economic, technical, and environmental reasons have led to the delay or cancellation of many LNG projects. In particular, the high cost of moving gas compared with liquid and solid fuels has made for a relatively low net price for gas at the wellhead. Hence, petroleum producing countries have typically pursued other uses for their natural gas. These include reinjection to maintain oilfield pressure, other field applications, extraction of natural gas liquids for domestic and export markets, and commercial applications in local markets, particularly energy-intensive industry, such as water desalination, petrochemicals, and aluminum smelting. The balance, still a substantial amount, is being flared or vented when it is produced in association with crude oil or is being shut in when it is nonassociated.

Today's international oil prices have greatly improved the economic viability of LNG projects. By the same token they have also improved those of other high-cost sources of energy in countries which are not energy-deficient. Investment decisions for LNG projects thus have to be justified in comparison to the possibilities for alternative energy supplies.

Deep conventional gas refers to accumulations found at depths greater than 15,000 ft. The potential for such gas accumulations may be much greater than has been assumed in the past and could well be one of the world's major fossil fuel sources. A major deterrent to deep gas exploitation is the increased cost of seismic surveys and drilling. In some countries, where oil reserves are plentiful, there is little incentive to drill if there is no ready market for large volumes of gas. The Middle East could have giant deep gas deposits, but it is doubtful that this potential will be tested until oil reserves begin to be depleted.

Worldwide, meantime, the addition of new reserves of gas currently exceeds consumption.

Section V is concerned with the important topic of oil from shale and tar sands, heavy oil, and enhanced oil recovery. Shale oil production is presently small, though the resources are huge. At present oil prices, some development has started and some projects may at last become economically viable. Environmental and technological problems hinder their exploitation on a commercial scale when using the method of surface mining, but this depends largely on geographic location. In the USSR, large quantities of oil shale are used as fuel in power stations. In this case, however, the oil is not first extracted; the rock is simply crushed, yielding a very high ash.

Heavy oil and tar sands are low quality oils probably existing on land and offshore in far larger quantities than conventional (light) crude. The main difference between heavy crude and light crude is the much greater viscosity of heavy crude, which makes its extraction from underground more difficult. It usually has no lifting mechanism provided by nature and thus the production costs are higher than those of light crude. A variety of technologies has been developed in recent years for the production of heavy crude, and with today's oil prices, heavy crude and tar sands are economically attractive. This is true even if heavy crude entails higher production costs and needs to be upgraded in quality in order to make it equal to a barrel of light crude. It is estimated that the cost of producing and upgrading a barrel of heavy crude from the Orinoco Belt of Venezuela would range between US $5 and $13 (in 1979 dollars). Moreover, the estimated quantity of recoverable heavy crude in the Orinoco Belt is in the vicinity of 500 BB, which is almost equal to the estimated global proven and recoverable reserves of conventional crude. One chapter reports on the progress of the underground tunneling method for the production of heavy crude and confirms earlier estimates of the costs of production of heavy crude in Alberta at Can. $7 per barrel using such methods. Tar sands, which are an even lower quality material than heavy crude are today profitably produced at two surface-mining projects in northern Alberta in spite of the difficult winter conditions there. Underground extraction of tar sands, which soon will become operational, is believed to cost less than surface mining under hard winter conditions.

The economic feasibility of enhanced oil recovery techniques has been improved by today's higher oil prices. Several types of recovery methods are discussed here, including: primary recovery from crude oil reservoirs by natural reservoir energy; secondary recovery by waterflooding or gas injection; tertiary recovery using micellar, thermal, carbon dioxide, hydrocarbon miscible, and chemical waterflooding techniques; and thermal techniques, i.e., steam injection and in-situ combustion. In addition, there are possibilities for using microbiological processes for enhanced oil recovery.

Section VI is comprised of a series of papers devoted to naturally occurring gas from other than conventional reservoirs. As a result of improved technology and higher energy prices, much attention is being given to finding and producing such deposits. Any estimates of such gas in place are highly speculative, as described in Chapter 4, but are not implausible. Even if the gas does exist, we cannot now predict that it can be produced economically in significant amounts in every country, though considerable quantities of some of these unconventional gases are already produced. Nevertheless, they must be included in a comprehensive survey of energy resources.

Underground coal gasification offers a relatively low-cost, environmentally sound method to produce clean fuels from coal. This opens up the opportunity to exploit coal which is not mineable by present-day techniques and could thereby enable vast and widely distributed resources to be reclassified as recoverable reserves. The Soviet Union has developed a commercial technology which some American interests have purchased. Field tests in the USA to date have confirmed the potential economic and environmental advantages of this process, and the process is being scaled up. This interest in the USA could lead to commercial production by the late 1980's.

Another source of gas is methane obtained from coalbeds by drilling. In small volumes it is already extracted from underground coal mines for commercial use, and through hydrofracturing and other methods, yields can be substantially increased. An additional benefit is that the extraction of gas from coal deposits before mining begins reduces the risk of gas explosions. Furthermore, coalbeds which are presently considered unmineable by reason of depth or thinness of seams can be tapped by drilling wells. It seems clear that natural gas from coalbeds is an attractive energy source, and its development should be pursued more vigorously.

Encouraging results were found in a study by Hydro-Quebec to substitute peat for diesel oil in small-scale electricity generation in remote areas of Quebec, through a process of gasification.

Gas dissolved in water can be placed into three categories: (1) gas dissolved in underground water near hydrostatic pressure, (2) gas in geopressured zones, and (3) gas in lakes. Also, the formation waters in many petroleum-bearing basins of the world may contain large quantities of dissolved natural gas. Gas in underground water is produced today in Japan, the United States, and Italy. Significant resources undoubtedly exist in many parts of the world, but little exploration has yet taken place. While this type of gas may be unable to compete economically with conventional gas where the latter is readily available, it could compete with imported gas.

Geopressured gas is now known to exist in the geopressured zones found in many parts of the world, both onshore and offshore. The economic feasibility of using this gas is a function of a variety of factors, including the use of by-products, such as water and salt, and of the geothermal and hydrostatic energy of the reservoir. At this time, however, the only active program in exploration and development of geopressured resources is taking place in the USA. Nevertheless, the potential of this type of gas is believed to be truly enormous, and it is consequently thought to be a major source of energy supplies in the future.

Lake Kivu, which is shared by Rwanda and Zaire, is presently the only known case of a large gas reserve in solution in a lake. There has been no systematic exploration for gas in other lakes, and it may be expected that in many of them, as in marshes, methane is being produced. However, the safe extraction of gas from a lake is still a matter of research.

Reserves of black (Devonian) shale rich in organic matter represent an important potential source of gas. Large quantities of methane trapped in and beneath thick layers of shale have been produced in significant volumes in the United States for decades. In the past, gas production per well was low, but well life was very long. In more recent years, hydro-fracturing has increased well productivity. As such shales are found in many parts of the world, the global resource potential, though unknown, is surely considerable. Novel approaches to raise production include the use of (1) chemicals to cause fractures and (2) bacteria to cause erosion of the shale. These methods are now the subject of experimental work.

Reports of the discovery in Siberia of natural gas in a frozen state--called gas hydrates--have aroused widespread interest and have led to the conclusion that hydrates may exist over large regions elsewhere in the world. Although significant deposits on land are limited to regions of thick permafrost, such as that which covers parts of Canada, Alaska, and the Soviet Union, many areas under the deep oceans also are favorable for the formation of gas hydrates, their existence in fact being confirmed by the deep sea drilling program.

Production of fully hydrated reservoirs does not look like an imminent possibility, because there is no known technology today to liberate the gas from their hydrate forms. The first efforts might be to produce gas from beneath a hydrate deposit. This could lower the reservoir pressure, causing adjacent hydrates to decompose and thereby maintain reservoir equilibrium. The process might be stimulated by injection of steam or hot water.

Initial studies are focused on finding and identifying hydrates, studying the engineering problems for their extraction, and gaining further understanding of their properties, particularly their decomposition. Petroleum scientists and engineers should be aware of the potential vastness of the hydrate resource and should be considering how to recover the gas. Indeed experts from the USSR estimate global gas hydrate resources to be very great. In the short term, production may not be economic, but the longer term outlook could be different, depending on technology and economics.

Gas is found in very large quantities in tight sandstones, which are low permeability structures requiring a fracturing of the host rock to allow the gas to flow freely to the well-bore. About 1 trillion cubic feet per year or almost 5 percent of the total conventional gas production of the USA is produced from tight sandstones. The amount of the recoverable quantities of gas from such tight sands in the USA is not known due to insufficient field data. Exploration has just begun in Canada, and the potential in many developing countries appears promising. There is little doubt that gas from tight sands is another significant resource in the future. Although gas production from tight sands is not the subject of a separate chapter, it is too important as a resource to not bear inclusion here. Resource estimates are included in Chapter 4.

The 11 chapters of Section VII are devoted to applications of solar energy. The economic aspects of solar water and space heating have been studied in many countries, but no generalizations are possible, because cost comparisons can only be made on a country-by-country basis, taking into account local conditions, including the price of alternative energy forms. For example, in North America, while solar energy systems normally cost more initially than conventional alternatives, there are economic merits in the longer term.

In assessing the feasibility of solar heating systems, the method of calculating costs is crucial: for instance, is the cost of maintenance or of the required storage facilities included? Is allowance made for rising fuel costs? What is the reliability of the system and how easily can spare parts be procured? Is a provision made for a back-up or supplementary system? Chapter 62 describes the development of solar and other energy resources in Tanzania, which has begun to develop renewable energy systems, including solar box cookers, water heating, and distillations systems. Efforts are concentrating on finding energy technologies applicable in rural areas, which are much more affected by the firewood crisis than the oil-import burden.

There are two methods of generating electric power from sunlight. One is the sequential conversion of sunlight to heat, mechanical energy, and, ultimately, electrical energy. The other uses solar (photovoltaic) cells to directly convert sunlight to electricity. The attraction of solar cells is that they require no moving parts, are nonpolluting, and have the potential to last a long time with minimal maintenance. Their major disadvantages derive from the intermittent nature and varying intensity of sunlight and, of course, their cost.

The technology of solar cells is well proven; they have been used successfully in space vehicles and at remote sites for several years now. The problem is that they are expensive: the present cost of silicon solar cells is about $10 per peak watt. Although silicon solar cells are the only type presently commercially available in large quantities, research on other materials is occupying a bevy of companies, large and small. The ideal material would combine high efficiency (gallium arsenide has the highest), an abundant supply, and a low cost of manufacture. Chapter 59 describes a promising amorphous material that may approach this ideal. With a goal of producing solar cells for about $0.50 per peak watt, one may voice optimism on their future role in solving energy supply problems.

Solar cells are already beginning to be introduced for rural applications in developing countries, typically through international or bilaterial programs of technical assistance. Indications are that it is becoming economically viable, particularly in areas where the cost of electricity is already high. An example is given in Chapter 58 for a location in Upper Volta.

Another method of collecting energy from the sun which has interesting possibilities is the solar pond, which collects and stores the sun's heat. Solar ponds are a potential source of low temperature heat in large quantities, unlike solar collectors which are typically small-area devices. The art is to avoid water at the bottom from heating and rising by convection, which would dissipate the temperature difference. In the nonconvecting pond, convection is prevented by making the water at the bottom heavier than the water above; this is done by dissolving salts in a higher concentration at the bottom than at the top. Heat is extracted by decanting the hot layer at the bottom.

Solar ponds realize certain advantages over regular solar collectors: (1) solar ponds can be made very large; (2) the water and soil below it act as built-in energy storage; (3) the ponds have no collector windows that need cleaning; and (4) the hot water can be transported easily to a central place of usage some distance away (e.g., 1 km or more). The solar pond may also prove to be a source of solar electric power in areas lacking an electricity grid, as the calorie cost is evidently sufficient to make low temperature turbines a practical application. Pond sites are limited roughly to the area within 40° of latitude.

Solar ponds are ideally suited for decentralized sites needing thermal energy and power generation. Among other end uses, which have been demonstrated in the field, are domestic hot water, swimming pool and greenhouse heating, space heating, industrial and agricultural processing, and boiler feed preheating.

The advantages which might be obtained by constructing a solar power satellite are enumerated in Chapter 60. According to this concept, a solar power satellite would convert solar energy into electricity and feed it to microwave generators. From these, the energy would be directed down a microwave beam of very low power to receiving antennae on earth. The microwave energy would then be reconverted to electricity and delivered to end users. The advantage to this system is that, because the satellite always faces the sun, it could collect energy and produce electricity 24 hours a day. This concept is envisaged as an option for the 21st century, but presently is a costly energy source.

Energy derived from exploitation of the biomass is detailed in the 22 chapters of Section VIII.

Biomass refers to the variety of plant and other organic materials which can be used as feedstocks for conversion to useful fuels and by-products. These include trees, forestry and agricultural residues, manure, and aquatic plants. Many countries both in the developed and the developing world are exploiting the energy potential of biomass. Sweden, which is the country most dependent on imported oil, has an active program, and 1 to 2 percent of the energy used in the USA is derived from biomass, a percentage which is expected to increase significantly over the next 40 years. This source of energy should be of particular interest for developing countries, because it can provide small-scale energy plants for use in cooking, electricity generation, and driving engines in rural areas.

In regions which are well endowed with forests, it may make economic sense to run thermal power stations on wood rather than other fuels (Chapter 71). Naturally a perpetual supply of forest biomass must be within economic distance. Forest management would have to be expanded to comprehend the production of the biomass for energy as well as for all other uses.

Trees and forest residues can be converted into useful energy forms by thermo-chemical methods including direct combustion, pyrolysis, gasification, and liquefaction. The pyrolysis of biomass in the absence of air produces char, condensable organic liquids, water, and noncondensable gases. It is a very old process and has been used since earliest times to produce charcoal, tars, pyroligneous acid, and gases. In recent years pyrolysis of biomass has received renewed attention, as it offers a means of reducing dependence on imported oil.

Research into the commercial utilization of seaweeds dates back to the beginning of this century, but seaweed production of fuels really began in the 1970's. The potential for oceanic biomass conversion to energy is now being studied at several places. The great advantage of the marine environment is the vast space available for biomass production and the constant availability of water. However, there are some disadvantages involving high capital cost estimates, crop damage from storms, and the relatively limited knowledge of aquacultural science compared with agronomy.

There has been significant development of processes for large-scale production of ethanol by fermentation from carbohydrate raw materials, which could replace a portion of gasoline in motor engines. The ethanol is often called power alcohol, or colloquially gasohol. All starch crops can be fermented to alcohol, but the most important ones are sugar (raw sugar and molasses) and starchy roots (such as cassava). Because these crops are relatively expensive in Europe, they cannot serve as cheap raw materials for ethanol fermentation. In contrast, they are usually inexpensive in developing countries. A number of countries throughout Africa, Asia, Latin America, and the Caribbean have the potential to grow sugar and molasses for ethanol fermentation. The major producers of starchy roots (Brazil, Nigeria, Indonesia, India, Thailand, and Zaire) also have this potential. The total ethanol potential from these three products is estimated to be nearly 7 million tons (135,000 barrels per day or b/d) by 1985. Brazil and Thailand account for some 40 percent of this potential, Cuba and India another 20 percent, then the Philippines, Nigeria, and Zaire in smaller amounts. In particular, Brazil's National Alcohol Program aims at producing 11 million m^3 (185,000 b/d) of ethanol by 1985. Chapter 74 presents a very interesting program in which alcohol from sugarcane and cassava is expected to reduce significantly Brazil's dependence on oil.

Brazil has a long tradition, dating back to 1931, of adding small amounts of alcohol to gasoline; at that time, the government ordered the addition of 5 percent ethanol. But now they will be going much farther, the aim being to eventually use alcohol to replace up to 20 percent of automotive gasoline and to gradually replace such petroleum products as diesel oil and certain petrochemicals. If all goes well, hydrated ethyl alcohol will entirely replace gasoline as the automotive fuel.

In addition to relieving the burden resulting from the fact that Brazil has to import over 80 percent of its fossil fuel needs, other benefits will flow from this project. For example, Brazil's sugar industry will not suffer from a decline in the world market price, and it will be possible to manufacture locally all the equipment required for the new distilleries to support the expanded alcohol production, thus stimulating Brazil's capital goods sector.

A similar approach has been adopted elsewhere. In Australia, the sugar crop has been doubled with the expectation of supplying enough ethanol for 12 and 15 percent of the country's fuel needs, while the Tate and Lyle Company of the United Kingdom has allocated $34 million for the development of new uses of sugar and other crops as sources of energy and chemicals.

The availability of arable land and competition between alternative uses of land for crops and animal feed are important economic and political issues. However, frequently little else can grow on land where sugar can be grown, and therefore use of land for ethanol crops or animal feed is not necessarily exclusive. Moreover, there are still a number of countries with large areas of uncultivated land which could be put into production for biomass materials.

Methanol, or methyl alcohol, can be synthesized from low calorific coals, natural gas, biomass, or any material with available carbon. Although not a primary energy source, it has value as a fuel, a solvent, and a petrochemical feedstock. With its ease of transportation, methanol has long been a valuable product, but now increased oil prices make it still more attractive, both as a fuel product (for use in power stations and in automobiles) and as an intermediate product in the production of gasoline. The result has been a sharp increase in methanol utilization; the outlook is for further increases.

During the last decade efforts have been made to develop systems for converting urban refuse into useful fuels. As fuel costs continue to escalate, the value of energy in solid residues generated by society also increases. The techniques being developed to extract energy from these residues range from mass burning with recovery of steam, to pyrolysis for production of fuel gases, to anaerobic fermentation for methane recovery.

There are also projects operating and more planned to extract methane gas generated in sanitary landfills.

Among various renewable energies, biogas is now receiving the greatest attention. It is produced from anaerobic digestors in which wastes are decomposed by the action of bacteria. This system improves the environment by disposing of such wastes, and allows the remaining sludge and effluent to be used as fertilizer. A household-sized plant can produce gas for cooking and lighting. The process is generally referred to as "biogas technology" (BGT) and has been promoted extensively in such countries as China and India. In China, approximately 20 million biogas plants are in use. On a farm in Kenya, methane from biogas has been the only source of fuel for cooking since 1955 and the residues have been providing fertilizer for the cultivation of coffee.

Although in some rural areas the expansion of biodigestors is slow, because of ample supplies of firewood, dung, or other energy sources, the use of biogas is expanding in both developed and developing countries. From all indications, biomass in its various forms can supply energy for a number of purposes. Its importance should be recognized and development encouraged to extract the maximum benefit from this renewable energy source.

Section IX is comprised of 11 chapters describing various applications of water to power generation. Hydroelectricity has many evident advantages. It is a continuously renewable and widely distributed resource. Generating plants can be expanded with relative ease and the power can be integrated into multi-use development. Hydroelectricity is nonpolluting in that it does not produce such by-products as thermal effluents and it is an end result of one of the more efficient energy-conversion technologies. Projects have a long life, low operating costs and no fuel costs, and the technology is well known. In addition, energy is storable as water in the reservoirs.

The use of hydroelectricity is expected to continue to grow during the next decades. Although new sites are becoming harder to come by in most developed countries other than Canada or Norway, there are still many opportunities in developing countries for hydro projects.

Hydroelectricity does have certain disadvantages. It is highly capital-intensive, the most promising sites are often far from markets, and the construction projects can have an undesirable environmental impact on the ecology. In addition, there are geo-political constraints associated with the flooding and harnessing of international rivers.

Small-scale hydroelectric plants have generally declined in number for many years, with emphasis being placed on generating electricity in thermal and large-scale hydro plants. However, this trend may now be changing, particularly in developing countries. A striking example is China, which has installed over 80,000 hydro units during the last three decades, of which 60 percent are less than 500 kW each in capacity.

In Greenland, the large volumes of water melting each summer from the ice shield close to the coast appear to offer favorable conditions for large-scale hydro development. The potential for such a project has received intensive study, and seems to be feasible, if highly capital-intensive. The project envisages that this potential source of power be supplied for local use and--for the most part--for export to North America and Western Europe by electric cable, or ship or pipeline after conversion into ammonia or hydrogen.

Energy can be created by the mixing of electrolyte solutions of different concentrations. The possibility of using this salinity energy has only recently been considered, and initial research and development has started in at least three countries.

The methods under consideration are based on osmotic pressure differences, ionic diffusions (reverse electrodialysis), and vapor pressure differences which occur between solutions of different concentrations. These methods, essentially the reverse of water desalination technologies, are still in the early laboratory research stage.

The oceans, which absorb most of the earth's solar radiation, are stratified by temperature and density, and in principle, wherever there is a difference of temperature, a heat engine can be built to convert part of the energy into useful work. Two systems which would convert the ocean's thermal energy are under consideration: the closed Rankine cycle and the open cycle. The Ocean Thermal Energy Conversion (OTEC) systems would produce electricity to be cabled ashore or used to make energy-intensive chemicals and metals on board the floating OTEC facility. There is still a range of technical and environmental problems to be solved, but though this source of energy is still at the research stage, a small pilot unit does operate offshore Hawaii.

Inventors have been intrigued by the possibilities of wave power for many years, but only recently have serious engineering studies been made on the cost effectiveness of such machines for large-scale power generation. Large-scale wave energy converters look technically feasible on the basis of largely existing technology, but they represent expensive ways of producing electricity and appear to have some significant flaws. However, further research may well show how to lower costs and redesign wave energy converters.

The Japanese have built a huge machine which is being tested in the Sea of Japan, and the Lockheed Co. is developing a wave energy converter in the USA. Wave Power Ltd. of the United Kingdom is working on a device called a "wave contouring raft," and similar research is underway in Norway. The government of Mauritius has initiated an ambitious wave energy project which is expected to provide 20,000 kWh of electricity a year. Nevertheless, it does appear that, if there is any application of wave energy technology for large-scale electricity generation, it will not be before the next century.

There has been a resurgence of interest in tidal energy in the search for new energy sources. The only modern, relatively large tidal power station in the world today is at La Rance, France. There are some small plants in China, one in the USSR, and an experimental one in India. Some sites elsewhere have received intensive study, including the Bay of Fundy in eastern Canada, but there are relatively few locations where the tidal range is large enough to justify consideration of tidal power development.

The seven chapters of Section X investigate the use of wind energy for power generation. Although used throughout much of history to drive ships and windmills, wind power has been neglected since the application of fossil fuels, and crude oil in particular, to energy production.

In recent years there has been a revival of interest in wind power, based on modern technology. Wind energy can be used to generate electricity or direct heat, to drive pumps, and to propel ships of increasingly large size. Prototype and experimental wind energy converters have been built in a number of countries as part of research or demonstration programs sponsored by private and government institutions. These programs include compilation of wind speed data, study of the performance and reliability of wind turbine generators, and investigation of their environmental acceptability and means for integration into electric supply systems. Similar units are available commercially.

In the past 5 years, firms in Denmark have sold and erected several wind energy generators which will produce electricity, and Australia plans to install 3,000 generators to provide two-thirds of the energy needs of western Australia. Tanzania has started to build and test low-cost wind turbines and has concluded that not only is wind power a technical possibility, but it is more economical than entirely diesel-based systems for village power stations. A 2,000 kW wind converter is in operation in North Carolina (USA) and even larger ones are being developed in Sweden and the Federal Republic of Germany.

The best sites for wind energy production are along ocean shorelines, in the mountains, or offshore, where the wind's velocity is unimpeded; favorable sites may exist onshore at special locations. Some national energy programs envisage that wind energy could be used to generate significant amounts of electricity by the mid-1980's. It might represent 10 percent of total electricity generation if there were no system for wind energy storage and more if there were. There is also a significant market for smaller converters to supply farms and isolated houses.

Section XI is comprised of 14 chapters addressing the complex subject of geothermal energy.

The natural heat existing under the earth's surface can be transmitted to the surface and made to produce energy. There are six kinds of geothermal energy resources: three of these are fully developed (dry steam fields, hot-water fields, and low-enthalpy--or low temperature--fields), and three are at the stage of research and development (magma energy, geopressured zones, and hot dry rocks).[2] Hot water fields are the most commonly exploited form of geothermal energy, but dry steam fields have the advantage that the steam can be directly used in generating electric power. Geothermal energy can also be used in horticulture and pisciculture.

The greatest efforts to exploit geothermal resources are in the high enthalpy range (more than 150°C), although the geothermal energy used for direct heat worldwide from resources of low to moderate temperature is about three times greater than that for electrical power generation (6,000 MW versus 2,000 MW). This is because low grade geothermal energy sources, which are economically submarginal for power production, can still be used for industry and agriculture as well as for heating or cooling.

Research is now being conducted on ways to capture the geothermal energy in hot dry rocks, that is, the technology is being developed to exploit the earth's heat in regions where steam or water are not naturally present as a means of bringing the energy to the surface. Successful tests have been run on a system in which two wells are drilled into a rock formation and the rock in between is fractured; water is then injected down one well and steam recovered from the other. Further development is needed to demonstrate the commercial feasibility of the system.

---

[2] Geopressured zones are described in the section on natural gas.

Magma chambers are underground occurrences of molten rock in the temperature range of 700°-1,200°C. There are many areas around the world with magma chambers within 10 km of the surface, and these represent a very large source of thermal energy. Research has overcome many of the problems involved in drilling into such hot material, including keeping the drill hole open and preventing the casings from being destroyed by the powerful corrosive forces of the magma. All such research is presently funded by governments, and if more funds were devoted to it, magma could develop into a significant source of geothermal energy in the near future. One research project funded by the United States Government has already proven a process, for which a patent has been taken out, whereby water and biomass are introduced into the magma chamber for the production of steam and methane--a sort of high temperature natural biodigester!

Geothermal resources are common and commercial nearly everywhere. Although the technology is mature and expertise is available, their commercial development until recently has been slow.

Development efforts of geothermal energy are impressive. Iceland is by far the leading country in the use of geothermal energy for space and greenhouse heating, while Japan is using such energy to supply hot water to hotels and dwellings, to heat greenhouses, to raise fish, and to provide air conditioning to buildings. Other countries now exploiting geothermal energy include China, El Salvador, France, Hungary, Indonesia, Italy, Mexico, New Zealand, the Philippines, the USSR, and the USA.

The United Nations has made a key contribution toward this development. For the past 28 years it has assisted member countries in evaluating their geothermal energy resources and has made available technical assistance of experts for training, funding, and executing major exploration projects in several developing countries. Worldwide, geothermal electric plants now have approximately 1,800 MW of installed capacity compared with 680 MW in 1969. By 1985, perhaps 10,000 MW could be in operation based on known exploration work and projects.

Substantial geothermal resources may also exist in some smaller islands, whose limited energy needs could be met by them. The greatest potential is in the Pacific Islands although, on a lesser scale there is also potential in the Caribbean, some Greek islands, the Azores, and the Canary Islands.

The four chapters which make up Section XII consider the principal factors necessary to evaluate nuclear power as a future energy source.

The outlook for nuclear power in industrialized countries is uncertain in all respects. Confidence in safety was badly shaken by the accident in March 1979 at Three Mile Island, Pennsylvania. The treatment of spent fuel elements is yet to be solved. Concern about proliferation of nuclear weapons has been revived by reports of tests and planned tests in a growing number of countries and has heightened fears of the possible diversion of nuclear material for military and terrorist purposes.

In general, nuclear power has not been demonstrating the clear economic advantages which earlier had been forecast, because of rising capital costs, high repair costs, long down-time for repairs and maintenance, and financing difficulties. At the same time, since the slowdown in economic growth commencing in 1973, the demand for electricity has grown at a lower rate than forecast in the developed countries, a reduction that has fallen disproportionately on nuclear power, which had been expected to account for a large part of projected increases in electric generating capacity. This has led to the deferral or cancellation of a large number of orders for new nuclear power plants.

Despite these radically lowered expectations, the outlook for nuclear energy suggests a gradually rising share of electricity generation in some industrial countries, and a few (particularly France and the Soviet Union) are moving ahead with breeder reactors. Plans in Britain, FR Germany, and Japan for breeders have been delayed, and those in the USA suspended.

The development of nuclear power in developing countries, according to one chapter, should not be precluded because of fears of proliferation of nuclear weapons, rather, economic and technological grounds should be the basis for that decision. Whatever position is taken on this assertion, the fact that nuclear energy produces electricity in large, centralized power stations must also be viewed with its social and economic implications. These impact most strongly in developing societies and should consequently be taken fully into consideration when deciding whether to embrace the nuclear option.

In another chapter, it is argued that nuclear power is desirable and can be made acceptable for industrial countries. Although technical improvements are forthcoming, institutional changes are still needed. A clustering of reactors in a relatively few, remote sites is advocated in order to reduce the risks associated with them. It contends that all energy options should be preserved, including nuclear energy and breeders. For the long term, in this view, only fusion and solar energy would be potentially large enough systems to provide an alternative to fission, and while they may one day prove technically and economically feasible, this is still not known. Caution would, therefore, dictate developing the breeder and deploying it in the event alternatives prove insufficient.

Possible future constraints on the use of fossil fuels as a result of the release of carbon dioxide (CO_2) into the atmosphere were also introduced. This argument is that the serious risk of climatic change in the direction of increased temperature from the increased concentration of CO_2 in the atmosphere dictates a prudent course of action, i.e., one that emphasizes energy sources producing no CO_2, above all nuclear energy. There is no consensus on this view, for some question the reliability of climate models in predicting temperature increases proportionate with an increase in CO_2 concentration. There is a general recognition that much work remains to be done on climate change. It also may be possible to take advantage of the increasing commercial value of CO_2 by capturing it from the combustion gases of power plants.

It may be that nuclear electricity generation is a most expensive way to generate electricity (if all costs are included) and, as every energy source can be used for electricity production, nuclear energy is superfluous.

The five chapters in Section XIII are related in that they suggest that electricity may provide a source of power which is of growing importance to developing countries, that its effectiveness is greatly enhanced through cogeneration, and that important new means of producing electricity are now under study. The point is made that the crisis in oil supply may prevent fossil fuel-poor developing countries from pursuing a course of development such as that followed by the industrialized countries, but two factors may mitigate this problem: many of the developing countries tend to be rich in hydroelectric and biomass resources. This program still will necessitate a conservative use of electrical power and the evolution of higher efficiencies in biomass utilization. However, just as the "wood crisis" in 16th century Britain forced a shift to coal and set into motion a chain of events that culminated two centuries later in the Industrial Revolution, the present "petroleum crisis" could provide the impetus needed to shift developing countries to a course of sustainable development (Chapter 124).

Two chapters deal with the very important topic of cogeneration or the combined production of heat and electricity. Although not a resource, this more efficient capture of available energy amounts to the same thing. In the United States and the USSR, for example, it is estimated that as much as 30 million tons of coal per year is saved through cogeneration. In many other countries, especially FR Germany, Italy, Poland, Romania, and Sweden, less but still significant savings are recorded.

Two chapters describe energy sources which serve as examples of those in an early stage of development. Research is now being conducted on the generation and conversion of electrostatic energy. The earth is a huge generator of electricity, as is evidenced by electric storms. Experiments are presently being carried out to explore whether this energy could be tapped by appropriate technology, including sharp-point or radioactive antennae.

An electric current can be produced in electrostatic generators by mechanical means such as rotating disks, cylinders, or belts. And although electrostatic generators are now economical where power output of 10 kW or less is needed, outside this range, they are usually inferior to electromagnetic generators. It is thought that with some research they could become superior over a much larger range of power output and voltages.

Electrostatic motors create motion as a result of electrostatic forces acting between electric charges. In contrast, conventional electric (i.e., electromagnetic) motors create motion as a result of magnetic forces acting on electric currents. Many different types and designs of electrostatic motors are possible. Indeed, they can operate from a much greater variety of sources than conventional electromagnetic motors; these sources include charged capacitors, high impedance chemical batteries, electrostatic generators, atmospheric electricity, and nuclear batteries. It appears that research into electrostatic motors may have great potential.

In conventional electric motors, motion is created by magnetic forces acting on electric currents. A new concept concerns the potential of the permanent magnet motor. In electric and permanent magnetic motors, a field imbalance is created wherein the field in focused and timed, and the magnetic leakage is controlled. In the concept of the permanent magnet motor, the timing would be built into the motor by the size, shape, and spacing of the magnets in the stator and armature. The focusing would be controlled by the shape of the magnet's pole length and the width of the air gap. The permanent magnet motor represents an electricity-generating unit requiring no fuel or other energy input except the very gradual using up of the magnets. Patents have been issued for such a motor and the first small units will become available within 2 years.

The final, but in many ways the most significant portion of this volume, Section XIV, is comprised of 28 chapters prepared by energy experts from specific developing countries. These contributions clearly reflect a variety of conditions from energy-rich to energy-poor, in different stages of industrialization, and with varying availability of technical people. Nevertheless, a few general observations may be made.

One phenomenon which developing countries are experiencing is a high rate of urbanization which implies a radical change in the supply base. While rural dwellers consume mostly noncommercial, locally produced energy forms and produce much of their own food, the basic needs of urban dwellers must be met from commercial (and not infrequently imported) supplies. Hence, urbanization will lead to a vast expansion in the demand for commercially traded energy.

A further point is the generally lower base line of energy-resource knowledge in developing countries as compared to industrialized countries. Even for conventional oil and gas, the total amount of exploration until now is clearly unsatisfactory, and as regards unconventional oil and gas, resource knowledge is practically nonexistent in almost all developing countries. In addition, knowledge of coal, lignite, brown coal, geothermal, and other underground resources is inadequate. In evaluating surface resources such as hydro, wind, and solar radiation, specific local measurements which are essential for project planning are often lacking.

In analyzing the appropriate degree of centralization and decentralization of energy supply, cost is regarded as the most crucial element for most nations in assessing the relative merits of centralized and decentralized systems. The growing role for renewable energy sources is recognized, though most solar installations in developed countries will continue to need back-up support from centralized systems. However, systems in many of the developing countries are already highly decentralized. Very likely, the efficiency of electric energy use could be considerably improved by decentralizing power generation, so that (a) increments in capacity are better matched to those in demand, and (b) opportunities for cogeneration can more easily be seized. The economic feasibility of solar energy was thought to depend on how it is integrated into nonsolar back-up systems.

By way of conclusion, the Conference itself endorsed the principles contained in the declaration of the Conference on the Human Environment approved in Stockholm in 1972. In particular, the present Conference stressed the need for rational management of resources in which (a) environmental protection is integrated with development, (b) the earth's capacity to produce renewable resources is maintained, and (c) nonrenewable resources are safeguarded against excessive depletion. Assistance should be given to developing countries to promote their development in a manner compatible with preservation of environmental quality.

In the second week of the Conference, a working group was formed to produce a paper for discussion by the full Conference. This paper summarized the proceedings succinctly, and from it the following conclusions:

> The papers presented and the discussion which followed made the Conference conclude that the world's energy potential is great, but the labor and capital requirements of exploiting new energy sources in sufficient quantities will be high, and the transition will take time. This emphasizes the importance of efficient use of energy, as well as the strategic need for every country or region to utilize the greatest possible range of energy resources. Also emphasized was the necessity of not degrading the global environment.
>
> An essential point is the role played by energy in development efforts, and its implications for the relations between industrialized and developing countries.
>
> Leaving to other forums suggestions regarding the imperative and urgent need for assistance to countries, particularly developing countries, to tide them over the serious problems of transition to a future based on renewable sources by appropriate international arrangements, the following suggestions are made:
>
> 1. There is a need to intensify the energy resource assessment effort in all countries, particularly in developing countries. Countries should take appropriate steps to recognize the existence of the energy resources available within their national boundaries and develop them for optimal utilization.
>
> 2. Countries (or regions) should evolve a national (or regional) energy policy that integrates energy into their economies with the aim of achieving optimum levels of self-reliance. Preferential emphasis should be made on the development of those available energy resources that could be most economically exploited within the context of the nation and on those nations not endowed with abundant resources of unconventional energy.

3. Countries should create appropriate national institutions where these do not already exist to coordinate energy planning, to promote orderly development of their energy resources, and to establish liaison with other countries and agencies involved in the energy sector.

4. Countries should encourage cooperation in sharing the benefits of energy technology and joint energy resources. Without in any way jeopardizing their national sovereignty, all countries should promote regionally combined development of such resources as hydro-power and solar energy.

5. Countries should promote the rational use of optimal energy systems, with attention paid to socio-economic, scale, and environmental factors as well as efficiency of conversion.

6. The Conference recognizes that potential social, economic and ecological changes have resulted and may in the future result from the introduction of various energy conversion systems: whilst in the main these changes are detrimental, beneficial effects have also been noted. The Conference recommends that particular attention should be paid to ascertain at the early planning stage the full range of such consequences of any energy projects.

7. Scientific and technologically-oriented meetings and conferences should be held at national, regional and international levels, at which the active participation of all countries in the developing world should be ensured.

8. Given the diversity of energy resources, of conversion and delivery systems, and of the prospects of their rapidly changing technological, financial, and economic profiles over the next decade and in the more distant future, the United Nations system could make a significant contribution through:

(a) monitoring new contributions in the energy field and reporting on such developments, e.g., through annual reports;

(b) drawing the implications of such assessments to the attention of international financing bodies; and

(c) suggesting areas for initiating coordinated efforts or new research, either through existing institutions or--if necessary--by creating the required bodies.

9. The Conference suggests the establishment of regional and/or national energy centres to constitute an essential infrastructure for indigenous technological development in energy. Such infrastructure should include: technology transfer mechanism, information base, energy recovery inventory, efficient Research and Development management systems, education and

training, and public support, as well as ways of financing. Personnel from the developing countries should be involved in all of them.

The conclusions and suggestions made in this document have been synthesized from a diversity of opinions voiced at the Conference.

While not committing in detail each of the participants, these conclusions and suggestions were largely shared by all the Conference participants.

Richard F. Meyer
Editor

LONG-TERM ENERGY RESOURCES

SECTION I. WORLD ENERGY OVERVIEW

CHAPTER 1

LONG-TERM ENERGY RESOURCES AND THE ENERGY OUTLOOK

Joseph Barnea[1]

The UNITAR Conference on Long-Term Energy Resources was designed to give a broad view of the energy resources that will be available in the future in order that energy policy can be framed on the basis of knowledge of the essential facts, rather than on sentiment or on impressions of temporary shortages which can easily distort the long-term perspective.

First of all it should be acknowledged that in the history of energy resources, no energy resource has ever become obsolete. As a result, there has been a steady increase in the number and type of energy resources. Historically, new resources were brought into production, because old resources were either insufficient to meet the demand or could not meet a new need. Thus, coal replaced charcoal in making steel, as there was not enough wood available for the growing iron and steel industry. Similarly, because hydro power was insufficient to meet electricity needs, other energy resources had to be developed. The history of energy resources has thus been one of a constant introduction of new sources of energy and consequently a steady expansion in the diversity of energy sources, a trend which will likely continue in the future.

Furthermore, although each new energy source has entered the production stage at a relatively high cost, the cost has normally declined once experience in finding and producing the source is acquired. For example, before World War I, the cost of gasoline in the USA was higher in real terms than it is today, and electricity in its early years was considered a luxury. This lesson of declining energy costs, however, must be applied with care in order to avoid supporting either an energy source (or an energy conversion technology) which is inherently expensive and does not lend itself to significant reductions in cost.

The gradual but steady advance in technology and the expansion in the number of energy sources, which together have resulted in a reduction in energy costs has changed dramatically in the seventies because of the sudden sharp

---

[1]UNITAR, New York.

increase in oil prices and the prolonged period of inflation. How has this unusual period affected the energy outlook?

The dramatic increase in oil (energy) prices--and the price rise was considerably higher than the rise in the cost of production--has brought about a sharp expansion of the energy resource base, because the high prices made it possible, indeed economically attractive, to produce energy resources which had previously been uneconomic. Prudhoe Bay, Alaska, the biggest oil field ever discovered in the USA, which entered production in 1977, would never have been exploited, if the price of oil had remained at $2 or $3 per barrel, because oil in northern Alaska cannot be produced and transported to consumers at such a low price. In addition, the high price of light oil has made high-cost oil such as heavy crude and tar sands, economic today. With this expansion of the resource base, there has been over the past 10 years an unparalleled increase in energy research. The United States Department of Energy, which is largely devoted to promoting energy research and development, has a budget of $10 billion, and large amounts of money are being spent by many other countries, and by companies, universities, and other institutions on energy research. Some of the results of their efforts are reflected in the Conference documents and demonstrate that it is an ongoing process, which will continue to bear fruit in the future.

The combination of higher energy prices and accelerated technological advances will not only result in the production of new sources of energy, but will also change the outlook for the known energy resources. Let us consider a few examples. In conventional oil fields, the worldwide average recovery rate is about 30 percent of the oil in place, and in some countries it is even lower. With the very much higher prices for oil and the introduction of new technology, there is every reason to expect that the recovery rate will sharply increase in the future, and if during the next 50 years, the recovery rate in existing fields goes up from 30 percent to 60 percent, as much or more oil would be produced than was produced over the past 120 years. If one also considers the additional oil that would be produced as a result of the anticipated higher recovery rates in conventional oil fields expected to be discovered in the future, then the total amount of oil likely to be produced due to higher recovery rates may be able to provide a large portion of oil needs in the next 50 years. But it is known that the world has not only conventional oil or light crude, but also heavy crude and tar sands, for which a whole new technology has come into being. As the recent UNITAR Conference on the Future Supply of Heavy Crude and Tar Sands showed, these unconventional types of oil, which were neglected in the past, can now be produced economically at present world market prices. Assuming continued technological improvement in the recovery efficiency of these two types of oil, and assuming they are fully developed, oil needs could be covered for a period far in excess of 50 years.

The new technology is not only significant for light and heavy crude, but also applies to other conventional energy sources as well. For instance, underground gasification of coal will allow the use of coal which cannot be produced economically by conventional methods. A paper to be given at the Conference will show that low-grade coal, lignite, and biomass can be converted to methanol and so make their contribution to energy needs in a variety of ways. Timber resources can be increased dramatically by tissue culture and crossbreeding, and the steadily increasing efficiency of solar cells gives hope for them to become attractive as generators of electricity in the near future. Most surprising, and perhaps even most significant, will be the impact of an old resource, namely natural gas. In the past, only associated natural gas (wet gas) and nonassociated gas (dry gas) have been considered as sources of methane, the almost ideal fuel. At the Conference, 14 different types of gas are going to be discussed, and the list is by no means exhaustive. There is, for instance, no paper on marsh gas, because the technology does not yet exist for exploiting it, although according to an estimate by the Max Plank Institute in Germany, marsh gas is naturally produced in very large quantities. Undeveloped and uncontrolled, marsh gas is a pollutant of the atmosphere. There is also no paper on natural gas in lakes, such as Lake Kivu in Africa, because the occurrence of gas in lakes has still not been explored, and the technology for its extraction has not yet been developed. But the 14 types of gas that will be discussed at the Conference under various headings make it clear that gas is going to become one of the main energy sources of the future, though most of the new types of gas have not yet been explored and are therefore unknown in most countries of the world. There will be a paper on gas from tight sandstones and one on gas from Devonian shale. These types are being produced in significant quantities in the United States; according to a recent press release from the White House, the production of gas in the United States from tight sandstones and Devonian shale has reached about 1 trillion ft^3 per year. Another paper to be presented at the Conference will provide the latest estimate of gas occluded in coal mines in the United States. The estimate is 765 trillion ft^3--indeed a very large reserve--and the estimated cost of producing gas from coal mines, as given in the paper, is 50¢ per 1,000 ft^3. Another paper, written by a staff member of the United Nations, analyzes the coming glut of gas in many developing countries, and recommends the construction of large liquefaction plants to allow for the export of that gas to industrialized countries. This glut is expected as early as 1985. I happen to believe that developing countries should, if they can, establish their own infrastructure for the local transportation, distribution, and consumption, especially in industry, of natural gas, because many of them could place natural gas in the centre of their long-term energy planning.

The paper predicting a glut of natural gas in a number of developing countries is also based only on the two conventional types of gas--associated gas and dry gas. There are many other types of gas in the future, but very little work has been done on them outside the United States, Canada, and, to some extent, India. I therefore approached a number of experts suggesting that they write a paper speculating on which types of gas could be found in the various countries of the world. In this way, the participants of the Conference would return to their respective countries with the knowledge that there are many new types of natural gas on which work has been done elsewhere, and they would be induced to consider whether some of these types of gas might exist in their own country. One group of experts at a university found this such an interesting idea that they decided to make it a major study, but the results of their work will not be available until a year after the Conference. Fortunately, at the last moment, Richard Meyer of the United States Geological Survey agreed to write the proposed paper.

The Conference will deal not only--not even predominantly--with fuels. The dividing line between fuels and other types of energy sources is becoming fuzzy, and this applies to the dividing line between renewable and nonrenewable energy resources as well. It is not clear at first whether gas from landfill areas is a renewable or nonrenewable resource, nor is it clear whether firewood is a fuel, though it is a renewable source of energy. Fortunately, the Conference will avoid questions of terminology and classification and will concentrate on the characteristics--including physical, technical, economic, and environmental--of every energy resource.

One classification however is important, and that is the classification by possible end use. This classification is largely dependent on the technology of end use as it exists today and probably will exist in the near future. For transport, and air transport in particular, liquid fuels are a vital necessity, whereas for the production of heat and electricity any source of energy can be used. This is an important distinction for two reasons: first, because modern industry cannot operate without electricity, and second, because electricity is today a necessity for our communications system and social life. Electricity, therefore, must play a key role in the energy planning of every country. As will be seen at this Conference, many of the new sources of energy are used exclusively for electricity generation. This applies to magnetic electricity, a new source which promises low-cost electricity generation in the future, and it also applies to other systems in the research and development stage, such as salinity gradient and magma energy, both of which will be discussed at the Conference.

The wide availability of so many sources and resources for electricity generation raises an important question: Is nuclear power needed for long-term energy planning? We believe that nuclear electricity generation--and nuclear energy can only be used for the generation of heat and electricity--is among the costliest methods of generating electricity, if the true cost of all aspects of nuclear energy production are included. We believe therefore that, given the many and increasing types of energy sources, the world may not need nuclear electricity in the future. This is certainly a point that deserves full discussion and might be an important conclusion of the Conference. A few words should be said at this point about carbon dioxide (CO_2), a by-product of fuel burning, since its adverse effects are often mentioned in conjunction with many fuel energy sources. First it must be noted that CO_2 is vital for agricultural production because of its role in photosynthesis. Further, it is a material of increasing industrial significance. Indeed, there is research presently being conducted on its use as an energy source. However, very often the point is made that the increase of CO_2 in the atmosphere will lead to a "greenhouse" effect, an increase in temperatures worldwide, and the melting of the polar ice caps. This theory is frequently mentioned, including in some of the papers submitted to the Conference, but it is a theory that is out of date. Recent work by Dr. George Kukla, the world-renowned climatologist, and his colleagues has been published in the British journal Nature indicating that the hypothesized heating effect of CO_2 may be incorrect and that, on the contrary, a sharp increase in the CO_2 content of the atmosphere may have a cooling effect. Unfortunately, Dr. Kukla was unable to attend the Conference, but I mention the latest research on CO_2 in order, if possible, to eliminate discussion of CO_2 from the Conference.

The large expansion in energy resources in a number of countries and the need to obtain a variety of end-products has led in the last few years to ever increasing expenditures on the conversion of primary energy products to the desired end-products. This applies, for instance, to the conversion of coal to high or low Btu gas or to oil, and to a variety of other conversion programs. In general these conversion processes have been excluded from the agenda of the Conference, because they do not deal with supplies of primary resources. However, there are a few exceptions, namely in those cases where conversion is necessary for the utilization of a resource, such as underground gasification of coal which otherwise cannot be exploited, and the conversion of unusable biomass to methanol. These conversion processes expand the economic concept of a resource and are included for this reason. However, all conversion processes of one energy product to another are expensive, being very often capital intensive, and involve considerable energy cost. As a rule therefore, governments and companies should make certain that the energy resource they need is not available in their country before embarking on capital intensive conversion projects. Nevertheless, there are cases where governments, for a variety of reasons, are proceeding with such energy conversion projects. This usually occurs when a country has vast quantities of one resource, which has been utilized for a particular purpose and is subsequently needed for a different use. New Zealand recently announced that it will invest close to $800 million in the conversion of natural gas, which it has in large quantities, to gasoline, because it presently has to import all its petroleum products. Often energy conversion projects involve difficult economic judgments, some of which will be discussed in some of the papers prepared for the Conference.

Another aspect which in general has been excluded from the Conference is the energy infrastructure required for the utilization of energy resources. Such infrastructure may involve pipelines systems, refineries, railways, transmission lines, and so on. This is a serious and important problem, especially for developing countries, as is discussed in one economic paper to be presented. In this context it must be realized that the restricted availability of energy infrastructure in so many developing countries justifies the development of small resources near consumers.

In areas that lack an infrastructure, a distinction has to be made between small resources and small units of production. Small energy production units are needed, and many of them are now available. For instance, the first 250 kilowatt geothermal power generating unit was installed in Indonesia in 1978 and is operating well. Small hydro power units are now found in many countries, and small coal mines are beginning to prove themselves very attractive economically. The UNITAR Conference held in Mexico in 1978 on the Future of Small-Scale Mining contained papers on the small-scale production of geothermal, coal, and oil shale resources. Small oil and gas fields were discussed in the first UNITAR energy conference, which took place in 1976 and was entitled the Future Supply of Nature-Made Petroleum and Gas. More recently, the UNITAR Conference on the Future Supply of Heavy Crude and Tar Sands discussed small deposits of heavy crude. The development of small units of production for rural areas is important, but it must not overshadow the even greater importance of small deposits, which will probably provide a large part of future energy resources.

Small deposits will, to an increasing extent, play a vital role in energy production not only because an appropriate technology for rural districts is being developed but also because with higher prices small deposits become too valuable to be neglected. Furthermore, as the energy infrastructure grows, it becomes possible to develop small deposits in rural areas in excess of the rural demand. Gradually for certain energy resources, small resources may become quantitatively significant. Unfortunately, there is little room in the Conference agenda to discuss small energy resources in detail. I would only like to point out that for some energy resources small resources have additional attractions. It appears that some of the small deposits involve smaller risks and may be more economical to develop than large ones. Small, shallow gas deposits in Ohio (USA) have, for example, been discovered with an 80 percent exploration success ratio, and in up-state New York where development is now underway the exploration success ratio for small gas fields is 90 percent.

Also omitted from the Conference agenda are water projects other than river hydro power. This includes for instance the hydro power potential of the Quattara depression in Egypt, the diversion of sea water to the Dead Sea, and the flooding of the Amazon Valley in Brazil and the Chaco Valley in Colombia.

(Both of the flooding projects have been proposed by the Hudson Institute.) The agenda also excludes energy dreams such as building a dam across the entrance of the Red Sea so that the drop in the Red Sea water level would allow large-scale hydro power generation, and older schemes to reduce the level of the Mediterranean. The only exception included in the Conference is the hydropower potential of Greenland, because it involves large-scale disturbance of neither sea traffic, the environment, nor established populations, and it utilizes fresh water.

Many energy sources, though by no means all of the new sources or resources in our future, will be discussed at the Conference. Some people would not agree to speak on some new energy sources, because lawyers objected to the early disclosure of secret or proprietary material, and in one or two other cases permission of governments could not be obtained. Nevertheless, a very large number of energy resources will be dealt with, making it possible for energy planners in governments or companies as well as for other experts to become aware of the many alternatives for energy planning in the future. The many alternatives will of course make decisionmaking in future energy planning more difficult. The Conference cannot provide an answer for an individual country as to which energy sources to select for full development. This will vary from country to country according to resource availability, energy infrastructure requirements, the type of energy end-products needed, capital availability, and a number of other factors. But the Conference is designed to provide governments with a knowledge of the alternatives so that they may better know what resources to look for in their own countries and to base their long-term plans on.

The Conference could also express a view on the new sources of energy which are at the research and development stage, and it could indicate, in one form or another, which of the resources should have priority in research and development and which are less promising. It is to be hoped that the Conference will overcome the pessimism which was created at the beginning of the 1970's by the Club of Rome in a study on "The Limits of Growth." It would be significant if, at the end of the 1970's, this UNITAR Conference was able to tell the world that there are a great variety of energy sources in our future, that there are large fuel resources as well as such new resources as biomass, solar energy, magma energy, magnetic energy, and various water-based resources, and that the waste products of yesterday can become the energy materials of tomorrow. The world therefore does not have to expect a decline in the availability of energy in the form in which it is needed. Neither is it necessary to expect long-term shortages in the future. The availability of energy in the future, for a higher standard of living, is a function of our work and our efforts, and it is on us that the future supply of energy for our children and grandchildren will depend.

CHAPTER 2

THE NEED FOR A DIVERSITY OF ENERGY SOURCES

Sir Hermann Bondi[1]

INTRODUCTION

If the aim of this book is "to provide policy makers with a comparative evaluation of the full range of future energy sources, taking into account their costs, availability, and environmental impacts," then it seems unfortunate that, in this introductory chapter I must argue that such a task is impossible. In my experience, no group of independent thinkers such as are gathered here can agree on the future cost of a single energy source, and effects on the environment are difficult enough to assess for energy systems which exist today, let alone those of the future. All planning must allow for uncertainty, particularly when so much is at stake. The man in the street puts his eggs in more than one basket, and we must in our turn keep energy options open wherever we can. There are two main reasons why the world needs a diversity of energy sources: the diversity of energy needs and the uncertainties surrounding each individual energy source.

THE DIVERSITY OF ENERGY NEEDS

Even if we were omniscient and could state now our exact future energy needs, we would find that to meet all of them at at least cost would require a diversity of sources, since these needs vary in at least three ways. These ways are thermodynamic nature, place, and timing.

Thermodynamics. Energy needs may be distinguished thermodynamically. "Heat does not of itself flow from a colder to a hotter body," and no amount of warm water will melt steel. Conventional energy sources can yield a sufficiently high temperature to meet most of our energy needs, so that in practice the temperature levels of energy needs usually turn out at present to be a guide to efficient use of fuels, rather than an absolute constraint on their choice. Looking to the future, diffuse solar energy and heat from geothermal aquifers are both lukewarm resources which seem best fitted to heating buildings

[1]Chief Scientist, UK Department of Energy.

and domestic water, where they can replace on a joule-for-joule basis fuels used at present below their full potential to give high temperature heat which is immediately dissipated. One way of making efficient use of high temperature energy sources for low temperature applications is the heat pump, which may be described as a thermodynamic lever. This requires, by the very nature of its two inputs, a "diversity of energy sources." It offers some very interesting possibilities for upgrading low temperature heat, either gathered from the environment or recovered.

Place. Another source of diversity in energy needs is the place at which they occur. Energy for transport contributes about 20 percent of world energy demand (R. J. Alden et al., 1977). Over 90 percent of energy for transport is derived from petroleum, since this easily yields portable liquid fuels. Transport is perhaps the most inflexible of all energy needs.

I welcome the emphasis on developing countries in this book of Conference reports. One of the main energy problems for these countries is the need to supply fuels to remote rural areas when local firewood becomes exhausted. Solutions appropriate to industrialized countries with their highly developed distribution systems may not be cost-effective for the impoverished village.

Timing. A third source of diversity in energy needs is the timing of demand. It seems that the more sophisticated our energy carriers become, the more difficult they are to store; woodpiles and coal bunkers were inconvenient for the individual user and have often been replaced by the wire or the gas pipe, but they did ensure day to day security of supply. Now this has become the concern of central utilities which, by forecasting the sum of many individual needs, can reduce redundancy in the energy system as a whole. Since capital-intensive facilities used at low load factor have high unit costs, the timing of energy demands exerts a strong influence on the primary energy source chosen to meet them. This limits the proportion of nuclear capacity in an electricity system and does so especially in a small one without many different patterns of individual demand to average out. The timing of demand thus leads to multiple sourcing, particularly in integrated systems. It also has implications for the potential of intermittent sources of energy, such as the winds, waves, and tides, since their output may be expensive to store for a long enough period of time to meet all demands. It seems inevitable that for most purposes a back-up supply must complement them. Here again we see a need for diversity of energy sources.

Energy Needs in Industrialized Countries

Table 2-1 compares the pattern of energy demand in the mid-seventies in the UK and the USA. It is clear that the USA, perhaps partly because of its lower population density, has a greater proportionate need for transport fuels than the UK. The distribution of temperature levels of demand appears broadly similar, although there is a greater need for air conditioning in the USA and for space heating in the UK. What the table does not show, however, is the fact that in many states of the USA peak electricity demand occurs in summer, for air conditioning, and therefore is in phase with solar supply. The UK is not so fortunate and indeed cloudy winter weather makes economic use of solar energy for space heating more difficult than for any other country of our latitude; in winter only about one-fifth as much heat is available as in summer. This demonstrates the importance of timing.

Energy Needs in the Developing World

By contrast, consider the current pattern of energy supply in India (A. V. Desai, 1978). Only about 50 percent of needs for useful energy are met by commercial energy; of the remaining needs, mostly in rural areas for agriculture and cooking, two-thirds are met by noncommercial fuels (mainly wood) and one-third by the muscular efforts of human beings and animals. Energy planners must never forget that hoped-for increases in the standard of living of the world's poor will result in increased energy consumption to replace physical labor and prevent the exhaustion of limited wood resources. It has been estimated (O. Brandon, 1979) that to provide a sustainable and tolerable quality of life for all humans would increase world energy consumption by 3 to 10 percent. This takes no account of population growth. Developing countries urgently need means of providing energy to rural areas without impossibly high capital investment. The density of rural energy demand per unit area is low, and this may offer scope for use of biomass on a renewable basis, preferably integrated with food production where fertile land is scarce.

So far I have concentrated on the varied nature of energy needs. Next I will consider some of the possibilities for meeting these in the future.

TABLE 2-1. End uses of primary energy in the USA and the UK in 1976

Use	Primary energy as percentage of total USA	Primary energy as percentage of total UK
Buildings	36	44
of which,		
Space and water heating	22	31
Lighting, cooking, and appliances	10	13
Air conditioning	4	0.2
Transport	26	19
Industry	38	37
of which,		
Motive power and electrolysis	9	11
Process heat <120°C	~ 7	6
>120°C	~ 19	16
Feedstocks	3	4
	100	100

NOTES: 1. Transport includes ships' bunkers.
2. Industry includes agriculture and off-road vehicles.

Sources: U.S. Energy Information Administration, "Statistics and Trends of Energy Supply" (1977); U.S. Office of Technology Assessment, "Application of Solar Technology to Today's Energy Needs" (1978); UK Department of Energy, Energy Paper 39, "Energy Technologies for the United Kingdom" (1979).

MATCHING NEEDS WITH ENERGY CARRIERS

Oil, the only major energy source which is at once capable of providing high temperature heat and is easily transported and stored, is no longer cheap. On grounds of both cost and availability, we will need to reserve it in future years for premium applications such as transport and chemicals manufacture, where replacements are proving difficult to find. Therefore, there is a pressing need to exploit other energy resources and to use energy more efficiently.

Natural gas has all the advantages of oil except ease of storage and marine transport. In countries with access to natural gas, it is likely to take up much of the heating load at present met by oil, up to the limit of gas supplies.

Coal is a relatively plentiful resource but at present it is used mainly for bulk heat applications. Extending its range of applications is presenting a challenge to several countries besides the UK. Ideally, one wishes to avoid the waste and expense of converting coal to a fluid, by instead using coal directly. Devices such as fluidized-bed combustors can help raise the intensity of coal combustion to that of oil, and at the same time dispose of some undesirable products of combustion. It seems to me that the most intractable aspect of coal's unpopularity with the small user is the bother of transporting the coal before use and the ashes afterwards. In a factory at present using oil there may be no room to install facilities for coal storage and transport. The inventor of a foolproof way of transporting coal cheaply in pipes without either soaking it in water or requiring more oil than coal would, I think, make a fortune. Meanwhile we must anticipate a continuing demand for clean-burning, portable fluids not just for transport but also for domestic heating and much of light industry.

Electricity is an energy carrier both versatile and easy to use. For applications such as lighting, electronics, and sodium manufacture there is no practical replacement. Electricity can be made from any energy source but the source chosen by many utilities is nuclear power, which is considered to promise a more stable price than conventional sources. If this hope turns out to be realized in practice there will be an incentive to examine ways of replacing scarcer energy carriers by electricity. The greatest challenge is the motor car. The market for electric vehicles will probably be restricted to urban short-haul use unless better acceleration and range can be achieved.

Hydrogen is a possible competitor with electricity for very long-range energy transport and with oil for air transport. However, it is not itself an energy source and appears to be most cheaply made by electrolysis, so its eventual function may be to complement electricity as an energy carrier.

Heat transport using steam or hot water is, of course, a vital part of current energy systems. Renewable heat sources or recovered heat from power stations could act as an input to present-day district heating networks. If such a network does not already exist the main barrier to replacing individual fossil-fuelled heating systems, where economic, is likely to be an institutional one, that of persuading potential users to join a suggested scheme. The individual's desire for a free choice of energy sources may work in such a case against the national interest.

An exhaustive list of energy carriers would be tedious; it is clear from those considered already that the nature of energy needs and carriers is such that more than one energy source will be needed to replace oil if the high costs of making synthetic fuels are to be minimized. The utilization technologies needed to match energy sources and needs have perhaps been neglected in the past in favor of the more glamorous field of energy production, and I am pleased to see that one section of this volume covers the important question of energy matching.

UNCERTAINTIES SURROUNDING ENERGY SOURCES

Beyond the need to diversify, which results from the structure of energy demand, there is a second need: to provide security of energy supply.

Conventional Energy Sources

Uncertainty dogs all conventional energy sources:

Oil and natural gas reserves were considered in a previous UNITAR conference in Austria in 1976. They are notoriously difficult to estimate; ultimately recoverable oil reserves have been variously estimated at between 1,200 and 4,000 billion barrels (Seidl, 1976), while at the end of 1978 published proved world oil reserves were 649 billion barrels (BP, 1979). These figures should be compared with world production of 23 billion barrels in 1978 (BP, 1979). Proved gas reserves are comparable to oil reserves, at 2,500 trillion ft^3 (BP, 1979), the equivalent to about 450 billion barrels of oil, of which perhaps 40 percent is associated with oil (Whiting, 1976). World gas production and consumption in 1978 was equivalent to 9 billion barrels of oil (BP, 1979), mainly in the country of origin, but consumption and trade is likely to increase. These figures are summarized in Table 2-2.

The extent of ultimately recoverable gas reserves is an open question; geopressured reservoirs, methane hydrates, and deep abiogenic methane might each multiply gas reserves many-fold. Taking into account also the possibilities for oil production from the deep ocean and from shales, the sparse nature

TABLE 2-2. Proved reserves and production (10^9 barrels oil equivalent)

	Oil	Gas
Reserves (end 1978)	649	450
Production (1978)	23	9

of historical oil exploration over much of the world's surface, and the incentive for exploiting small reservoirs at current oil prices, it is quite conceivable that, as Professor Odell often reminds us, we may have oil and gas coming out of our ears for many years to come--at a price. However, no country would wish to rely exclusively on resources outside its own control to meet needs of such strategic importance as energy, and the need for security of supply will probably be a sharper spur to diversification of energy source than questions of ultimate resource.

Coal is plentiful in the ground; the Workshop on Alternative Energy Strategies in 1977 estimated that there are about 700 billion tonnes of economically recoverable proved reserves of coal in the world, and potential reserves may be as high as 12,000 billion barrels of oil equivalent. World consumption in 1978 was 13 billion barrels of oil equivalent. However, extracting it will require heavy investment, perhaps $120 billion in the USA alone to increase annual production from 500 million tonnes now to 2,000 million tonnes by the year 2000, according to Bechtel Corporation. The capital demands of coal conversion are on the same scale. Labor supply for deep mining may also be a problem in some countries unless robotic techniques are further developed.

Environmental issues raise the greatest uncertainties for the future of coal as a global energy source. Both open-cast and deep mining interfere with land use. Sulphur and trace elements present in the coal and carried over in waste gases may, at high enough concentrations, have harmful effects on plant and animal life. In the long term an undesirable side effect of using any fossil fuels may be the influence of carbon dioxide in the atmosphere on global climate. Recoverable fossil fuel reserves contain at least 10 times as much carbon dioxide as the atmosphere at present (Chadwick, 1978). Climatic modelling suggests that a doubling of the atmospheric carbon dioxide content would raise average world temperatures by 1-2°C, and temperatures at high latitudes by as much as 10°C owing to increased absorption and downwards reradiation of heat; effects on local climate and rainfall are unknown. By comparison the direct heating effect of fossil fuel use is negligible on the global scale. But sources and sinks of carbon in the oceans and the biosphere are poorly understood, and climatic models are rudimentary compared to the complexity of the real world, so that this must remain an open question. By the end of the coming decade we may have a clearer picture; meanwhile this "greenhouse effect" stalks in the undergrowth of long-term energy planning.

Nuclear fission using thermal reactors holds promise as an electricity source of stable cost but uranium supplies are limited. Reserves of uranium in the noncommunist world costing less than $80/kg are equivalent to about 125 billion barrels of oil if used in thermal reactors substituting for coal-fired power stations (based on 1977 OECD estimates). However, this is a small fraction of a much larger resource base. The energy available from uranium could in principle be increased by two orders of magnitude by using breeder reactors, but only when their capital costs are better known will it be possible to estimate the uranium price at which they could become viable. For the nuclear industry as a whole, public concern about safety and proliferation is an observed fact with impacts on the energy policies of many nations. To rely on this option alone and foreclose other lines of energy research and development would be premature.

THE IMPORTANCE OF ENERGY CONSERVATION

As costs of energy increase and secure supply becomes less certain, both individual users and governments are paying greater attention to the many possibilities for saving energy. Certainly within the UK savings of at least 30 percent of energy in most areas of use have been identified as technically feasible. Energy conservation can probably ease the strait jacket of constraints on global energy supplies and we must strive to achieve it. But energy-saving measures which have been cost-effective for years remain to be fully implemented by individual users--loft and water tank insulation spring to mind--so that while our task is to reduce unnecessary energy use to a minimum by providing information and technologies to back this up, the extent to which this will succeed is unknown.

RENEWABLE ENERGY SOURCES

Uncertainties in the balance of global energy needs and supplies can be reduced by diversifying energy sources, since any additional contribution which does not divert resources from conventional energy production will help to ensure adequate energy supplies. Renewable energy sources will be needed to replace conventional fuels as these run out, and indeed the sum total of these sources is in theory more than adequate to meet the world's foreseeable energy needs. In all questions of energy choice, reducing the cost of the system required to meet each need is a primary objective. Choice of discount rate bears heavily on the renewable sources, and this, together with the investment risks associated with unproved technology, is the reason for the predominance of government rather than commercial interest in researching these sources. Even at commercial discount rates some of their applications are already cost-effective, but their diffuse nature and often their intermittent supply is likely to limit the impact they can make on the energy needs of urban societies. Despite this many countries, including the UK, are performing R&D to assess the potential of renewable energy sources and where possible to develop them as an additional option for energy supply. Many speakers including my colleagues from the UK will be considering these sources at greater length in subsequent chapters.

CONCLUSION

To sum up, the end uses of energy are most efficiently met from several sources. The uncertain nature of the future means that we must diversify energy sources, and perform R&D on many technical options. Let us not prejudge the results of such research by defining priorities too tightly now, but instead allow choice for future generations.

REFERENCES

British Petroleum Company, 1979, Statistical review of the world oil industry 1978.

Brandon, O., private communication.

Chadwick, A. T., 1978, Human activity and the carbon dioxide level in the atmosphere: Energy Technology Support Unit Report R5.

Desai, A. V., 1978, India's energy consumption: Energy Policy, v. 6, no. 3, p. 217-230.

Eden, R.J., and others, 1977, World energy demand: World Energy Conference, 10th, Technical Report, p. 56.

Seidl, R. F., 1977, Implications of changing oil prices on resource evaluations, in Meyer, R. F., ed., The future supply of nature-made petroleum and gas; UNITAR/IIASA Conference, Technical reports: New York, Pergamon Press, p. 113-137.

Whiting, R. L., 1977, Gas from conventional gas fields, in Meyer, R. F., ed., The future supply of nature-made petroleum and gas; UNITAR/IIASA Conference, Technical reports: New York, Pergamon Press, p. 277-293.

CHAPTER 3

THE SOLAR TRANSITION: COMPATIBLE CONVENTIONAL ENERGY SYSTEMS

Barry Commoner[1]

INTRODUCTION

It is widely recognized that the relationship between energy and the economic system which it must support is in a state of crisis. In order to define the nature of the energy crisis and to evaluate possible plans for solving it, it is useful to examine the efficiencies of the thermodynamic and economic processes that comprise the relationship.

The relationship between the economic system and the production and use of energy is outlined in Fig. 3-1. Certain of its features are consequences of the Second Law of Thermodynamics:

a. That work (W) can be done only as a result of the flow of energy from some suitable source (E);

b. That because the processes which yield the products (P) of industry, agriculture, transportation, and communication do not occur spontaneously, they, as well as the economic output which they yield (O_G), depend on work;

c. That since the acquisition of energy (e.g., producing petroleum and natural gas, mining coal, or manufacturing a solar collector) is itself a nonspontaneous process, it also requires that work be done.

Figure 3-1 also illustrates a feature of the relationship which depends on the First Law of Thermodynamics--that any energy (E_L) which is simply lost from the work-requiring system, (e.g., heat lost through inadequate insulation) fails to contribute to the overall process of production and economic output. Finally, there is an economic feature--that a certain fraction (O_E) of the gross economic output (O_G) must be devoted to the acquisition of the needed energy, leaving the remainder, net of the cost of energy production (O_N), for uses in consumption and in the accumulation of capital for further production.

---

[1]Director, Center for the Biology of Natural Systems, Washington University, St. Louis, Missouri.

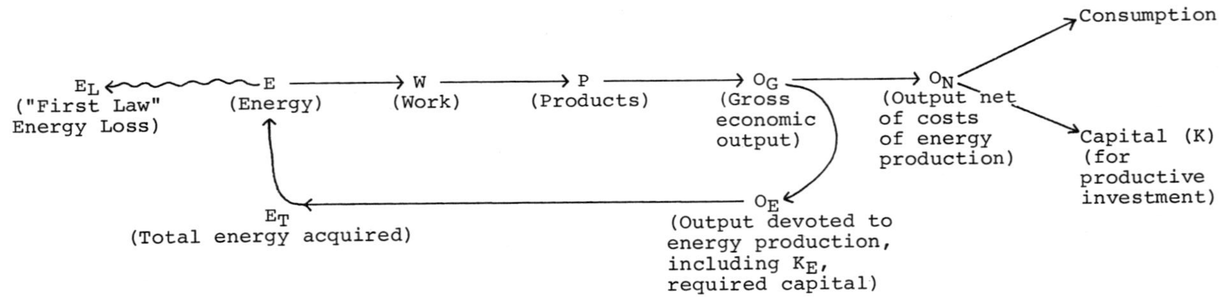

Figure 3-1. Relationship between the economic system and the production and use of energy

It follows from these relationships that economic output will depen[d on] certain thermodynamic efficiencies:

a. "First Law" efficiency, i.e., the ratio of the amount of energy that is converted to work (thus contributing to the production system) and the total amount of energy acquired, or E/E_T (where $E_T = E + E_L$).

b. "Second Law" efficiency, i.e., the amount of work accomplished per unit of energy used (W/E), and the amount of product yielded per unit of work (P/W), which for the purposes of this discussion may be combined in the form of P/E (the productivity of energy, analogous to the productivity of labor or capital).

The relevant economic efficiencies are the productivity of economic output used in the production of energy (E/O_E) and, in particular, the productivity of the capital devoted to energy production (E/K_E).

As a result of these relationships, the efficiency with which economic output is used to produce energy (i.e., the real cost of energy), and the efficiency with which energy is converted to work and economic output, will determine the ability of the economic system to support consumption and capital investments in further production. And it also follows that the degree to which these efficiencies can be improved--for example by means of "First Law" or "Second Law" conservation measures--will determine the degree to which improvements can be made in the ability of the economic system to meet human needs.

THE EFFICIENCY OF ENERGY PRODUCTION

The Effect of the Nonrenewability of Energy Sources

In most countries a very large part (96 percent in the U.S.) of current energy sources (oil, natural gas, coal, and uranium) on which the production depends is nonrenewable. The numerous separate deposits of these energy sources vary considerably in their accessibility, and therefore in their real cost of production. In the natural course of events the most accessible, least costly, deposits are exploited first; as each deposit is exhausted, the next most accessible (and therefore more costly) deposit is exploited, and so on. Thus, there is a strong interaction between the process of finding and producing a nonrenewable resource and the structure of the "crop" of accessible deposits which remain to be found. This interaction leads to an exponential negative relationship between the efficiency of production (E/O_E) of a given energy resource and the total amount of production (i.e., the cost of producing a unit of energy rises exponentially, over time, with production). Data regarding this process, in the case of U.S. domestic oil and natural gas production, are shown in Fig. 3-2, which is taken from a report by the U.S. National Petroleum Council (NPR, 1973). The analytical methodology used to generate the exponential curve which describes the projected costs of production between 1971 and 1985 is summarized in Fig. 3-3, which is also taken from the NPC report.

Also shown in Fig. 3-2 are the actual selling prices (in 1970 dollars) of U.S. domestic oil since 1971 (DOE, 1979). They correspond quite well with the projected price curve (based on the computed production cost and several alternative rates of return on assets), taking into account that in 1974 the U.S. oil companies raised their "targeted" rate of return from 15 percent to 20 percent (Blair, 1976). This is evidence, then, that the recent economic behavior of U.S. domestic oil production conforms to expectations predicted from the relevant geophysical and economic factors. It suggests that this natural relationship, rather than the oil embargo of 1973, is the cause of the present rising course of oil prices. Indeed, it seems likely that the ability of OPEC producers, beginning in 1973, to set out on an exponential course of price increases may have resulted from the prior evidence, in the NPC report (which was published in 1972 in draft form) that a major consumer of OPEC oil, the United States, would be forced, in the natural course of exploiting this nonrenewable source, to pay progressively more for its own domestic oil (thus enabling OPEC countries, despite their much lower production costs, to sell their oil at comparable, rising prices).

The nearly constant price of U.S. oil in the period 1955-1971 is only an apparent departure from this exponential relation, for it is a consequence of industry-sponsored state and federal laws that established "market demand prorationing" of production, which according to Allvine and Patterson (1976) "...permits prices to be maintained above a free market level." This can be seen by extending the projected exponential curve of Fig. 3-2, backward from 1971 to 1955. It is then evident that the actual, constant, price in that period was well above the price (computed from the NPC analysis) at which oil should have been sold to maintain a 15 percent rate of return.

A particularly important consequence of the exponentially rising cost of producing a depleting resource such as oil or natural gas is that the

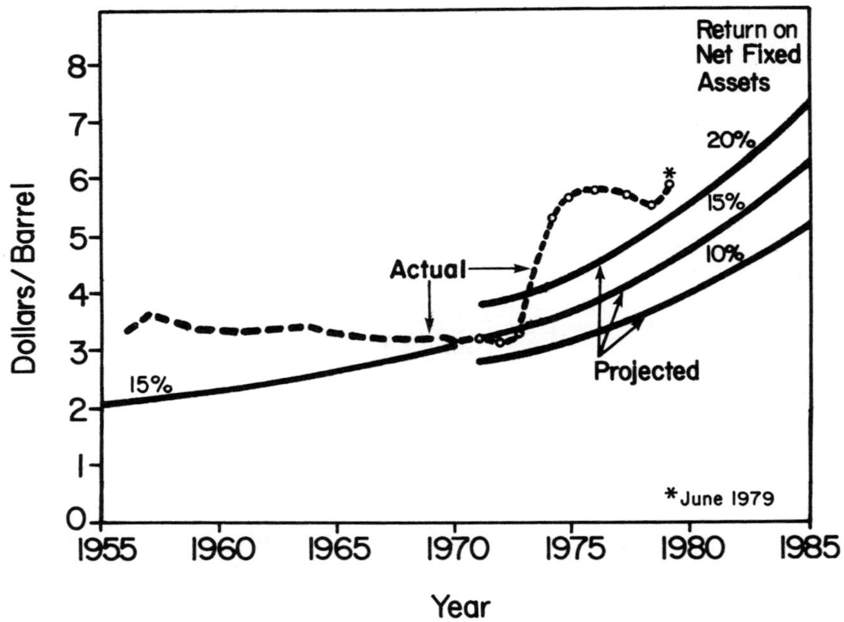

Source: National Petroleum Council (NPC), U.S. Energy Outlook: Oil and Gas Availability (Washington, D.C.: NPC; 1973).

Figure 3-2. Projected (National Petroleum Council) and actual crude oil prices (1970 dollars)

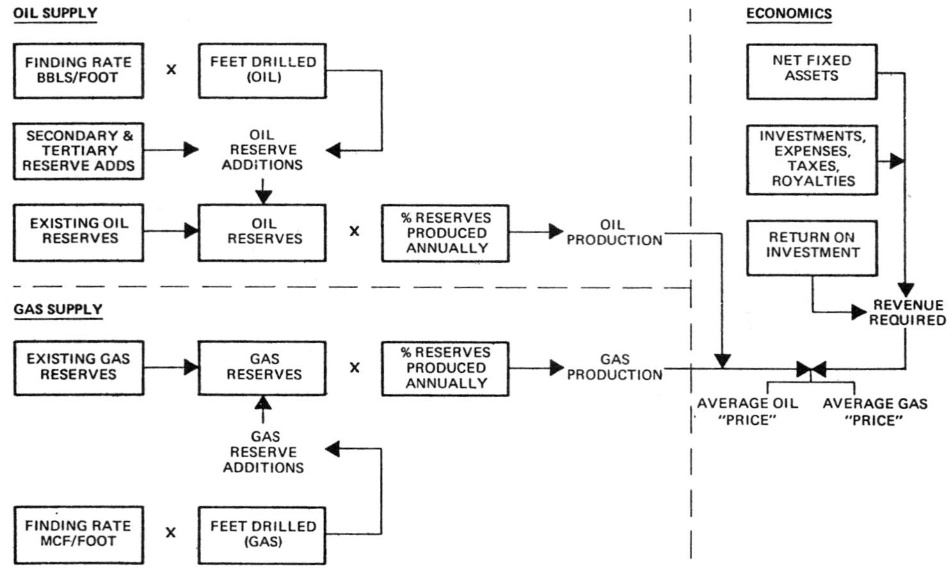

Source: National Petroleum Council (NPC), U.S. Energy Outlook: Oil and Gas Availability (Washington, D.C.: NPC; 1973).

Figure 3-3. Oil and gas supply--Economic methodology

productivity of the necessary capital (to drill the progressively deeper, more numerous wells and to use more complex, expensive methods to extract the fuels from their deposits) also declines sharply with production. For example, according to a study reported by the Federal Energy Administration in 1974 (FEA, 1974), the productivity of capital invested in oil production (i.e., the amount of oil that can be produced per dollar of invested capital) is expected to fall by 70 percent by 1988.

Thus, on both theoretical and empirical grounds it can be seen that, at least in the case of the best-known deposits of a nonrenewable energy source, U.S. oil, the efficiency with which economic output is used to produce the energy falls exponentially as the resource is exploited. This defines the most crucial feature of the energy crisis--that dependence on nonrenewable energy sources, with continued production, diverts an exponentially larger fraction of total economic output into the production of energy. It means that the energy sector will, in effect, cannibalize the economic system which it is supposed to support.

The Relative Efficiencies of Different Modes of Nonrenewable Energy Production

In order to compare the thermodynamic and economic efficiencies of the production of different forms of energy, one needs to determine the relative energetic and economic costs involved in primary production of the sources, in their transport and distribution, and in the operation of the end-use equipment. The overall consequence of such a comparison can be evaluated conveniently by computing the total capital required for all of these steps, per unit of energy delivered to the work-requiring task by the end-use equipment. Such computations have been made recently by the technical staff of the American Gas Association and are summarized in Table 3-1 (AGA, 1978). For given end uses the efficiency figures vary over ranges as great as four-fold. In general, the most efficient source is natural gas (used directly for heat); liquid and solid fuels used for the same purpose are somewhat less efficient. The efficiency of producing electricity is, in general, significantly lower than direct use of the fuels considered. This reflects the very large Second Law inefficiency which is inherent in the conversion of heat energy into mechanical work. As a result, only about one-third of the energy represented by the power plant's

TABLE 3-1. Average capital investment estimates for new domestic energy (1977 billion $ per added annual quad)

Delivered fuel	Residential/commercial spaceheating: Conventional	Residential/commercial spaceheating: Advanced	Premium industrial use	Large industrial boilers
Natural gas				
Lower 48	$42.3	$ 62.9	$19.3	$ 8.9
Alaskan	56.2	71.2	32.0	25.9
Coal gasification				
High-Btu gas	58.5	70.1	30.4	N.A.
Medium-Btu gas	N.A.	N.A.	N.A.	11.4
Coal liquids				
Electric	89.7	97.8	88.6	N.A.
Liquid	N.A.	N.A.	33.4	21.1
Coal electricity				
Power plant near end users	89.0	97.9	85.1	N.A.
Power plant near coal mines	98.7	104.6	97.2	N.A.
Combined cycle	74.2	87.1	69.0	N.A.
Coal/direct-fired				
SO_2 scrubbers	(Not applicable)			13.5
Fluidized-bed				12.0
Oil shale				
Electricity	90.1	98.6	89.2	N.A.
Distillate	57.4	N.A.	28.2	22.0
Nuclear electric	94.6	100.8	94.2	N.A.

N.A. = Not applicable or not included in this analysis.

NOTE: Terms "conventional" and "advanced" refer to end-use spaceheating equipment (e.g., furnaces vs. heat pumps).

Surface coal mining is the only mining cost assumed.

Source: American Gas Association, "A Comparison of Capital Investment Requirements for Alternative Domestic Energy Supplies" (Arlington, VA: AGA; May 18, 1978), 1978-9.

fuel is recovered as electricity. From Table 3-1 it can be seen that, in terms of required capital, the least efficient source of energy is nuclear power.

The Effect of Centralization on the Efficiency of Energy Production

If energy production is centralized and hence distributed to users over appreciable distances from the central location, transmission and distribution costs--both in energy and capital--are incurred. Table 3-2 summarizes these costs for different forms of energy. It shows, in general, that centralization most sharply reduces the efficiency of the production and delivery of coal and electricity, and has a lesser effect on the use of oil and natural gas, particularly with respect to capital productivity.

In the case of electrical power, efficiency is also affected by centralization in another way. If power is produced by a few very large power plants rather than more numerous small ones, it is impossible to achieve a close match between capacity and demand (see Fig. 3-4). Demand is made up of a very large number of small power-operated devices, so that overall demand rises with time in a smooth, continuous way. In contrast, when it is based on a few very large plants, capacity rises in successive large increments, as each new power plant is connected to the system. Inevitably, for a time the system has much more capacity than it needs. In that period the capital invested in the power plant is, to a degree, wasted. (If the excess power is sold out of the system's demand area, efficiency nevertheless falls because of higher transmission costs.)

THE EFFICIENCY OF ENERGY END-USE

The final step in the transformation of an energy source into economic output is the production process, in which energy, along with other necessary inputs such as labor and capital, is used to generate the product. Here, a guiding idea that derives from the Second Law of Thermodynamics is important--i.e., that the efficient use of energy depends on coupling the work-requiring task with a source of energy which is in a form appropriate to that task. Thus it is wasteful of energy to use electric resistance units for space heat, since electricity is a "high quality" source, produced at the expense of the two-thirds of the original fuel energy that is lost at the power plant. Fuel directly applied to space heat is used much more efficiently for that purpose.

TABLE 3-2. Energy and capital requirements for transporting various forms of energy

Form of energy	Mode of transmission[3]	Energy productivity[1] Btu-miles per Btu used	Capital productivity[2] million Btu-miles per year per dollar invested
Oil	Pipeline	87,000	2,369
Natural gas	Pipeline	25,000	4,121
Coal	Rail	37,000	572
Electricity	Power line	3,400	788

[1]The number of Btu's of energy that can be transported 1 mile using 1 Btu of energy.

[2]The number of Btu's of energy that can be transported 1 mile using $1 of capital investment.

[3]Modes considered are long-distance bulk energy transmission, and do not include distribution system costs, for ease of comparison.

Sources: Office of Technology Assessment, U.S. Congress, Application of Solar Technology to Today's Energy Needs (Washington, D.C.: U.S. Government Printing Office, 1978). Eric Hirst, "Energy Intensiveness of Passenger and Freight Transport Modes, 1950-1970" (Oak Ridge, Tenn.: Oak Ridge National Laboratories, 1973). Barry Commoner, "Testimony before the U.S. Senate Transportation Subcommittee, Hearings on the Future of the Highway Program" (St. Louis, Mo.: Center for the Biology of Natural Systems, 1975).

When comparisons are made among different processes that yield the same product one can express the efficiency with which energy is used to generate fixed amounts of the product (i.e., energy productivity), for example transport of people or freight. As shown in Table 3-3, there is considerable range of variation in such estimates--for example, a nearly six-fold range of energy productivities among alternative ways of transporting passengers over land and a nearly four-fold range of variation with respect to overland freight transport. The variations in capital and labor productivity are equally large. Generally speaking, bus travel is the most efficient means of transporting passengers over land, while for freight, railroads are the most efficient mode

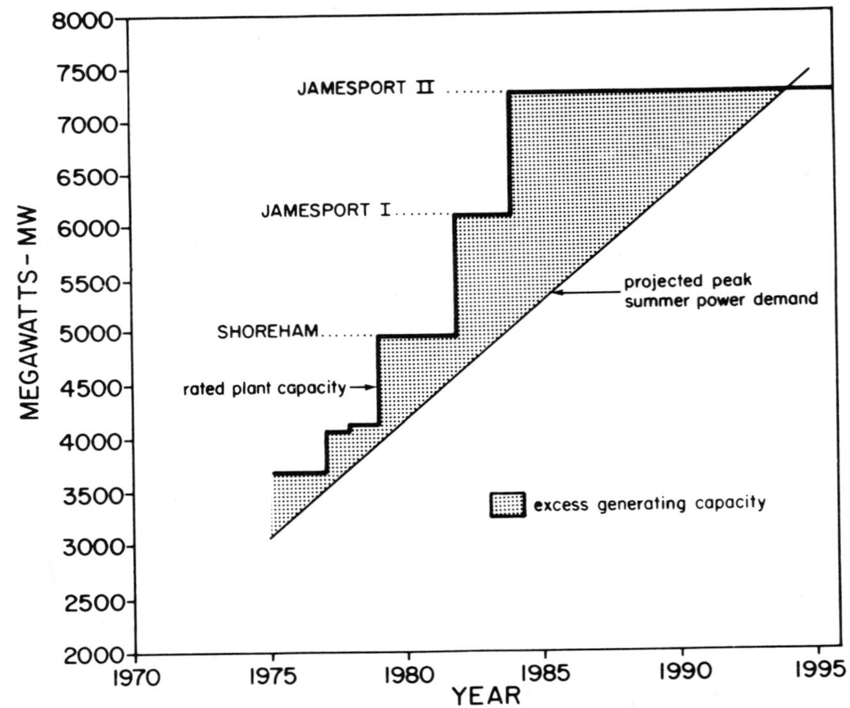

Source: Dubin-Mindell-Bloom Associates, P.C., "A Study of Existing Energy Usage on Long Island and the Impact of Energy Conservation, Solar Energy, Total Energy and Wind Systems on Future Requirements" (New York: Dubin-Mindell-Bloom Assoc., P.C.; 1975).

Figure 3-4. Long Island Lighting Company (LILCO): Projected peak summer power demand and installed generating capacity, 1975-1995

TABLE 3-3. Transportation productivities and total travel

	Energy				Capital				Labor		Share of total travel*	
	Pass.-miles per 10,000 Btu		Ton-miles per 10,000 Btu		Passenger-miles per year per dollar invested		Ton-miles per year per dollar invested		Pass.-miles per man-hour	Ton-miles per man-hour	Percent of all pass.-miles	Percent of all ton-miles
	1960	1970	1960	1970	1960	1970	1960	1970	1970		1970	
Auto	1.8	1.1			7.44	6.43					86.6	--
Truck (for hire)			3.4	3.6			9.13	6.70		93.7	--	21.3
Bus	6.7	6.3			23.31	30.30			294.1		2.1	--
Air	1.4	1.2	0.3	0.2	7.45	8.10	0.72	0.84	188.7	18.7	10.0	0.2
Rail												
Estimate 1	3.4	3.4	16.7	13.0	7.19	7.18	15.56	19.33	231.5	625.0	0.9	39.8
Estimate 2	3.4	3.4	16.7	13.0	1.49	1.48	22.42	22.67	47.8	731.4	0.9	39.8
Pipeline			100	52.6			60.61	99.40		11,904	--	22.3
Water			16.1	14.7			34.02	40.94		1,219	0.3	16.5

Sources: Barry Commoner, "Testimony before U.S. Senate Transportation Subcommittee, Hearings on the Future of the Highway Program" (St. Louis, Mo.: Center for the Biology of Natural Systems, 1975). U.S. Department of Commerce, Statistical Abstract of the United States 1978 (Washington, D.C.: U.S. Government Printing Office), 1978.

(apart from pipelines and water transport, which are more limited with respect to the materials that can be carried and the routes followed). Obviously, other considerations also enter into such comparisons--for example, that air transport, although inefficient in its use of energy and capital, is uniquely rapid. The distance traveled is also an important consideration, which leads, for example, to the conclusion that modern rail transport can be competitive with air transport, with respect to door-to-door speed, for distances up to several hundred miles.

It is also evident from Table 3-3 that the more efficient modes of transportation are relatively little used in the United States, so that a shift of intercity passenger traffic from autos and aircraft to railroads, and of freight from trucks to railroads, would greatly increase the productivity of energy (and also of capital) in transport.

When comparisons need to be made between production processes that yield different products, they must be based on economic value--a procedure which inevitably involves factors other than thermodynamic efficiency. An example of such data is shown in Table 3-4, again only to illustrate the wide range of values between energy-intensive industries such as steel and chemicals and an industry such as leather production, in which little energy is used relative to sale price. Such comparisons may also suggest substitutions--such as leather for plastic--which can provide a similar good at a considerable improvement in energetic efficiency.

SOLVING THE ENERGY CRISIS; THE TRANSITION TO RENEWABLE ENERGY

General Considerations

From a number of recent studies, it is evident that there is considerable room for improving the efficiency with which energy is used in an industrialized society such as the United States by reducing "First Law" energy losses (for example, through improved insulation). A Bonneville Power Administration study concluded that residential energy consumption could be cut by 14 to 48 percent through the application of energy-saving measures (BPA, 1976). It is also possible to reduce energy consumption by measures which diminish the resultant good relatively slightly (for example, driving more slowly, or reducing building temperatures). Such measures are, of course, useful in that they reduce energy expenditures and thereby relieve, to a degree, their pressure on the

TABLE 3-4. Labor, capital, and energy productivities for selected industries (1971)

Industry	Labor productivity Value added ($) per production worker man-hour	Capital productivity Value added ($) per dollar of fixed assets	Energy productivity Value added ($) per million Btu's
Petroleum and coal products	28.43	.34	1.90
Chemicals and allied products	27.75	.80	4.85
Stone, clay, and glass products	11.59	.82	7.39
Primary metal industries	11.54	.49	4.42
Paper and allied products	11.40	.58	7.64
All industries average	12.43	1.13	14.42
Leather and leather products	6.25	3.64	62.04

Source: The Census of Manufactures.

economic system. However, such conservation measures do not in fact address the qualitative problem that is at the heart of the energy crisis--i.e., that dependence on nonrenewable energy sources, over time, diverts an exponentially larger part of economic output to energy production, thus eroding the system's capability to support consumption and investment in new production. The energy crisis can be solved--as distinct from being temporarily ameliorated--only by a transition to renewable sources of energy.

Such a transition could conceivably be based on two existing renewable sources: solar energy and breeder-based nuclear power. While there are numerous substantive reasons to favor the solar alternative over the breeder, it suffices here to mention only one, overriding consideration. Solar energy sources can be introduced at once, thus beginning immediately to combat the escalating cost of energy which is the main effect of the energy crisis. In contrast, even the most optimistic estimates conclude that a significant introduction of breeder-based power is unlikely in less than 25 years or so, a period of time in which the economic impact of rising energy prices would be intolerable.

The Link Between Improved Energetic Efficiency and the Solar Transition

It follows from the above considerations that the answer to the exponential decline, with time, in the efficiency with which economic output is used to produce energy is to replace the present nonrenewable energy sources with solar ones. In accomplishing such a transition it is necessary, as well, to correct the considerable inefficiencies, such as those outlined earlier, in the present ways in which energy is produced and used. Given the capital-intensive nature of most solar processes, the solar transition will be very costly under the best of circumstances. Unless the present inefficiencies are corrected the cost of the transition might be so large as to render it infeasible.

From a review of the previous section it is apparent that there is a generic fault in the present energy system which explains most of its inefficiencies and which suggests how they can best be corrected in ways that support the solar transition. That fault is the failure to integrate the various parts of the system into an efficient whole. Thus a chief reason why electric power production is so wasteful of both energy and capital is that the energy output of the power plant is not effectively linked to the thermodynamic characteristics of its customers' energy-requiring tasks. Since the power plant produces both heat and electricity, its output could be used to meet both of these needs. Such an integrated system for the cogeneration of electricity and heat would use both energy and capital much more efficiently than conventional power plants, which produce only electricity. Similarly, in transportation, energy and capital could be used efficiently by operating electrified trains powered by cogenerators that also supply heat to nearby residences and commercial buildings. And another, more general, reason why electric power is used so inefficiently is that the growth in capacity is so poorly linked to the growth in demand.

From these considerations it follows (a) that the efficiency of energy use could be improved considerably by decentralizing power production, so that increments in capacity are matched to increments in demand--for example, by constructing local power plants sized to match a new or rehabilitated housing development; and (b) that efficiency can also be increased markedly by using cogeneration rather than conventional power production. Taken together with the fact that transport of heat from a cogenerator is efficient only over relatively small distances, these considerations urge the installation of decentralized cogenerator power plants, sized to meet local demand for both electricity and heat. As will be shown below, such nonsolar steps to decentralize power production and to derive it from cogenerators not only conserve energy and reduce costs, but also facilitate the entry of solar sources.

The efficiency, and therefore the economic feasibility, of solar energy also depends on how it is integrated into nonsolar "back-up" energy systems. For example, a number of studies have shown that the efficiency of providing space heat can be optimized by integrating both the conventional back-up and solar sources into a single system. The integrated solar/conventional system is more flexible than either a purely solar or a purely conventional system. When sufficient sun is available relative to the concurrent outside temperature, the system is able to derive all the needed energy from the sun. On the other hand, when sunlight is scarce and/or the temperature is low the system can operate on conventional fuel. Because of this flexibility the complex system is able in the first situation to substitute cost-free sunlight for expensive fossil fuel and in the second one to substitute fossil fuel for a costly expansion of the solar system.

Similarly, the high cost of energy storage, which is often necessary in solar technologies, can be reduced by incorporating the solar source into an energy-_using_ system which itself has a built-in storage capability. Thus, if a photovoltaic system is used to provide power for an electric car, which must have a battery in order to operate (regardless of whether the electric source is conventional or solar), the high cost of storage normally needed in a photovoltaic system is shared with the cost of the vehicle.

In designing a transition to renewable, solar energy, then, we need to be guided by the following progression. First, for the work-requiring tasks of the economy, we must determine that form of energy which is most appropriate to the given task, both in terms of Second Law considerations, and in terms of practical considerations, such as mobility. Second, solar sources capable of yielding energy in a form optimally suited to these tasks must be identified. Third, where possible, the solar source must be capable of providing energy on the scale in which it is needed locally (i.e., avoiding wasteful transport). Fourth, where transport is necessary, solar energy must be transported in a form which can be handled efficiently, both in transport and in end use. Fifth, since disruption in energy supplies is economically disastrous, a gradual transition must be effected, which steadily transforms the present, nonrenewable system into an ultimately wholly solar, renewable one. To accomplish this transition it is important to begin the process by improving the energetic efficiency of the existing nonsolar systems, changing designs (usually to achieve "Second Law" savings) in ways that facilitate the subsequent entry of solar sources.

Table 3-5 summarizes the efficiency and practicality of various forms of energy in supporting the chief work-requiring tasks, in the present nonsolar system. Certain crucial relationships emerge:

a. The most efficient energy source for space heat is a gas fuel, or the heat yielded by a cogenerator.

b. The most efficient source of energy to operate stationary machines (e.g., motors) is cogenerator-produced electricity.

c. The most efficient source of energy for the operation of land vehicles is cogenerator-produced electricity, available either directly, as in an electrified railroad, or indirectly in the form of battery-stored power, as in an electric car.

d. The most efficient source of energy for the operation of aircraft (and to a lesser extent, ships) is a liquid fuel.

Table 3-6 shows that in the U.S. energy system, the energy sources that are most efficiently suited to the various tasks represent a rather small part of the total use of energy.

Existing solar technologies can readily provide energy in forms that are suitable for their optimal use in these various work-requiring tasks:

a. Where the local solar flux is sufficient, space heat and hot water (as well as cooling) can be effectively provided by various types of solar collectors. Where the local insolation is inadequate these tasks can be performed by transporting into the area (at optimal efficiency) a gaseous fuel that is readily derived from solar sources such as manure, sewage, or other forms of organic matter.

b. Electricity needed to operate stationary machines can be produced directly from sunlight by means of devices such as the photovoltaic cell, or, in the instance cited above, from a cogenerator driven by solar methane.

TABLE 3-5. Efficiency and practicality of various forms of energy

Task	Energy source				
	Solid fuel	Liquid fuel	Gaseous fuel	Electricity (without cogeneration)	Electricity (with cogeneration)
Heating (space heat, hot water, process heat, cooling)	Relatively efficient (if local)	Relatively efficient (transportable)	Relatively efficient (highly transportable)	Inefficient (TD loss at power plant)	Very efficient
Stationary motors	Inefficient, impractical (TD loss) (fixed position)	Inefficient, impractical (TD loss) (fixed position)	Inefficient, impractical (TD loss) (fixed position)	Inefficient, practical (TD loss at power plant) (flexible)	Very efficient, practical
Land vehicles	Inefficient (fuel weight, TD loss)	Inefficient (TD loss)	Inefficient (TD loss)	Inefficient (TD loss)	Very efficient
Aircraft, ships	Impractical (fuel weight)	Relatively efficient (suitable weight/volume relation)	Impractical (fuel volume)	Impractical (battery weight)	Impractical (battery weight)

TD = Thermodynamic, i.e., "Second Law," loss of useful energy.

TABLE 3-6. Most efficient energy sources for selected tasks

Task	Most efficient and practical energy source	Percent of energy from most efficient source (U.S., 1978)
Heating	Gaseous fuel	52.0
Stationary motors	Electricity with cogeneration	15.0
Land vehicles	Electricity with cogeneration	0.1
Aircraft	Liquid fuel	100.0

Source: Energy Information Administration, U.S. Department of Energy, End Use Energy Consumption Data Base: Series 1 Tables (Washington, D.C.: U.S. Government Printing Office, 1978), DOE/EIA-0014.

c. Electricity needed to operate electrified railroads can be derived from cogenerator power stations driven by solar methane. Electricity needed for battery-operated cars could also be derived from such local cogenerator stations, or from arrays of photovoltaic cells.

d. Liquid fuel, needed to operate aircraft (and probably ships and large land vehicles such as heavy tractors and trucks) can be produced, in the form of ethanol, from a common solar source--crops and other forms of biomass.

The overall strategy of the solar transition involves matching such solar sources to the work-requiring tasks which they can most efficiently support, and to facilitate the entry of the solar sources into the overall energy system by means of appropriate nonsolar steps. An example of how this approach can be carried out is outlined in Figs. 3-5, 3-6, and 3-7. These diagrams show how, on a national basis, the energy demands of the residential and commercial sectors of the U.S. economy are now met; how the introduction of gas-fired cogeneration can conserve energy and save money; and finally, how such a system can facilitate the introduction of solar energy.

At the present time, on a national basis, the residential/commercial sector demands 9.2 quads for space heat. Most of this energy is supplied by direct burning of oil and natural gas, but some is also provided by coal and uranium, which is used to produce the electricity that powers some space heat

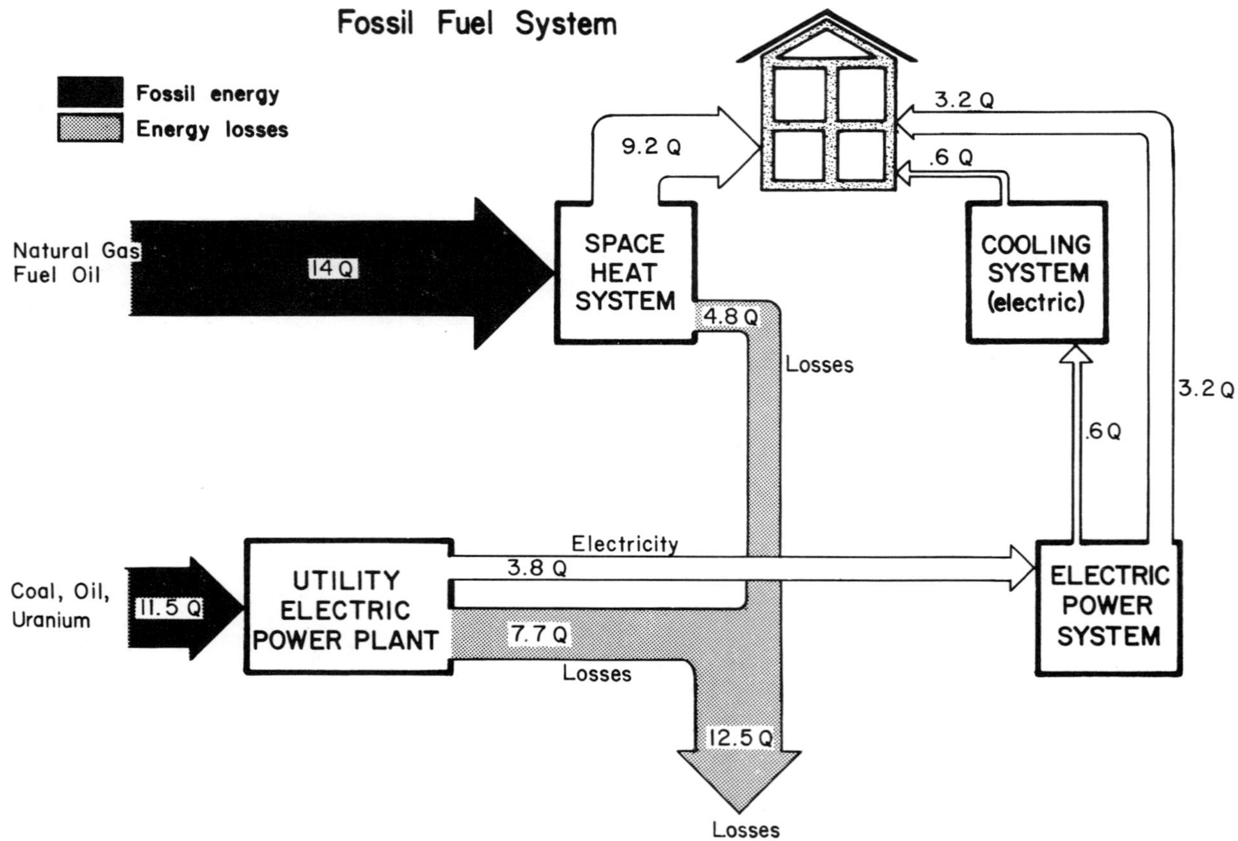

Figure 3-5. Conventional generation of space heat, cooling, and electricity

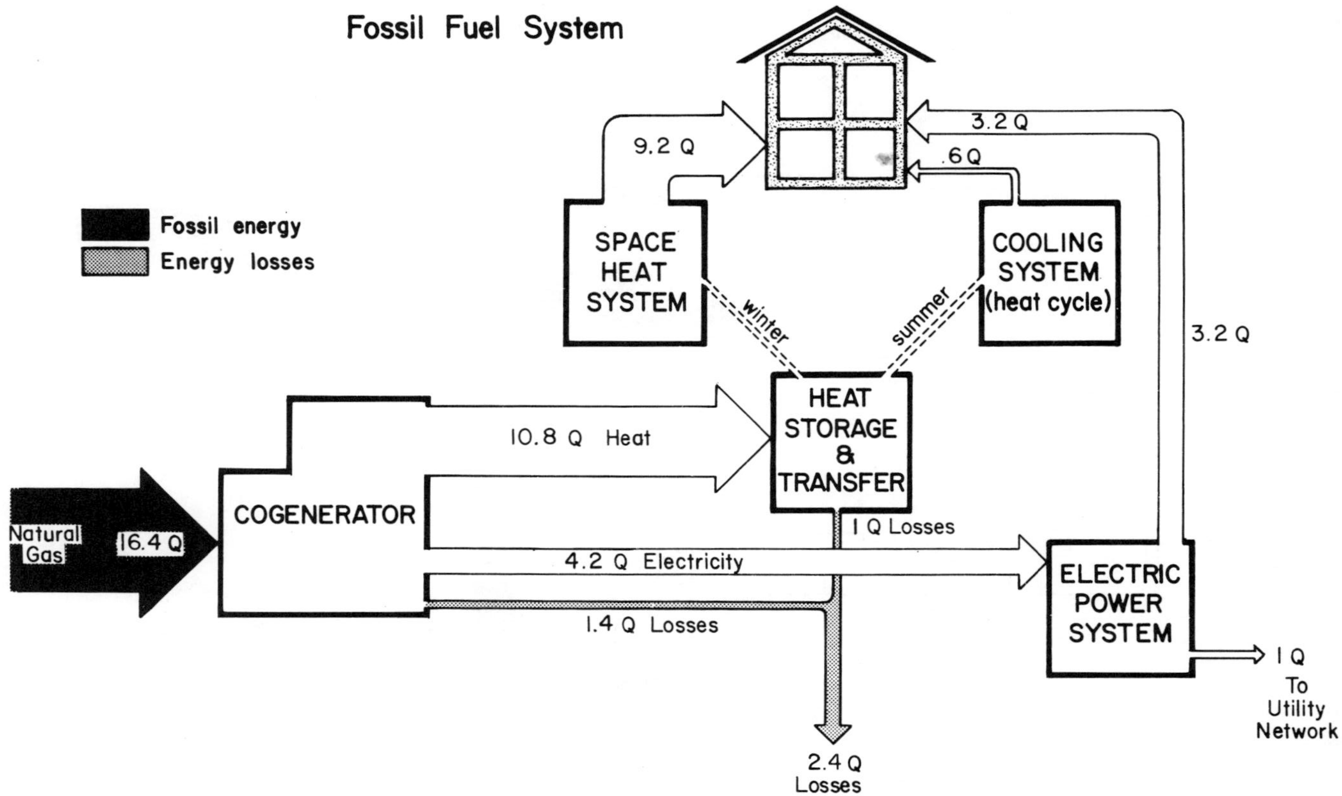

Figure 3-6. Cogeneration of space heat, cooling, and electricity: Fossil fuel system

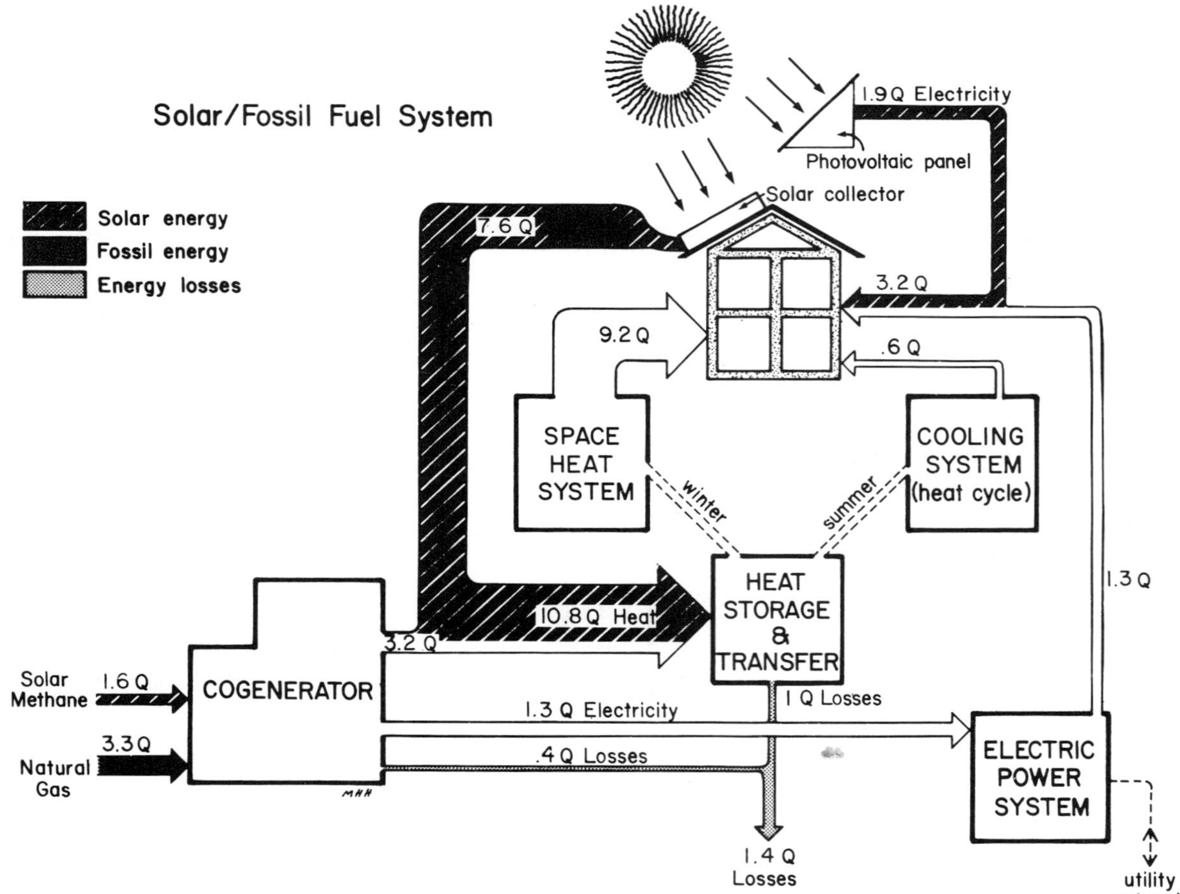

Figure 3-7. Cogeneration of space heat, cooling, and electricity: Solar/fossil fuel system

systems. The sector demands an additional 3.8 quads of energy in the form of electricity, of which 0.6 quads is used for air conditioning. These demands are met by consuming 14 quads of fuel (largely oil and natural gas) for space heat and 11.5 quads of fuel (coal, oil, some natural gas, and uranium) to produce the required electricity. The electric utility systems' losses of "waste" heat amount to 7.7 quads, while 4.8 quads are lost in space heating. Thus, of a total input of 25.5 quads, 12.5, or nearly half, is wasted (see Fig. 3-5).

As shown in Fig. 3-6, if the system were based on cogenerators, only 8.5 percent (instead of about 66 percent) of the energy input into the power plant would be lost. The plant can provide the system with 3.2 quads to meet the need for electricity, 9.2 quads of recovered heat to meet space heat needs, and 0.6 quads of recovered heat to operate air conditioners, a net energy savings of one-third of the energy now used.

Once the system is transformed in this way, three different sources of solar energy can be readily introduced with relatively small changes in the system itself. This is shown in Fig. 3-7. Solar collectors can be readily introduced by feeding the heat which they collect into the system's existing heat storage and transfer system. A major part of the heat energy needed for space heat and air conditioning (7.6 quads of the 10.8 quads required) can be met by this solar source. At the same time, panels of photovoltaic cells can be installed to provide 1.9 quads of the total of 3.3 quads of electricity needed. Now the cogenerator needs to provide only 3.2 quads of heat and 1.3 quads of electricity, which can be accomplished with a total fuel input of only 4.9 quads--an 81 percent saving over the present fuel expenditure. Moreover, methane from a solar biomass source can be substituted for at least part of the natural gas that fuels the cogenerator. Based on expected availability of solar methane, this could reduce the input of fossil fuel to 3.3 quads, or a savings of 87 percent of the fossil fuel now used to supply the residential/commercial sector.

Fundamental to this scheme is the crucial role of natural gas. Because cogenerators must be close enough to residential and commercial consumers to efficiently supply them with heat, cogenerators must be compatible with an urban environment. Only a cogenerator fueled by natural gas, which releases only carbon dioxide and water when it is burned, can meet this requirement.*

---

*A gas-fired cogenerator would produce nitrogen oxides, which can be controlled with presently available technology.

Natural gas is also the only fossil fuel which can be replaced by a solar fuel--methane generated from biomass--with no change in the energy-using equipment. It therefore facilitates the gradual introduction of solar energy without excessive disruptions and costs. (It is worth noting that this process is already under way in Chicago, where the gas utility is buying methane produced from manure at a cattle feedlot in Oklahoma at $1.97 per 1,000 ft^3. The solar methane is pumped into the Texas-Chicago natural gas pipeline and delivered to Chicago consumers as a very small proportion of the gas that they are now burning in their furnaces and kitchen stoves.)

Methane is also a form in which solar energy can be stored on a large scale. Like all processes that depend on sunlight (such as farming), the output of a national solar system would depend on the weather. We could not afford to run out of energy in a bad year, so that a standby energy source is essential. Again, natural gas is the leading candidate since it could be quickly produced from capped-off wells, distributed in the pipelines used for solar methane, and burned in the equipment that uses the latter with no change.

This scheme is described not as a concrete blueprint but only to illustrate the principles which must govern the solar transition: an integrated transformation in which the solar technologies are gradually brought into an overall system which is itself--for sound energy- and money-saving reasons--transformed to facilitate the process. It reminds us that an effective solar transition demands an understanding of the opportunities for improving the efficiency of the existing system in ways which not only alleviate its catastrophic impact on the economy, but which, at the same time, facilitate the transition to the only source that can solve the energy crisis--solar energy.

REFERENCES

Allvine, Fred C., and Patterson, James M., 1973, Competition, Ltd.: The Marketing of Gasoline: Bloomington, Ind., Indiana University Press.

American Gas Association, 1978-79, A Comparison of Capital Investment Requirements for Alternative Domestic Energy Supplies: Arlington, Va., AGA; May 18, 1978.

Blair, John M., 1976, The Control of Oil (New York: Pantheon Books).

Bonneville Power Administration, 1976, Electric Energy Conservation study, prepared by Skidmore, Owings and Merrill: Portland, Ore., BPA.

National Petroleum Council, 1973, U.S. Energy Outlook; Oil and Gas Availability: Washington, D.C., NPC.

U.S. Department of Energy, 1979, Monthly Energy Review: October.

U.S. Federal Energy Administration, 1974, Project Independence Blueprint Final Task Force Report on Oil; Possible Levels of Future Production: Washington, D.C., U.S. Government Printing Office.

CHAPTER 4

SPECULATIONS ON OIL AND GAS RESOURCES IN SMALL FIELDS AND UNCONVENTIONAL DEPOSITS[1]

Richard F. Meyer[2]

INTRODUCTION

Niccolò Machiavelli tells us (Discourses, Book 2, Chap. VII) that "...[The Romans] judged that it was not the extent of land, but its good cultivation, that made wealth. It is necessary also that each colony should have public pastures for their cattle, and forests to supply wood for fuel, without which no colony could exist." Two millennia after the Romans, we still struggle for food and fuel.

At the beginning of 1980, the world is heavily and, in the short term, inextricably bound to oil and gas, but it will not, and indeed cannot, satisfy indefinitely its needs for energy from conventional deposits of gas and mostly medium to light crude oil (20° gravity API and above). In the near future, increasing dependence on a variety of unconventional sources will come into play. Some unconventional sources relate essentially to technology and, therefore, economics. Others are unconventional in that they require new methods of exploration and exploitation.

In due course, the world must find substitutes for petroleum for such things as heating and transportation; coal may substitute for petroleum as a base for petrochemicals. Even petroleum substitutes for heat and power will not be found, developed, and implemented at once. In considering the shift to alternatives, we should speculate on all the ways in which oil and gas might continue to contribute to energy supply without our resorting to technologically complex manufacturing processes to produce fuels in liquid or gaseous form. Naturally occurring petroleum supplies, although not perfect with respect to environmental effects of their exploitation, are far more benign than most of the proposed energy alternatives, such as coal and nuclear fuels.

[1]Preliminary draft, report presented to United Nations Institute for Training and Research Conference on Long-Term Energy Supply, Montreal, December 1978; to be published by McGraw-Hill, 1980.

[2]U.S. Geological Survey, Reston, Virginia.

Definitions are required for the various resource terms used in this report. In general these definitions follow the usage recommended by the U.S. Bureau of Mines and the U.S. Geological Survey (1980):

Resource. A concentration of naturally occurring solid, liquid, or gaseous material in or on the earth's crust in such form and amount that economic extraction of a commodity from the concentration is currently or potentially feasible.

Original Resource. The amount of a resource before production.

Identified Resources. Resources whose location, grade, quality, and quantity are known or estimated from specific geologic evidence. Identified resources include economic, marginally economic, and subeconomic components.

Reserves. That part of the reserve base which could be economically extracted or produced at the time of determination. The term reserves need not signify that extraction facilities are in place and operative. Reserves include only recoverable materials; thus, terms such as "extractable reserves" and "recoverable reserves" are redundant and are not a part of this classification system.

Undiscovered Resources. Resources, the existence of which are only postulated, comprising deposits that are separate from identified resources. Undiscovered resources may be postulated in deposits of such grade and physical location as to render them economic, marginally economic, or subeconomic.

In this report, the amount of the original resource is not given. Instead, the data are for reserves or identified resources. The latter implies that the resource is technically recoverable but not at present economically recoverable.

BASINS OF THE WORLD

Basic to any speculation on ultimate production is an understanding of the geologic setting in which oil and gas are found. Klemme (1977) compiled much information on the known worldwide occurrence of oil and gas, especially with respect to field size and type of basin. Table 4-1 was compiled from data in Klemme's review. The extracontinental basins--those marginal to small ocean basins--contain the most conventional oil and gas, accounting for 50 percent of the world total. Examples of such basins are those in the Middle East Arabian/Persian Gulf, the U.S. Gulf of Mexico, and eastern Venezuela. The intracontinental basins account for another 25 percent of oil and gas and include the Alberta-Western Canada Basin, the great basins of Siberia, and the Libyan

TABLE 4-1. World petroleum basins

Type of basin	Producing basins: Containing giant fields	Containing nongiant fields only	% World reserves	Nonproducing basins: Tested	Poorly tested
Interior	2	7		16	13
Intracontinental	24	15	25	13	13
Rift/graben	6	7	10	17	11
Extracontinental	12	11	50	20	8
Stable coastal	2	2		33	22
Intermontane	13	18	10	35	30
Delta	3	1	4	10	10
Total continental	374				
Total deep offshore	29				
Total	403				

Source: Klemme, 1977.

basins. The small intermontane basins are confined mostly to the circum-Pacific borderlands of the Americas and Asia, whereas the rift basins are typified by the basins in the North Sea and northern Africa. Rift and intermontane basins each account for about 10 percent of oil and gas. Adding 4 percent for deltaic environments--a surprisingly low figure--means that only 1 percent is in continental interior areas, such as the U.S. midcontinent, and stable coastal (continental shelf) basins, exemplified by the circum-Atlantic borders.

In summary, then, among the seven basic types of basins are 62 basins containing at least one giant field that has recoverable reserves of 500 million barrels of oil or 3 trillion ft^3 of gas; 61 productive basins that contain no known giant fields; 144 basins that are nonproductive but that have been extensively tested; and another 107 basins that have been poorly tested, if at all. To this total of 374 may be added perhaps 29 virtually untested deep-ocean basins. This count of basins can easily be expanded. The conterminous United States can be divided into 95 easily defined geologic basins and provinces, as well as many offshore areas, and 22 more onshore basins in Alaska

(Meyer, 1970). Obviously, most countries can be so divided but this expanded number of basins neither alters the validity of Klemme's work nor adds to greater understanding of petroleum occurrence.

These data tell us a great deal about the favorable environments for conventional oil and gas but much less about the most likely places to look for unconventional petroleum deposits. When one speaks of conventional oil and gas reserves worldwide, it is almost invariably in terms of giant fields--those containing more than 500 million barrels of oil or its natural-gas equivalent of about 3 trillion (million million) ft^3. This is reasonable, for giant fields possess impressive credentials indeed: They contain 75 percent of known recoverable oil reserves, or about 820 billion (1,000 million) of a total of about 1,000 billion barrels of reserves plus cumulative past production. Further, more than half the world's known oil has been found in only 33 fields of at least 5 billion barrels each (Nehring, 1978).

SMALL OIL AND GAS FIELDS

An overwhelming proportion, about three-quarters, of all wells drilled in the world for oil and gas have been drilled in the United States, according to Grossling (1977). The proportion is about the same for exploratory wells alone. In terms of drilling density, the same source gives the United States nearly one well per mi^2, the next best, 0.15, and the Middle East, for contrast, only 0.0083. Yet the Middle East ultimately will produce three or four times as much as the United States. The reason for this is that the Middle East is uniquely prolific because of its particular geologic setting, including favorable reservoir rocks and giant structures.

Nehring (1978) has estimated that more than 20,000 fields, each smaller than 100 million barrels, will ultimately produce a total of about 100 billion barrels or 10 percent of the 1,000 billion barrels of presently known (not undiscovered) oil (reserves and past production).

My own data, wherein small fields are defined as those that have less than 50 million barrels of recoverable oil equivalent, support this proposition for the world as a whole. However, for the United States, where the data are comprehensive, they indicate that although 97 percent of the fields are small, they include 26 percent of the known ultimate oil equivalent. For the balance of the world, 71 percent of the recorded fields are small, but they contain only 3 percent of the oil equivalent.

More precise figures could be obtained by means of an improved oil- and gas-field data base not available when my work was done and by compiling oil-field and gas-field data separately rather than as oil-equivalent units. Unfortunately, the necessary data for gas fields simply are not available, but I believe that enough is known to suggest strongly that the ultimate resource significance of small fields is greatly underestimated.

In the United States, more than 26,000 small fields account for 52 billion barrels of oil equivalent, whereas the 717 larger fields account for about 150 billion barrels (Table 4-2). Although a field that contains 50 million barrels of recoverable oil is considered a very good field for the United States, where fields of less than 1 million barrels may be produced, such a field is marginal in most parts of the world in 1980, for economic reasons. This will not always be so as world oil prices continue to rise and as production and, consequently, exploration mature elsewhere. Doubling the proportion of reserves of small fields could thus easily double Nehring's (1978) figure of 100 billion barrels, even though his estimates are based on a definition of the term "small" as being 100 million rather than 50 million barrels.

TABLE 4-2. Relative number of small and large oil and gas fields in the world

[10^6 barrels oil equivalent (BOE)]

	Small fields (<50 x 10^6 B)		Large fields (>50 x 10^6 B)		All fields	
	Number	BOE	Number	BOE	Number	BOE
United States	26,612	52,477	717	147,294	27,329	199,771
Alberta, Canada	958	5,169	97	19,693	1,055	24,862
Others	1,545	14,903	627	451,899	2,172	466,802
World total	29,115	72,549	1,441	618,886	30,556	691,435

Source: U.S. Geological Survey Petroleum Data System; Petroconsultants' Field Records File.

Obviously, the data shown in Table 4-2 are not precise. The totals should be much larger, as indicated earlier. In particular, the Middle East is grossly understated in the data base used to compile the totals. Furthermore, information on natural-gas production is available for only a few areas. Nevertheless, the table is very useful in showing relative proportions of ultimate production, 90 percent of which is accounted for by the 5 percent of large fields.

The reason for understanding the relative proportion is twofold. First, it is highly unlikely that drilling around the world will simply cease when the giant fields have been exhausted because petroleum will always be a valuable commodity, and the search for it will continue even when field sizes diminish. The United States contains large numbers of small deposits. My studies indicate, for example, that 73 percent of South American fields are small, but these fields contain only about 3 percent of known recoverable oil equivalent. In Europe, the figures are 84 percent and 12 percent, respectively. I believe that the growth in the amount of reserves credited to small fields as the result of future drilling will be in addition to presently counted reserves--even those now attributed to small fields--because most present estimates of future discoveries are based upon large fields only.

Second, available information from studies of actual exploration behavior supports this view. Root and Drew (1979) examined the U.S. Permian basin in detail and found that as of 1974, from a total of 4,014 fields, the 140 fields that are larger than 50 million barrels oil equivalent accounted for 76 percent of the hydrocarbons. The trend in size of discoveries over time is important (Table 4-3). For the United States as a whole, the finding rate shows a sharp increase just after World War II, as does the finding rate for the Permian basin. An indefinite period of a nearly constant discovery rate of small fields followed the sharp break. Although this suggests that few large fields have been overlooked, we have no information that indicates an end to the small discoveries for a long time to come.

Even if many small discoveries of the sort described above were made, conventional petroleum supply cannot continue indefinitely at past rates. In many parts of the world, however, continued finding of small fields will extend well into the next century, the cumulative impact of those discoveries will be large, and their ameliorating effect on the economies of many countries will be significant.

TABLE 4-3. Trend in average size of oil and gas fields discovered in the Permian basin, 1921-1974

Time period	Number of exploratory holes	Number of fields discovered	Average field size (10^6 BOE)
1921-38	2,015	111	121.2
1939-46	1,937	171	27.7
1947-50	2,334	277	22.7
1951-52	2,122	251	8.4
1953-54	2,352	322	6.5
1955-56	2,671	377	4.5
1957-58	2,430	334	2.8
1959-60	2,228	309	2.3
1961-62	2,102	327	3.3
1963-64	2,029	362	4.1
1965-66	2,100	304	2.5
1967-68	1,804	271	2.4
1969-71	2,157	309	3.5
1972-74	2,059	288	1.4

Source: Root and Drew, 1979.

HEAVY CRUDE OIL

Heavy crude oil, defined here as oil of 20° gravity API or heavier, is conventional oil in conventional reservoirs produced at least in large measure by means of unconventional technologies. Most of this oil is heavier than 20° gravity API, contains large amounts of trace metals and sulfur, is highly viscous, and retains sufficient mobility to flow, especially when given thermal or other treatment. We know that the world contains much heavy oil--as much as 15 billion barrels of reserves in fields under production or in reservoirs known with a considerable degree of certainty (Table 4-4; Meyer and Dietzman, 1980). Already very substantial amounts of heavy oil are in production, probably as much as 1.2 billion barrels annually, and at least 45 billion barrels have been produced in the past (Table 4-5). Increasingly, more heavy oil is coming to market as reserves of lighter crudes are depleted. How much heavy crude oil remains to be discovered or, in fact, how much is already identified

TABLE 4-4. Summary of known world recoverable heavy-oil reserves and resource
(million barrels)

Area	Identified resource	Reserve recoverable	Total
Canada	211,840	2,669	214,509
United States	30,065	1,754	31,819
Middle America	--	331	331
South America	500,006	5,743	505,749
Europe	1,406	736	2,142
Africa	1,289	260	1,549
Middle East	--	3,528	3,528
USSR and Asia	31	122	153
Oceania	--	--	--
Totals	744,637	15,143	759,780

Source: Meyer and Dietzman, 1980.

TABLE 4-5. Heavy-oil production compared with world total production
(million barrels)

Area	Annual production (1976)		Cumulative production (1979)	
	Heavy oil	Other oil	Heavy oil	Other oil
Canada	N.A.	489	197	8,535
United States	245	2,976	8,183	103,536
Middle America	27	328	1,712	6,431
South America	626	1,269	25,033	43,094
Europe	36	462	443	8,512
Africa	41	2,139	431	26,207
Middle East	164	8,117	2,425	104,226
USSR and Asia	77	5,235	6,782	97,737
Oceania	0	157	0	1,167
Totals	1,216	21,172	45,206	399,445

Source: Meyer and Dietzman, 1980.

but not yet counted in reserves is not known, but current estimates of the total amount are highly conservative.

A major unknown at this time is the future of enhanced oil recovery above and beyond that obtained by primary and secondary methods. Best results have been obtained through the various thermal methods applied to heavy oils alone but room for improvement remains: Overall recovery rates average only about a third of total oil in place. A technological breakthrough comparable with that of hydraulic fracturing in the 1950's would add immediately to reserves. If recovery of all oil could be improved from 33 to 50 percent of original oil in place, as much as 100 billion barrels of additional oil could be produced from currently known fields. Improvement of heavy-oil recovery alone from 15 or 20 percent to 50 percent of oil in place could increase heavy-oil reserve estimates twofold, estimates that were conservative to begin with.

Other heavy oil is in natural bitumen-impregnated rocks, either sandstone or limestone; these occurrences are commonly called tar sands or oil sands but chemically the oil is indistinguishable from heavy oil. Therefore, all heavy oil, including natural bitumen, is included in Tables 4-4 and 4-5. The Athabasca (Alberta, Canada) tar sands are generally considered to be the prototype of this sort of petroleum accumulation. The Orinoco heavy oil belt of eastern Venezuela is usually placed in this category as well, but its oil is more properly considered to be heavy oil because it retains some reservoir mobility.

Large new deposits of natural bitumens probably will not be found, at least at the surface, and for most of the known occurrences, the amount of bitumen in place is fairly well known. Production efforts are underway in Alberta as well as in Venezuela, but the production problems in the two places differ and center on the best means of improved recovery. Most of the deposits are so large that even a few percentage points of recovery improvement could be the equivalent of the discovery of several giant oil fields. I emphasize that the tar-sand deposits included in Table 4-4 are of likely technical recoverability; therefore, they represent only 5 to 10 percent of the bitumen in place.

SHALE OIL

Resources of oil shale in the world are extremely large. In Table 4-6, the resources listed as recoverable are those found in shale at least 10 feet thick and having an average grade of 10 gallons of oil per ton. The marginal

TABLE 4-6. Identified shale-oil resources of the world land areas
(10^9 barrels)

Continents	Recoverable under present conditions	Marginal and submarginal
Africa	10	90
Asia	24	84
Australia and New Zealand		1
Europe	30	46
North America	80	2,120
South America	50	750
Total	194	3,091

Source: Donnell, 1977.

and submarginal resources refer to lower grade oil shales or those of higher grade that are more difficult to mine as a result of depth or of other factors. Obviously, an enormous difference exists between the amount of recoverable shale oil and the amount that is in place, as reflected by the marginal and submarginal categories. This difference in recoverability is caused by the conventional recovery methods, in which the shale must be mined and retorted at the surface, meaning that about half the shale in place must be left as pillars. Even if the shale is recovered by in-situ thermal methods, part (as much as 50 percent) of the oil is lost in the combustion process, the amount of recoverable oil, that is, 190 billion barrels, is very large. If the bulk of it could be recovered by in-situ thermal processing, it would to a large extent alleviate the environmental and technological problems involved with mining and retorting, which are very large undertakings. However, the in-situ processing methods have not yet been perfected, and such methods appear to entail at least a certain amount of mining and fracturing by explosives to create a rubble zone that has the necessary permeability for the production of the fluids released by heating. Alternatively, it may be possible to fracture the shale rock hydraulically through wells. The heart of the recovery problem is that the kerogen in the rock is intimately intermixed with the rock matrix; each ton of shale mined or processed yields only about 350 pounds of organic material, or about 17.5 percent of the original weight, and of this, only about 270 pounds represents oil and gas (Prien, 1974).

Although serious recovery problems are involved, the indicated recoverable resource of about 190 billion barrels is very large. As petroleum demand provides additional incentives, both economic and institutional, for fresh petroleum supply, we can expect that improvements in shale-oil-recovery technology will be made to permit not only exploitation of shale oil but perhaps even an improvement in productive capacity, which is at present a serious handicap to development.

In addition to the oil resources from the Colorado Plateau represented by the preceding data, very large quantities of oil are contained in the black shales of the eastern United States. These black shales are discussed below in relation to their natural-gas content, but they may be an important oil source as well. The Institute of Gas Technology in the United States is conducting research that suggests a recovery of 400 billion barrels of oil, by means of a proprietary process.[3]

METHANE DISSOLVED IN WATER

A large amount of methane is dissolved in subsurface brines. Estimates for the United States Gulf Coast region alone, where the reservoirs are geopressured, range from 1,000 trillion to 50,000 trillion ft^3 (Kuuskraa and Meyer, 1980). A principal difficulty in the estimation process is determining the solubility of methane in the brine, which makes possible an estimation of quantity. Furthermore, the brines are under various pressure regimes, generally exceeding hydro-static pressure. The final difficulty is determining the size of the individual reservoirs under consideration, that is, their thickness and areal extent. At this time, we simply do not have the requisite basic reservoir engineering information on which to base rational estimates of future recoverability of methane from overpressured reservoirs. Dorfman (1977) estimated that, on the basis of assumptions of gas solubility in water and pressure gradients, that the amount of natural gas that might be recovered from the U.S. Gulf Coast is approximately 250 trillion ft^3.

The Gulf Coast is an extracontinental basin (Table 4-1) of the type that accounts for about 50 percent of world petroleum reserves. Fifty-one such

---

[3] Written communication, F. C. Schora to J. Barnea, January 28, 1980.

basins are identified in the world, of which 23 are presently productive. Only the U.S. Gulf Coast basin has been critically examined, most recently by the National Petroleum Council (1980a), for its potential for producing methane dissolved in overpressured reservoirs. Should the estimate by Dorfman of 250 trillion ft^3 for this basin be even approximately correct, then other similar basins elsewhere in the world could easily contain 10 times this amount, or 2,500 trillion ft^3. This is indeed highly speculative, yet this 2,500 trillion ft^3 value includes such areas of the world as the Gulf of Mexico, eastern Venezuela, the north slope of Alaska, the Arabian/Persian Gulf area, the Indus River area of India, the North Sea, and the Black Sea/Caspian Sea region of the USSR.

One of the other less spectacular but possibly significant occurrences of natural gas in water is the estimated 2.2 trillion ft^3 of methane dissolved in the bottom waters of Lake Kivu, Africa (Borgniez, 1960). The waters of the lake are strongly stratified, and the gas is restricted to a stratum of water averaging 600 feet and occurring below the 900-foot level in the lake. Initial efforts to produce the gas were successful (Oil and Gas Journal, 1959), and plans were made to produce the gas as fuel for a local brewery (Oil and Gas Journal, 1962) but have not as yet been carried out. The methane-bearing waters also contain about four times as much carbon dioxide as conventional methane, significant nitrogen and hydrogen sulfide, and large amounts of dissolved salts and phosphate. Whether this is a unique occurrence is not known, but its existence suggests that the possibilities for similar occurrences elsewhere should be examined.

That natural gas has been produced from shallow subsurface brines in Japan for some years (Marsden and Kawai, 1965) and that other occurrences are known in Italy, the USSR, China, and the Philippines is of great importance because such productive possibilities have undoubtedly been overlooked in association with many of the world's conventional oil and gas fields. To estimate a potential supply of such gas is virtually impossible until the necessary investigations concerning the origin of the gas have been made. Such investigations will require examination of records not only of oil and gas wells but of water wells drilled the world over.

NATURAL GAS IN ORGANICALLY RICH BLACK SHALE

Since the early 1970's, interest in the extensive organically rich black shale deposits of the eastern United States has increased markedly as natural-gas reserves and production have declined and as various fuel embargoes and crises have been imposed during the past decade. Extensive investigations are underway into the resource potential of this shale, spurred by the fact that the 2,000 mi^2 Big Sandy gas field has been producing from such a reservoir in Kentucky for more than 50 years. This field is credited with ultimately recoverable reserves of about 2 trillion ft^3.

The black shales have long been of interest to geologists because of the possibility that they may be a source for several energy commodities. They are relatively rich in uranium and, like the oil shales of the western United States, they have been considered as a potential source of oil, although their lithology differs greatly from that of the western shales. Although the shales could be retorted to extract oil as well as gas, this expensive recovery method would be applicable only to those parts of the shale that are very rich in organic matter.

Most current attention is directed to the shales as a source of gas-well gas produced from fractures in the shale. Most current research, in fact, is directed to technologies for improving both the gas recovery and the productive capacity from the shales, commonly through some form of artificial fracturing. Because these shales are very widely distributed (Table 4-7), they must be given serious consideration as a potential gas source. This table, derived from Duncan and Swanson (1965), includes only the relatively rich shales containing 10 to 65 percent organic matter. The shales of lesser organic content, although their total amount of contained gas is much greater, are not considered because they are too lean to be considered exploitable in the foreseeable future. The richer shale, as may be seen from the table, is another matter entirely, although the gas resource given for the United States may be as much as three times too high in terms of potentially recoverable natural gas, in the light of current research. A recovery of about 25 trillion ft^3 of gas from the shale is considered reasonably certain (National Petroleum Council, 1980b), but recovery may be as much as 100 trillion feet. The table indicates an estimate by Duncan and Swanson that the world total from shale of this richness might be about 4,000 trillion ft^3. Even if this estimate is reduced to one-third of that number, in keeping with the U.S. experience, it still amounts to more than

TABLE 4-7. Estimate of total stored energy in organically rich shale of the United States and principal land areas of the world
(Estimates and totals rounded)

Continent or country	Approximate area underlain by sedimentary rocks (10^{12} mi^2)	Shale containing 10 to 65 percent organic matter		
		Shale in deposits (10^{12} short tons)	Minimum organic content (10^{12} short tons)	Gas resource (10^{12} ft^3)
United States	1.6	120	12	310
Africa	5.0	370	37	960
Asia	7.0	500	50	1,300
Australia	1.2	90	9	230
Europe	1.6	120	12	310
North America (including United States)	3.0	220	22	570
South America	2.4	180	18	470
World total (rounded)	22	1,600	160	4,150

Source: Duncan and Swanson, 1965.

1,000 trillion ft^3 of gas, which is indeed a large quantity and one that could have an enormous effect on the economies of the nations in which it is located. Obviously, there are large quantities of natural gas from conventional gas deposits that could be available to the world market, but the deposits are not presently exploited because of transportation difficulties and institutional problems. The transportation and handling of gas from black shale are the same as for gas from conventional deposits. Some countries have black shale resources but not conventional gas. The black-shale gas thus could become a significant national asset requiring careful consideration wherever it occurs in the world.

NATURAL GAS IN TIGHT SANDS

Significant occurrences of natural gas in so-called "tight" sandstones are well known and well documented (see, for example, Meyer, 1977; National Gas Survey, 1978b). The problem of exploiting such reservoirs, which owe their tightness to exceptionally low permeability, is much more extensive than the mere estimation of resource potential. Because the permeabilities of these tight sands are generally 1,000 times less than those of conventional reservoirs, the tight host rock must be extensively fractured by injecting large quantities of fluid under very high pressure into the formation through the wellbore, the fluid carrying an entrained propping agent. When the pressure at the wellhead is relieved the fluid is forced back out of the formation, and the propping agent is left to maintain permeability paths along the fractures to the wellbore, which serves as a pressure sink. An outstanding example of tight reservoirs is found in the basins of the Colorado plateau of the western United States. The tight gas sands underlie, to a large extent, the same formations that contain the well-known oil-shale deposits of the Green River, Unita, and Piceance Creek basins. The tight gas resource estimate for these basins is as much as 600 trillion ft^3 of gas, but only about 250 trillion ft^3 is identified by reasonably close well control (National Gas Survey, 1978b). Therefore, 250 trillion ft^3 serves as a conservative yardstick of the resource for this area. Other examples of tight sands (National Gas Survey, 1978b) are found in the Denver basin, to the east, which contains as much as 1 trillion ft^3 in tight formations; east Texas; the Sonora subbasin of the Permian basin; the San Juan basin of New Mexico, 63 trillion ft^3; the northern Great Plains in Montana, 285 trillion ft^3; and New York State.

The U.S. tight sands fall into the category of cratonic intracontinental-composite basins (Table 4-1), a geological environment that also includes the western Canada sedimentary basin, the Appalachian basin of the eastern United States, North Africa, northern Europe, and the great broad basins of the Siberian part of the USSR. Consequently, the tight sands of the small western U.S. basins probably are not unique occurrences but are symptomatic of very widespread geological environments. Undoubtedly, many tight gas-sand formations have been drilled in the past but have been overlooked for any of a number of reasons. First, because the formations are of low productivity without stimulation, they are easily overlooked in the drilling of a well, even an exploratory well. Second, even if identified, the reservoirs probably would not be exploited if they were within the same province as fields and reservoirs of normal productive capacity in areas that have a surplus of natural gas. Under present conditions, however, the technology for transporting natural gas by means other than pipeline is advancing rapidly, and exploration of the resource potential of all gas sources is increasingly worthwhile. Additional tight sands probably will be found in either producing or scarcely explored basins along the Andean trend of South America and in the highly structured area along the northern border of India. I surmise from this that a resource of at least 2,000 trillion feet probably remains to be identified in tight-sand reservoirs in basins throughout the world outside of the United States.

METHANE OCCLUDED IN COAL BEDS

For as long as coal has been mined underground, explosive methane gas has led to the death and injury of countless miners. In order to minimize explosions in underground coal mining operations, the U.S. Bureau of Mines instituted a program of mine degasification. One means of degasifying the coal seams before mining operations begin is to drill ordinary gas wells into the seams and to produce the gas to exhaustion; this was done as far back as 1905. Attempts are now being made to perfect this technique in the United States, and, in fact, small amounts of the gas are being utilized commercially. Because wellhead pressures are low, the gas at this time is exploited only near the well site.

While conducting the degasification work, we naturally attempt to make estimates of the amount of recoverable gas that might be present as free gas occluded in fractures in the coal seams. Estimates of 300 (Deul, 1976) to 850 (National Gas Survey, 1978a) trillion ft^3 of gas in U.S. coal have been made; these estimates include both identified and hypothetical coal beds, but, of course, not strippable coal. At the same time, a total of 3.2 trillion tons of coal was estimated for the United States, about evenly divided between identified and undiscovered (Table 4-8). Total coal in the world was estimated at about 13.6 trillion tons, most of which occurs in Asia, including the European part of the USSR (Averitt, 1973). If these numbers are approximately correct and if the proportional amount of gas occluded in U.S. coal beds is about the same as that for the world, roughly 3,500 trillion ft^3 would be available in the world from this source.

TABLE 4-8. Estimated total original coal resources of the world, by continents[1]

(10^9 short tons)

Continent or country	Identified resources[2]	Hypothetical resources[3]	Estimated total resources
Asia[4]	7,000[5]	4,000	11,000[6]
United States	1,581	1,643	3,224
Other North America	139	1,237	1,376
Europe	620	210	830
Africa	80	160	240
Oceania	60	70	130
South and Central America	20	10	30
Total	9,500	7,330	16,830

[1]Original resources in the ground in beds 12 inches or more and generally less than 4,000 ft below the surface, but includes small amounts between 4,000 and 6,000 ft.

[2]Identified resources: Specific identified mineral deposits for which extent and grade may or may not be evaluated, and whose contained minerals may or may not be profitably recoverable under existing technology and economic conditions.

[3]Hypothetical resources: Undiscovered mineral deposits, whether of recoverable or subeconomic grade, that are geologically predictable as existing in known districts.

[4]Includes European USSR.

[5]Includes about 6,500 billion short tons in the USSR.

[6]Includes about 9,500 billion short tons in the USSR.

Source: Averitt, 1973.

The exploitation of this particular gas resource has salient advantages over other unconventional sources. Most important of these is that methane production leads to a degasification of the coal seams, making later underground mining of the coal safer. In addition, much of the coal is in areas where the natural gas could be utilized to the benefit of the local communities.

GAS HYDRATES

Gas hydrates are solids that resemble ice (Katz, 1971) in that gas molecules are incorporated in the molecular structure of the water. The formation of gas hydrates depends on temperature, pressure, and salt content of the water. For example, at 30° F, methane in pure water forms hydrates at 350 pounds per in^2 absolute (psia), whereas at 30° F in water containing 20 weight-percent NaCl, a pressure of 1,000 psia is required.

Gas hydrates are well known to the natural-gas industry because their formation can plug high-pressure flow lines if the gas is not dry. Aside from such practical problems, gas hydrates are known to occur widely in areas of permafrost as well as in the upper layers of sediment beneath the sea floor. Thus, two potential sources of new gas supply present themselves.

The phenomenon appears to be almost universal below the sea floor outside the equatorial region. The presence of gas hydrates has been determined from deep-sea cores and inferred in many studies that are based upon ocean seismic profiling (Tucholke, Bryan, and Ewing, 1977, in the western Atlantic, and Grantz, Boucher, and Whitney, 1976, in the Beaufort Sea, for example). Although a means to exploit such deposits is very difficult to visualize at this time, the resource appears to be large enough to require further study. In addition, Trofimuk (1973) has made the very significant observation that, because hydrates render sediments impermeable to natural gas, the upper sedimentary layers of the ocean floors, and their gas hydrates, may form a cap rock beneath which oil and gas could be trapped.

The amount of natural-gas hydrates in permafrost areas may be about 500 trillion ft^3, as documented in a number of sources (Cieslewicz, 1971; Trofimuk et al., 1977; Bily and Dick, 1974), but the means of exploiting them are not clear. Trofimuk et al. (1977) have suggested that a definite correlation exists between areas of former continental glaciation and the presence of large natural-gas accumulations. The evidence given by them (see Tables 4-9 and 4-10) indicates that this possibility is too important to overlook. Table 4-9

TABLE 4-9. Hydrate-formation zones and ice sheet distribution on the earth
(Average thickness of hydrate-formation gases on continents, 700 m; on ocean floor, 300 m)

Age	Continental glaciers			Hydrate-formation zones (HFZ) Continents (Northern Hemisphere)		
	Area (10^6 km^2)	Area (%)	Volume (10^6 km^3)	Area (10^6 km^2)	Area (%)	Rock volume in HFZ (10^6 km^3)
Carboniferous and Permian	70	--	175	--	--	--
Quaternary (temperature minimum)	45	30	60	75	50	52
Holocene	15	9	24	40	27	28

Source: Trofimuk et al., 1977.

TABLE 4-10. Distribution of world oil and gas resources relative to Quaternary glaciation

	Continental glaciation (N. lat)	Hydrate-formation zones (N. lat)	Percentage of oil and gas in region that is associated with glaciation	
			Oil	Gas
North America	38-46°	35°	95	40
West Europe	49°	50°	95	90
East Europe	--	47°	--	40
West Siberia	59°	56°	80	98
Total Northern Hemisphere	50°	45°	60* (8**)	60

*Exluding the USSR.

**Exluding the USSR and Athabaska tar-sand deposit, Alberta, Canada.

Source: Trofimuk et al., 1977.

shows the relationship between ice coverage by continental glaciation of different ages and extent of zones of gas hydrate formation. Table 4-10 relates the distribution of continental glaciation to present-day oil and gas accumulation. Such deposits have been estimated to occur over half the area of the USSR and may represent reserves of as much as 350 trillion ft^3 (Cieslewicz, 1971). Bily and Dick (1974) described the presence of gas hydrates in the permafrost of northern Canada and drew attention particularly to the drilling problems that the hydrates present when warmed by the drilling fluid. They pointed out the production problems that may be involved if we attempt to exploit the hydrate resource, particularly the heat requirement and consequent water release during decomposition.

UNDERGROUND COAL GASIFICATION

The concept of exploiting coal by burning it underground and recovering the released gases is 100 years old; significant efforts have been undertaken to recover gases by this method since the 1950's (Dinsmoor et al., 1978). The basic purpose is to generate methane in situ rather than in a synthetic natural-gas plant after coal has been mined. The methodologies for such in-situ combustion are well known and have been reviewed by Schrider and Whieldon (1977) and Buder and Terichow (1978).

In the United States, the coal resource base is large (Table 4-8) and even these 1973 data are understated compared with more recent estimates. Buder and Terichow (1978) reported estimates for the contiguous 48 States of the United States at 50 percent recovery, of 9,000 trillion ft^3 of gas that is potentially extractable by underground gasification from coal beds not otherwise minable.

In general, the resulting gas is low in heat value--70 to 150 Btu per cubic foot--and useful mostly for fuel-gas and power generation. It could, of course, be upgraded to a high-Btu gas by the addition of hydrogen. Furthermore, the in-situ combustion process is most practical for the western U.S. subbituminous and lignitic coals, which are nonswelling, thick, and low in rank.

The coal bed is destroyed when this method is used; in the future the bed could be neither mined nor degasified for production of occluded gas. In addition, the decrease in volume of near-surface coals could result in serious subsidence problems. Nevertheless, the procedure is significant, and perhaps the gas supply is one of the most available new sources in any area containing coal resources.

SUMMARY AND CONCLUSIONS

Conventional oil and gas fields have been exploited for more than a century. Although remaining resources of oil and gas are large, their geographic distribution and other constraints on their use are leading to a careful examination of alternative sources. Such naturally occurring deposits have been the subject of this chapter. A summary of world oil and gas resources in small fields and unconventional deposits is as follows:

Oil

Small fields	100 billion barrels
Heavy-oil (reserves in known fields)	15 billion barrels
Heavy-oil (resource in deposits (tar sands))	745 billion barrels
Undiscovered heavy-oil resources	300 billion barrels
Shale oil	190 billion barrels
Total	1,350 billion barrels

Natural gas

Geopressured reservoirs	2,500 trillion ft^3
Lakes	20 trillion ft^3
Shallow Japanese-type brines	100 trillion ft^3
Shale rich in organic matter	1,000 trillion ft^3
Tight sands	2,000 trillion ft^3
Occluded in coal	3,500 trillion ft^3
Hydrates in permafrost	500 trillion ft^3
Underground coal gasification	13,000 trillion ft^3
Total	22,620 trillion ft^3

It has been my purpose, in common with other chapters in this volume, to suggest that large amounts of oil and gas remain to be exploited throughout the world. The timing of such exploitation will be a function of demand, economics, and technology. These factors are, of course, intricately interwoven and much remains to be learned.

One example of the attempt to come to grips with economics is the study of unconventional gas by the United States National Petroleum Council (1980c). The Council estimates for cumulative gas reserve additions (trillion ft^3), using only the 10 percent rate of return case (15 percent for tight gas) follow:

	Gas price, constant 1979 dollars, per 1,000 ft^3		
	$2.50	$5.00	$9.00
Coal seams	5	25	45
Black shale	7	20	27
Geopressured brine	0	0.1	0.6
Tight sands	116	175	204
Total	128	220.1	276.6

The amounts thus indicated to be recoverable are but a fraction of the technically recoverable resource but do suggest the important relationship between price and availability.

Studies of the size of the resources contained in the kinds of deposits described here are being conducted in many places. Schrider and Wise (1980) describe the efforts by the United States Department of Energy, in conjunction with the United States Geological Survey and industry, relative to tight sands, black shales, geopressured gas reservoirs, and methane from coal beds. The International Institute for Applied Systems Analysis (IIASA) held a Conference on Conventional and Unconventional World Natural Gas Resources in Laxenburg, Austria, June 30-July 4, 1980. Earlier, IIASA and the United Nations Institute for Training and Research held a conference on The Future Supply of Nature-Made Petroleum and Gas (cited in Meyer, 1977). In 1979, UNITAR conducted conferences on Heavy Crude and Tar Sands (cited in Meyer and Dietzman, 1980) and on Future Energy Sources. Doherty (1980) describes the research efforts on unconventional gas sources by the United States Institute of Gas Technology. The results of all such activities will be a firmer understanding of the nature and resource potential of unconventional deposits and the economics of their development.

REFERENCES

Averitt, Paul, 1973, Coal, in Brobst, D. A., and Pratt, W. P., eds., United States mineral resources: U.S. Geol. Survey Prof. Paper 820, p. 133-142.

Bily, C., and Dick, J. W. L., 1974, Naturally-occurring gas hydrates in the Mackenzie delta, N.W.T.: Canadian Petroleum Geol. Bull., v. 22, p. 340-352.

Borgniez, G., 1960, Donnees pour la mise en valeur du gisement de methane du Lac Kivu: Acad. Royale Sci. Coloniales Sec. Sci. Tech. Mem. in 8° N.S., v. 13, no. 1, 110 p.

Buder, M. K., and Terichow, O. N., 1978, Underground coal gasification can reach unminable energy: Oil and Gas Jour., v. 31, no. 6, p. 54-61.

Cieslewicz, W. J., 1971, Some technical problems and developments in Soviet petroleum and gas production: Mines Magazine, v. 61, no. 11, p. 12-16.

Deul, Maurice, 1976, Natural gas from coal beds, in Natural gas from unconventional geologic sources: National Academy of Sciences, Washington, D.C., p. 193-205.

Dinsmoor, B., Galland, J. M., and Edgar, T. F., 1978, The modeling of cavity formation during underground coal gasification: Jour. Petroleum Tech., v. 31, no. 5, p. 695-704.

Doherty, M. G., 1980, Unconventional gas: IGT GaScope, no. 49, p. 2-8.

Donnell, J. R., 1977, Global oil-shale resources and costs in Meyer, R. F., ed., The future supply of nature-made petroleum & gas: New York, Pergamon, p. 217-260.

Dorfman, M. H., 1977, The supply of natural gas from geopressured zones: engineering and costs in Meyer, R. F., ed., The future supply of nature-made petroleum & gas: Pergamon, p. 873-888.

Duncan, D. C., and Swanson, V. E., 1965, Organic-rich shale of the United States and world land areas: U.S. Geol. Survey Circ. 523, 30 p.

Grantz, Arthur, Boucher, Gary, and Whitney, O. T., 1976, Possible solid gas hydrate and natural gas deposits beneath the continental slope of the Beaufort Sea: U.S. Geol. Survey Circ. 733, p. 17.

Grossling, B. F., 1977, The petroleum exploration challenge with respect to the developing nations, in Meyer, R. F., ed., The future supply of nature-made petroleum & gas: New York, Pergamon, p. 57-69.

Katz, D. L., 1971, Depths to which frozen gas fields (gas hydrates) may be expected: Jour. Petroleum Tech., v. 23, p. 419-423.

Klemme, H. D., 1977, World oil and gas reserves from analysis of giant fields and basins (provinces), in Meyer, R. F., ed., The future supply of nature-made petroleum & gas: New York, Pergamon, p. 217-260.

Kuuskraa, V. A., and Meyer, R. F., 1980, Review of world resources of unconventional gas: Report presented at IIASA Conference on Conventional and Unconventional World Natural Gas Resources, Laxenburg, Austria, June 30-July 4, 1980, 66 p.

Marsden, S. S., and Kawai, Kozo, 1965, "Suiyosei-Ten' Nengasu, a special type of Japanese natural gas deposit": Amer. Assoc. Petroleum Geol. Bull., v. 49, no. 3, p. 286-295.

Meyer, R. F., 1970, Geologic provinces code map for computer use: Amer. Assoc. Petroleum Geol. Bull., v. 54, no. 7, p. 1301-1305.

Meyer, R. F., 1977, The resource potential of gas in tight formations in Meyer, R. F., ed., The future supply of nature-made petroleum & gas: New York, Pergamon, p. 645-656.

Meyer, R. F., and Dietzman, W. D., 1980, World geography of heavy crude oils, in Meyer, R. F., and Steele, C. T., eds., The future of heavy crude oils and tar sands: New York, McGraw-Hill.

National Gas Survey, 1978a, Nonconventional natural gas resources--Methane in coal: U.S. Dept. Energy DOE/FERC-0010, p. 1-36.

National Gas Survey, 1978b, Nonconventional natural gas resources--Gas in tight formations: U.S. Dept. Energy DOE/FERC-0010, p. 47-103.

National Gas Survey, 1979, Nonconventional natural gas resources--Gas dissolved in water: U.S. Dept. Energy DOE/FERC-0029, 217 p.

National Petroleum Council, Committee on Unconventional Gas Sources, 1980a, Geopressured brines--unconventional gas sources: Washington, D.C., National Petroleum Council, variously paged.

National Petroleum Council, Committee on Unconventional Gas Sources, 1980b, Devonian shale--unconventional gas sources: Washington, D.C., National Petroleum Council, variously paged.

National Petroleum Council, Committee on Unconventional Gas Sources, 1980c, Interim summary--unconventional gas sources: Washington, D.C., National Petroleum Council, variously paged.

Nehring, Richard, 1978, Giant oil fields and world oil resources: Rand Corporation Report R-2284-CIA, 162 p.

Oil and Gas Journal, 1959, Weird Belgian Congo gas field is tapped: Oil and Gas Jour., v. 57, no. 17, p. 82.

Oil and Gas Journal, 1962, Brewery will tap big African gas bubble: Oil and Gas Jour., v. 60, no. 11, p. 83.

Prien, C. H., 1974, Current oil shale technology: A summary: Rocky Mountain Assoc. Geologists 1974 Guidebook, p. 141-150.

Root, D. H., and Drew, L. J., 1979, The pattern of petroleum discovery rates: Amer. Scientist, v. 67, no. 6, p. 648-652.

Schrider, L. A., and Whieldon, C. E., 1977, Underground coal gasification--a status report: Jour. Petroleum Tech., v. 30, no. 9, p. 1179-1185.

Schrider, L. A., and Wise, R. L., 1980, Potential new sources of natural gas: Jour. Petroleum Tech., v. 32, no. 4, p. 703-716.

Trofimuk, A. A., 1973, [Accumulation of natural gases in zones of hydrate-formation in the hydrosphere]: Akad. NaukSSSR, Doklady, v. 212, p. 931-934.

Trofimuk, A. A., Chersky, N. V., and Tsaryov, 1977, The role of continental glaciation and hydrate formation on petroleum occurrence, in Meyer, R. F., ed., The future supply of nature-made petroleum & gas: New York, Pergamon, p. 919-926.

Tucholke, B. E., Bryan, G. M., and Ewing, J. J., 1977, Gas-hydrate horizons detected in seismic-profiles data from the Western North Atlantic: Amer. Assoc. Petroleum Geol. Bull., v. 61, p. 698-707.

U.S. Bureau of Mines and the U.S. Geological Survey, 1980, Principles of a resource/reserve classification for minerals: U.S. Geol. Survey Circ. 831, 5 p.

SECTION II. ENERGY POLICY AND PLANNING

CHAPTER 5

APPLES, ORANGES, AND THE COMPARISON OF ALTERNATIVE ENERGY TECHNOLOGIES

Alvin Kaufman[1]

INTRODUCTION

The selection of alternative energy technologies in a scientific manner is a matter of considerable concern to all those responsible for energy development. This is crucial in view of the large amounts of capital required to fund both the research and commercialization phases of these new technologies. This large capital requirement in a world facing limited economic growth, means that choices between technologies must be made. Not every technology can be funded.

The question of how to compare these then becomes vital, since the method can bias the result. In view of the importance of the subject, this paper will deal with the methodology of comparing alternative technologies rather than with an actual comparison.

AN OVERVIEW

In the United States and Canada, comparative processes are generally evaluated in terms of economic factors, including the cost of environmental and social regulation with the final choice dependent on the estimated return on investment believed probable from the project. In other countries, however, economic viability is not the only, or even the prime, determinant. The choice, by and large, will be based on the contribution of each process to the social, economic, and political development of the country. That is, the predominant considerations will be such items as employment potential, contribution to infrastructure development (highways, ports, plants), the ability to encourage other industrial development, and national security considerations, such as making the nation self-sufficient. This latter system of comparison has been making inroads into the more conventional form of analysis because of steadily increasing governmental interference in the economy, even in the so-called mixed economies. Therefore, in our discussion of methods for comparison, we

[1]Congressional Research Service, Library of Congress, United States.

will deal with the latter, although the return on investment procedure still forms the base methodology.

In any consideration of alternative technologies one can usefully develop a model, which is simply an organized way of thinking about a problem. It can be expressed in words or in equations. The equations may be solvable on a computer, a calculator, or through simple mathematics. In short, the equations can range from complex to highly simplistic.

Commonly considered models are those dealing with process and with economics. The process models are based on detailed engineering studies, and as a consequence incorporate information on future technology and probable structural changes. The econometric models, on the other hand, deal with consumer behavior and economic activity. The two types are often merged into a single integrated major model. These models, however, do best as a means of testing or assessing alternative policies. For example, such models can be useful in determining whether to build additional nuclear plants or not. As Hamilton and Manne (1978) have indicated, public debate tends to focus on low probability-high consequence events, but the high probability-low consequence events may be of greater importance. Integrated models are useful in predicting these events and thus may be able to produce the impact data needed for our technologic comparisons, but generally they will not be suitable for the selection of alternative technologies. These models tend to be too aggregated and oriented toward the answer to "big" questions such as "should we tax or subsidize." There is, however, no reason why suitable integrated models could not be developed for the task of technology selection. Lacking access to such a sophisticated tool, however, we have elected to approach the topic in a simple manner.

Before discussing our models, it will be useful to consider the problems that must be faced by any model designed to evaluate alternative technologies. These problems include those connected with data and those dealing with uncertainty.

Data Problems. In any modeling effort, there are problems with data, but these are compounded when one is dealing with new technologies. These generally lack a data history in the sense that the available cost data are based on engineering estimates rather than actual operations. In most instances, the technologies being compared have not operated at commercial scale. As a consequence, the cost data are based on laboratory or pilot plant activity which may be at variance with the future commercial cost experience. Further, comparison of these estimates against data for established technologies will result in a biased answer. That is, the data for the established technology will be weighted downward by economies of scale, while the new technology will be weighted upward by the relatively primitive stage of technologic development. In any case, the new technology will probably follow the rule of two and one-half. That is, the costs will be double and the benefits half of the estimates.

Uncertainties. In comparing alternatives we are faced with uncertainty in predicting environmental, social, and resource impacts. A new technology is uncertain in its causes and effects because there is no experience upon which to base estimates, and because our estimating techniques are poor. The impacts include environmental and health risks, catastrophic accidents, political problems (nuclear proliferation), and such land use effects as strip mining. Such impact uncertainty is compounded by an inability to predict which will be mitigated by future technology and which will not (Ramsay, 1979).

Thus, our knowledge of impacts will be greater in the future than it is at present. Therefore, the early cancellation of a technologic option based on imperfect knowledge may foreclose that option even though it may turn out to be preferable in the future because its impacts could, at that time, be more easily mitigated as a result of the march of technology. Given then our imperfect knowledge, the selection of alternative technologies becomes essentially a value judgment. That judgment must encompass an evaluation of an acceptable level of risk; the importance of the preservation of health, resources, and the environment; as well as such equity values as income, regional, and intergenerational effects. The weights given these various items will be different for different people. Risk perception is thus integrated with the goals and priorities selected by each country. For example, a country setting economic development as a high priority goal may view the risk of environmental damage as acceptable. A more fully developed economy, on the other hand, may perceive the environmental risk as too great compared with the benefits of further economic development.

From the above it is apparent that any evaluation of alternative technologies is, to a large extent, a subjective exercise.

COSTS

In any comparison of alternative technologies we must deal with two kinds of cost, those that are directly related to the process, and those that are external.

Direct Costs. These costs would include the materials, capital, and labor involved in the creation and operation of the plant, as well as those incurred to meet existing environmental and social regulations. For example, the fixed and variable costs of a cooling tower built to meet water effluent standards would be considered as a direct cost. In most cases, the direct costs will be computed as engineering estimates based on expected materials flows and balances. As a consequence, these will be subject to change as additional information on construction and operational requirements becomes available, and as the technology advances up the learning curve.

External Costs. External costs are those related to noneconomic transactions. That is, an economic transaction between a buyer and a seller tends to have its impacts restricted by the marketplace to the parties to the negotiation. The market, however, will not reflect the value of multiple service resources such as air and water. These are not only useful in a particular process, but are required by others for life sustenance, recreation, waste disposal, and so forth. The pollution or damage of these resources and the consequent impact on others will often not be reflected in the prices set by the marketplace. This occurs because air and water are generally regarded as almost free goods, while subsequent damage to others is difficult to evaluate. Generally, it will not be possible to pinpoint either the identity of those who cause the problem, or the value of the effect in a manner sufficient to determine cost. In those instances where it is possible to readily pinpoint the offender, the externalities can be internalized by forcing the cost of abatement on those who cause the damage. For example, if the hot water effluent from a nuclear plant is believed damaging to the environment, those units can be required to install heat dispersal mechanisms.

The reduction or elimination of external costs will often entail other costs, sometimes of a rather staggering magnitude. The cost of abatement must be measured against the benefits in order to determine a trade-off point between abatement and suffering (Kaufman, 1976). Kahn (1975) has noted that "...environmental protection, like conservation, is an economic good, and, like all other economic goods, we cannot have unlimited quantities except at unlimited costs."

Lave and Silverman (1979), however, have stated that the benefit-cost procedure tends to focus on economic efficiency as the goal, but other goals also need to be considered. This form of analysis also has an important bias toward the overestimation of the costs of internalizing external costs, and the underestimate of the benefits. The costs are generally overstated because technological change is usually not accounted for in the procedure. The benefits are generally understated because hard-to-value items, such as an attractive view or better income distribution are usually left out.

The cost-benefit procedure also puts considerable emphasis on the importance of the timing of costs and benefits. That is, present costs or benefits are more important than those accruing in the future because money not expended or collected today can be invested to produce more money tomorrow.

A DISCUSSION OF METHOD

There are a number of methods that can be used to evaluate alternative technologies. Of these, however, there are two I would like to discuss.

The Point Method. This method is used to compare technologies at a point in time, such as the year 1990. As a consequence, some adjustments must be made to the data in order to assure that all costs are included. In making our comparison, it is best to keep all money figures in real terms, that is, based on prices in a common year, such as 1979. This avoids the problem of attempting to divine the future pace of inflation.

The comparison needs to be undertaken in two steps: the first in terms of process cost; the second in terms of external costs. The process costs would be computed in the usual way, based on engineering estimates for each technology, and including costs incurred to meet government regulations and standards. In doing so we must keep in mind that it is the incremental cost that counts. That is, the cost for the last unit on the line provides a measure of the human, capital, and natural resources that will be needed. A comparison of incremental cost for each alternative yields a measure of the economic resources that need to be foregone in order to produce energy by each method. Incremental cost comparisons have other advantages. They avoid the problem of apples and oranges. Comparing a new technology against an existing technology on an average cost basis is comparing apples and oranges. The established technology will have a heavy portion of fixed features with change occurring on the margin. Its average cost will also be heavily weighted by economies of scale not yet apparent in newer technologies. The comparison of the marginal unit of the established technology versus the last increment of the newer technology assures that comparable things are being reviewed. The use of incremental cost also has the advantage of making the comparison more dynamic in the sense that the marginal plant for a newer technology may well represent second or third generation technology.

In any case, these costs should be computed without regard for government subsidies so that true costs are compared. In addition, costs not incurred until far into the future, such as nuclear decommissioning costs, should be annualized and added into the calculations for the year in question. The same needs to be done for such costs as nuclear waste disposal, which occur at intervals rather than annually.

The comparison, however, only tells us whether the process is economic; it does not reflect such external costs as possible mining fatalities, acreage disturbed, life style changes, or boom-town upheavals. To include these items requires a second-phase computation. This addresses external costs and must resolve three major problems: the difficulty of determining what items should be included; the diversity of units in which the impacts are measured; and the difficulty of assigning a money value to many of these items. In large measure, the weight given to the various external cost items is dependent on the importance of various goals. Therefore, the analyst must make a judgment as to whether goals such as health preservation, resources, environmental protection, or equity are more important than other external costs and in what order. The risk to be taken and the weight to be assigned each item then can be determined. Those costs that have been internalized, such as water cooling towers for nuclear plants or scrubbing units for coal plants, will be included in the normal operation costs and need not be considered here other than to note that the costs of abatement can be used as a proxy for the damage done. This assumes that the regulation requiring such a device was based on some kind of a cost-benefit analysis.

The difficult-to-measure costs pose a more formidable problem. These are not only hard to measure, but often are measured in a variety of different units, making comparison arduous. For example, land disturbed by strip mining might be measured in acres or injuries in number of accidents. Even within the same cost element there are problems. A million acres disturbed in barren scrub country will carry a lower external cost than in a fertile farm area. In this particular example one could apply to the disturbed acreage the average regional yield per acre, in monetary units, to derive a value for the damage. If reclamation is required, this cost to return the land to use could be used as the value of damage. The external costs should be annualized and added to the process costs.

For those costs that cannot be measured, however, a more subjective approach is required. In this case, it is suggested that a matrix be developed to rank the various technologies in regard to the cost and benefits that would accrue for the various social items believed to be important, as well as an indication of the importance of each item. The matrix allows one to evaluate, in an orderly manner, the relative value of the various technologies. When used in conjunction with the earlier cost analysis, a complete comparison is available. The highly subjective nature of this process, however, tends to make it less desirable than others. Regardless of the method selected, there is no way to avoid a large measure of subjective judgment.

The Cost-Benefit Method. The cost-benefit method is somewhat more dynamic than the point method. In preparing this analysis one would create a framework in which costs and benefits are inserted in the year in which these would be incurred without annualizing, amortizing, depreciating, or otherwise reflecting payoff periods, and then discounted. This system gives direct recognition to the time value of money. A technology with heavy front-end costs but with benefits far off in the future would be less desirable than one with equal costs but with benefits accruing in the short term.

Two procedures are available. In one case, the two streams of money (costs and benefits) could be netted and the result discounted to arrive at a present value. The project with the highest present value is expected to yield the greatest benefits relative to costs, and considering the time value of money.

Alternatively, the costs and benefits could be netted and a rate of return computed that would result in a zero present value for the net stream of money. The rate is known as the internal rate of return. The higher that figure the greater the benefits that would be derived from the project in relation to the costs and therefore, the more desirable the project. The technology with the highest internal rate of return would be the "best." This method avoids the need to select a discount rate.

If, however, a discount rate is selected it could be varied for each technology to reflect the nonmeasurable costs and benefits. For example, one can start with the long-term bond rate as the nominal discount rate. This would then be adjusted up or down to account for perceived, unmeasurable costs or benefits. The adjusted discount rate, when applied against the data, would result in a present value reflecting the measurable and nonmeasurable costs and benefits.

CONCLUSIONS

It is apparent from the foregoing discussion that alternative new technologies can be compared and evaluated. The data problems and uncertainties, coupled to the inability to quantify some of the external costs, make such a comparison subjective.

The need to select technology in the light of national goals and priorities, however, makes this subjective element a useful way of injecting those goals and priorities into the analysis. As long as one recognizes that judgment is an integral part of the analysis, and as long as that judgment is prudent, the subjective nature of the technique should not be a deterrent.

REFERENCES

Hamilton, L. D., and A. S. Manne, 1978, Health and Economic Costs of Alternative Energy Sources: International Atomic Energy Agency Bulletin, p. 44-58.

Kahn, A. E., The Recognition of Environmental Costs: Iowa State University Regulatory Conference, Ames, Iowa, p. 7.

Kaufman, A., 1976, Electric Power: Regulation of A Natural Monopoly, chapter in Energy Supply and Government Policy: Cornell University Press.

Lave, Lester B., and Lester P. Silverman, 1979, Environmental Pollution and Energy; An Economic Evaluation: Materials and Society, v. 3, no. 2, p. 115-116.

Ramsay, W., 1979, Unpaid Costs of Electrical Energy: Johns Hopkins Press for Resources for the Future.

CHAPTER 6

COSTING METHODOLOGIES FOR ENERGY SYSTEMS

Jack Allentuck, Vinod Mubayi, and Ann W. Reisman[1]

I. INTRODUCTION

The ability to make comparisons of costs among energy sources and delivery systems is a sine qua non for long-term energy planning and for decisionmaking in establishing research and development strategies. Such comparisons are especially difficult for new energy sources and innovative technologies. An additional dimension of complexity is added by the general failure of monetary costs to coincide with social costs. Moreover, time is a crucial variable in any estimate of costs, especially when new sources and innovative technologies are not yet commercially available.

In this chapter, we address the problem of devising a methodology (Section II) for arriving at costs of such systems which may be used to compare alternative sources. We also examine the basic elements of such a methodology and treat such diverse subjects as resource supply curves, experience and learning, and the cost of environmental pollution. In Section III we consider specific conventional (hydroelectric) and nonconventional (renewable) technologies. In Section IV conclusions arising from our analysis are summed up and briefly discussed.

II. ENERGY RESOURCE AND TECHNOLOGY COSTING METHODOLOGY

We start with the premise that the basis for comparison of costs among energy sources and technologies is the cost of energy delivered to an end-use device. This requires consideration of capital and operating costs for each process as appropriate: extraction, cleaning, beneficiation, transportation, conversion, transmission, distribution. In addition, for each process, technological characteristics, such as efficiency and environmental residuals, must also be taken into account. Process modifications required to make the quantity

---

[1]National Center for Analysis of Energy Systems, Brookhaven National Laboratory, New York. Prepared under Contract No. DE-AC02-76CH00016 with the U.S. Department of Energy.

and form of residuals acceptable may have significant impacts on costs. The costs of foregoing benefits which might be derived from alternative uses of land required for energy systems must also be accounted for.

In Fig. 6-1 we have diagrammed information flows for an Energy Technology Information System (ETIS). Within the system we have established a hierarchy of technology data, primary and derived. Outside the ETIS we require data on resource characteristics and market allocations while data algorithms transform primary and derived data into a computerized data base of unit energy costs. Unit energy-costs data are combined with market penetration models to produce scale-of-output information which is used to further modify the estimation of cost factors at the primary data level. Cutting across all data are considerations of regional specificity. In fact, for certain technologies site-specific data are required. These are included in the resource characteristics data base and are inputs to the formulation of resource supply curves. In order to formulate long-term strategies and establish cost comparisons which are to be valid 10 to 25 years into the future, global, national, or regional scenarios of price levels in the energy sector and elsewhere are required. Scenarios must also address the state of technological progress, specifically the availability of new technologies and their costs. Government subsidies to energy resource exploration, development, and production or to research and development of energy technologies must be considered.

Having identified the principal components of an ETIS upon which we base our costing methodology we describe the derivation of energy unit costs. First we review the information called for under the "primary" data category of Fig. 6-1, foremost among which are the general technical data. These would include process information for various extraction, refining (conversion) transmission, and distribution processes, environmental residual produced, control-specification data for each process, as appropriate, and global, national, and regional scenarios of demographics, economic conditions, and technology availability. Scenarios are self-consistent assumptions of future conditions. Items of cost of delivered energy are likely to be extremely sensitive to such assumptions and scenario design offers an opportunity to incorporate changes in these assumptions arising from alternative economic or political policy variables. In addition, primary data must include market conditions: estimated demands and alternative supplies. Also, cost factors for each of the processes included in the general technical data base are required.

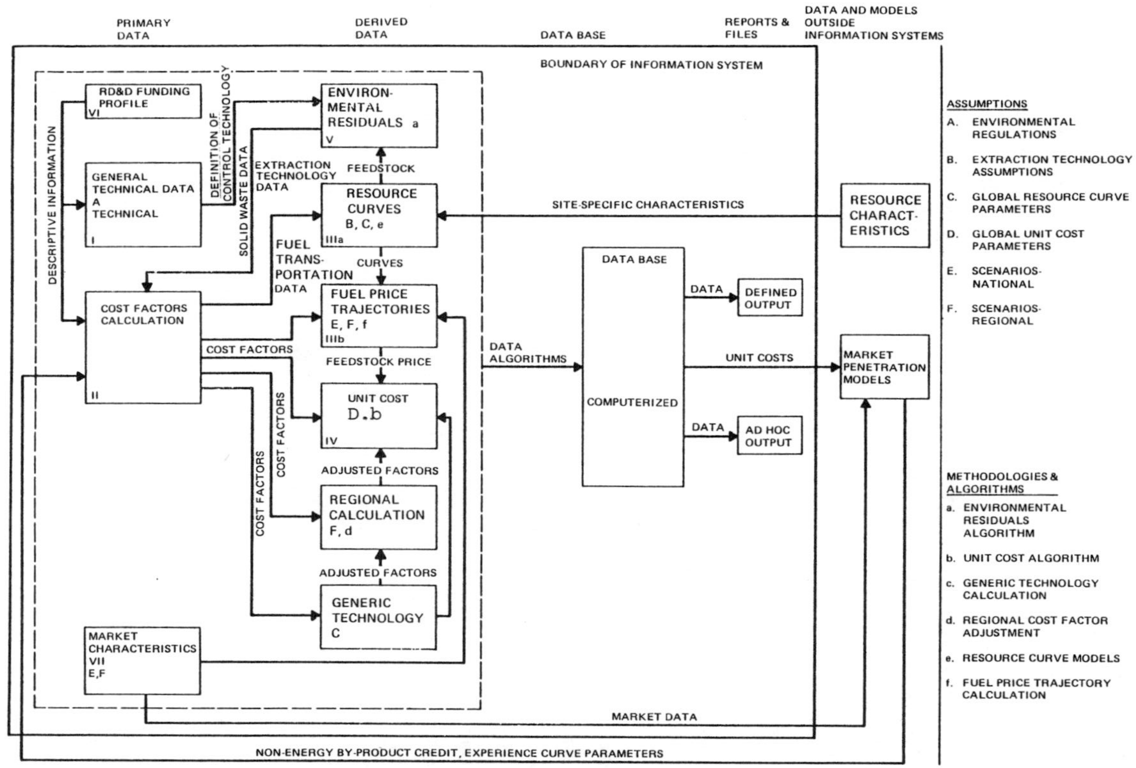

Figure 6-1. Energy technology information system

These cost factors include, for example, resource exploration costs, drilling or tunneling costs in various strata, capital and operating costs of conversion facilities of various sizes, capital costs of transmission systems of different types, waste disposal costs, or residuals abatement costs, etc.

Government subsidies, whether in the form of tax exemptions, direct financial support of exploration, development or production, or of research and development of energy technologies, will affect the cost of energy delivered to the consumer. Such subsidies may have as their objective certain development goals, as for instance, regional industrialization or perhaps political objectives such as "energy independence." They may serve to promote otherwise irrational investment decisions. In our view, the estimated cost of energy which should be used in comparisons is the social cost. (We recognize that it may not be easy to obtain this in real life.) Subsidies should be included in the construction of resource supply curves, since the date of availability of energy supplies may be advanced and the quantities available increased through subsidization. The costs of subsidies should then be added to obtain a "true" cost of the unit of delivered energy.

In continuing our description of data requirements we turn first to resource supply curves. These are central to the estimation of unit energy costs derived from extracted energy sources, e.g., coal, oil, gas, uranium. We start by outlining the construction of a resource supply curve of oil from new fields. The concept is typical and may be applied to other resources as well. Such a curve is shown in Fig. 6-2. Production must be preceded by exploration and development. The first of these discovers the heretofore undiscovered resource and converts it to <u>oil in place</u> (OIP). Development converts OIP to reserves while production extracts the resource. The process is diagrammed in Fig. 6-3.

The estimation of the cost of energy from a specific resource base requires site-specific data. These, when combined with the appropriate process data and associated cost coefficients into resource supply curves, are used to derive fuel price trajectories and to estimate the unit cost of energy inputs to an energy system.

Associated with each of these overlapping (in time) processes--exploration, development, and production--are a large number of activities, the costs of which enter into the cost of a delivered unit of energy. (These activities are listed in Table 6-1.) The decision to undertake these activities is rational

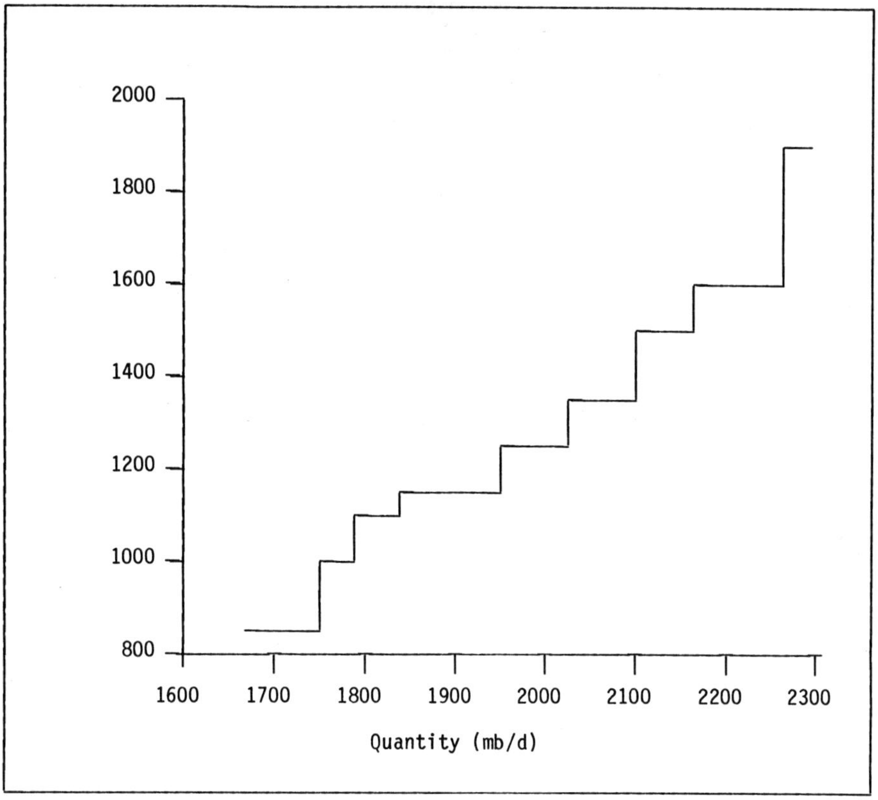

Source: Reference 7.

Figure 6-2. 1985 oil supply, Western Gulf Basin

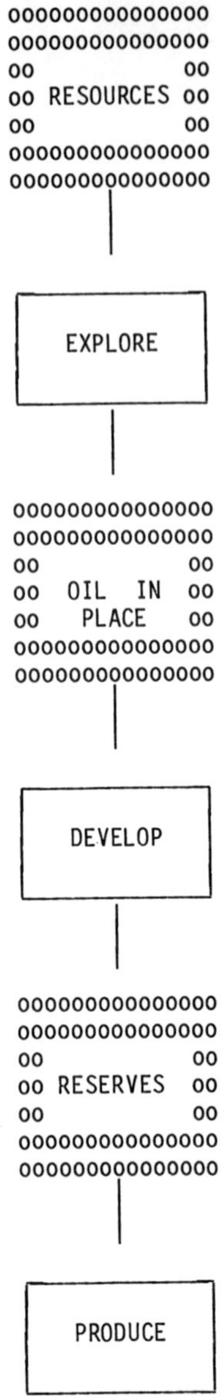

Source: Reference 7.

Figure 6-3. Activities and events of one reservoir

TABLE 6-1. Oil supply activities

EXPLORATION

- Preproduction geological and geophysical work
- Leasing
- Wildcat drilling
- Delineation drilling

DEVELOPMENT

- Construction of substructures
- Development drilling (producing wells)
- Equipping producing wells
- Construction of surface equipment
- Construction of gas plant
- Service well drilling
- Construction of enhanced recovery equipment
- Injection of enhanced recovery materials (capital items)

PRODUCTION

- Operation of producing wells and lease equipment
- Workover wells
- Replacement of worn-out equipment
- Operation of gas plant

Source: Reference 7.

when expected prices for the product are sufficient to yield a return which exceeds that on alternative investments of equivalent risk. (The lowest price which meets these requirements is the minimum well-head price.) This decision process is at the heart of resource supply curve construction. (Before proceeding to a review of the technical aspects of constructing such a curve we note that the supply curve calls on all the data resources of our ETIS for its derivation.)

The decision to invest in oil supply subsumes the allocation of such an investment between the development of known reservoirs and the discovery of new reservoirs. The supply curve depends on this allocation and in describing its derivation we shall elaborate on this dependence and indicate how the expected price for the resource (derived from scenarios of economic conditions) may in turn affect the allocation.

Consider first the definition of various kinds of reserves of the resource. (These are presented diagrammatically in Fig. 6-4.) "Measured reserves" have been discovered and a significant portion of the investment in their development has been made. Since we are concerned with the cost of energy from a new reservoir, measured reserves requiring no investment in discovery and little in development are not of interest to us here. "Indicated" reserves provide a resource which may be recovered from known reservoirs only if investment in fluid injection wells and equipment is made. (Primary recovery is defined as the unaided flow of oil to the surface under the force of the natural pressure of the reservoir.) "Inferred reserves" are extensions and revisions of known reservoirs. These arise from exploratory diggings which provide additional information on the size of a known reservoir. "Undiscovered resources" are the difference between the resource base and the sum of the previously identified reserves.

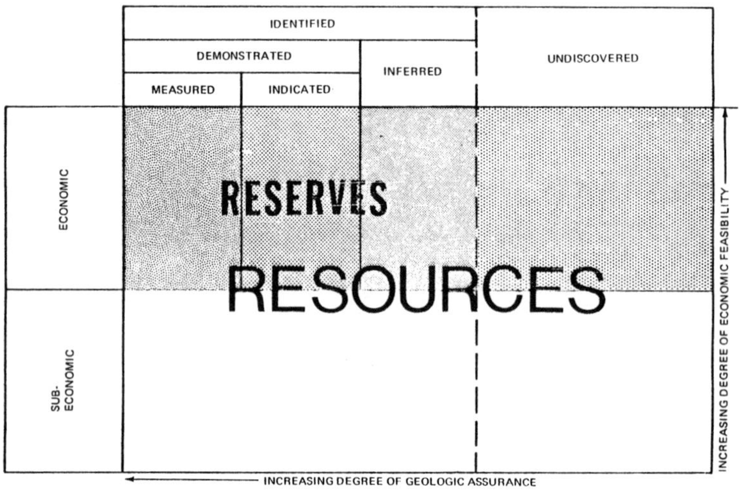

Source: Reference 7.

Figure 6-4. Reserve and resource definitions

The modeling approach leading to an estimation of resource supply curves is diagrammed in Fig. 6-5. The drilling submodel estimates drilling activity based on the available capacity of drilling rigs, leasing schedules of areas to be surveyed and a perceived trajectory of oil prices. In regions where exploration has already occurred, drilling activity is converted into inferred reserves using historical finding rates. Where historical finding rates are unavailable, as, for example, in marine environments or other unexplored regions, another approach is called for. This approach specifically accounts for the uncertainty in the resource data.

The estimation of a supply curve of oil from a specified tract in such a regime requires the following information:

1. The total resource base underlying the specified tract. This estimate is based on seismic data of structural traps which are present. Prior to drilling however it is unknown whether or not such traps contain hydrocarbons.
2. The quality of the resource base. This includes the field size distribution and factors affecting the producibility of the resource.
3. Physical conditions affecting exploration and production activities. These are especially important for off-shore environments and include weather conditions and water depth. Such conditions affect drilling and production technology choices and costs associated with these processes.

Since these data are likely to be uncertain to varying degrees, a probabilistic approach to their use is indicated. Large uncertainties are likely to surround the extent of the resource base, the field size distribution, and resource producibility. Uncertainties affecting technology choices and hence the capital costs associated with production are likely to be less.

A method of combining the probabilities associated with 1 and 2 above into a frequency distribution of cumulative production is presented in Reference 1. These data combined with data on capital and production costs and a required rate of return on investment lead to a minimum well-head price, again in probabilistic terms. (The cost of oil in domestic use however should be accounted at the world-market price.)

This minimum well-head price is the breakeven price which equates the present value of total yearly revenue with present value of total yearly outlays.

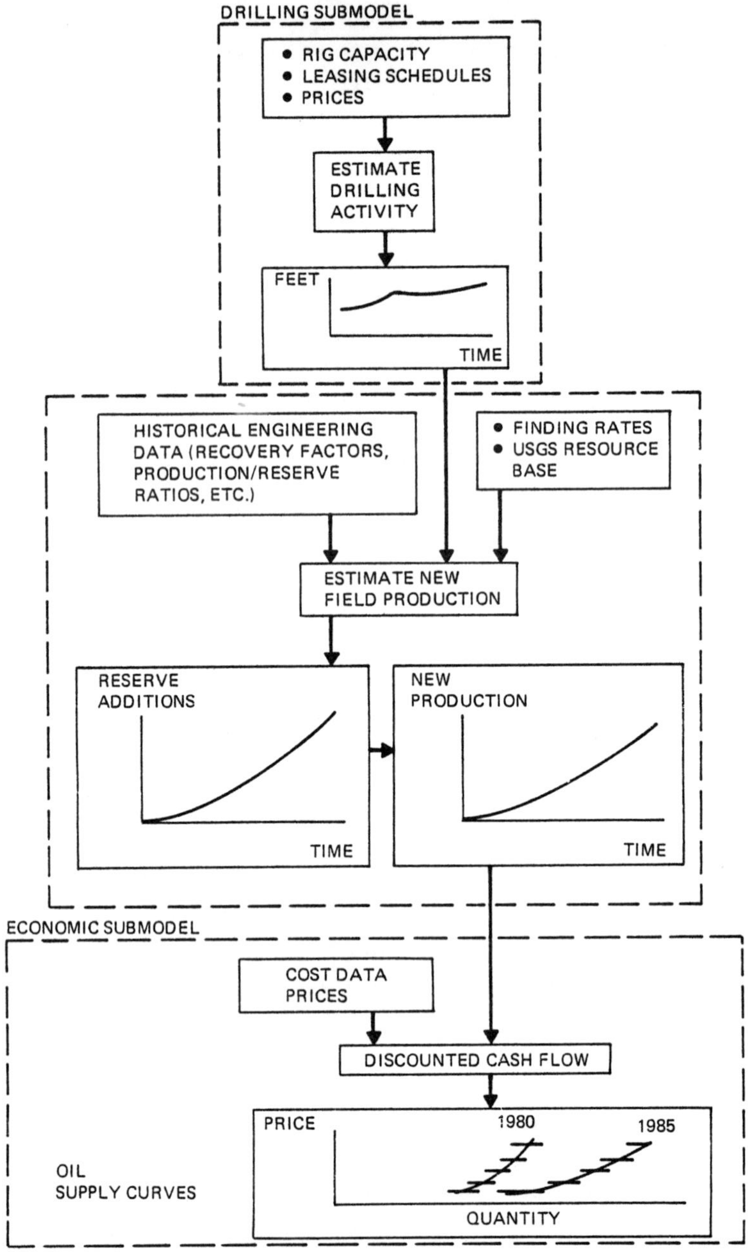

Source: Reference 7.

Figure 6-5. Modeling approach to resource supply curves

To achieve increased production from a given resource base requires increased exploration, more capital investment, and incurs greater production costs as producibility declines--hence greater revenues. The supply curve (Fig. 6-2) is a trajectory of minimum well-head price versus total annual production.

The assumptions in the scenario of a trajectory of world oil prices provides on a year-by-year basis the greatest well-head cost which may be met by market prices. Alternatively it provides a value for oil (as opposed to a cost of production) produced in a given year and hence an appropriate accounting price for its consumption, or depending on prevailing interest rates, an optimal production schedule.

The effect of the distribution of revenues between exploration and development may be discussed in terms of the computation outlined above. The lag between expenditures on exploration and the receipt of revenues from production is likely to be substantially greater than between expenditures on development and revenues from production. Thus, given a fixed expenditure budget, production will be less as the proportion of expenditure on exploration, with its resultant increase in the minimum well-head price, is increased.

The effect of taxation or royalties is similar: an increase in the minimum well-head price. Thus, if prices remain constant, taxation may have the effect of leaving smaller fields, or resources which are more costly to exploit, undeveloped.

A consideration of all the factors involved in the cost of oil (or gas) leads to the inescapable conclusion that estimates of such costs must be made in the context of a specific scenario of national and world economic conditions and government policy.

Having estimated the supply curve, the cost of the resource entering the energy system is available. What remains to be done is to estimate the cost at each step of the refining and delivery process up to its receipt by end users. This procedure--essentially an accounting methodology--may be organized as a reference energy system (RES) such as the Brookhaven National Laboratory (BNL) RES shown in Fig. 6-6. BNL's Energy System Network Simulator (ESNS) is a computerized version of this system. The oil subnetwork of ESNS is shown in Fig. 6-7. Space does not permit us to elaborate on ESNS but suffice it to say that using a data base of process costs and environmental residual coefficients for each mode of the process network a cost per 10^6 Btu of energy (or other

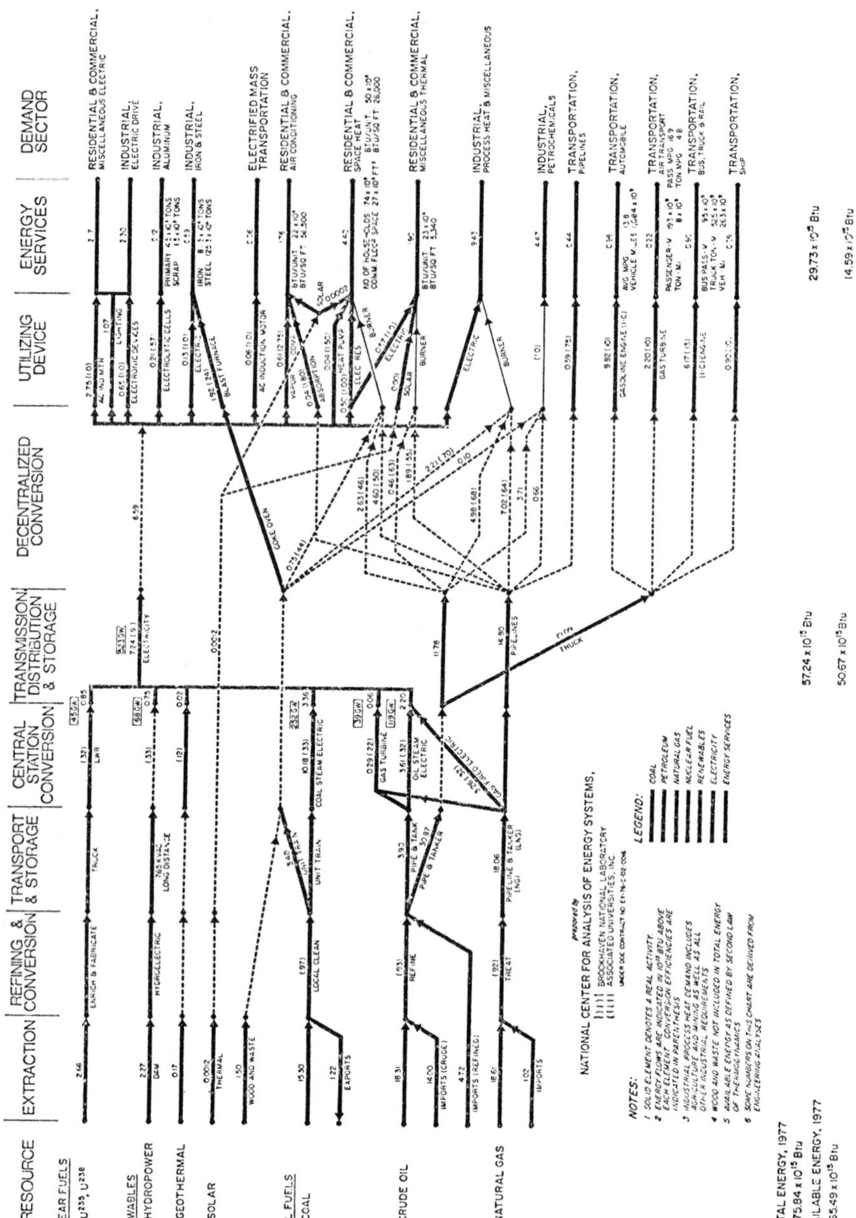

Figure 6-6. Reference energy system, year 1977

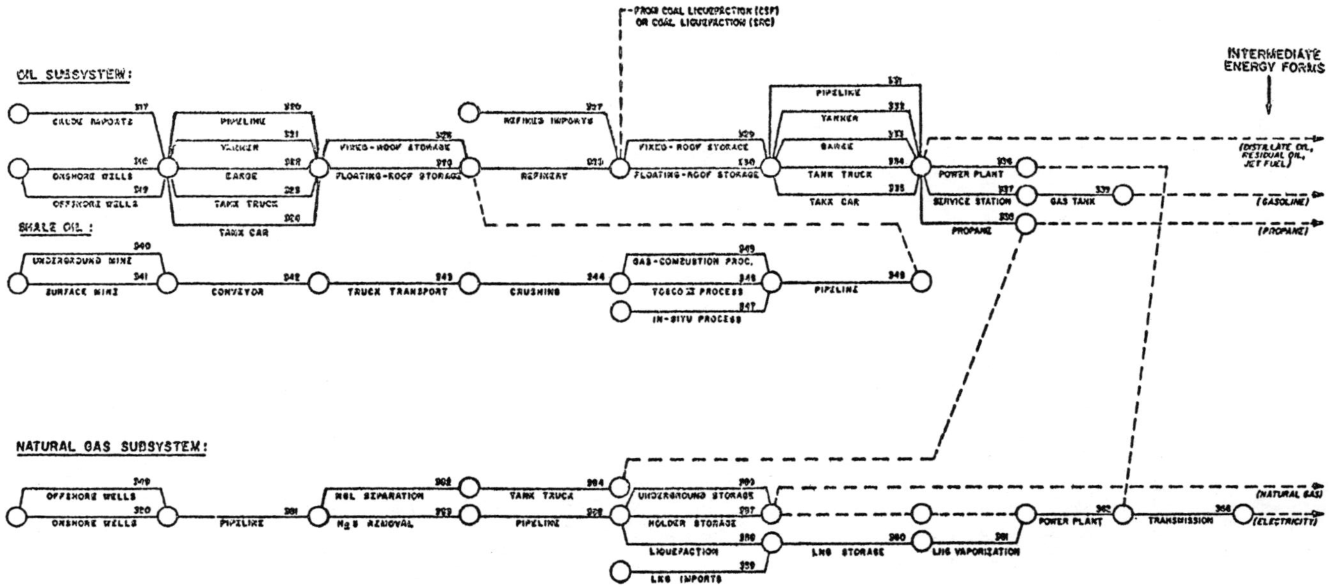

Source: Reference 8.

Figure 6-7. National energy system supply process network (oil, natural gas)

appropriate unit) delivered to a final use is estimated along with an estimate of residuals categorized by type.

Process cost data in the developed countries are well known. In developing countries it is possible to estimate such costs on the basis of costs in developed countries once problems such as the valuation of indigenous labor in the context of underemployment and of imported capital equipment procured with scarce foreign currency are resolved. This matter will be examined in greater detail when we discuss hydroelectric development. Costs of environmental degradation from process residuals are difficult to estimate in any context. One approach is to assume that processes will be permitted only to the extent that residuals per unit of output are socially acceptable. The costs of the product then will include capital and operating costs of control technology. This approach in a sense begs the questions since it implies that there are no environmental costs as long as residuals are socially acceptable as interpreted by government and incorporated in a set of "environmental regulations."

A more sophisticated approach to accounting for the cost of environmental residues is the estimation of costs imposed on society by these residues. A major portion of such costs may arise from adverse health effects of environmental residuals. At Brookhaven National Laboratory, health damage functions have been estimated for several energy conversion processes (Reference 2). Since these functions depend on the interaction of toxic residuals with populations, their use requires the estimation of the transport of residuals from a source to a populated area, along with chemical and physical processes which may take place along the way. Finally it remains to associate a monetary cost with such effects. Thus the direct estimate of social costs of residuals from energy processes may indeed be so complex as to be impractical to undertake.

III.A. EFFECT OF LEARNING AND TECHNOLOGICAL PROGRESS ON COSTS

An issue which arises in the consideration of energy costs from new sources or innovative technologies is how experience or learning will affect such costs. Is learning or experience likely to reduce the cost of wind-energy by reducing the cost of manufacture of propellor blades or the cost of construction of supporting structures? Are there conceptual differences between learning and experience? What operative significance do these differences have? Learning, while it directly affects labor (measured in physical units) input into producing the xth unit is not only learning by direct labor but also learning by management. The latter is reflected in the organization of the production process as a consequence of which requirements for direct labor inputs are reduced. Reduction in direct labor hour input does not necessarily mean an analogous reduction in labor costs. This follows from the likelihood that labor will succeed in capturing in increased wages a portion of its increased productivity.

The experience curve relates to the cost of producing the xth unit as a function of cumulative production. Its mathematical formulation is identical to that of the learning curve. Despite this identity in formulation, in concept it is quite different since it refers to unit cost, involving as it does the costs of all inputs as opposed to physical labor inputs as in the case of the learning curve. It is assumed that technological knowledge is held constant but that with experience management learns to combine inputs and to organize production so as to effect a reduction in production costs.

The technological progress function is concerned with improvements in some figure of technological merit such as efficiency or horsepower/unit weight.

It may be argued that for the exploitation of energy sources involving heavy construction for conversion processes as, for example, central electricity generation, the advantages derived from learning are unlikely to be captured by investors and costs should not include effects of learning. While there may be effects of experience on conversion technology costs they are indeterminate as to sign and therefore should not be included.

For new sources and technologies depending for their exploitation on repetitive manufacturing processes, such as photovoltaics, heliostats, or propellors, the effects of learning on production processes may be estimated by detailed analysis. Absent production data, it is unlikely that such estimates will prove very reliable. In any event, their consequences for total costs are likely to defy analysis since their effect depends on the ability of labor to capture increases in labor productivity in the form of increased wages.

Experience for technologies of this type may or may not have an advantageous effect on costs. Estimating this effect appears to be impossible except by comparison to other technologies.

As regards technological progress a case-by-case analysis is required.

III.B. HYDROELECTRIC ENERGY COSTS

The approach to costing energy derived from a specified hydroelectric facility involves estimating the electric energy which it will produce over its lifetime and the cost of producing it. This involves determining the economic dispatch of generation on an integrated system at each period over the life of the proposed facility taking into account costs of energy production from each generating unit. It follows that the cost of electric energy from a hydroelectric source cannot be considered in isolation.

The estimation of energy to be derived from a particular hydroelectric facility involves information on the one hand of the source under consideration and on the other hand of the system and the load which it must meet. Data of the hydro-source itself include its seasonal (or monthly) capability expressed in megawatts as well as capital and operating costs. Seasonal capacity factors for base and peak load operation are required as well. Forecast load data and data of the balance of the system, including thermal efficiencies, outage rates, and maintenance schedules must be sufficient to estimate the rank of the hydro unit in the system loading order. From these data an optimal system expansion plan is derived and the amount of energy produced by the hydro unit in each period is estimated. Hydroelectric facility investment is highly site-specific. For initial surveys, cost estimates which employ programs based on generalized cost data are available. These convert site-specific data such as head and flow into preliminary cost estimates.

In developing countries, estimating investment in hydroelectric facilities presents difficulties which arise from the necessity of importing heavy equipment which is not available domestically and employing foreign workers having skills in scarce supply in the local labor force. In order to take these factors into account, the fraction of equipment, material, and labor available locally has to be initially estimated. Such a review may, in fact, lead to a revision of construction techniques to take full advantage of local conditions. For instance, if various kinds of local labor are underemployed, estimates based on social costs may lead to construction techniques which are more labor-intensive than would be employed if the importation of high-priced foreign labor was required.

III.C. RENEWABLE ENERGY SOURCES

The analysis of the costs of energy from renewable energy sources such as direct solar and wind, whose availability is inherently stochastic in nature, involves additional dimensions of complexity compared with the conventional sources discussed earlier. It is useful conceptually to separate this discussion into two areas: (a) the operation of solar/wind devices in the context of an integrated energy generation and distribution system, such as an electric utility, whose other components are conventional, (b) "stand-alone" or decentralized mode of operation. We will consider the introduction of the latter in developing countries, where their use in rural areas unconnected to any centralized energy delivery system has aroused a great deal of interest and discussion.

In the case of (a), the complexity arises when energy generated from a random, fluctuating source has to interface with the energy produced from conventional facilities. In this setting, a comparison based simply on the unit costs of energy generated from a single facility, the yearly capital, operating and maintenance costs divided by the annual energy generated, is insufficient since it does not take into account questions of compatibility and reliability. An electric utility, for example, employs several types of generating plants to meet a dynamically varying load, with generally predictable daily, seasonal, and annual variations, at least possible cost with a given reliability. The introduction of an intermittent source, whose variations in output can be expected to range, as in the case of wind, on a time scale of seconds to months, involves a careful analysis of the techno-economic characteristics of utility operation before a comparison of the costs (and value) of, say, wind-generated and conventionally generated power can be made. In general, the introduction of solar/wind devices may be expected to lead to a decrease in both conventional capacity as well as fuel consumption. However, as has been recognized in studies of solar heating of buildings, the fact that solar heating requires conventional back-up systems creates several problems for utility operations such as a peaking effect, a low-load factor effect, and question of storage and conservation. A large demand for peak power could be created during times of solar outage (such as prolonged cloudiness over the utility service area). This could either create problems of lack of reserve margin and greater risk of outage or, by lowering the system load factor, force the utility to invest in large blocks of conventional peaking capacity leading to uneconomic operation of the system as a whole.

A careful and concise analysis of the numerous issues involved in integrating the operation of solar/wind technologies with conventional generation on a utility grid has been recently made by Kahn (Reference 3). A similar analysis devoted to wind-turbine electric generation has been made by Davitian (Reference 4). It is clear that costing methodologies devised for renewable energy sources in the context of centralized energy delivery systems have to take into account the techno-economic operations of utility systems as a whole and cannot be solely based on the technical performance or costs of individual solar/wind devices alone.

The costs of energy from isolated or decentralized renewable energy systems can be tackled, to a first approximation, in a more straightforward manner by evaluating the delivered energy cost to a final user in terms of annualized capital, operating, and maintenance costs, including provision for storage, if any. However, such systems are likely to see widespread use not in the industrialized countries, where centralized energy delivery networks already exist on a widespread scale, but particularly in the rural areas of developing countries where such networks are not available. Interest in analyzing the economics of such systems has been growing, following the "oil crisis," although the subject is still in its infancy. One of the energy technologies on which attention has been focused in the literature (Reference 5), is the production of methane by anaerobic digestion of animal and agricultural wastes at the village level. The gas can be used directly as a fuel or converted to electricity. As a by-product of the process, a slurry is produced which generally preserves the nutrient value of the original material and can be used as an organic fertilizer. The costing methodology which can be applied in this particular case is that of social cost-benefit analysis (Reference 6) which analyzes the returns to an investment at a microlevel in the context of a set of macroconditions in which decisions have to be evaluated. Biogas has a set of direct and indirect benefits associated with it such as the provision of fuel and fertilizer, an increase in self-sufficiency at the local level, possible increase in employment, and probable health benefits (which may be difficult to quantify) due to the more sanitary disposal of waste material. Its direct costs are the capital costs of plant construction and operating and maintenance costs associated with plant operation, including labor and provision of secondary inputs such as water. (The valuation of labor, particularly, when unemployment or underemployment in labor-surplus rural areas exists is a difficult issue and several analysts have argued for a zero "shadow-wage" rate.) However, there are a number of indirect costs also associated with biogas ranging from management and institutional problems (especially of waste collection) to the social issues involved in "commercializing" through new technology a heretofore "noncommercial" resource. It is clear that a valuation of costs and benefits of biogas will have to consider giving greater importance to local social and environmental issues, some of which may be difficult to quantify precisely in monetary units, before deciding on whether the net present value of a particular system compares favorably with conventional alternatives.

IV. CONCLUSIONS

In the brief discussion of costing methodologies for various energy resources and technologies that we have presented earlier, several important issues have been identified. Among these are:

1. Delivered costs of energy to final end use should be the basis of economic comparison between alternative energy forms. Treatments of subsidies and/or taxes should be explicitly considered insofar as they affect the time frame of development of new energy sources. However, their costs should be added to the final cost of delivered energy.
2. In the case of fossil fuels, such as oil and gas, different physical characteristics and geological conditions impose a different structure of capital and operating costs for development and production of energy. An example is oil derived from the outer continental shelf, shallow off-shore or on-shore facilities. The costs derived for energy have to reflect these differences.
3. While the quantity of residuals from energy processes can usually be estimated in a straightforward manner, their effect on costs depends on value judgments which are much more difficult to quantify. In effect, this amounts to placing a monetary value on human health and life. Even if the physical effects can be calculated, which are themselves matters of intense controversy, their translation into monetary units involves a considerable degree of subjective judgment.
4. When intermittently renewable sources, such as solar and wind are interfaced with conventional generating plants, costs have to be considered in the context of integrated operations of the generation, transmission, and distribution system as a whole. The economic costs

of decentralized renewable resource systems can be analyzed more directly, but, in developing countries, where they will largely be implemented a wider range of social and institutional issues have to be considered.

5. In developing countries attention has to be placed on obtaining appropriate shadow prices for labor and imported technologies and integrating them into the cost analysis. In particular, these factors should be an important consideration in the choice of technologies.

6. In general, the energy costs imposed by national political or social policies and goals such as "rural electrification" or "energy independence" should be explicitly taken into account in the analysis.

REFERENCES

1. F. W. Mansvelt Beck and K. M. Wiig, 1977, The Economics of Offshore Oil and Gas Supplies: Lexington Books, Lexington, Massachusetts.
2. S. C. Morris, K. M. Novak, and L. D. Hamilton, Databook for the Quantification of Health Effects from Coal Energy Systems: Brookhaven National Laboratory, Upton, New York, in preparation.
3. Edward Kahn, 1979, The Compatibility of Wind and Solar Technology with Conventional Energy Systems: Ann. Rev. Energy, 4, p. 313-352.
4. Harry Davitian, 1978, Wind Power and Electric Utilities: Wind Engineering, v. 2, no. 4, p. 234-255.
5. Andrew Barnett, Leo Pyle, and S. K. Subramaniam, 1978, Biogas Technology in the Third World: International Development Research Center, Ottawa, Canada.
6. Ramesh Bhatia, 1977, Economic Appraisal of Bio-Gas Units in India, Economic and Political Weekly, v. XII. For a recent discussion of social cost-benefit analysis see L. Squire and H. G. van der Tak "The economic analysis of projects," World Bank, 1975.
7. U.S. Department of Energy, Division of Oil and Gas Analysis, 1978, Technical Memorandum, Washington, D.C., TM/ES/79.
8. A. Reisman and R. Malone, 1978, Less Developed Countries Energy Network Simulation LDC-ESNS; A Brief Description: Brookhaven National Laboratory, Upton, New York.

CHAPTER 7

COGENERATION AS AN IMPORTANT WAY FOR ACHIEVING AN ENERGY CONSERVATION POLICY IN ROMANIA

Dr. Călin Mihăileanu and Dr. Nicolae Armencoiu[1]

SHORT PRESENTATION OF ROMANIAN ENERGY POLICY

The Socialist Republic of Romania lies in central-eastern Europe, has an area of 237,500 km^2, and a population of about 22 million, of which 47.8 percent live in towns. There are about 240 towns and over 13,000 villages in Romania. Among the big towns, 18 have a population of over 100,000; Bucharest, the capital city of the country, has about 1.8 million inhabitants.

Romania's electric power requirements are assured from the national power system for all consumers. The thermal and hydropower plants, the substations, and electric networks operate within a single system. For the hot water required by urban heating and domestic consumption, as well as for the steam required by industry, the necessary heat quantities are produced in thermal plants as well as in combined heat and power generating plants (CHPP). In 1978 the CHPP's delivered over 45 percent of the total heat requirements as hot water or steam.

Before the Second World War, Romania's economy was especially based on agriculture and timber production, on light industry, and on certain branches of the extractive industry, especially mining; energy was not quite developed, although there is a tradition in this field:

- Coal and ore extraction began hundreds of years ago.
- Electric lighting was introduced at Timisoara and Bucharest in the years 1882-1884, simultaneously with the appearance of the first power plant in the world.

Cogeneration was practiced in our country even before the Second World War, when the national system of power plants and networks was not yet formed. Industries were developed during that period which consumed both steam and electric power and for which CHPP's were built, equipped first with piston-type steam engines and then with small units with backpressure turbines.

---

[1]Central Energy Research Institute, Bucharest, Romania.

Romania has various energy resources and all types of fuels: coal/anthracite, pit coal, lignite, and bituminous shales; oil and natural gas; hydropower; and geothermal, wind, and solar energy; but all of these are in limited quantities.

After the Second World War, as the national economy had to be rapidly developed under conditions of limited energy and fuel resources, special attention was granted to the energy problem and its planned development was assured, so that the available resources could be rationally and efficiently utilized. Under these conditions, cogeneration development was possible as an integral part of Romania's energy policy.

Even before the occurrence of the oil crisis, Romania's future development was concerned with efficient utilization of energy.

At the same time, at the national economy level, a special stress was laid on energy conservation and efficient utilization. The following aspects have been considered: the modification of the industrial production structure in order to reduce energy consumption, improvement of manufacturing technologies, recovery of the secondary energy resources, and introduction of new energy resources in the primary energy balance.

Within this policy, cogeneration has an important position.

COGENERATION--A WAY OF ENERGY CONSERVATION AND ENVIRONMENTAL PROTECTION

Cogeneration--combined heat and power generation--is the most efficient way of energy conservation; fuel savings of about 15 to 20 percent being obtained as compared to separate generation.

The fact which leads to fuel savings is that the steam, after being expanded in the turbine and thus producing electric power, is utilized for the heat supply to the consumers instead of being sent to the condenser, where heat is removed through the cooling water.

Under the new energy circumstances of restricting hydrocarbon utilization and increasing the role of coal, another advantage of cogeneration is that of better environmental protection due to the fact that, in the case of large capacity units, efficient installations can be provided for ash retention and sulphur dioxide dispersion.

Being located far from residential areas and offering the possibility of efficient utilization of some low-grade primary energy resources (lignite, bituminous shales) the CHPP's contribute to the present need to maintain the ecologic balance by improving primary energy source exploitation or by phenomena of a different type, like the reduction of pollution or the improvement of social conditions.

At the same time the problems connected to plant operation, coal transport, and slag and ash discharge are more easily solved in the case of a large CHPP than in case of several small thermal power plants located near the consumption center.

The combined heat and power generating plants were developed in Romania as sources for supplying with heat and electric power the large industrial areas and urban agglomerations, as well as for supplying important industries simultaneously with urban consumers.

The planned development of the economy makes it possible to know the evolution of the heat consumption in each center (i.e., the extension of existing industries, new industries implementation, the number of new flats to be built), for a longer period of time (5 to 10 years), leading to the possibility of establishing the optimum heat supply solution in due time.

COGENERATION EVOLUTION AND ITS PRESENT DEVELOPMENT STAGE IN ROMANIA

The Main Stages of the Cogeneration Evolution in Romania

As we have already mentioned, cogeneration began in Romania even before the Second World War, several refineries and paper factories having their own CHPP's, but systematic studies and research for energy conservation began with the first 5-year plan, 1950-1955, when the first guidelines for the country's energy basis were drawn.

It was during this stage that, both by studies and research and by in-service checks, the bases of the cogeneration were laid.

The main problems to be solved were the following:

- Research on the type, magnitude, and variations of the heat requirements to be covered by urban and industrial cogeneration;
- Establishing theoretical conditions and background for cogeneration application in our country, determining the basic elements of heat consumption and of the cogeneration system;

- The simultaneous solution of important demands for heat and electric power due to industry development and to the occurrence of new industrial complexes with technological processes requiring large quantities of heat;
- The rationalizing of fuel consumption in certain areas where the latter was utilized with low efficiencies and the adopting of solutions leading to a minimum consumption of fuel resources.

In 1950 the total installed power in the CHPP's was 127 MW with an electric power output of 1,870 TJ (520 kWh) and a heat supply of 7,500 TJ (1.8 million G cal).

In the first part of the period 1950-1955, units of 6 to 12 MW were commissioned, with steam parameters: 3.4 M Pa (35 at), 450°C, where industrial steam consumers were to be found, then the first 25 MW units with steam parameters 9.8 M Pa (100 at), 510°C for the new industrial areas. In 1960 a total installed power of 472 MW was reached in the CHPP's, with a heat supply of 35,600 PJ (8.5 million G cal) and an electric power output of 5,760 PJ (1,600 GWh).

After 1960, cogeneration was introduced and extended in the big industrial centers, first aiming at industrial consumption and second, at suppressing the old industrial thermal power plants with low efficiency; where conditions were created from the urban point of view, district heating was also taken over.

The Present Stage of Cogeneration in Romania

At the end of 1978 the installed power in the CHPP's was 3,806 MW, out of which the backpressure industrial cogeneration units represented 31.6 percent, industrial cogeneration units with condensation and adjustable bleeding, 52 percent, while the district heating units represented 16.4 percent (Fig. 7-1).

The annual heat quantity supplied by the CHPP's in 1978 was 312.8 PJ, the maximum heat consumption assured from a single plant being about 700 MW for steam (about 1,000 t/h) and respectively about 2,600 MW for hot water (about 2,200 G cal/h).

Table 7-1 presents the evolution of the main cogeneration characteristics in Romania.

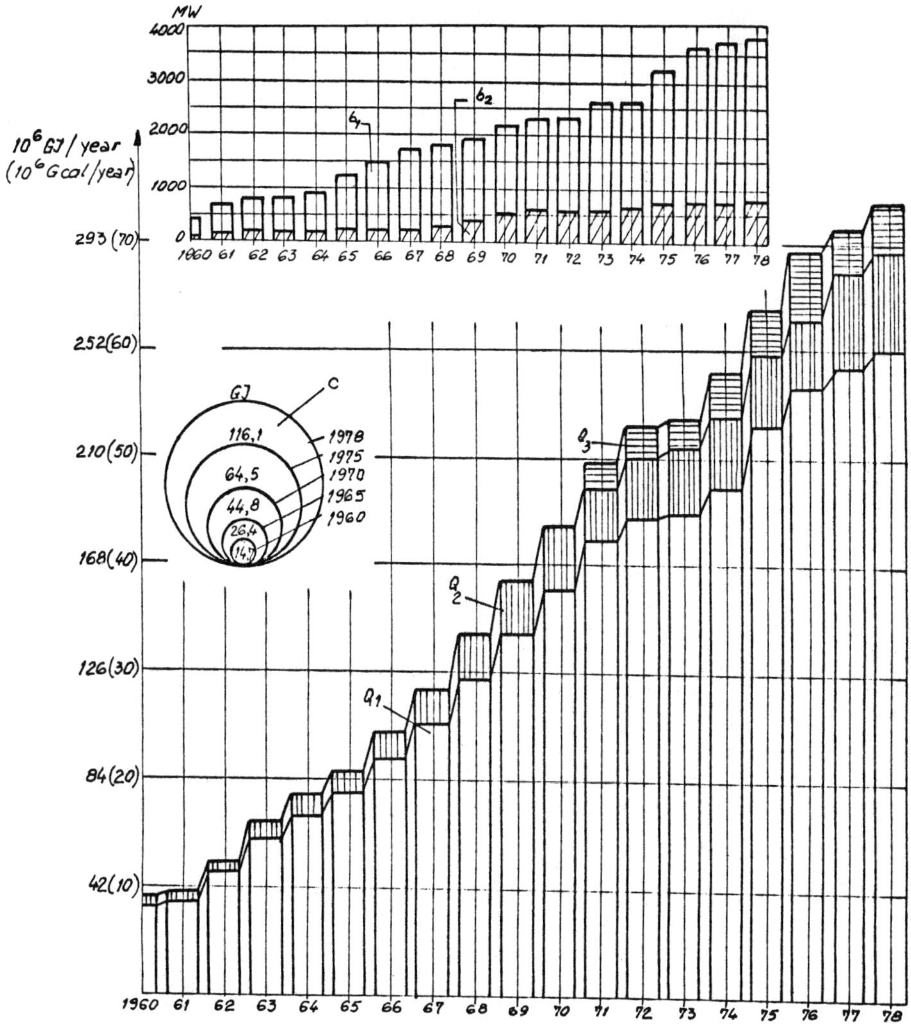

Figure 7-1a. Heat supplied by heat and power plants: Q_1 - industrial consumption; Q_2 - district heating consumption; Q_3 - greenhouse consumption.

b. Capacity installed in combined heat and power plants: b_1 - back pressure; b_2 - extraction-condensing

c. Fuel savings - 10^3 t.c.e.

TABLE 7-1. Characteristic elements concerning the cogeneration evolution in Romania

Elements		1960	1965	1970	1978
Installed power in combined heat and power generating plants as compared to the total installed in thermal power plants	MW	472	1203	2165	3806
	%	30.2	33.6	36.3	37.4
Heat supplied from combined heat and power generating plants	PJ	35.6	85.8	182.6	312.7
Unit power of the largest CHPP unit in operation	MW	25	50	100	150
Length of the heating networks	km	12.2	220	450	900
Maximum diameter of the pipes	m	0.5	0.8	1.0	1.2
Fuel saving obtained by cogeneration as compared to the separate generation	PJ	14.7	26.4	44.0	79.2
	10^3 t.c.e.	500	900	1500	2700

The share of the installed power in the CHPP's within the total installed power in the thermal power plants increased from 30.2 percent in 1960 to 37.4 percent in 1978.

The fuel savings achieved by cogeneration, as compared to the separate generation of electric power and heat, increased from about 14.7 PJ (500,000 tons of coal equivalent) in 1960 to about 79.2 PJ (2,700,000 t.c.e.) in 1978.

The heat transmission and distribution networks were mainly designed and built with radial pipes in closed circuit, safety being increased by connections between the main transmission networks. The heat supply control is made by modifying the outgoing temperature function of the air temperature and the wind velocity. In 1979 there were 30 public heat transmission and distribution systems as compared to only 17 systems in 1970.

The total length of the public heat transmission and distribution systems was about 950 km in 1978 and the maximum heat transmission distance was about 15 km for hot water and 10 km for steam.

MAIN ACHIEVEMENTS IN THE FIELD OF COGENERATION IN ROMANIA

General Concept on the Development of the CHPP

Concerning the CHPP having high unit powers, the first units were condensing and adjustable bleeding types, due to the evolution of heat consumption with time, approximately related to their magnitude.

For a more efficient use of the investments in cogeneration, initially LP and HW boilers are installed; the next step being the installation of the heating units; these boilers serve to take over the heat peak and as a reserve.

As a general rule, when developing CHPP's, at least three units are to be installed in a plant, so that when one of the units is under revision or undergoes a repair and another unit fails, at least one unit should be in operation to provide the minimum requirements of the industrial process.

To enhance heat supply reliability to the industrial consumers, the power reserve is provided by LP high-output steam boilers and by the reducing-cooling stations.

The incomplete return of the condensate from the steam supplied to the industry raised some problems related to the requirements of feedwater to the boilers. Solutions for the condensate treatment, with heat content recovery, were applied.

Some Cogenerating Systems Developed in Romania

At the end of 1978, in the Romanian CHPP's (Table 7-2), 31 units of 50 MW each were installed, with the parameters 13.7 M Pa (140 at) 555°C to 570°C; the units are condensation, adjustable bleeding, and counterpressure types, to meet particularly the industrial consumption requirements; six 100 MW units of the parameters 13.7 M Pa (140 at), 555°C, condensation and adjustable bleeding type, for urban heating; four 125 MW units, condensation and adjustable bleeding type for urban heating, at the parameters 14.9 M Pa (152 at), 540°C, with reheating at 540°C; and one 150 MW unit, Romanian made, 19.2 M Pa (196 at), 540/540°C.

One of the most interesting achievements in the cogeneration field in Romania is the cogeneration system in the Ploiesti-Sud-Brazi area, fed by the CHPP Brazi.

The basic idea in the conceptual design of this system was to supply both the existing refinery at Brazi and the Petrochemical Complex to be built on

TABLE 7-2. Important combined heat and power generating plants existing at the end of 1978

Art. no.	Power plant	Installed power MW	Heat supply capacity MW (G cal/h)	Achievements in 1978			
				Supplied electric power		Supplied heat	
				GWh	TJ	10^3 G cal	TJ
1.	Bucharest-Sud	550	2650 (2280)	3536	12730	5448	22810
2.	Brazi	460	1960 (1690)	2196	7910	4478	18750
3.	Galati	325	880 (760)	2334	8400	1598	6690
4.	Borzesti	335	1670 (1440)	1904	6855	7126	29840
5.	Bucharest-West	250	950 (820)	1745	6280	1456	6095
6.	Govora	208	1040 (900)	1198	4315	3811	15960
7.	Craiova	205	1065 (920)	6068	21850	3020	12645
8.	Oradea	205	720 (620)	1378	4960	2640	11055
9.	Grozavesti	125	640 (550)	732	2635	1026	4295
10.	Iasi	100	720 (620)	623	2245	2337	9785
11.	Palas	100	790 (680)	688	2480	1976	8275
12.	Pitesti-Sud	136	620 (530)	612	2205	2534	10610

that side, and the refineries in the southern side of Ploiesti from a single, economic heat source located at Brazi. Lately, in addition to the high industrial consumption, the power plant also took over the urban consumers of the Ploiesti town as well as a large vegetable greenhouse complex.

For heat transport there is a steam network of 1.56 M Pa (16 at) where the maximum length is 9.5 km for the refineries in the southern side of Ploiesti and 14 km for the supply of hot water to the town, at the parameters of 150t/70°C.

The CHPP has at present 460 MW installed power and is equipped with five 50 MW units (three condensation and adjustable bleeding units and two counterpressure units) which supply the technologic steam, and with two 105 MW units at the parameters 13.7 M Pa (140 at) and 550°C which supply heat to the town and industry.

Another interesting achievement is the cogeneration system of the city of Bucharest, where urban consumption is prevailing. It is grouped into large dwelling quarters where several industrial areas with technologic steam and hot water requirements do exist.

The solution adopted was that where the heat supply to the city and industries concentrates into three large CHPP's: Grozavesti, Bucharest-Sud, and Bucharest-West, the technologic scheme required in industry was grouped into zonal heating plants (Danubiana, 23 August, Pipera, Gtivita Rosie, and Militari) equipped with 4 and 6 MW counterpressure units summing an installed power of 36 MW except for the southern industrial area where two 50 MW industrial heating units were installed in Bucharest-Sud for the new steam consumers.

The cogeneration in the city of Bucharest began by providing the condensation power plant, Grozavesti, with one 25 MW unit and later with two 50 MW urban heating units.

The second CHPP is the Bucharest-Sud, with an installed power of 550 MW and a great number of hot water boilers (HWB) which take over heat-peak consumption.

The Bucharest-West, commissioned recently, has a 250 MW installed power and several HWB.

The urban heating network is the tree-like type; the maximum diameter of the mains reaching 1,200 mm; the total length being over 400 km.

Other CHPP's of importance are:

- CHPP Galati with a 325 MW installed power, which burns furnace gas and coke and supplies heat to the steel works and to Galati.
- CHPP Borzesti with a 335 MW installed power, designed initially to burn gas and crude oil and extended later to burn coal. It is designed to supply heat to the large industrial area Borzesti (refinery and petrochemical industry) and to Onesti.
- CHPP Govora with a 208 MW installed power, designed to supply heat to some large chemical enterprises.
- CHPP Craiova (105 MW) and Oradea (205 MW) are built to use the lignite in the area and to supply some industrial enterprises in the area and urban consumers.
- CHPP Palas with a 250 MW installed power, burning crude oil; the first 150 MW unit is Romanian-made and is designed for urban heating.

SOLUTIONS AND PROSPECTIVE TRENDS IN THE DEVELOPMENT OF THE COGENERATION IN ROMANIA

The examination carried out for prospective purposes outlined that combined heat and power generation will continue to have an important development, as in Romania there are concrete economic and technical possibilities to provide at least the same development rate as in recent years.

The cogeneration is to be accomplished in the power plants running on solid fuels with low-heat value: lignite (low-heat value of 6,700 KJ/kg) and on bituminous shales (3,800 KJ/kg) marking an important turn as against the present conditions when the power installed in the coal-fueled CHPP's represents only 12 percent.

In the future the solid-fueled CHPP will be prevalent, resulting both from the building of new CHPP running on low-grade fuel mentioned above and from the conversion of some existing power plants burning oil, to coal burning.

Upon using some indigenous reserves aiming at maximum energy savings, the use of the low-grade fuels in the cogeneration of heat and electric power represents one of the important ways of attaining the energy targets proposed for the next decade in Romania, namely, self-sufficiency (independence from the fuel and energy points of view).

The cogeneration of heat and electric power will develop along two lines: the building of new CHPP and the conversion of some condensation plants with heating characteristics, for heat supply.

The CHPP's will be designed to operate with 50 MW units, particularly for industrial heating, and with 150 MW units for the urban heating. At the same time, besides the large CHPP's equipped with the above-mentioned units, power plants with units having less than 3, 4, 6, and 12 MW units for low-heat consumptions; these units will be gradually replaced with larger units.

The 50 MW industrial heating units are usually provided with counter-pressure units, with initial steam parameters of 13.7 M Pa, 540°C. The condensation and adjustable bleeding units will continue to be used only in special cases, such as obtaining of a high degree of reliability in the supply of the consumers in the area with electric power. These power plants which operate interconnected with the power system are designed for separate and insular operation in case of a failure in the system, thus providing continuity of heat and electric power.

The 150 MW urban heating units can supply a quantity of heat in the form of hot water, 215 MW (185 G cal/h). These units have initial steam parameters 19.2 M Pa, 540°C and reheating at 540°C and a very good efficiency in the generation of the electric power in a condensation scheme, close to that of the units having a maximum unit installed power of 330 MW.

Both in the industrial and urban heating plants provisions are made for peak and safety installations: LP steam boilers with maximum 28 kg/s unit output and HWB of maximum 116 MW, respectively.

The equipment for the CHPP's is manufactured by the Romanian industry, in a wide range of programs.

The conversion of the condensation plants into cogeneration plants aims at minimizing heat losses in the condensers. This can be obtained by providing installations which take over the steam from the condensation 200 and 300 MW turbines and by modification of the condensation turbines, at the units with lower power.

For the same purpose of minimizing the heat losses in the condenser, the high vacuum operation is used for separation of one section of the condenser or the use of the built-in bundles for the heating of water in the heating networks or in the makeup circuits of the heating networks.

If by now the maximum heat transport distance in Romania were about 10 km for steam and about 15 km for hot water, in the future the steam transport is expected to extend to about 35 km. The extension of the area supplying heat to the consumers by the CHPP has favorable effects on the concentration of consumption and on reaching a better efficiency by using specialized heating units with high unit powers.

The increase of the heat transport systems is also one of the future concerns related to the development of zonal systems to which heat will be supplied by NPP's, far into the future.

At the same time, the cogeneration development in the future stirs concerns for scientific and research activities. The main directions are:

- The improvement of technologic heat-consuming processes, by minimizing consumption, the use of heat with lower potential, and the return of the condensate in its initial state.
- Minimizing heat losses by better building insulation.
- Building of high efficiency power equipment (steam and hot water boilers), turbogenerator units, and auxiliary units (fans, pumps, etc.).
- The study of possibilities to store and stock the heat generated in the CHPP's.
- The study of fluids, other than water and steam, for long distance heat transport and minimizing of heat and pressure loss.
- Study of the heat supply conditions from the NPP for prospective requirements from high power units and small specialized nuclear reactors, and deeper insight into the specific problems of heat generation and transport.

Under the more and more restrictive conditions facing the future energy supply to domestic and industrial consumers, the achievements and programs in the cogeneration field are, in the authors' opinions, an example of an efficient way to be studied for the conservation of energy and environmental protection.

CHAPTER 8

GOVERNMENT POLICIES AND SUBSIDIES TO ENERGY

Milton Russell[1]

It is fitting, I believe, that this section be devoted to the criteria to be imposed on long-term energy planning. The sections of the book that follow examine particular energy processes and opportunities, but here we seek to put those opportunities into perspective. For my part, I would emphasize three truths that each of us sometimes overlooks in our understandable enthusiasm to get on with whatever projects we are pursuing. First, developed and developing nations alike have limited resources to devote to energy. Second, every process that works, in the sense that it produces energy, cannot--and should not--be pursued. Each must also pass the economic tests of producing energy more cheaply than alternatives and less expensively than doing without. Finally, and especially appropriate in the context of a United Nations report, nations do not stand alone. Autarchy in energy research and development, as well as in energy supply, is an expensive and ultimately elusive goal. While slavish dependence on others for energy is unpalatable, efforts to go it alone may be just as destructive to economic and social values. Thus this book can contribute to breaking down barriers between researchers and thus speed the results of research into application to the benefit of everyone, which ultimately is the goal of all of our work.

More specifically, I wish to consider the criteria for subsidizing potential energy sources and to indicate the nature of the government policies which can most efficiently and effectively promote development of those sources which are decided upon. Let me note at the outset that conservation, doing more with less, is similarly important, and the principles relevant to subsidizing energy production mostly can be applied equally well to conservation. My remarks on energy production should be interpreted as encompassing energy conservation where applicable. I have kept my comments general so that they may be useful

---

[1]Director, Center for Energy Policy Research, Resources for the Future, Washington, D.C. The views expressed in this paper are those of the author and should not be attributed to RFF or his colleagues.

in the context of different national policies and diverse ways of organizing economic activity.

ENCOURAGING ENERGY: BEYOND BUSINESS AS USUAL

The increase in oil prices over the past 6 years, and such pressing problems as deforestation and desertification of vast tracts of land, have created a large menu of unexploited energy production and conservation activities that are economic and desirable *now*. The energy reality to which the peoples of the world must respond has changed, but behavior lags. Investments have not been made, skills have not been learned, and consumption and production patterns have not been altered. In this sense we are living now in the "long term" found in the title of this book. Because change has come so swiftly in the energy arena, there is a gap between what is, and how we act, that must be bridged, just as when looking to the future, we see a gap we wish to close between where we are and where we want to be. I will concern myself with both these long terms by addressing the issue of what actions it is and is not wise for governments to foster *now* due to past and prospective changes in the world oil price.

The special reason for energy subsidies, and the concern with government policies toward energy, is the result of four factors. The first two, noted above, are the very rapid rise in the price of oil and the widespread crises in the supply of fuelwood and other dispersed indigenous fuel sources; the third is the accident of nature which concentrated easily accessible petroleum in a few areas of the world; and the fourth is the political reality that the nations controlling these deposits can vary supply and hold prices above the long-run replacement cost of energy. Additionally, major oil supplies are located in a politically unstable region from which oil supplies may easily be disrupted.[2]

The desire for relief from the insecurity of imported oil is a powerful motive leading governments to foster indigenous energy supplies even if those supplies are more costly than imported oil. In an analogous way, uncertainty about future availability of energy offers a powerful incentive to invest now

---

[2]These factors are analyzed in Sam H. Schurr et al., *Energy in America's Future* (Baltimore, Johns Hopkins University Press for Resources for the Future, 1979), Chapter 16; and in *Energy: The Next Twenty Years*, A Study Group Report sponsored by the Ford Foundation, Hans H. Landsberg, Chairman (Cambridge, Mass., Ballinger Publishing Co., 1979), Chapter 6.

in developing energy sources even if they are more expensive than conventional supplies. I will concentrate on the economic rather than the more political aspects of the rationale for energy subsidies, starting with the fact that current world energy prices may be influenced by the level of demand for oil, just as future prices will be affected by how much we use up in this generation.

Oil Demand and Oil Price

Oil shares with most other commodities the characteristic that higher levels of output engender higher unit production costs as known deposits are worked more intensely and more expensive reservoirs must be found and produced sooner. Higher rates of production also lead to increases in the value of oil in the ground, representing the higher cost at the margin of future sales foregone. And, finally, with greater demand, both the lower and upper limits of the range within which oil exporters exercise discretion over prices are also lifted. Thus it appears likely that with greater oil demand, the price of oil flowing in world trade will rise.

The effect on consumers of an increase in the price of oil due to an increase in requirements for imported energy is multiplied because the higher price is paid for all oil, not just for the increase in imports. Other energy prices may rise as well. If demand of an extra million barrels per day raises the price of oil by, say, $1 per barrel, that extra million barrels adds its cost <u>plus</u> $1 per barrel for each of the millions of barrels of previously imported oil to the consumer's oil bill. This relationship makes each government's policy toward energy production and use a matter of concern for all members of the international community. Displacement of imported oil by alcohol in Brazil and by domestic oil in Pakistan, installation of nuclear power plants in Korea, use of solar cookers rather than kerosene in India, opening of hydropower facilities in China, and use of more fuel-efficient automobiles in the United States all tend to hold down the price of oil to everyone. In the same way, broadening the base of energy production and reducing consumption of depletable oil lessens its price in the future. Oil is today the incremental source of energy supply, and its price serves as the bridge between the energy policies of all countries of the world, and between present and future generations.

International Aspects

But what does this imply about national energy policies, and specifically about subsidies to energy production and conservation? Most importantly, it suggests that while the resource costs of reducing the demand for imported petroleum are borne by the country which acts, the benefits are shared with all net importers of energy. And those benefits in some measure are the result of transfers from oil producers. Lower oil imports by any country set in motion forces which alter the international distribution of wealth. It also may make the current generation worse off while benefiting those living in the future.

Here we come upon a difficult and troublesome international aspect of energy policy. Whose welfare is to be sought? Energy choices would be relatively simple if we were concerned solely with maximizing the welfare of the world as a whole, and if a world authority existed to implement decisions taken. Energy sources would be exploited in turn, the cheapest first, and research and development would be directed to develop new energy sources as they were economic, and not before. When the cost of producing the old sources rose above the prospective cost of exploiting new techniques or deposits, the latter would begin to be produced and the question of subsidies would never arise.

If all net energy importers were brought together under the single goal of minimizing the cost of energy, again the prescription would be clear: Alternative energy sources will be brought on stream, least expensive first, until the extra cost of more indigenous energy was equal to the extra cost of importing more oil, including the higher price for all oil imported that importing more caused.

But neither of these conditions exist, nor are they likely to hold for the foreseeable future. It is unfortunate, but we are unlikely to see credible agreements between energy exporters and importers as to how the benefits of more rational energy development would be divided. Similarly, agreements among energy importers as to what share of the costs of extra production and conservation each will bear in order to gain the benefits of lower oil prices are not in prospect. And realistically, the likelihood of nations acting knowingly in the interests of others and against those of their own people is as slim with regard to energy as it is in other matters.

The Special Case of Energy: Reasons for Subsidies

Let us analyze, then, the circumstances under which a country acting solely in its own interest should subsidize energy production or conservation. First of all, added production would have some influence on the world price and thus would be worth more to the nation as a whole than the price of the imported oil it displaced. For most countries acting alone, however, the impact of its import savings on price would be very small.

Second, the added security of an indigenous source of energy may be worth a considerable premium. Energy represents a small portion of the value of most countries' total consumption, but the energy input is vital. Thus, a substantial premium may be worthwhile to assure that in the event of an oil supply disruption, economic activity and, for some, life itself will not be seriously affected.

Third, indigenous energy supplies, so long as they do not require too much imported capital to create, do not cause the drain on foreign exchange which has been so destabilizing a factor in the international financial operations of oil-importing countries.

Fourth, some alternative forms of energy are labor intensive, and for countries bedeviled with a chronically underemployed labor force, some premium over imported oil prices may be justified. In effect, the subsidy to energy production and conservation is partially offset because the resources going into its production would otherwise have had no useful employment. As an adjunct to this argument, those unemployed or underemployed who are newly drawn into the labor force have an opportunity to learn new skills and to experience the discipline of wage labor. In effect, the new energy industry may create cadres of trained workers acclimated to industrial tasks. Those cadres can be drawn upon as development proceeds.

Fifth, the nation may be able to concentrate energy production in the least developed portions of the country and have added value for that reason. There it can serve the three functions of tying such regions more closely to the rest of the nation, raising the level of living in the depressed area, and providing growth centers around which industry and other economic activity can develop.

Finally, energy production and conservation activities may be subsidized for traditional economic reasons: They may fit the category of infant industries which, as they reach an adequate size, can compete on even terms in the world market. They may have otherwise unaccounted for spillover benefits of value to the nation. Or again, the energy produced at present may be more costly than imported oil, but may promise not to increase much in cost while imported oil can be predicted to do so.

Each of these reasons for subsidizing energy production and conservation is framed to provoke the rigorous test that the subsidy leaves the nation itself demonstrably better off than before. This test forms a defensible basis for a national energy policy with staying power--one that can survive shifts in fads, changes in governments, and redirections of priorities. It also means that some technically feasible and intellectually exciting projects will not be supported.

The plain fact is that in some nations there are few opportunities to enhance, wisely, indigenous energy production. The natural resources and human energies of such nations can better be used doing something else, with the full knowledge that this means continuing to depend on ever more expensive and possibly unreliable imported oil. By the same token, even for the best endowed nations, opportunities are limited. Just because subsidizing one type of energy production or conservation is attractive at some level of output does not mean that doing more is necessarily better or that other energy activities are necessarily wise.

This point deserves elaboration. Unfounded promises, based on faulty cost comparisons, are sometimes made for conservation and for alternative energy technologies. What is ignored is that costs rise as the best locations, applications, and raw materials are exploited. For example, a U.S. study shows that the average cost for different waste material feedstocks in producing ethanol varies by a factor of four; when differences <u>within</u> each feedstock at different levels of use are taken into account, the variation is even greater. If virgin undamaged raw material were considered, the range becomes a factor of seven or eight. The projected <u>difference</u> in feedstock costs alone, even among waste inputs, is only a little less than the current <u>cost</u> of petroleum.[3] The point is that some ethanol may be economic at only slightly higher petroleum prices, but costs rise rapidly as more and more ethanol production is brought into being. The same principle holds for conservation. As the number rises of

---

[3] U.S. Department of Energy, <u>The Report of the Alcohol Fuels Review Group</u>, June 1979 (NTIS, DOE/PE-0012), p. 12-13.

Btu's saved, the cost per Btu escalates until further energy saving would not be worthwhile.

To summarize, there are good social and economic reasons for ignoring "business as usual" criteria in energy policy. A number of justifications for subsidies to energy production and conservation exist. But we should all beware of programs defended solely on the basis of such vague catch-all criteria as that energy should be saved or produced because it is scarce, because it is important, because the world will some day run out of oil, or because it is any nation's responsibility to others.

ENCOURAGING OIL: A DIGRESSION

Before I leave the analysis of the role of subsidies to address the question of government policies to implement them, mention should be made of oil exploitation in relatively unexplored regions of the world. There is a regrettable tendency, when dealing with the long term, to discount petroleum and natural gas in favor of more exotic and nonconventional energy sources. Conventional oil and gas are not ignored in this volume but they are given relatively little attention in consideration of the fact that they will probably remain the most important source of energy throughout this century and perhaps beyond. Much oil remains to be found in the promising petroleum basins which exist in all regions of the world. The World Bank in its survey of 70 developing countries concluded that 23 of them had prospects for substantial oil development, and another 15 probably had meaningful deposits.[4]

It seems to me important that we examine means by which promising petroleum regions, especially in developing countries, may be explored and developed in a timely way. The World Bank has facilitated this work, but much remains to be done. Technical difficulties abound, but national attitudes, unrealistic expectations, financial and institutional barriers, and the understandable reluctance of oil companies to venture where political risks are high are perhaps more serious constraints. They deserve careful attention in international forums such as this one. There might be a substantial payoff in devoting some of our attention, in the corridors if not in the meeting hall, to these issues.

---

[4] World Bank, A Program to Accelerate Petroleum Production in the Developing Countries, January 1979, p. 17.

ENCOURAGING ENERGY: PRINCIPLES FOR POLICY

Nations have limited resources but many pressing social and human needs, of which energy production and conservation is only one. This means that it is important to consider carefully how much of a nation's wealth of talent and resources is to go to energy projects, to choose those projects which will have the greatest long-term benefit to the nation, and to see to it that each project is pursued efficiently and abandoned if it turns out to have been a mistake. These criteria lead to a few widely applicable principles for selecting and administering energy projects.

It is important to keep in mind that energy projects form only one section of a nation's agenda, and each project is only one item among a large array of potential energy activities. Thus, it is not sufficient merely to demonstrate that an energy project is worthwhile. It is also necessary to show that it is more worthy of support than competitive endeavors which it will not be possible to pursue if the energy project goes forward.

In evaluating energy and other projects the timing and certainty of expenditures and returns is important. Clearly, the earlier the benefits of the later the costs the better. Certainty of costs and benefits is also a factor. Some projects that promise very high returns may also be very risky, and if in addition the benefits are far in the future, a project with far less dramatic potential may be the better choice. Formal analysis leading to precise calculation of present expected net benefits from alternative projects may not be either practical or useful, but efforts at formulating alternatives in this way can discipline choices and help prevent serious errors. Reevaluation of programs is also important. As one moves ahead it may make another less valuable because both will serve the same end. Similarly, if an inferior technology can be deployed now, it may make pursuit of a better process uneconomic. The cost advantage of the superior process may not pay for the expenditures needed for its development. Choices among energy processes are always required, and wise decisions are not easy to make in the best of circumstances.

Decisions which promote the national well-being can best be assured when institutions created to administer programs have incentives which approximate those of the nation as a whole. An important principle in organizing a national energy program is to keep decisions about energy (and other projects) out of the hands of partisans of any one energy form or process. Otherwise, the incentive is to continue the pursuit of programs even if they prove uneconomic or otherwise impractical. There is a tendency to underestimate difficulties and even dangers and to overestimate results when the professional future and perhaps livelihood of the decisionmaker depend on continued government support of a particular process or technology. In practice, this means that agencies and laboratories should not be narrowly specialized--that it is not wise to establish a bureau for exploitation of the water hyacinth. Instead, set up a bureau to investigate the use of a range of renewable and other resources. In general, it is wise to pursue a range of energy activities through the early and cheap exploratory phase <u>so long as</u> there can be ruthless pruning of those which do not prove useful. The likelihood that necessary pruning will be done is greater when those affected (and especially those doing the pruning) know that other opportunities are open to them when one project is abandoned.

Another principle is that necessary subsidies should be designed to maintain the incentive for efficiency at the operating level. In practice this means that subsidies should be such that each reduction in cost or increase in output goes directly to the producer as added profits. Similarly, each loss in efficiency or failure should come directly from the return to those doing the producing. For example, subsidies should be directed toward paying part of the capital costs but not of the operating costs of a process. It is always unwise for subsidies to be structured in such a way as to make up the losses of a project. Under such a scheme the incentives are all wrong--they lead to activities and expenditures which enhance the ends of those doing the work. They remove the pressure to produce energy at as low a cost as possible.

Finally, there is always the danger that technicians and scientists are more interested in the game than in the score. There is the understandable tendency to seek the path which satisfies scientific curiosity, which offers the elegant solution or which yields the applause of colleagues at home and abroad. Only accidentally does such a path coincide with the one which produces the most practical benefit to the nation. Perhaps the most serious temptation is to play to the plaudits of countrymen by surrendering to nationalistic pride by attempting to do everything at home. Instead, every effort should be made to borrow or buy technology and processes abroad rather than to develop them more expensively domestically. There is much work to be done in the world--too much to waste scarce talents and skills in duplicating developments available elsewhere. This leads to the further point that specialization and then cooperative exchange is as useful in energy activities as it is in others. Conferences such as the one which led to this volume of reports can foster such exchanges and thereby contribute to lessening the cost of energy and raising the level of living of the peoples of the world.

CONCLUDING REMARKS

One is astounded by the vast array of processes for extracting and producing usable energy suggested by the various chapters included herein. Some of these processes are economic now, some only need development to make them practical, others would be attractive if the costs of alternative energy rose substantially, and some may never prove usable at meaningful scale. The key questions are which processes fall into which categories and how might subsidies for promising processes best be provided.

It is important to each of us that we be as insightful in rejecting alternatives as soon as they prove fatally flawed as we are in discerning new opportunities to produce and conserve energy. In this way we can make the world's resources go further. We can save scarce skills and talents to be applied more productively elsewhere. And for the long run we can foster the confidence of our governments and peoples that will lead to continued support for our work.

CHAPTER 9

PERSPECTIVES ON DECENTRALIZATION

Milton Searle, Blair Swezey, and Chris Whipple[1]

INTRODUCTION

Energy has become increasingly difficult and expensive to obtain in the United States during the last decade. This is partially a result of the depletion of domestic oil resources, but in addition, measures to protect the environment, the public, and workers have acted to restrain the growth of domestic energy supply, as has growing opposition to nuclear power. These factors were at work in varying degrees throughout the industrialized world. In addition, the 1973 oil embargo and subsequent increases in oil prices, coupled with political problems in exporting nations, have completely changed the conditions of oil supply. The oil situation has impressed on the world the ultimate finitude of conventional energy resources and the need to consider long-term solutions to the world's energy problems.

The response of the world to perturbations in its energy supply has been manifold. Energy conservation has become very important. The world-wide search for new energy technologies has intensified with increasing emphasis on the use of very large, previously uneconomic mineral resources such as oil shales and tar sands and on renewable resources such as solar and biomass. Research and development (R&D) on breeder reactors continues. This technology represents a long-term option which the world can ill afford to neglect.

On the demand side, fundamental questions have been raised concerning the energy-economy relationship. Historically, U.S. energy consumption and economic growth have had a strong positive correlation. Recently, many have questioned this relationship and maintain that historical economic growth rates can continue with reduced energy consumption. Some extremists question the need for any further energy growth at all, and some even question the need for any further economic growth.

---

[1]Energy Study Center Staff, Electric Power Research Institute, Palo Alto, California.

Numerous ideas for improving the structure of the world's energy systems have been advanced by energy professionals, politicians, and concerned citizens. We believe that these new ideas require scrutiny. The physical and economic aspects of new systems need to be favorable. Thermodynamic and physical laws must hold, net energy conditions must be satisfied, and there are limits, however difficult to quantify, on the rate of introduction of new technologies. Moreover, research and development results are uncertain and consumer needs and preferences difficult to predict.

In this chapter we seek to provide some perspectives on one aspect of the energy problem which is frequently viewed within the context of broad economic and social reform--the appropriate degree of centralization and/or decentralization of energy supply.

The centralization/decentralization debate has often taken on a philosophical air, with decentralists developing scenarios of energy systems which would presumably bring about a more humane, better integrated, socially responsible and peaceful future, more in harmony with our natural surroundings.

But the main thesis of the decentralists seems to be that increased energy production and consumption using conventional technologies degrades the environment and also leads to more centralized institutions. In the extreme, they view our energy problems as unsolvable without broad social and economic reform.

In examining the subject of decentralization we will adopt a pragmatic approach.

ATTRIBUTES OF DECENTRALIZED SYSTEMS

To begin with, what are the characteristics of decentralized technologies which set them apart from centralized technologies? Attributes identified in the literature, and which we consider below, pertain to: unit size, system boundaries, ownership, operation, maintenance, control, and fuel supply.

The unit size of decentralized systems is small and usually restricted to one social or economic unit. However, a unit which supplies several social and economic entities may be thought of as decentralized. Centralized facilities, on the other hand, are large with size being based mainly on economies of scale. Historically, economies of scale in the commercial energy world have favored centralization.

System boundary considerations are very important in defining the existence and degree of decentralization. Many technologies which are decentralized in use may be economically viable only if their components are mass produced in centralized facilities. Thus, centralized production at an earlier stage may be an underlying requirement if decentralized end use is to be economically feasible. Consumer acceptance may also depend on quality guarantees that only large manufacturers are able to offer.

There is little discussion in the literature as to where system boundaries should be drawn for decentralized systems. Should a system that is decentralized in application, such as the personal automobile, be considered truly decentralized if it depends on centralized production of components, centralized distribution, and centralized fuel supply?

Failure to consider centralization effects outside of the arbitrarily determined system boundaries may compromise the broad social objectives claimed for decentralization. This is a very important point because it can be persuasively argued that centralization of many key societal functions has actually promoted decentralization in end use. For example, highly centralized electricity generation has permitted a high degree of decentralization in the operation of small industrial and commercial establishments which could not have afforded the high costs of self-generation. Centralized electricity generation also permits the decentralization of numerous services within the household, such as clothes washing, appliance operation, and home entertainment.

A decentralized installation is generally considered to be owned, operated, and maintained by the end user. Thus, the end user possesses a feeling of control over energy production. At the opposite extreme, centralized facilities are controlled by businesses and thus are removed from the decisionmaking power of the individual.

Fuel supply is critical to the operation of any energy installation. For decentralization it should be under the control of the end user. Solar energy is an abundant resource available in varying degrees in all regions and meets this criterion. However, its availability is independent of the degree to which a solar installation is decentralized.

BROAD SOCIAL OBJECTIVES

It is clear from the literature on decentralized systems that a major consideration concerns the social ends that these technologies serve. Such matters are quite subjective. This does not make them less important, only more difficult to deal with. We next look at some of the values that have been suggested.

National welfare is said to be enhanced through decentralized systems because they employ indigenous energy resources, which reduce the demand for foreign exchange, increase national security, and provide greater political independence. However, it is the nature of the energy source itself that is the decisive factor and not its degree of centralization. Centrally generated power based upon any indigenous resource, including solar, would appear to satisfy these objectives just as well as decentralized systems.

For the less developed nations, excluding those exporting oil, use of indigenous resources is particularly important. However, deciding on the proper balance between indigenous fuel sources and imports is not a simple matter. Spending a disproportionate share of national income on developing indigenous energy resources may be worse than importing more energy and exporting more of some other commodity.

Mineral resource depletion. The 1973 oil embargo and subsequent events continually remind us that world fuel resources are finite, that prices for them are likely to be rising, and that access to resources is less assured than previously believed. This has produced a desire to turn away from mineral fuels and toward renewable resources, particularly solar. The physical characteristics of these resources often make them more amenable to use in decentralized installations, such as solar space and water heating facilities, than in highly centralized facilities, such as those used with coal and nuclear energy. By and large, though, these resources can be utilized in either central or decentralized installations.

But we must not forget that the depletion of fuel resources of world commerce--oil, natural gas, coal, and uranium--is not imminent. Although we are more conscious now than we were a decade ago of the threat of depletion, the world's remaining mineral fuel resources constitute a large total resource. And rising energy prices are likely to result in still further increases in these resources.

Rough estimates show that world crude oil resources are perhaps 80 times annual world consumption, with from 1 to 10 times more in oil shales and tar sands; conventional natural gas resources are perhaps 180 times current consumption; coal over 1,000 times current consumption; and low-cost uranium resources well over 100 times current consumption even without breeder reactors.

By contrast, some renewable resources now in widespread use, such as fuelwood in South America, Africa, and Asia, are now in short supply. In any case, the question of encouraging decentralization in many of these areas is a non-issue; the systems are already extremely decentralized.

Protection of the environment provides a strong motivation for the use of renewable resources--particularly solar. However, it should be noted that environmental impacts often result more from the nature of the energy resource that is being utilized than from the degree of system centralization.

Nevertheless, it is true that centralized installations have a potential for overwhelming their immediate environment. The environment is more likely to absorb the scattered adverse effects of decentralized units than the concentrated effects of centralized installations. This factor, however, may be offset by the lower cost and easier regulatory burden of controlling pollution in a few big facilities than in many small ones. Indeed, this is becoming more apparent as many U.S. homeowners are switching to wood for home heating and creating local air pollution problems. Attempts by government to control many small decentralized installations could result in an onerous increase in direct government intrusion into the lives of private citizens.

TECHNICAL CONSIDERATIONS

Improvement in technical performance is one of the arguments presented for preferring small decentralized systems. It is claimed that such systems will be more reliable, will provide better thermodynamic matching of end use to source, and will be more efficient in the use of fuel. These arguments are generally made in comparing decentralized systems with centralized electric systems.

There is little reason to believe that decentralized electric systems will be more reliable than centralized systems in the industrial nations. In fact, U.S. electric systems are often criticized for having excessive reliability.

The relative reliability of centralized and decentralized systems will vary with the source of system failure. By definition, decentralized systems are not subject to central equipment malfunction. Both systems are subject to failure as a result of catastrophic weather. However, in such a situation central system service would in all probability be restored to the average user more rapidly than would be the case with decentralized systems. While central systems stock spare equipment and have trained personnel to deal with major outages, decentralized systems, by their very nature, do not make provision for dealing with disasters beyond their equipment boundaries. In the case of local outages, central station electric service is generally restored in a matter of hours. However, it often takes days for a homeowner to obtain the services of a plumber or an electrician. There is no reason to believe that the homeowner would obtain better service for his decentralized electric system.

Finally, there is the matter of design reliability. Decentralized electric systems, such as solar, will need to either incorporate sizeable energy storage facilities or have backup systems capable of providing energy requirements during extended periods of overcast. Such long-term storage will be quite expensive. In most cases the most economic alternative to adequate storage is reliance upon centralized backup systems.

Claims of greater reliability for decentralized systems need to be more specific as to the circumstances under which the greater reliability is claimed and as to the costs of achieving such reliability.

Energy is never used with 100 percent <u>efficiency</u>. In any energy conversion facility some low-temperature energy is wasted because it is not economic to exploit. For example, the warm cooling water discharged from power plants is rarely used as a source of energy, although this water contains large amounts of low-temperature heat. Uses for low-temperature heat are not generally located near power plants and because of the low temperatures and large water volumes it is not ecomomic to move the water to potential users.

Decentralization is often thought to provide an opportunity to use waste heat because the decentralized energy facility can be located close to both electric and heat loads. In practice however, there are difficulties in matching the load patterns of demand for electricity and space heat on both a daily and a seasonal basis. In a shopping center, for example, it may be necessary to heat early in the morning to warm the buildings before customers arrive. At this time electrical load is relatively low. Also heating loads may be low in spring, summer, and fall relative to electricity demands.

Waste heat from decentralized power generation facilities may also be used for industrial process heat. However, the same problems in matching electrical demands and process heat demands may exist. In certain industries, where the electricity and process heat demands are reasonably well matched, there is an opportunity for cogeneration of electricity and process steam in a common facility. If tied to a central grid, any mismatch between the heat and electricity load can be accommodated by buying from or selling power to the grid. Interest in such cogeneration facilities is growing.

At sites where there are compatible electricity and heat loads, decentralized installations can increase fuel efficiency. However, fuel efficiency is not an end in itself. Costs of equipment, fuel, maintenance, and energy backup must also be considered. In some cases greater fuel efficiency is obtained only at a greater total energy expense.

In addition to considerations of efficiency in energy conversion, it is sometimes argued on thermodynamic grounds that it is improper to use a high quality energy source, such as electricity, to perform low quality work, such as space heating. Just how important it is to preserve this high-temperature potential for performing high quality work depends upon the scarcity of this high-temperature potential. Since both nuclear and solar energy can be used to produce high grade energy, and face little prospect of resource depletion, the thermodynamic argument does not appear to be a compelling one. To date, American society has chosen to give greater weight to such factors as convenience, cleanliness, availability of relatively inexpensive utilization equipment, and ease of distribution.

ECONOMIC PERFORMANCE

The economic costs of any system, centralized or not, must, in the long run, play a large role in its viability. Such costs represent the demands which the technology makes on a nation's productive resources. In principle, these costs should also include the effects of national resource depletion and other externalities, but such all-inclusive costing is difficult to realize in practice.

Decentralized electricity production with solar energy is currently very expensive. Small wind systems, rated at 1 to 2 kW, average about $4,000/kW installed. Larger wind systems, approaching 15 kW, average about $1,700/kW installed. The rated output of these machines is usually achieved only at a wind speed greater than 20 mph. However, in actuality, most urban areas have mean annual wind speeds of less than 10 mph. The average output of these machines at 10 mph may be as little as 20 percent of the rated output. Even assuming a mean annual wind speed of 12 mph, the life cycle cost of the electricity generated by these systems is very high, from 14¢ to 22¢ per kWh (Massachusetts Energy Office, 1978). This compares to an average U.S.

residential electricity price of about 4½¢ per kWh. Furthermore, these wind generated electricity costs are for systems using a synchronous inverter which allows electricity generated in excess of the load to be fed back into the utility grid and for electricity to be drawn from the grid when the load exceeds the wind system output. A dedicated battery storage system would add 10¢ or more to the cost of each kWh generated by the wind machine. Clearly, stand-alone systems incur large economic penalties due to the storage system.

The cost of photovoltaic (PV)-powered installations is also very high, ranging from $15,000 to $30,000 per peak kW of capacity, several times that of the wind systems. These costs represent currently available small power systems for remote applications and U.S. Department of Energy-funded test systems. Decentralized residential PV systems have yet to be commercially installed. A truly decentralized residential unit would, like the wind system, require a large storage system with the attendant extra costs. Cost reductions are imminent for solar cells but the high costs of the total PV system will restrict any appreciable penetration of photovoltaic technology to remote, nongrid-connected applications for some time to come (Swezey, 1979).

Solar hot water systems are generally considered to be the most economically competitive of the solar technologies. Commercially installed residential solar hot water systems generally range in cost from about $2,000 to $3,000 for a system expected to provide up to 70 percent of the annual water heating load. Combined residential hot water and space heating systems may range in cost from $5,000 to $15,000 for a system expected to provide about 70 percent of the annual water heating load and about 50 percent of the annual space heating load.

The most recent analysis available to us shows that, on the average, solar water heating becomes economic when the costs of alternative fuels reach $8 to $12 per million Btu. Active solar space heating systems become economic at $15 to $30 per million Btu for alternative fuels (Bezdek, et al., 1979). For comparison, in early 1979, U.S. heating oil prices averaged about $4 per million Btu, natural gas averaged $3 per million Btu, and electricity averaged about $12.50 per million Btu (U.S. Department of Energy, 1979). It should be emphasized once again that these systems will require a backup energy source. This problem is discussed in greater detail in the next section. It is evident that large reductions in installed solar system costs and large increases in conventional energy costs will be necessary before solar can make a significant penetration into the heating market.

MANAGING TRANSITIONS

Transition Mechanisms. For those technologies and those situations where it makes sense to move from a more centralized to a less centralized technology, there are still questions of how the transition will take place. If the decentralized technology is a clear economic choice, then normal economic forces may do the job with the transition time depending on the margin of economic superiority, the retirement rate of existing equipment, and the overall growth of the market. At the present time even the most advanced decentralized technologies are barely economic in the United States and then only in particularly favorable situations. Where consumer economics are not sufficiently favorable to bring about a socially desirable transition, then other mechanisms such as subsidies, tax credits, or legal requirements for installation must be involved.

Role of Backup Systems. During any transition to greater reliance on decentralized systems there will be complex questions concerning the relationship of the decentralized systems to the centralized systems. Most decentralized systems will need to use centralized systems for backup. Exorbitant collection and storage costs would be required for decentralized solar systems to provide service equivalent to centralized systems without backup. The cost of supplying all of a customer's heating or electrical requirements from a decentralized system may never be economic on a large scale compared to using central systems for backup.

The solar water heating systems previously considered had only 1 day's storage capacity. In some regions at least 5 day's storage would be required to provide service equal in reliability to that of a central system. Since 1 day's storage capacity may be 20 percent of total system costs, the costs of 5 day's storage plus some additional cost for collectors to charge the storage could easily double total system costs. And since each day's added storage will be needed less than the prior day's storage, marginal costs will rise rapidly. For example, whereas the basic 1 day's storage in a hot water system may, on an annual basis, provide 70 percent of the total water heating load, the 5th day's storage may provide only 1 percent of the load. Thus, the marginal cost of the 5th day's storage would be about 70 times the basic storage cost. The cost effects are similar for space heating systems. For hot water systems some decrease in present reliability standards may be tolerated in order to save on costs; for space heating in cold areas the maintenance of reliability may be crucial.

In view of the high marginal cost of supplying 100 percent of the required energy from solar systems there are advantages to long-term reliance on mixed systems where consumers have the option of using centralized systems for all, part, or none of their energy supply as their preferences and economics dictate. Central electric systems may be particularly well adapted to this role because of the versatility of electricity. Electricity can provide backup for solar water heating, solar space heating, and solar electricity all at the same time. Indeed, mixed systems may turn out to further many of the values advocated by those who favor decentralization while retaining many of the economic benefits of centralization, and at the same time giving the consumer maximum flexibility of choice.

Net Energy Balances. Another important transitional question to be considered is the net effect on energy supply of the rapid production of new decentralized technologies. It requires energy to produce and install each unit of new energy technology, such as a solar cell. If the energy to produce each unit of the new technology is high, then rapid growth may produce a net energy drain on the economy for a period of time. For example, it has been shown that a growth in solar heating installations which might provide 14.68×10^{15} Btu cumulatively between now and the year 2000 would produce a net energy gain to the economy of only 2.68×10^{15} Btu in this period. This assumes it requires 5 years for a unit to produce an amount of energy equivalent to the quantity used in its manufacture. Estimates of the energy (not economic) payback time in the literature tend to be in the 3- to 9-year range depending on the technology and the geographic location in which it is used. If the payback period is 7 years, the economy would actually experience a cumulative deficit of 2.12×10^{15} Btu. A slower rate of growth in solar installations would provide almost as much net gain at a 5-year payback and a small gain even at a 7-year payback (Whipple, 1979).

These net energy effects occur because the energy required to manufacture a unit is expended in a single year while the energy produced is spread over many years. If the payback period is 5 years then on the average a unit is producing in any 1 year only one-fifth of the energy it took to manufacture it. It thus takes five units in operation to produce the energy to manufacture one new unit. If the yearly rate of production is greater than one-fifth of the number of units in operation, then the system has an energy deficit for that year. And the effects are cumulative. Of course, when the rate of production of new units slows or ceases then the operating units will produce a large surplus of energy.

The role of solar energy in displacing oil imports could be more important in the next few decades than its role as a net energy producer. If solar equipment were produced using domestic fuels (e.g., coal or nuclear) and if the solar output served end uses which would otherwise be supplied by imported oil (e.g., oil-fired space heating) there could be fuel substitution gains even if the net increase in energy was small.

SUMMARY AND CONCLUSIONS

The classification of energy systems as centralized or decentralized can be very arbitrary. Many attributes of energy systems vary along a whole spectrum of values. Whether the system is considered centralized or decentralized will depend upon where the system boundaries are established.

The broad social advantages of enhanced national welfare and less environmental impact attributed to decentralized systems are often due more to the nature of the energy source than to its use in decentralized installations. Thus, the advantages of decentralization per se may be overstated.

In some cases, such as with cogeneration facilities, decentralized systems have proven more energy efficient than centralized systems. However, for the majority of cases, the greater reliability and efficiency claimed for decentralized systems have not as yet been demonstrated. It may be possible to achieve better thermodynamic coupling of load and source with decentralized systems but the extent to which such systems represent a major market also remains to be demonstrated.

Worldwide mineral energy resources are relatively large and fear of their imminent depletion is not a reason for rushing to renewable resources. However, concern about the terms on which mineral energy resources will be available is a reason for increased emphasis on domestic resources, renewable or not.

For most prospective users, economic costs remain the most crucial element in assessing the relative merits of centralized and decentralized systems. Presently, economics do not generally favor decentralized systems.

Finally, there is a growing role for the use of renewable energy resources, whether in centralized or decentralized installations. By focusing on the relative merits of each of these extremes, vis-a-vis the other, we lose track of the immense potential of these resources for providing a significant portion of our energy needs on all scales. Centralized and decentralized energy production, together, can be compatible, and the challenge that lies ahead is to concentrate on the quick and efficient integration of renewable energy systems, centralized or decentralized, into the existing energy supply structure.

REFERENCES

Bezdek, R., and others, 1979, Economic Feasibility of Solar Water and Space Heating: Science, v. 203.

Massachusetts Energy Office, 1978, Wind Report.

Swezey, B., 1979, An Assessment of the Potential for Photovoltaics as Envisioned in the National Photovoltaic Program Plan: EPRI Energy Study Center Information Paper.

U.S. Department of Energy, 1979, Monthly Energy Review.

Whipple, C., 1979, The Energy Impacts of Solar Heating: Science (in press).

CHAPTER 10

ENERGY DEVELOPMENT THROUGH INDUSTRIAL COOPERATION

Jean d'Herbes[1]

The many activities of ONUDI, CNUCED, and the recent CNUSTED meeting in Vienna show that the success of the new international economic order now in embryo depends largely on a felicitous solution to what is called the problem of transfers of technology. There is general consensus as to the complexity of the problem, starting with the matter of vocabulary. Indeed, what do we mean by the term technology? Are we talking about patents, licenses, or industrial know-how? These are obviously quite different things. Do we mean basic research or tricks of the trade? The solutions cannot be the same.

It is even more difficult to derive a precise definition for what we call adapted technology. If we consider that most major techniques were developed in industrialized nations for their particular uses, it is obvious that, like fish in an aquarium, such techniques can only bear full part within the specific industrial fabric that allows for this growth.

They are necessarily crippled if transplanted as such in a different environment with no prior effort to rethink them and adapt them to that particular environment. Dr. Barnea (Chapter 1) calls attention to the importance and the future role of natural gas, even when produced in small fields. He is 100 percent correct. But here again, careful consideration must be given to the distribution costs involved.

For instance, the cow-dung used as fuel for cooking rice, the staple diet of the Indian population, represents the equivalent of some 25 million tons of petrol, and there are no costs for setting up facilities. How much would it cost to set up a gas distribution network?

In my opinion, questions of this nature demand our utmost attention, and I would like to share with you some of my ideas on the matter.

My example raises some unusual questions.

We are all in the habit of reasoning in financial terms: "pay out, rate return, intercalated interest," and so on.

[1]Total Coopération Industrielle, Compagnie Francaise des Petroles, Paris, France.

This would be logical if we had unlimited energy resources and more, particularly hydrocarbons. But we know this is not the case.

Therefore, wouldn't it be better to put the question in terms of the worldwide energy budget?

Obviously, the development of a new technology or the exploitation of new energy deposits (whatever form they may take) requires a more or less substantial investment of additional energy.

But can we draw up a complete budget between anticipated contributions and investments granted? Certainly, such a balance is very difficult to establish since we would have to reckon with the energy required for manufacture of the equipment, transport, and the livelihood of the people who implement the new processes.

Aren't there instances in which the balance is negative?

In such cases, are we not justified in asking whether it contributes to stoking worldwide inflation? This is one hypothesis.

It would probably require searching consideration. It would certainly offer the advantage of purely technical and thus objective reasoning with regard to the energy balance by eliminating all political or philosophical interference.

We clearly perceive here the idea of <u>risk</u>. This is a very difficult notion to grasp, and I would like to use the oil industry as an example. This notion of <u>risk</u> appears at all levels:

- Exploration,
- Development,
- Exploitation,

of a hydrocarbon field. It represents the counterpart of nature's own intervention. In this area, perhaps more than in any other, nature brings her full force to bear. Remember that, barring very rare exceptions, no one in the world has ever actually seen an oil or gas field. This already shows the evident risk involved in exploration, since it is tantamount to finding a relatively small needle in a haystack. This is generally the only risk that pays off.

There may be an even greater risk involved in development since most of the investments earmarked for development purposes (wells, transformation, and transportation facilities) must be decided on the basis of 3, 4, or at most 10 bore-holes made in a large rock-mass and giving direct access to only 100-billionth of the existing mass.

Fortunately, by applying the sciences of mathematics and fluid physics we can attenuate the risk. But when we talk about optimum recovery of existing reserves, the degree of risk involved depends on the methods used.

Taking a deposit of 1 billion tons of oil, one single point of difference in the final recovery represents $2 billion at today's prices! These figures may appear phenomenal but unfortunately we are looking at very practical problems. Likewise, there is considerable risk when it comes to exploitation. This point needs explaining.

Let's take a field that produces 10 million tons per 100 active wells. We will say that the field is operating at 100 percent efficiency if all the wells, presumed to be identical, are operating at the optimum 24 hours a day, 365 days a year.

Everyone who has dealt with the oil industry up close or from afar knows that this would be utopia, for just like humans, wells are subject to more or less serious ailments (the introduction of water, gas, salt, asphalts, among other substances). We will therefore have a coefficient of less than 100 percent.

But let's see what a difference of 10 points, say between 85 percent and 95 percent, means for the same field, since we can consider these two figures as sound management coefficients. To maintain production at the same level of 10 million tons a year, we would need 10 percent more wells, meaning a minimum of $10 to $100 million of additional investments each year, depending on whether we are dealing with a field on land or at sea. This is actually a useless over-investment that could be avoided depending on how the field is managed; it is probably the easiest to minimize.

These are all real risks that entail major expenditures: ranging from about $100 to $500 million for exploration, $1 billion for development, about the same for exploitation.

I would repeat that only the risk of exploration is taken into account as such, which I find abnormal.

This brings us to the industrial cooperation policy which the Compagnie Francaise des Pétroles has spearheaded for many years. What is this policy? Quite simply, to help those who ask us to minimize the risks I have described. So you see, this is not some new gadget to deck out technical aid or the supply of services out in modern colors. The basic idea behind the policy is that only an experienced industrialist who has coped directly with these risks can pass on appropriate answers to others facing the same problems.

Certainly, there are many highly qualified experts and consulting, engineering, or service firms that can provide partial answers to the problems raised. But only a similar company, in other words a company with the same concerns and facing the same kind of decisions, can fully grasp the weight of the risks incurred and the seriousness of the necessary decisions.

Reciprocal confidence is obviously supposed by such cooperation, which has already been instituted in countries as diverse as India, Abu Dhabi, Sudan, Algeria, where it was first set up, the Soviet Union, Argentina, the Ivory Coast, and Cameroon, and in fields of activity as different as exploration, development, field exploitation, refining, distribution, and hydrant systems for air fields. This type of cooperation also supposes a frame of mind unlike the attitude which guides traditional supplier-customer business relations, for it implies that both parties involved are making a joint effort to solve a given problem as effectively as possible.

Although the burden of investment may rest on only one of the parties, it should be understood that the other risks its world image and its technology in the broad sense of the word. Actually, it is no longer a question of technology transfer in the narrow sense, but instead of the transfer of industrial know-how. Such know-how, which cannot be described in any manual of procedures but which is widely present within a company, has been acquired over many years of research and exploitation.

To cite an example, in the space of 20 years the Compagnie Française des Pétroles has invested upwards of \$10 billion in one manner or another in the specific area of research, development, and marine oil exploitation.

What price tag should be put on the transfer of this experience? Nothing? \$10 billion? Neither one nor the other, evidently, but rather some figure between the two that basically should take into account the result expected by the partner.

We also should realize that industrial know-how is dependent not only on the mastery of specific techniques, but also, and perhaps especially, on the mastery of what I would term "the human tool" in the noble sense. I am referring to the management, training, and job environment of the labor force which we consider to be the key to success.

Indeed, nothing of any real value can be accomplished without human input. In sum, it is a matter of genuine cooperation supported by mixed teams that are animated, directed, and supervised by multi-level cooperation committees, entrusted with sustaining and fostering further development of cooperation as a means of guaranteeing a genuine transfer of know-how.

Since this is a new attitude, I would go so far as to say that a new type of contractual relations should be considered, relations patterned more after a marriage contract than the international commerce code. Our experience along these lines has convinced us of the important place of the industrial companies of developing countries in the current dialogue.

This motivated our strong desire to play an active part in the creation of an institute for international cooperation and development, located near Nice, France, an institute in which jurists and companies from both hemispheres can talk together, examine problems, failures, and successes, determine their causes, and make objective attempts to find remedies.

The Compagnie Française des Pétroles works through this industrial cooperation policy and lays no claim to philanthropy, and here I anticipate your questions on this point. It is concerned with fulfilling its specific industrial world role in all energy-related fields of activity and obtaining fair remuneration for its efforts and its experience, while aiding in the solution of the many problems raised at the world level.

In conclusion, I would like to quote part of the address of Mr. Aigrain, Secretary of State for research and head of the French delegation to the CNUSTED meeting held this past August in Vienna:

> Is it not on the foundation of such principles that an assembly like ours could resolutely base the action of the international community, bearing in mind that there is no finer co-operation than that which respects the independence of the partners? In this connection, I have strived to confirm the orientation of the scientific and technical co-operation policy which I currently head, a policy that is designed primarily for negotiated co-operation. Essentially, negotiated co-operation means the expression and the mutual consideration of the concerns and needs of the partners.

CHAPTER 11

SOME CONSIDERATIONS REGARDING ENERGY RESOURCE DEVELOPMENT IN AREAS WITHOUT ADEQUATE INFRASTRUCTURE

G. V. Subba Rao[1]

The purpose of this chapter is to outline a few broad considerations in the development of energy supplies in areas which are without adequate physical infrastructure in the oil-importing developing countries.

The conclusion is that there are no really fixed energy sources and that, as technology, energy price levels, development patterns, and locations change, the basic parameters for energy resource exploitation are significantly altered. Unprecedented opportunities now exist for energy exploration and development in the many unexplored regions of the developing countries on the basis of the exploitation of local energy sources.

Some of the greatest developmental challenges and opportunities lie in the expansion of the resource base of both conventional and nonconventional energy resources. The developing countries have yet to undertake vigorous programs of exploration of underground energy resources (oil, coal, oil shale, tar sands, natural gas, geothermal energy, uranium) and of utilization of their vast potentials of renewable sources of energy, particularly hydropower. Even the use of currently available technologies is believed to result in an extraordinary expansion of their energy base. It has been estimated that 50 to 60 developing countries have either offshore or on-land petroleum potential; 30 to 40 countries have coal potential; an equal number of countries have oil shale and tar sands potential; and at least 25 to 30 countries have geothermal potential. There are vast hydropower potentials in developing countries, only a small fraction of which is at present utilized. Nearly half a dozen countries have tidal power potential and many more may utilize solar and wind energy. The wide spectrum of rich energy resource potentials awaiting development in developing countries requires both centralized and decentralized technologies and investment patterns.

---

[1]The author is a senior staff member dealing with energy work in the United Nations; the views expressed here are his own and should not be attributed to the organization he serves.

ENERGY RESOURCE DEVELOPMENT STRATEGIES

In planning energy resource development in the developing countries, the first step is to survey the potential energy resources, their size, characteristics, and location in relation to market centers, and the infrastructure of investments and facilities that will be necessary to utilize them. In this process, a resource will be developed to the extent that markets open up locally, regionally, or abroad. Gradual development of the infrastructure will facilitate the linkage of dispersed centers of demand and will give an impetus to regional industrial development. In this process, it will be necessary to develop both centralized and decentralized technology approaches--the former for the satisfaction of the needs of the large concentrations of urban populations and the latter for making energy available to rural and remote areas. Where there are large-scale energy resources, such as hydropower, to combine with large-scale demand for industrial and other uses, centralized electricity generation can be efficiently provided at low cost; to the extent that the centralized electricity-generation system can be extended through a grid system, impetus can be given to rural electrification, which will permit the operation, as an example, of water-lifting pumps for irrigation from tube wells. By and large, the process of industrialization and urbanization requires the development of centralized, large-scale supplies of energy.

Outside of urban areas, there are many obstacles to energy resource development in developing countries, including lack of infrastructure and very limited and dispersed markets. Energy resources may exist either in large- or small-scale dimensions, but because of the lack of nearby markets their development is delayed. Several examples can be given.

Ethiopia has a number of geothermal energy deposits but many are located far from populated areas which could provide markets for electricity or hot water. Zaire has one of the largest hydropower potentials in Africa. The lack of domestic and regional markets to utilize even a fraction of this potential means that this valuable energy resource will remain unutilized for a long time.

There are also examples of local energy sources which are not exploited for lack of infrastructure in terms of transport and other development and utilization facilities. Large-scale flaring of associated gas occurs in many oil-producing countries for lack of outlets for local gas utilization. Natural gas utilization requires an infrastructure of pipelines and distributing systems to reach consumers. These facilities are heavily investment-intensive and expensive gas-distribution systems and pipelines can be justified only if there are large industries to utilize the resource or if there are dense human settlements with adequate purchasing power to provide a concentrated market. Low-grade fuel deposits, which until recently were not economical to exploit, have now become attractive in the light of the increased prices of imported energy. However, if these deposits are located far from market centers, they are likely to remain unexploited for a long time unless new technologies are found for their economic utilization or new industries are set up nearby.

Appropriate strategies for the development of energy resources which are location-bound have to be devised initially in terms of the needs of the areas and regions where they are located and, after initial development, on a country-wide basis. Industries may be developed where large-sized energy and other natural resources exist.

Large-scale centralized systems of power generation become appropriate to feed the energy demands of urban complexes; however, for dispersed low-demand centers in rural and remote areas, decentralized energy resources development and decentralized generation systems will be more appropriate. Thus, questions of centralization or decentralization in energy resources development have to be decided in terms of the location of resources in relation to markets, the size and scale of such resources, and the size of investment and infrastructure requirements for their exploitation.

In the past, the preoccupation with large-sized projects in exploration, power generation, refining, or marketing at the national level has resulted in the neglect of numerous opportunities for the development of small-scale activities in energy as much as in other fields. The time has come for fixing attention on small-scale energy resources and technologies and on decentralizing energy systems and services for the benefit of the rural populations which constitute a majority of the developing world.

It is now necessary to evolve a new strategy of energy resource development to fit the needs of areas lacking electricity grids, distributive infrastructure, and large, concentrated market centers. The secret of small-scale development in energy lies in the exploration (or identification), development, and utilization of economically attractive or locally significant small-scale deposits of fuels and power.

Small hydropower systems and micro-hydropower projects; small-scale fuel deposits and power-generator systems utilizing locally available low-grade fuels, such as bagasse; biodigesters utilizing vegetable or animal waste for the generation of methane gas; small-scale solar-energy applications for space heating, water heating, ice-making, or refrigeration; small-scale wind-power generation and geothermal power plants; geothermal hot-water applications; small oil, coal, and oil-shale deposits--all these are examples of possibilities of small-scale development suitable for local use to satisfy limited markets in developing countries.

It is not entirely incorrect to assume that for every large fuel deposit there are, generally speaking, a much larger number of small deposits. These may be distributed over a wide area and hence may not necessarily be economical for a centralized approach to energy exploitation. However, these dispersed and small deposits may be quite significant in local terms and adequate for local utilization. In any case, a step-by-step approach to gradually expand the circle of development is possible, depending on the growth dimension of the local and regional markets for energy.

While small-scale development is more the rule than the exception in agriculture and farming in practically all developing countries, and while small-scale industrial development has increasingly gained support in practically all countries, small-scale development in energy resources has been neglected in almost all countries, with a few such exceptions as small-scale oil operations and coal enterprises in North America.

The exploitation of small-scale power and fuel sources can give suitable impetus to local development and provide gainful employment by making energy available to poor populations. Furthermore, the financing and management of small-scale energy projects is less complicated than for large ones. The lead time for their fruition is also significantly shorter.

The challenge of small-scale energy resource development must be met by the energy planners in developing countries by instituting a package of energy extension services in a manner similar to agricultural extension or small-scale industrial services. The modernization of small-scale energy enterprises by the infusion of technology, management skills, links to markets, provision of infrastructure facilities to the extent possible, credit supply, and training programs must become an important element of national energy resources development planning.

ENERGY RESOURCE DEVELOPMENT OPPORTUNITIES IN DEVELOPING COUNTRIES

Opportunities for energy resource development are governed by a number of factors: the level of energy prices, the state of technology development, and, above all, the location of the particular energy resource in relation to markets and the time dimension for its exploitation. Many of these parameters are subject to considerable change.

Oil and Gas

The changing circumstances of the new energy situation have substantially altered the economically exploitable energy resource base as well as the number and variety of energy resources, technologies, and systems. With delivery costs of imported crude oil running from $25 to $30 per barrel, it has become quite attractive to find and exploit even comparatively high-cost crude oils, to employ costly secondary and tertiary oil-recovery technologies, and to explore and develop a large variety of nature-made, nonconventional resources on land and offshore, including deeper structures or even small oil deposits in locations distant from markets or in hostile climatic conditions. For a similar reason, it has now become attractive to exploit what were previously considered marginal or small-scale or deep coal deposits. With the new increases in oil prices, the economics of coal exploitation have changed considerably, and it is now possible to state that coal costing as high as $70 per ton can be competitive with imported crude oil at now around $150 per ton. Similarly, with fuel costs in oil-fired electric power stations rising to as high as 50 mills per kwh, even expensive hydropower, which might have been considered uneconomic earlier, could become attractive. It must be remembered that only a few years ago a cost level of 10 mills per kwh was viewed as the uppermost limit for the exploitation of hydropower.

Large areas in many developing countries consist of sedimentary basins which await land and aerial surveys for oil and gas exploration. Furthermore, the history of oil discovery by field size is partly a matter of biases in exploration and development, which are concentrated on larger and shallower oil structures before moving to smaller and less accessible ones. The increased level of oil prices and advances in exploration technology have made smaller fields economically more attractive today than previously.

In regard to the question of exploratory activity in oil-importing developing countries, exploratory drilling took place in 37 countries out of a total of 113 in 1976. Seismic surveys have been done in less than 20; more than half did not experience any significant exploration activity in 1976. During 1967-1976, in only 16 countries was exploratory drilling intensive enough to result in significant discoveries. These statistics on petroleum exploration activity in the developing countries reveal the vast scope for accelerated oil exploration.[2] In some instances, in spite of requisite surveys, exploration has not led to significant discoveries of hydrocarbons. This does not necessarily exclude the possibility of discovering oil and gas deposits if deeper drilling is undertaken. In fact, experience has shown that the incremental increases in costs of deeper drilling have been offset by sizeable discoveries of petroleum in areas where success has not been encountered at shallower depths.

A number of smaller countries which are either land-locked or in the interior continental systems have discovered reserves of a magnitude much larger than their own foreseeable requirements, for example, Chad. In such cases, appropriate marketing arrangements as well as distributive systems have to be devised in order to give an impetus to their development.

Coal

The exploration of many known coal-bearing areas is far from complete, and therefore additional reserves may be discovered as a result of intensified exploration, particularly by the use of such new technologies as seismic surveys. At this stage, it is not inappropriate to state that coal resources are poorly known at best; vast areas of the developing world have not yet been explored for coal, and large areas even in the countries now producing coal are not fully explored. Furthermore, the recent advances in coal exploration are such that, even in such classical coal countries as the United Kingdom, substantial new discoveries are being made using new tools of exploration. Developing countries which are known to have substantial reserves of coal include Botswana, Brazil, Colombia, India, Indonesia, Madagascar, Nigeria, Peru, Swaziland, and Zambia. Many more countries are known to possess minor reserves which may, nevertheless, make a significant contribution to their economies if developed and utilized. The exploitation of existing coal deposits, even if some of

[2] E/C.7/108.

them are not of a very high calorific value, will enable a number of countries, which depend heavily on oil imports to meet their electricity needs, to shift to domestic coal, provided modern mining methods are introduced and technologies of coal utilization are implemented. The technology of fluidized bed-burning presents opportunities for the use of small-scale coal deposits, even if they have a high ash content, for the generation of electricity. Such electricity would have to be generated close to the mine in order to avoid both high coal transportation and electricity transmission costs.

Oil Shale

Oil shale deposits have been reported in many developing countries: Chile, Cyprus, Ecuador, Egypt, El Salvador, Gabon, Jordan, Lebanon, Madagascar, Malaysia, Mali, Mexico, Morocco, Niger, Pakistan, Panama, Peru, Somalia, Syria, Thailand, Trinidad and Tobago, Tunisia, Turkey, Uganda, Uruguay, Yemen, and Zaire. The comparatively high investment cost as well as technical problems which are still encountered in exploiting oil shale combine to keep its development at a very slow pace. However, oil shale could be used for the generation of electricity by burning it as a boiler fuel. Projects for the conversion of oil shale to oil have been initiated in a small number of countries since the rise in the price of imported crude oil. The ash remaining after the oil shale has been burned can be converted to bricks or used as an additive to cement or for the production of heavy-weight aggregate for concrete. The potentialities of utilizing small-scale oil shale deposits for power generation, brick making, and other uses can be realized if the oil shale mine is not too far from market.

Tar Sands

These are sands impregnated with heavy, viscous oil. Such sands are known to be widely distributed in the world: Albania, Canada, Colombia, Ecuador, Ghana, Ivory Coast, Malagasay, Nigeria, Peru, Trinidad, and Zaire. The availability of tar sands deposits in many countries, the improvement in techniques now under way in oil recovery, and the comparatively high level of petroleum prices indicate that the production of heavy crude oil has arrived at the point of economic viability. Production costs of less than $10 per barrel for synthetic crude oil from tar sands in remotely located and very severe climatic conditions, such as in the Athabasca region of western Canada, have been reported. It is believed that comparable costs could be achieved in more developing countries with the higher ambient temperatures. It is also possible that exploitation could commence with a relatively simple and unsophisticated plant provided that some means of transporting the product to market were available.

Geothermal Energy

Geothermal energy is the natural heat of the earth, usually extracted in the form of hot water or steam. Geothermal activity is widely distributed in the volcanic areas of the world, particularly in the Circum-Pacific "circle of fire." Volcanic activity is also prevalent in the African rift-valley system. In addition, large areas of the world have a high average thermal gradient, and hot water could be extracted from wells for use in space heating and for agricultural and industrial applications. While the pioneer countries in the application of geothermal energy were Italy, Iceland, New Zealand, and the United States, a number of developing countries have actively pursued geothermal development, including Mexico, El Salvador, Nicaragua, Chile, Turkey, the Philippines, and Kenya. At least 25 to 30 developing countries are believed to have significant geothermal potential.

Geothermal energy has many economic advantages provided the exploration costs do not exceed those of oil. In such cases, geothermal generating stations can produce electricity at a lower operating and capital cost than coal or nuclear stations. Another attractive feature of geothermal electricity development is that plants can be built in the 10-1,000 megawatt range economically and can be expanded as demand grows; this contrasts with the large sizes for economic coal or nuclear plants. Geothermal energy can be used for electricity generation or as process heat in any industry requiring low-grade heat. Examples are pulp and paper making and food processing. The principal advantage of geothermal heat over more conventional sources is its extreme cheapness once the exploratory costs have been amortized. This advantage will be accentuated as the price of conventional fuels rises.

Hydropower

Hydroelectric power has many advantages. It requires no fuel and is perennially replenishable; it can be combined with other aspects of development, such as irrigation and flood control; maintenance and operating costs are low. However, there are some disadvantages also. The capital costs per kilowatt installed can be very high, depending on the site; the hydroelectric power sites are generally located in remote areas requiring costly electrical transmission systems and involving energy losses in the process; fluctuations in river flows affect the firmness of electrical supplies.

There are a number of developing countries which possess vast sources of hydropower which it is technically possible to develop but economically unjustifiable until a very large energy market has been assured. There are a number of cases of somewhat premature development of large-scale hydropower involving large capital investments, but whose utilization has been delayed and severely impeded by lack of power markets. Such projects have become "white elephants." Since foreign loans have been taken to finance such large blocks of investment, the country is burdened with large debts whose servicing has affected its balance of payments. In certain circumstances, it is possible to develop run-of river schemes, which are generally cheaper in capital outlay but have the disadvantage that only a fraction of total available energy from a river may be utilized.

With the rise of imported energy prices and technological developments, the relative economics of small hydropower plants have improved. The potential capacity of small hydropower dams, which can be defined as those less than 100 feet in height, with less than 1,000-acre feet of reservoir storage capacity, and with a potential capacity of less than 5 megawatts, can be quite substantial and can meet local needs in areas not connected with centralized electricity-generating systems. While it can be contended that large, centralized generating stations can provide cheaper electricity, the heavy investments necessary in transmission and distribution lines can be avoided with microhydropower stations serving local areas.

Large hydropower plants involving considerable civil engineering work on dam construction will be ill-suited to the supply of electricity to small rural communities. In some locations, it is possible to install small or microhydro units of limited capacity to exploit the power of nearby small streams or rivers. Such hydropower production utilizing small low-head facilities can supply power to small communities in the process of developing small industries or needing power to lift water for irrigation. Accurate quantitative assessment of the potential of micro-hydro energy resources is difficult; however, there are some estimates of potentials as, for example, in India, where it is considered that mini- and micro-hydro, canal drop, and other low-head methods could yield economic hydro potential equivalent to an annual energy generation of about 25 terawatt hours.

Wind and Solar Energy Utilization

These two energy sources have been utilized in many countries for a variety of limited uses--solar distillation, heat or air-conditioning, pumping water supply--but they have not been taken seriously in the past because of their limitations. Solar energy is too diffused and wind energy often too sporadic. Both sources thus are intermittent and need a supplementary back-up or storage system in addition to the generation plant. But both have several advantages, namely, that they are indigenous resources and that the technology of their utilization can be labor-intensive. Solar energy is at present being utilized in many small installations, mostly in the sunnier parts of the world, for domestic water heating, domestic space heating, and production of fresh water from brackish and saline water. In favorable climates, both domestic hot-water systems and solar stills for fresh water have proven successful technically and in some instances economical. New and innovative approaches for generating electric power from solar heat and storage of such electricity are in the process of development.

The use of wind energy is growing in developing countries for direct pumping of water from low-level sources. Wind energy may be harnessed on a small scale in a number of locations in developing countries where wind conditions are favorable. Before any development of this type is undertaken, areas with suitable wind conditions must be located and detailed meteorological data compiled.

Draft-Animal Power

A neglected but extremely important source of traditional rural energy systems is draft-animal power for plowing and farm operations as well as for drawing vehicles in rural and semi-urban areas.[3] A number of developing

---

[3] Based on information provided by Mr. N. N. Ramaswamy, Director of Indian Institute of Management, Bangalore, India.

countries depend on animal power and will continue to do so for a number of years to come for a variety of reasons: fragmentation of farm holdings so that big tractors are uneconomical to use, lack of road transportation, and small loads. It is broadly estimated tht there could be as many as 200 million draft animals in the developing countries. The draft-animal power system represents an investment of as much as $20 billion at the rate of $1,000 per draft animal. The power capacity of these animals at the rate of ½ horsepower per animal comes to as much as 100 million horse power, which equals approximately 80,000 megawatts and may very well exceed the present installed capacity of electrical power in all the poorer developing countries combined. The replacement costs of the mechanical power represented by these draft animals can be estimated at $100 billion, at the rate of $1,000 per horse power. For example, in the case of India, two-thirds of energy input to the farms is through draft animals; two-thirds of rural transportation is represented by animal-driven vehicles. While bullocks predominate in many Indian Ocean countries, in China water buffalo are more widely used--about 50 million. And camels predominate in the arid regions of North Africa and the Middle East. It is believed that the replacement of animal energy by alternate sources will take at least 25 years, perhaps longer.

The main point to emphasize is the potential for increased productivity of draft-animal power by effecting improvements in the efficiency of harnessing devices, efficiency of plows, carts, and other farm implements, and improving the animal breed and strength through improved animal health care and nutrition--all these improvements are expected to lead to better transport capability, more extensive agriculture, and longer working life for animals. Until such time as the transport infrastructure is improved and alternative sources of energy are available, draft-animal power will coexist with more advanced systems of transport and mechanical power. Animal power is in the decentralized and nonorganized sector, but preliminary studies made with respect to this source of energy in rural areas indicate that the modernization of draft-animal power systems by the manufacture of improved designs of farm implements and carts, training of artisans, and standardization will lead to significant benefits to rural energy systems.

Noncommercial Fuels and Rural Energy Economy

In many developing countries, a significant amount of energy is provided by the utilization of noncommercial fuels which do not normally reach the local energy markets. Of these fuels, the two most important are wood and dried animal dung. Waste vegetable matter is also used and is illustrated by the utilization of bagasse, which can be found in areas growing sugar cane. The efficiency of utilization of these fuels is often extremely low, as wood and dung are burned in open fireplaces. In addition, their use is frequently accompanied by undesirable side effects, such as loss of valuable manure where dung is burned and deforestation accompanied by soil erosion where wood is used. Fumes and smoke, which result from the poor combustion conditions, may also give rise to health and fire problems.

Considerable thought is being given to the possibility of supplying the energy needs of rural communities from indigenous renewable energy resources. This concept involves the application of known and well-tried technologies to conditions in the rural environment. Such an integrated rural energy economy would utilize solar energy for providing hot water, wind power for pumping water and possibly also producing electricity. Animal dung previously burned as fuel would be deposited in anaerobic fermentation plants to produce methane gas for cooking, heating, and lighting. The resulting sludge from the digester plant would form a valuable organic fertilizer for agricultural purposes and thus reduce the need to import chemical fertilizers.

The concept of supplying rural energy needs from local energy sources is relevant to all developing countries experiencing comparatively high delivered costs of improved energy.

In conclusion, the alternative energy technologies which appear most promising to rural areas are:

a. Solar energy for supplying low-temperature heat, for water-heating, and post-harvest operations;
b. Biogas plants to supply heat, provide mechanical energy, and meet lighting needs, either with gas mantles or through electricity;
c. Producer gas plants based on charcoal to run mobile equipment; and
d. Micro-hydro stations to provide electricity.

Combinations of various sources of energy to supply the needs of rural or remote areas must be tested through demonstration projects in a variety of local circumstances in different parts of the developing world.

CHAPTER 12

ENVIRONMENTAL HAZARDS FROM ENERGY PRODUCTION AND USE

Torgny Schütt[1]

SUMMARY

Extensive use of cheap energy, mainly from oil, hydropower, and nuclear fission, has brought welfare to industrialized society like Sweden. Networks for oil and electricity are nationwide and penetrate the whole country.

In Sweden imported oil accounts for around 70 percent of the total energy supply, which makes the country sensitive to external changes in oil prices and restrictions in supply.

The extensive and increasing use of energy also includes some drawbacks. Environmental hazards from the use of oil and nuclear fission energy have become apparent during the last decade.

Increasing oil prices, disturbances in oil supply, foreseeable oil shortages, and environmental hazards from the use of oil and nuclear fission energy, have caused Sweden, like many other countries, to search for alternatives--long-term, renewable, and preferably domestic energy resources with low environmental hazards and acceptable costs.

The environmental hazards that go together with conventional energy resources and with some new alternatives are discussed in this chapter.

ENVIRONMENTAL ENERGY HAZARDS

Because of limitations in physical, technical, human, and organizational performance, economy, and knowledge, our energy systems have, like all other systems, unwanted impacts on the environment together with their designed main performance. These impacts include land and water demands, mechanical impacts, emissions of heat, chemical, and radioactive substances to air, soil and water, acoustic noise, radiation, and many others.

Such primary impacts on the environment sooner or later will affect people, near or far away, through physical, chemical, biological, ecological, or other

---

[1] Swedish Energy Research and Development Commission, Stockholm.

mechanisms. The impacts can be internal to the plant or process and affect employees or external and affect local, regional, national, or global territory and the public. The impacts can be temporary or last for a long period of time and they can strike selectively.

Most systems have a characteristic environmental impact profile of their own. Normally it is practical to distinguish between two profiles, one for the system during normal operation and one for severe and normally rare accidents.

Beside the plant itself also the manufacturing and the scrapping of the plant give rise to environmental impacts which should be considered in connection with the plant.

The final impacts on people can be fatal or nonfatal but cause disability and illness. Impacts on the physical, ecological, or cultural environment can also degrade the support systems of people and their well being. We can thus distinguish between impacts on peoples' security, on their health, and on their environment. This goes for employees and the public as well.

Because it is not possible to determine completely the impacts of a plant--the possible accidents, when they will happen, what the consequences will be, and whom they will strike--the impacts are uncertain and therefore hazardous. Occupational and other accidents usually can be predicted only in a probabilistic way but the probabilities are normally uncertain. This is by necessity the case for rare accidents. The knowledge is usually best and the uncertainties smallest for the early steps in the impact-chain, such as flue-gas emissions, and for frequent impacts, as with many occupational injuries. Comparisons between alternatives are therefore restricted to frequent primary impacts, whenever this is appropriate.

A good starting point for a discussion of environmental impacts of a national energy system like that of Sweden, is the present energy system and its impacts. Then its future developments, improvements, and impacts can be discussed more fully.

THE ENERGY SYSTEM AT PRESENT--SWEDEN, 1978

The present energy balance for Sweden, the production of energy and its use in different sectors are shown in Tables 12-1 to 12-3. The tables show that:

TABLE 12-1. Total energy supply--Sweden, 1978

	PJ	TWh	%
Coal and coke	65.5	18.2	4.0
Oil products	1.135.5	315.4	70.0
Domestic fuels	132.2	36.7	8.1
Electricity from			
Hydropower	207.4	57.6	12.8
Nuclear power	85.9	23.9	5.3
Net import	-3.6	-1.0	-0.2
Total	1.622.9	450.8	100.0

TABLE 12-2. Electricity production--Sweden, 1978

	PJ	TWh	%
Hydro	207.4	57.6	62.1
Nuclear	86.0	23.9	25.7
Heat (oil, gas, etc.)	40.6	11.3	12.2
Total	334.0	92.8	100.0
Net export	3.6	1.0	
Own use and distribution	39.6	11.0	
Losses			
Total supply	290.8	80.8	

TABLE 12-3. Energy end use--Sweden, 1978 (PJ)

	Industry	Transports	Residential and commercial	Total
Electricity	138.1*	7.6	145.1*	290.8
Coal	9.6	-	0.1	9.7
Coke	30.9	-	0.3	31.2
Wood, bark, liquors, wastes	125.0*	-	<1	126.0
Gasoline	-	155.3*	-	155.3
Diesel oil	7.3	63.1*	16.0	86.4
Light oils	-	24.8*	2.2	27.0
Fuel oils--light	27.0	2.2	255.2*	284.4
Fuel oils--heavy	172.8*	1.0	90.4[1]	264.2
Gas	19.0	-	4.1	23.1
District heat (steam and hot water)	8.5	-	79.8*	88.3
Total	538.2	254.0	594.2	1.386.4

[1]Includes consumption for hot water deliveries from mining, quarrying, and manufacturing industries.

*Dominating use.

- Sweden imports about 80 percent of its energy demand, 70 percent as oil, 4 percent as coal, and 5 percent as uranium.
- Oil dominates and is used mainly for residential heating and for production of industrial process-heat.
- Wood, bark, and liquors are used almost entirely for production of process-heat in the pulp and paper industry.
- Gasoline and diesel oil are almost entirely used for transportation, where they dominate.
- Electric power is generated mainly from hydropower and nuclear power; only a small fraction derives from fossil fuels.
- Electricity is used mainly in the residential sector and in industry. In each sector it amounts to about 25 percent of the total energy use in the sector.

ENVIRONMENTAL IMPACTS AT PRESENT

Because Sweden imports 80 percent of its energy demand, most impacts from mining and extraction of energy resources lie outside Sweden, in the exporting countries. Environmental energy impacts inside Sweden are caused mainly by the end use.

The most severe environmental impact of the energy system at present is air pollution, resulting from emissions from oil combustion, and causing lung cancers, bronchitis, and other public diseases, ecological degradation, and severe corrosion and erosion. Occupational and public injuries due to accidents within the energy system are few in number.

Table 12-4 shows the present emissions to air of the whole Swedish energy system in 1975. The figures are uncertain but show the right order of magnitude of the emissions, which impact both public health and the environment.

Public Health Impacts

Many of the emissions, like carbon monoxide, oxides of sulphur and nitrogen, particulates, polycyclic aromatic hydrocarbons (PAH), and heavy metals, cause public health problems, if people are exposed long enough to air in which

TABLE 12-4. Emissions to air from the energy system--Sweden, 1975
(tonnes)

Pollutant	Industry	Transports		Residential and commerce	Power plants
		Gasoline	Diesel oil		
Carbon dioxide	20 000 000	10 000 000	7 000 000	25 000 000	8 000 000
Carbon oxide	2 000	1 300 000*	75 000	5 000	200
Nitrogen oxides	120 000*	100 000*	120 000*	15 000	30 000
Polycyclic aromatic hydrocarbons (PAH) [1]	500	4	3	100*	6
Particulates	10 000	4 000	15 000*	15 000	1 000
Sulphur oxides	300 000*	2 000	15 000	200 000*	70 000
Heavy metals					
Lead	1	1 600*	10	2	1
Cadmium	0.1	-	-	-	-
Mercury	0.05	0.01	0.01	0.05	0.01
Vanadium	100	-	-	150	50*

[1] Benso(a)pyrene constitutes about 10 percent of PAH, which is a part of polycyclic organic matter (POM), which is a part of the unburned hydrocarbons in the flue gases.

*Dominating and critical emissions.

the concentrations of these substances are high enough. About 2,000 new cases of lung cancer are registered annually among Sweden's 8.2 million inhabitants. About 200 of these cases are estimated to be caused by exhaust gases from cars and 50 by emissions from residential heating and other heating plants (1 lung cancer case for each 20 PJ thermal energy produced). PAH and heavy metals are under suspicion as causes. The number of man-days lost because of air pollution is not known but have been estimated to be around 300 for each thermal PJ produced.

About 1,500 of the registered cases of lung cancer are estimated to be caused by smoking in connection with the present air pollution. Chemical pollutants in our environment are believed to cause about 90 percent of the 31,000 new cases of cancer of all kinds which are registered annually in Sweden.

Environmental Impacts

In low air concentrations, pollutants can, by accumulation over many years, cause damage to the natural environment. Acidification of lakes and land ecosystems by emitted sulphur dioxide (SO_2) from oil combustion is at present the most severe environmental problem in Sweden. More than 10,000 lakes are already "dead" or badly disturbed. And this process is accelerating. At present about 800,000 tonnes of SO_2 are annually emitted and also deposited in Sweden. About 75 percent of the emissions are exported over national boundaries and the same amount is also imported by transportation over long distances from different European countries.

Acidic lakes act as sinks for airborne mercury. Together with additional amounts in the lake, the mercury is partly methylized by the microorganisms in the bottom sediments of the lake, dissolved in the acid lake water, taken up by the lower organisms of the lake, and finally accumulated in the fish, which then become toxic for humans. With more than 1 ppm MeHg in fish the lake is "black-listed" and the fish is forbidden for sale. This has happened to many lakes in Sweden. If the pH-value is increased in such a lake, by adding limestone to its water, the mercury will settle out and the lake will recover.

The emitted sulphur also contributes to the acidification of coniferous forest land and is under suspicion of retarding the growth of the forest. This is the subject of present research. Arable land also is affected; the use of limestone for agricultural purposes is steadily increasing. The costs of the corrosion due to acid rain have been estimated at $100 to $200 million a year at present.

The emissions of nitrogen oxides from industry and transportation are also under suspicion of contributing to the acidification problem and are the subject of research.

Recently negotiations have taken place within the U.N. Economic Commission for Europe to solve the regional and international sulphur emissions problem. Some agreements will be signed this month.

Emissions of carbon dioxide from the combustion of fossil fuels contribute to the worldwide emissions of this gas, which increase the atmospheric concentration of carbon dioxide. How this will influence the climate of the world is not yet known. An intensive international research is going on to penetrate this problem.

Emissions of vanadium from large oil-fired plants are under suspicion of retarding growth of plants in the vicinity of the plant, where the vanadium is deposited. This also is the subject of ongoing research.

The problems above were mentioned in order of importance.

DEVELOPMENT OF THE SWEDISH ENERGY SYSTEM

In 1975, the Swedish Parliament approved a number of new measures in the field of energy policy on the basis of a comprehensive Bill presented by the Government. The main objectives of the new policy were to reduce the growth of total final energy consumption and to minimize dependence on imported oil. To these ends, new or reinforced programs in the field of energy conservation and energy R&D were adopted. The Bill emphasized the importance of ensuring improved security of supply in the long term as far as electricity was concerned. The Bill further stressed the need for a more active oil policy and for international energy cooperation.

One of the main tasks of the Government that took office in October 1978 was to elaborate new proposals with regard to long-term Swedish energy policy. A Bill to this effect was introduced in March 1979.

The Bill underlies the imperative need to increase security of supply by reducing dependence on imported oil. It stresses the need to minimize environmental and safety problems associated with all kinds of energy conversion. The importance of energy conservation is further emphasized, as well as the need for an active oil policy and for further efforts to develop durable, preferably renewable and indigenous sources of energy with a minimum of environmental impact.

The Bill confirms the high priority attached to conservation in Swedish energy policy. It is assumed that total final energy consumption will increase at an annual rate of only 0.4 percent to 1 percent during the period ending in 1990, to a large extent due to conservation efforts in the residential and industrial sectors.

As far as the residential sector is concerned, a 10-year program to reduce energy consumption in existing buildings was introduced in 1978. The objective of the program is to achieve, by 1988, savings of 25 to 30 percent, corresponding to 3.4 to 4.1 million tonnes of oil a year.

Grants and loans have been available to houseowners since 1974 for improving insulation and the efficiency of heating systems. New building regulations, establishing minimum thermal efficiency standards for all new buildings, were introduced in 1976.

Due to the measures taken, energy consumption in the residential/commercial sector is actually expected to decrease in the period up to 1990.

Around 25 percent of total heating requirements are presently covered by district heating. The Bill presented in March 1979 stresses the importance of continued district heating expansion.

In the industrial sector a number of measures have been introduced aimed at energy savings. The National Swedish Industrial Board is responsible for information and training in the field of conservation technologies, for state grants to promote the construction of prototypes and demonstration plants, and for the system of state grants to promote increased utilization of more energy-efficient technologies or the substitution of solid fuels for oil in existing industrial facilities. This latter grant scheme has been in operation since 1974. The savings achieved are estimated at 800,000 tonnes of oil equivalent per year.

In addition to these measures, the Government has legal authority to request that energy conservation measures be taken before permission is given for very large industrial investments.

The measures taken in the transport sector include speed limits, progressive taxation according to the weight of the car, energy labelling of all new cars, guidelines for fuel consumption of new cars, and continued measures to support public transportation.

In the Bill a further strengthening of the 3-year energy research and development program adopted in 1978 was proposed. This program is divided into six subprograms, covering industrial processes, transportation, buildings, energy production, general energy system studies, and basic research. The cost of the whole 3-year program has been estimated at 842 million kronor. The biggest subprogram is the one dealing with energy production, where the wind energy program and the program for developing indigenous fuels receive the largest shares. A special effort has been launched to develop solar heating and environmentally acceptable coal technologies.

The Bill envisages increased use of coal in the period up to 1990. Today, very little coal is used in Sweden for energy purposes. It is estimated that 6 to 9 million tonnes may be used in 1990 in industry, in hot water and district heating plants, and in condensing power stations. The use of peat is also planned to increase.

The Energy Bill rejected two preliminary agreements, one of which concerned LNG, for the introduction of natural gas in Sweden. However, there is an interest in examining closely the possibilities of deliveries of gas from the North Sea.

The Bill establishes a target for the expansion of hydroelectric power of 65 TWh in 1990. No change of policy is proposed with regard to the four rivers in Northern Sweden which Parliament has decided to protect.

As far as nuclear energy is concerned, the Bill proposes that the nuclear power program should be limited to 12 reactors, i.e., the units already operating, completed, or under construction.

If the energy balances contained in the Energy Bill were to be achieved, annual oil imports could be reduced from approximately 30 million tonnes in 1979 to around 20 million tonnes in 1990.

The Energy Bill was discussed in Parliament in early June 1979. However, following the accident at the Three Mile Island nuclear power plant, agreement had been reached among the political parties to hold a consultative referendum on the role of nuclear power in the Swedish energy supply system. Against this background, it was decided to defer decisions on the parts of the Bill related to nuclear power until after the referendum. A majority in Parliament was in favor of holding the referendum in March 1980.

Parliament generally approved the measures proposed by the Government in the areas of inter alia conservation, oil, coal, peat, forest residues, natural gas, research and development, and nuclear safety.

Table 12-5 summarizes the energy balance for 1990 according to the estimates contained in the Bill.

TABLE 12-5. Energy balance for Sweden, 1990
(TWh)

	Higher level of consumption	Lower level of consumption
Oil and petroleum products	279-238	239-198
Coal and coke	45-70	45-70
Natural gas	-	-
Hydroelectric power	65	65
Wind power	0-1	0-1
Solar heat	1-3	1-3
Bark, liquors	40	40
Timber waste, energy forests, etc.	12-20	12-20
Waste heat	2	2
Peat	5-10	5-10
Nuclear power (electricity)	55	55
Nuclear power (heating)	6	6
Total supply	510	470
(of which conversion and distribution losses)[1]	50	45
Total final consumption	460	425

[1]Including refinery losses.

We shall in the following focus on such development of the energy system that is also desirable from an environmental point of view.

IMPROVEMENTS OF THE EXISTING SYSTEM

Of highest priority is the development of more efficient and cleaner cars--lighter cars, more efficient engines, fuels without lead, and better exhaust-gas filter systems. Administrative means may be used to prevent high traffic concentrations in cities.

Next comes change to low-sulphur oils. This can be done quite soon. The prescribed standard value for sulphur in light oils will in Sweden be decreased from 0.5 percent at present to 0.3 percent in 1980. For heavy oils, allowed sulphur will be 1 percent in most parts of the country, which corresponds to 240 mg per MJ thermal energy of the fuel, and for coal, 0.8 percent by weight.

Decreasing the hazards of oil transportation is urgent but technical development and international legislation are necessary.

Emissions of heavy metals from oil combustion, e.g., vanadium, should be studied and permissible levels lowered if the metals are found toxic for the ecosystem.

The security of nuclear power plants can be improved by technical and other means.

NEW ENERGY SOURCES

Increasing price and scarcity of oil will sooner or later make the production of gasohols, like ethanol and methanol, economic. The simpler chemical structures of these lead-free fuels will make the exhaust-filtering problem easier to solve. The high emissions of nitrogen oxides from the engines of today thus will be lowered.

New energy systems in the residential sector will be solar heating systems and heat pumps. These new systems involve no significant environmental problems. Increased use of wood and peat for residential heating will probably take place and will involve some environmental hazards. We will discuss these in connection with generation of electric power below.

In industry the use of coal, wood, and peat will increase. To keep the flue-gas emissions at acceptable levels, development of new techniques for cleaning the fuels, for combustion, and for cleaning of the flue-gases must be developed. The most difficult problem perhaps will be to clean the flue-gases of gaseous mercury, a special problem for Sweden. Also the PAH-emissions must be kept low by efficient combustion and other means, without increasing the emissions of nitrogen oxides. Handling of coal and disposal of ash, slag, and the sludge from scrubbers also involve severe environmental problems, which have to be solved. Coal mining, logging, and transportation of coal, wood, and peat involve substantial occupational risks. These will be discussed in connection with power generation, below.

Electric power constitutes about 20 percent of total energy use in Sweden. To generate electric power about 22 percent more hydropower is needed and almost three times as much chemical and nuclear energy as is produced in fuel-fired or nuclear power plants. New energy sources for power production in Sweden are coal, wood, peat, natural gas, and wind power. A comparison of these with conventional energy sources for power generation is presented in the next section, from an environmental point of view.

ENVIRONMENTAL IMPACTS FROM ELECTRIC POWER GENERATION

When generating electric power each energy source has its own characteristic environmental impacts. Any fair comparison of the different sources must involve the whole production line, from mining or extraction to power generation, and also the manufacturing industry for the different plants and processes along the production line.

In the following comparison a standard size of the power plant is chosen for all the energy sources--a plant of 1 GW electric power, with an annual production of 6 TWh (21.6 PJ) electric energy. The manufacturing industry mentioned earlier is excluded from the comparison.

The Power Plant

Common for all power plants is that all except windmills need land and water. The fuel-fired plants emit flue-gases and produce waste products (Table 12-6). The nuclear plant produces radioactive solid waste. The fuel-fired plants have all the same kind of impacts and are because of this quite simple to compare with each other, whereas comparisons with other plants are tricky. From Table 12-6 we find the following characteristics:

- o The hydro- and wind-power plants both use large areas of land, whereas the nuclear power plant uses a small area.

TABLE 12-6. Impacts on the external environment of power plants-- 6 TWh annual electricity production

	Fuel							
	Oil	Coal	Peat	Wood	Natural gas	Nuclear	Hydro	Wind
Fuel consumption, ktonnes/yr	1 500	2 300	5 700	3 400	1 350	0.3	-	-
Specific heat, KWh/tonne	11 500	7 500	3 000	5 000	12 900	6 000 000	-	-
Land use, km^2	3	4	4	4	2	0.5	100*	10*
Emissions to air, tonnes								
Carbon dioxide	5 000 000	7 000 000	6 000 000	6 000 000	5 000 000			
Carbon oxide	200	1 200	10 000*	10 000*	20			
Polycyclic aromatic hydrocarbons	3	10	100*	100*	0			
Nitrogen oxides	20 000	20 000	20 000	20 000	20 000			
Particles	100	1 200	800	200	400			
Sulphur oxides	3 000	7 000	10 000	2 000*	15			
Heavy metals, kg								
Arsenic	-	400*	-	-	-			
Lead	-	2 000*	-	-	-			
Cadmium	-	100*	-	-	-			
Mercury	-	200*	-	-	-			
Vanadium	10.000*	-	-	-	-			
Solid wastes, tonnes								
Ash and slag	2 000	230 000*	200 000*	70 000	-	-	-	-
Desulphurization sludge	250 000*	350 000*	-	-	-	-	-	-
Radioactive wastes	-	-	-	-	-	-	-	-
Low and medium, m^3						500		
High, kg						800*		

[1]The oil contains 1 percent sulphur. Flue-gas cleaning removes 99.5 percent of the particulates and 90 percent of the sulphur.

[2]The coal contains 10 percent ash and 1.5 percent sulphur. Flue-gas cleaning removes 99.5 percent of the particulates and 90 percent of the sulphur.

[3]Flue-gas cleaning removes 99.5 percent of the particulates.

[4]No flue-gas cleaning.

[5]Uncertain values.

[6]Cleaning the flue-gases from sulphur gives a sludge of 10 to 20 percent of the consumed fuel. Here the value 15 percent is used.

*Special characteristics.

- The nuclear power plant produces a relatively small amount of waste but it is strongly carcinogenous, high-level radioactive waste (about 100 million fatal doses), whereas the fuel-fired plants, except the gas-fired, produce large amounts of waste products but these are not radioactive.
- The coal-fired plant produces the largest amounts of waste products. The ash contains among other things about 50 tonnes of such substances as heavy metals that constantly will stay carcinogenous.
- The coal-fired plant also emits large amounts of heavy metals, which are hazardous for both people and the environment. Cadmium and mercury are perhaps the most hazardous for the Swedish environment. Cadmium enters the food-chain through the crops, mercury through fish.
- The oil-fired plant emits vanadium, that is under suspicion of retarding the growth of plants.
- The peat- and wood-fired plants are under suspicion of emitting large amounts of carcinogenous polycyclic aromatic hydrocarbons (PAH)--annually about 100 million fatal doses. Also if only a small fraction of this is emitted and reaches a populated area the consequences can be severe.
- The gas-fired plant is almost free from environmental problems, except for the emission of nitrogen oxides and carbon dioxide, which all fossil-fired plants have in common. Wood does not have this emission. In the case of wood the carbon dioxide is recirculated.

The occupational hazards are small for all plants (Tables 12-7 to 12-14).

Rare disaster-like accidents can happen only with nuclear power plants and with hydropower dams. By using the well-known Rasmussen methodology, which a recent Swedish governmental commission on nuclear safety has found no reason to criticize, the hazards of some Swedish installations have been calculated (Fig. 12-1), from which the mean-values used in Tables 12-12 and 12-13 are calculated. The problem of low probability for very severe consequences is too complex to discuss in this paper. By technological and other developments and measures the hazards can be reduced significantly compared to those shown in Fig. 12-1.

TABLE 12-7. Main hazards in producing electricity from oil

Process	Public hazards					Occupational hazards[1]
	Normal plant oper.		Rare accidents			
	Health	Environ.	Security	Health	Environ.	66/0.7
Extraction						Blow-outs, etc. 20/0.2
Transportation of crude oil			Oil harbor operations			Tanker accidents 15/0.3
Refining						
Transportation of fuel oils						
Power production	Flue-gas emissions					
Transportation of desulph. sludge						Transp. injur. 20/0.08
Waste disposal						

[1]The figures of the table show the number of injuries/number of fatalities of these, when 6 TWh (21.6 PJ) electric energy is produced.

TABLE 12-8. Main hazards in producing electricity from coal

Process	Public hazards					Occupational hazards
	Normal plant oper.		Rare accidents			
	Health	Environ.	Security	Health	Environ.	1000/4.6
Mining		Mining				Injuries 734/3.6
Transportation of coal						Injuries 237/0.8
Power production	Flue-gas emissions					
Transportation of wastes						
Waste disposal	Leakage of heavy metals					

TABLE 12-9. Main hazards in producing electricity from peat

Process	Public hazards					Occupational hazards
	Normal plant oper.		Rare accidents			
	Health	Environ.	Security	Health	Environ.	264/0.8
Extraction		Extraction				
Transportation of peat						Injuries 230/0.7
Power production	Flue-gas emissions					
Transportation of wastes						
Waste disposal						

TABLE 12-10. Main hazards in producing electricity from conventional wood

Process	Public hazards					Occupational hazards
	Normal plant oper.		Rare accidents			
	Health	Environ.	Security	Health	Environ.	1540/3.9
Forestry	Land use and forestry measures					Injuries 1300/3.1
Transportation of wood						Injuries 230/0.7
Power production	Flue-gas emissions					
Transportation of wastes						
Waste disposal						

TABLE 12-11. Main hazards in producing electricity from LNG

Process	Public hazards					Occupational hazards
	Normal plant oper.		Rare accidents			
	Health	Environ.	Security	Health	Environ.	24/0.3
Extraction						Blow-outs, etc. 20/0.2
Transportation of LNG						
Conversion NG-LNG-NG			Gas explosions			
Power production		Nitrogen-oxides emis.				

TABLE 12-12. Main hazards in producing electricity from uranium

Process	Public hazards					Occupational hazards[1]
	Normal plant oper.		Rare accidents			
	Health[2] -/0.1	Environ.	Security	Health	Environ.	39/0.08
Mining		Mining				Injuries 13/0.05
Enrichment						
Nuclear fuel production						
Power production			Core melt down -/0.001	-/0.12		22/0.03[3]
Final waste disposal	Leaching of radioactive substances					
Transportations all						

[1]Occupational injuries from radiation are not included. For the entire production line these are 0.1 fatalities per 6 TWh.

[2]Emissions of carbon 14 from reprocessing are assumed to be negligible.

[3]Injuries during the construction phase are included as 11 accidents and 0.03 fatalities.

TABLE 12-13. Main hazards in producing electricity from hydropower

Process	Public hazards					Occupational hazards
	Normal plant oper.		Rare accidents			
	Health	Environ.	Security	Health	Environ.	18/0.15
Construction		Flooding				Injuries 10/0.15
Power production			Dam failure -/0.12			

TABLE 12-14. Main hazards in producing electricity from wind power

Process	Public hazards					Occupational hazards
	Normal plant oper.		Rare accidents			
	Health	Environ.	Security	Health	Environ.	30/0.2
Construction		Tower				Injuries 20/0.2
Power production						

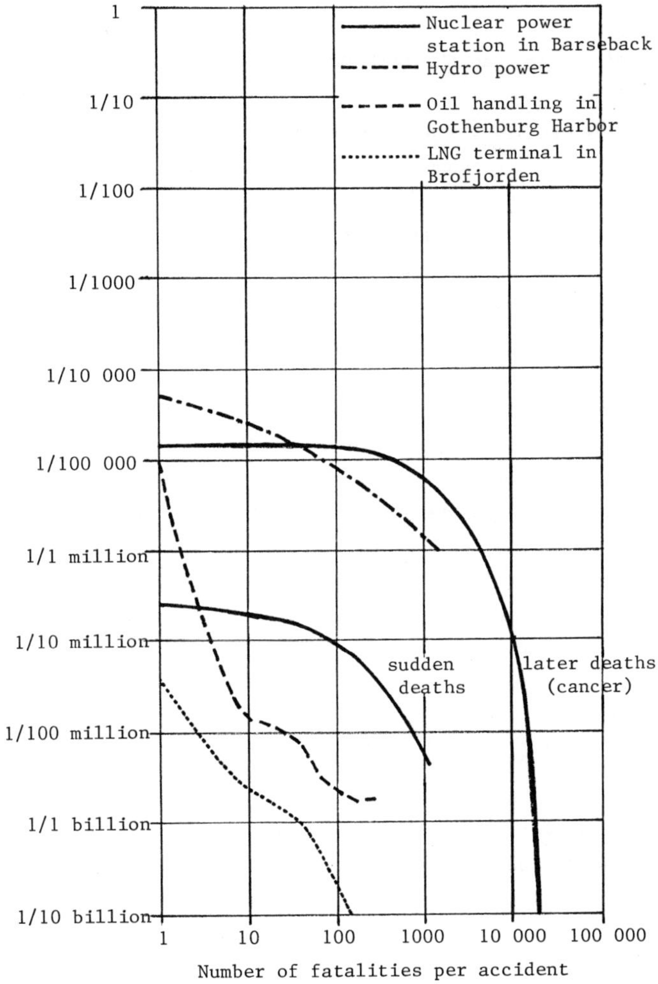

Figure 12-1. Probable fatalities in connection with big accidents

THE WHOLE PRODUCTION LINE

The production lines are different for the different energy sources, and so are the environmental impacts (Tables 12-7 to 12-14). The tables, which are only partial and uncertain descriptions of the corresponding systems, give the following picture:

- Wind power has the smallest environmental impact and few occupational injuries.
- Also hydropower, nuclear power, and gas have small occupational hazards. The risks in connection with LNG-explosions are negligible. The hazards in connection with nuclear power and hydropower are considerably higher but still quite low. The hydropower plant has substantial impacts on the river valley where it is situated.
- Oil has reasonably low occupational hazards, no risks for accidents that threaten the public, but environmental impacts from oil spills and SO_2 emissions.
- Peat has severe flue-gas emissions, is not low occupational hazards, and the extraction of peat has considerable impact on the natural environment.
- Wood from conventional forestry and coal both have high occupational hazards. In the case of coal the hazards lie outside Sweden. Coal has severe emissions. Both energy sources require large land areas for logging (400 km^2 annually), mining (1 km^2 x 2 m annually), and for waste disposal (coal).

CONCLUSIONS

In a complete evaluation of different energy alternatives environmental impacts are only one aspect among many others. In most decisions concerning the energy system environmental impacts are considered more as restrictions than main goals, and can usually be properly defined, modified, and met by research, technical development, or higher accepted costs. For Sweden and many other countries a mixture of different energy sources and systems seems to be a feasible solution, each one with its own advantages and drawbacks. Table 12-5 illustrates the situation.

CHAPTER 13

THREE ALTERNATIVE ENERGY SCENARIOS FOR BRAZIL: 1980-2000

Edgar E. M. Rechtschaffen, Arnaldo V. de Carvalho Jr., Leonardo Goldstein Jr., and Artur Obino[1]

INTRODUCTION

Petroleum importing countries have an urgent need to identify and implement energy policies that ultimately will reduce the dependence on foreign supply of oil.

The processes for future production and utilization of energy vary over time. The dynamics of energy patterns make evident the need for time-phased energy modeling as an analytical basis for energy policy evaluation.

The major features of the Dynamic Energy Model (DEM) are described. DEM is the outgrowth of several methodological efforts (Centro de Tecnologia Promon, 1978) aimed at modeling energy systems in which petroleum-related sectors are adequately detailed.

DEM has been extensively used in a study (Centro de Tecnologia Promon, 1979a) for the Companhia Energética de São Paulo (CESP) (São Paulo Energy Company) and, also, in another study (Centro de Tecnologia Promon, 1979b) for the Ministry of Industry and Commerce (FTI).

In this chapter DEM is applied to analyze three future energy scenarios for Brazil, year by year, for the time horizon from 1980 to 2000.

Each scenario is based on assumptions that realistically incorporate viable energy policies for Brazil, from the economical, technical, and political points of view.

The impact of each scenario, in terms of the cost of displacement of petroleum is also evaluated.

DYNAMIC ENERGY MODEL (DEM)

DEM has been developed to be used as a tool in the assessment of alternative energy policies and for the analysis of future energy scenarios. The

[1]Centro de Tecnologia Promon, CTP, Rio de Janeiro, Brazil.

model is based on a linear time-phased optimization algorithm and is structured to make it flexible.

For the Brazilian case, the internal structure of the energy system incorporates, among others, the following seven major sectors:

- Gasoline
- Diesel oil
- Fuel oil
- Crude oil refining
- Ethanol
- Methanol
- Political-strategic.

As such, some sectors relate to production, some to demand, and others to both.

Major fuels or energy sources considered are petroleum, ethanol, methanol, coal, hydroelectricity, wood, and vegetable oils. Each sector includes its own technical coefficients, equations, and restrictions (limitations). Thus, for the specific analysis of future energy scenarios for Brazil, DEM includes the parameters shown in Table 13-1. The internal structure of DEM is completed by the incorporation of equations that interrelate the different sectors. Some of the most important results provided by DEM for each scenario and over the 21-year time span are listed in Table 13-2.

EXAMPLE OF APPLICATION--BRAZIL

In this chapter DEM is utilized to analyze three alternative future energy scenarios in Brazil. For these applications, the objective function adopted minimizes the yearly accumulated petroleum consumption over the time period under analysis: 1980-2000. Thus, the model reflects a "gradual struggle against the barrel." Major weapons in this struggle were ethanol, methanol, hydroelectricity, coal, wood, and conservation measures.

BACKGROUND

Brazil's present energy supply is heavily dependent on imported petroleum (36 percent of primary energy), with a yearly price tag of about U.S. $4.2 billion in 1978. Estimates for 1979 indicate that the oil import bill may top the U.S. $7 billion figure. Besides strategic considerations, this dependence causes a severe burden on the balance of payments. In 1978, one-third of

TABLE 13-1. Parameters built in DEM

- Profile of the automobile fleet growth
- Profile of the bus and truck fleet growth
- Profile of the demand for diesel oil (in addition to the diesel oil demand of buses and trucks)
- Profile of fuel oil demand
- Performance attained with the different fuels
- Blend of ethanol (up to 20 percent in volume) with gasoline
- Blend of additive (up to 10 percent in volume) with methanol
- Blend of vegetable oil (up to 20 percent in volume) with methanol
- Blend of vegetable oil (up to 30 percent in volume)
- Blend of diesel oil (up to 15 percent in volume) with gasoline
- Profile of the maximum available supply of the alternative sources of energy: coal, wood, hydroelectricity, ethanol, methanol, and vegetable oil
- Efficiencies and improvement of the engine utilization efficiencies with time
- Energy conservation measures
- Predetermined market penetration limits for the new technologies
- Restrictions on the production of neat alcohol automobiles, not to decrease over the years
- Restrictions on the evolution of the neat alcohol car fleet, not subjected to unreasonable increases
- Restrictions on the penetration of coal, wood, and electricity not to decrease over the years
- Upper limits for oil consumption, starting at any given year
- Oil consumption profile
- Minimization of oil derivatives imports; preference given to importation of crude oil
- Oil derivatives available for export
- Stability of production and consumption of alcohols

TABLE 13-2. Results obtained by DEM

- Gasoline refined in the country
- Gasoline consumed by Otto trucks
- Number of Otto engine trucks
- Lack of gasoline
- Surplus of gasoline in a given sector
- Gasoline blended with diesel oil
- Excess of gasoline available for export
- Diesel oil refined in the country
- Surplus or lack of diesel oil
- Fuel oil refined in the country
- Surplus or lack of fuel oil
- Ethanol blended with gasoline
- Ethanol used in automobiles
- Number of automobiles running on neat ethanol
- Ethanol plus additive in buses and trucks
- Ethanol blended with diesel oil
- Ethanol in the diesel sector
- Methanol consumed by automobiles
- Number of neat methanol automobiles
- Methanol plus additives in trucks and buses
- Vegetable oil blended with diesel oil
- Coal, wood, and hydroelectricity penetration in the fuel oil sector
- Number of barrels of oil refined in the country
- Total ethanol consumption
- Total methanol consumption
- Refining scheme profile

Brazil's exports earnings were spent for petroleum imports. In 1979 this might climb to one-half.

A vigorous effort is therefore needed to increase oil, coal, and uranium ore domestic reserves, along with developing nonconventional energy sources, notably biomass-derived fuels, due to their particular importance and suitability to Brazil, considering her favorable climate, available land and water, and social conditions.

One major step toward this objective was the establishment of the "Brazilian National Alcohol Program-PROALCOOL," in November 1975, to promote ethanol production for use in gasoline blends. In 1978, anhydrous ethanol added to gasoline reached 1.5 million m^3, representing 10 percent of the gasoline market and 64 percent of total ethanol production. Estimates for the current year (1979) indicate a total production of over 3 million m^3 of ethanol.

PROFILE OF OIL PRODUCT CONSUMPTION

For substantial savings in oil imports, action should be taken to reduce consumption of the three main products with roughly equal weight in the market, i.e., fuel oil (31 percent), diesel oil (27 percent), and gasoline (25 percent). This is equivalent to a struggle against the whole barrel of crude and not simply the gasoline fraction. Ethanol substitution of gasoline has led to gasoline surpluses instead of a net crude oil savings, since diesel oil and fuel oil demand have increased at higher annual rates (about 10 percent and 8 percent in the period 1975-1978, respectively) relative to gasoline (about 1 percent in the same period).

In order to increase the yield per barrel of diesel oil and fuel oil while diminishing the gasoline yield, switching to heavier crude oil imports, combined with changes in current refining schemes, have been employed with limited success.

CHARACTERIZATION OF SCENARIOS

In Table 13-3, we present the major assumptions that characterize each of the scenarios under study. Table 13-4 indicates the characteristics common to all scenarios.

TABLE 13-3. Fuels for major petroleum sectors

Scenarios	Sectors: Gasoline sector	Diesel oil sector	Fuel oil sector
I	Gasoline Gasoline + ethanol (20%) Ethanol (hydrated)	Diesel Diesel + gasoline (15%) Ethanol + additive (10%)	Fuel oil Coal Wood Electricity Ethanol
II	Gasoline Gasoline + ethanol (20%) Ethanol (hydrated)	Diesel oil Methanol + additive (10%)	Fuel oil Coal Wood
III	Gasoline Gasoline + ethanol (20%) Ethanol (hydrated)	Diesel oil Diesel + gasoline (15%) Gasoline + ethanol (20%) (for Otto trucks)	Fuel oil Coal Wood Electricity

TABLE 13-4. Characteristics common to all scenarios

- Refining flexibility is limited and changes over the years.
- Ethanol added to gasoline does not exceed 20 percent of the blend (1:4).
- Consumption of coal, wood, and hydroelectricity, in fuel oil sector, may not decrease over the years.
- Fleet of vehicles and equipment using straight alcohol shall not decrease over the years.
- Importation of petroleum derivatives is undesirable and will occur only as a last measure.

For Scenario I:

- The fleet of trucks with Otto engines decreases, vanishing after the year 1984.
- Gasoline added to diesel oil does not exceed 15 percent of the blend.
- Ethanol added to diesel oil does not exceed 10 percent of the blend.

For Scenario II:

- The fleet of trucks with Otto engines decreases, vanishing in year 1984.

For Scenario III:

- Gasoline added to diesel oil does not exceed 15 percent of the blend.
- Fleet of trucks with Otto engines decreases until 1984, but may penetrate the market starting 1985.

RESULTS: ENERGY-MIX

Figure 13-1 illustrates, for each scenario, petroleum consumption and the penetration of alcohols, coal, wood, and electricity. Barrels of petroleum displaced, at any year, are obtained from the difference between the dotted and full lines. Thus, in Scenario I, there is an accumulated petroleum displacement of 3,870 million barrels, over the 21-year period. In Scenario II, the accumulated petroleum displacement amounts to 2,420 million barrels and in Scenario III the displacement equals 1,920 million barrels.

Figures 13-2, 13-3, and 13-4 illustrate the profile of production (refining) and consumption of major petroleum derivatives and alcohol penetration in the fuel market, for each of the scenarios.

COST IMPACTS

A "cost of petroleum displacement" comparison was made for the three scenarios. This is the aggregate cost to be paid for the alternative fuels displacing imported petroleum. For that purpose, the cost range of the alternative fuels (as of October 1978) was considered to be (with Exchange rate: U.S. $ = CR $30):

Ethanol: U.S. $ 0.23/liter	-	U.S. $ 0.30/liter
Methanol: U.S. $ 1.13/liter	-	U.S. $ 0.20/liter
Coal: U.S. $ 40/ton	-	U.S. $ 47/ton
Wood: U.S. $ 10/ton	-	U.S. $ 17/ton
Electricity: U.S. $ 27/MWh	-	U.S. $ 33/MWh

The following figures were obtained:

Cost of petroleum displacement:

Scenario I:	U.S. $ 31/barrel	- U.S. $ 39/barrel
Scenario II:	U.S. $ 32/barrel	- U.S. $ 43/barrel
Scenario III:	U.S. $ 35/barrel	- U.S. $ 45/barrel

Petroleum: U.S. $ 27.50/barrel (OPEC FOB price + transportation + refining)

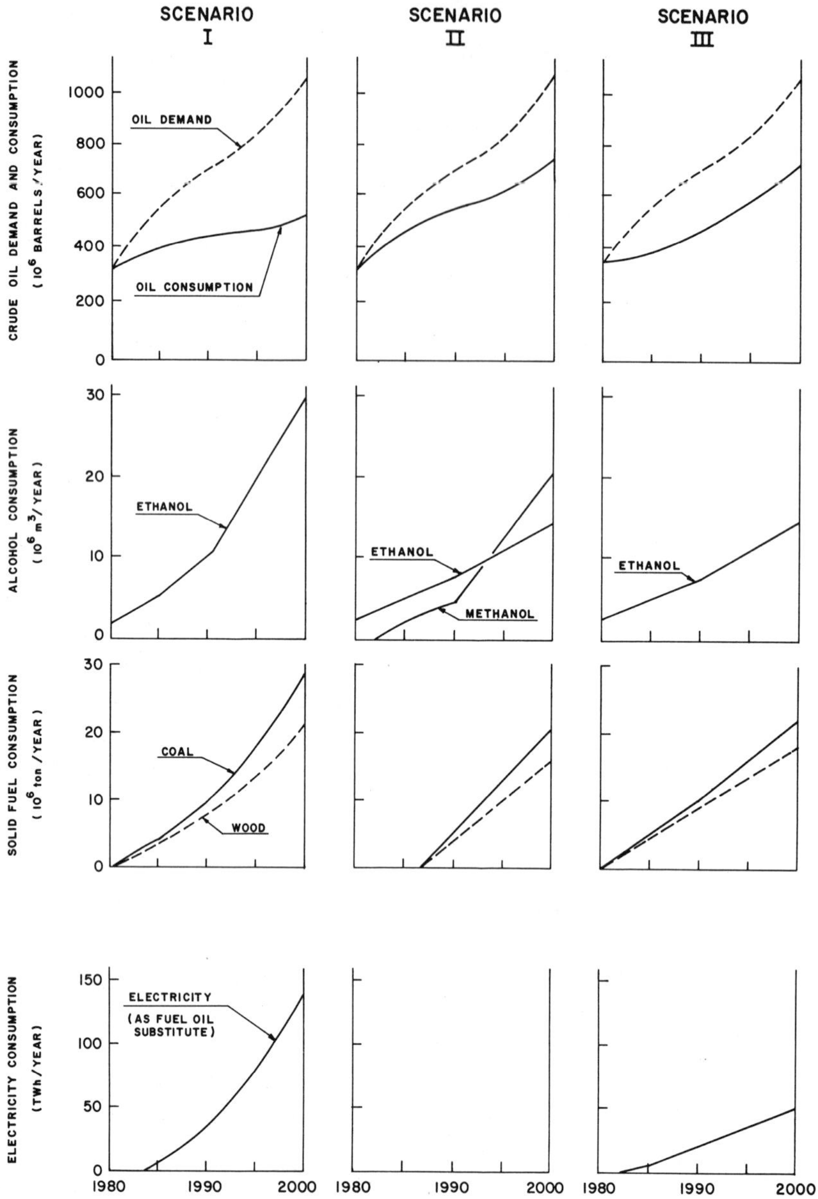

Figure 13-1. Refined petroleum and major substitutes

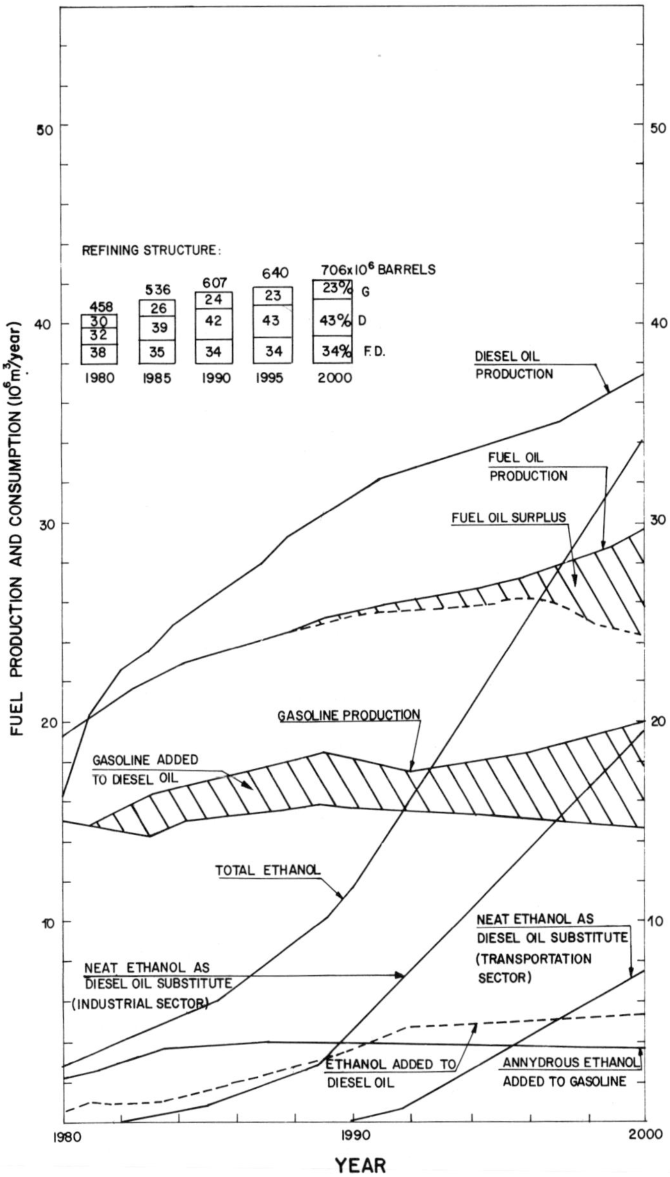

Figure 13-2. Scenario I

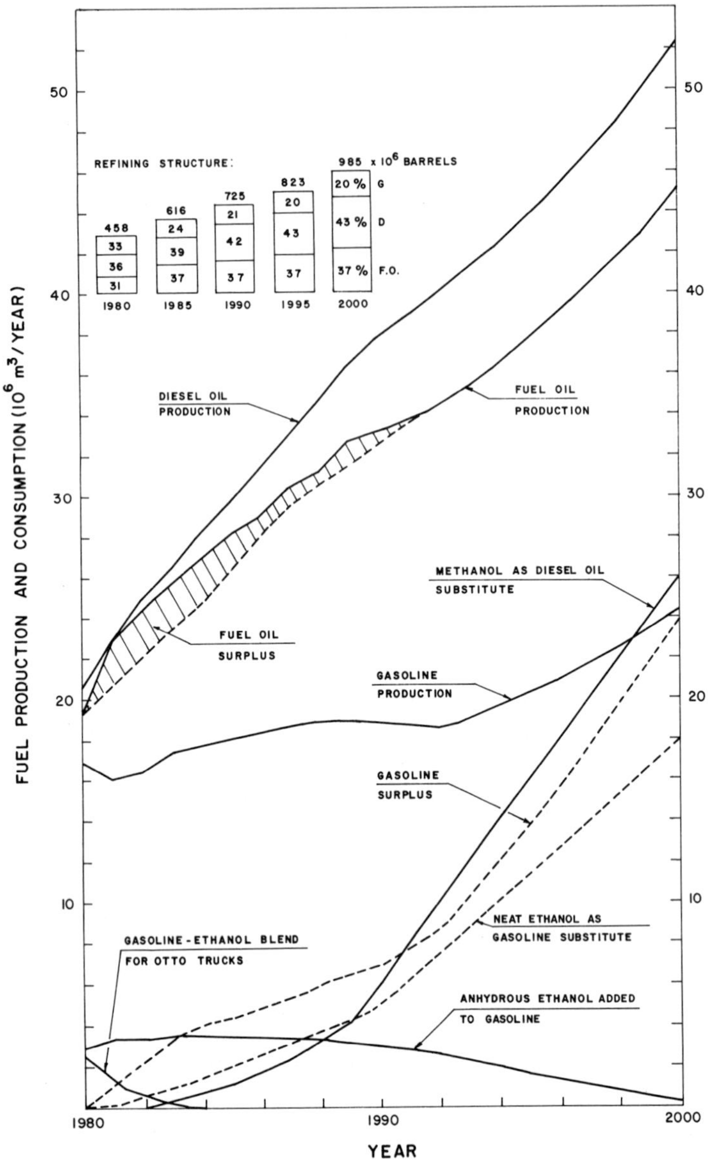

Figure 13-3. Scenario II

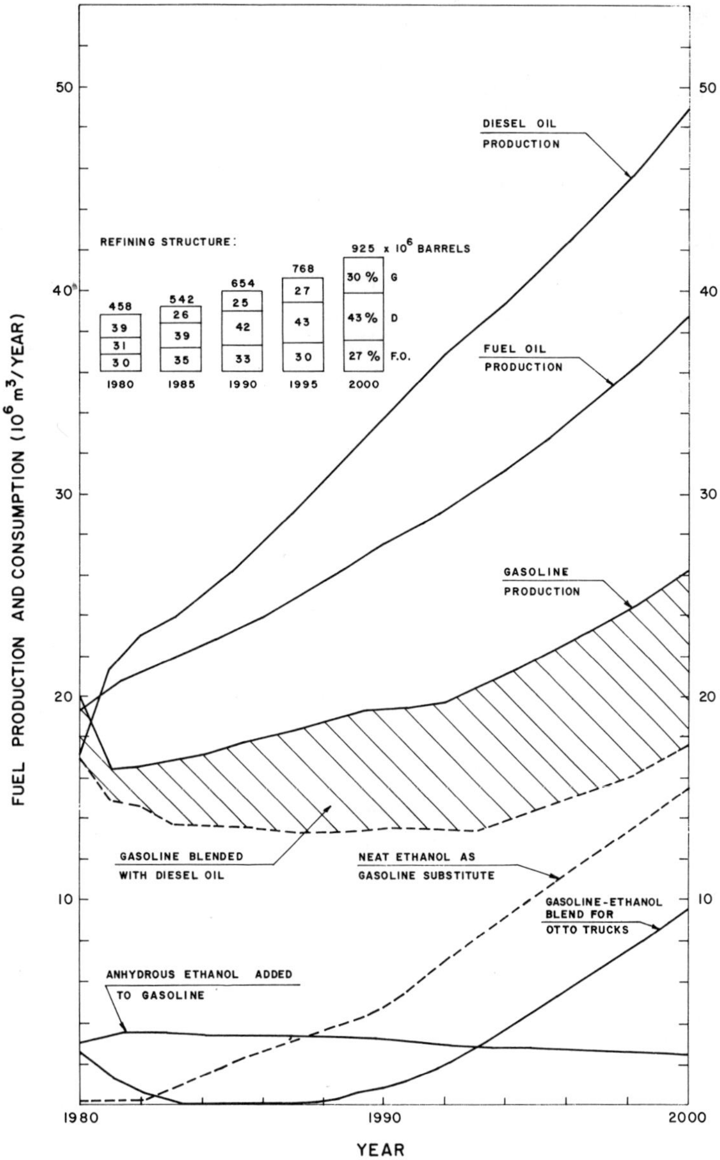

Figure 13-4. Scenario III

SCENARIO CONCLUSIONS

For Brazil, the displacement of petroleum by alternative indigenous sources is on the verge of economical viability. Scenario I analysis indicates that petroleum may be displaced at an aggregate cost varying from U.S. $ 31/barrel to U.S. $ 39/barrel.

COMPUTER SYSTEM

A computer system was developed to perform the analysis of the energy scenarios. This system is comprised of three computer programs:

- CONSISTE: Program to prepare the data for the energy model, in COBOL language for the HP 3000. It also creates an optional file to run the PLIN program.
- PLIN: Finds the optimum energy-mix of the model. Written in FORTRAN language and run on a CDC-6600 computer.
- DOUTPL: Creates graphic outputs from the results obtained by PLIN. It is written in FORTRAN for the HP 3000.

REFERENCES

Centro de Tecnologia Promon, 1978, Sistema de Análise e Acompanhamento de Projetos sobre fontes e Formas Alternativas de Energia: Ministério de Minas e Energia--SETEC.

Centro de Tecnologia Promon, 1979a, Estratégias de Penetração do Metanol no Mercado Energético Brasileiro: Companhia Energética de S. Paulo--CESP.

Centro de Tecnologia Promon, 1979b, Diagnóstico do PTE/BIOMASSA--Programa Tecnológico Industrial de Alternativas Energéticas de Origem Vegetal: Ministério de Indústria e Comércio (FTI).

CHAPTER 14

LONG-TERM ENERGY PLANNING FOR ALGERIA

Entreprise Nationale Sonatrach[1]

INTRODUCTION

Every concept of development seeks to ensure the fulfillment of the human being. Such fulfillment is achieved through the improvement of living conditions as well as through social and cultural progress. These goals are of necessity achieved through substantial economic development. It is for that reason that Algeria is continuing and intensifying its efforts to develop a strong, integrated, and independent economy.

Consequently a policy of intensive, overall industrialization is being followed in order to facilitate Algeria's speedy emergence from a state of underdevelopment and to give substance to the concept of economic independence.

Agricultural development is also an important goal to the extent that it will permit Algeria to achieve a higher nutritional level. This objective will be reached through an increase in the area planted as well as by soil improvement, mechanization, and the choice of appropriate techniques.

While development strategy therefore stresses the means for laying the foundations of a strong and self-sustained economy, it is equally true that emphasis must be laid on actions designed to improve the living standard of the population and to promote continuing social and cultural progress. This is done best through broadened education and professional training, and through such social means as the introduction of free medical care, and the development of cultural and sporting activities.

Accordingly, the existence of substantial resources of hydrocarbons makes it possible for a developing country like Algeria to utilize this resource as a powerful lever in achieving all these goals, which depend to a large extent on the development of an appropriate energy policy. Such a policy, implemented by means of optimum energy planning, will make it possible for the hydrocarbon sector to be an effective driving force in the development of the overall economy.

---

[1]Marketing Division, 46 Boulevard Mohamed V, Algiers.

This energy policy must therefore be based on three main principles:

- The need to build up financial resources through a production and valorization policy for hydrocarbons.
- The need to ensure the country's energy supplies in the short, medium, and long terms at minimal costs, while bearing in mind the needs of a rapidly growing economy.
- The need to ensure that hydrocarbon reserves are managed in the best possible manner by means of an appropriate exploration and conservation policy, so as to ensure development for the benefit of future generations.

However, the realization of these goals for the energy sector will not be possible without a new framework to replace the structure inherited from the colonial period. This is the goal on which Algeria will concentrate at first, with a view to developing and consolidating new infrastructures for achieving the established objectives.

HYDROCARBONS AS A SOURCE FOR THE FUTURE

The employment of the country's natural wealth for development and economic independence requires, as first priorities, the retrieval of such wealth, the improvement of the terms of trade, control of the means of production, and the valorization of these resources.

Resource Recovery

Without full and complete sovereignty over its natural wealth, a country cannot hope to achieve development and progress; accordingly, the cornerstone of Algeria's development strategy has always been its policy for the recovery of national wealth.

This has been realized by:

- Nationalization in 1971 of hydrocarbons and operations concerned with their exploitation.
- Distribution monopoly since 1968 for oil products and other liquid or gaseous hydrocarbon derivatives.

The recovery of natural wealth without control over the price mechanism would make such recovery an illusion without significance. Moreover, the lack of national mechanisms for retrieval would also be a serious impediment in the way of the effectiveness of such measures. Algeria therefore has always attempted to implement these two principles in order to strengthen further its control of the sector.

Price Control

The pricing mechanism employed in the past made it possible to transfer the bulk of the wealth of the developing countries to the developed countries. In that connection, action taken by Algeria in OPEC sought to ensure that organization's control over the fixing of prices as well as the continuing revalorization of the basic price of hydrocarbons. This policy could not be separated from action for the development of national wealth and indeed gave such action its full significance.

Control over hydrocarbon prices thus became an essential element in commodity revalorization and consequently a prime force for improving the terms of trade.

Introduction of National Mechanisms for Development

The achievement of the goals fixed by Government policy in the energy sector was made the responsibility of two State enterprises, under the aegis of the Ministry for Energy and the Petrochemical Industries:

- The national entity for exploration, production, transportation, conversion, and marketing of hydrocarbons (Sonatrach).
- The national entity for electricity and gas (Sonelgaz).

Achievement of the established goals was nevertheless conditional on the availability of a sufficient number of trained personnel capable of taking charge of all these activities.

The highly technical character of the energy field and the pressing need for a large number of qualified staff at all levels led to the adoption of a continuing, intensive training policy. Apart from the training afforded by the universities, a variety of mechanisms has been brought into being under the aegis of the Ministry for Energy and the Petrochemical Industries:

- Establishment in 1964 of the National Hydrocarbon and Chemical Institute (INHC).
- Establishment in 1965 of the Algerian Petroleum Institute (IAP).

- Establishment and development of technical schools within the national entities.

Hydrocarbon Valorization

Algeria has abundant hydrocarbon resources, consisting basically of crude oil, gas, condensate, and liquefied petroleum gas.

Crude Oil

Algeria's proven reserves reached 1 billion tons on 1 January 1979. Production during the year 1978 reached almost 55 million tons. Continuing exploration hopefully will lead to a substantial increase in reserves.

Currently the bulk of crude oil production is exported while the remainder is processed at the three refineries in Algiers, Arzew, and Hassi Messaoud, with a total capacity of 5.5 million tons, which will rise to 20 million tons in 1979, once the Skikda refinery is completed.

Future crude oil production prospects must take into account:

- The maintenance of refinery production.
- The reduction of crude exports in step with the growth of refinery capacity and lower production.
- Substitution of crude exports by exports of refined products once refinery production exceeds national demand.

The development plan for crude oil will therefore permit the product's valorization within the country in the form of refined products which will not only satisfy national demand but will also be a source of foreign exchange from exports.

At this point it should be emphasized that, for Algeria, oil is not only an export commodity but also the chosen instrument which will help fulfill aspirations for development of the country. Oil will not only produce the financial means for development but will also be a source of energy. It is for that reason that the price level, which establishes national income level, must be viewed in terms of the country's purchasing power. In that connection the discriminatory price policy for capital goods and services applied to the OPEC countries must be remembered, in that the index of import prices has reflected 25 to 30 percent inflation since 1974, with the consequence that development efforts have been totally negated.

For that reason Algeria has always supported a progressive price policy based on the maintenance of purchasing power and the recovery of the cost of substitute forms of energy. Such a policy has the merit of avoiding sudden adjustments resulting from crises, as has happened in the recent past.

Such an approach would make a major contribution to solving the crisis and would facilitate the better absorption of the economic impact of such an increase by consumer countries, as well as better planning of the price of energy and a decisive incentive to the development of alternative energy sources.

Accordingly, the development concerns of the producer countries, the preoccupation of the industrialized countries with regard to energy supplies, and the development of replacement sources of energy could be resolved within the framework of a worldwide agreement which could not fail to lead to the evolution of a fair and equitable price.

Natural Gas

Natural gas represents Algeria's most substantial hydrocarbon resource. Proven reserves are estimated at over 3,000 billion m^3 and have made it possible to consider an ambitious development program. Such a program covers both exports (nearly two-thirds of reserves) as well as the country's internal needs, which of course will escalate. Exports are expected to take the form of both liquefied natural gas (LNG) and natural gas and will require a substantial infrastructure, from processing units in the field to the installation of liquefaction plants on the coast and construction of international pipelines.

These developments will require enormous investment by the country but they should also obtain for the country the resources needed for development and its economic takeoff. The current low price level for gas, however, must be emphasized. In terms of heating equivalent, the current price of crude oil is between double and triple the FOB price of natural gas. The discrepancy is even greater in the case of gas extracted from coal.

There is no question but that from the present time until the end of the century, gas will make an important contribution to world energy consumption, bearing in mind the declining share of crude oil in such consumption. The delays which will be required before gas can be made available to the consumer must be stressed. The contract for the delivery of gas to El Paso, signed in 1969, only became operational during 1978, almost 10 years later. From the standpoint of the supply of gas, decisions taken at the present time will only be implemented within the framework of the next decade. A lesson must be learned also from the past evolution of prices for crude, insofar as the price of gas must rise progressively as from the present time in order to avoid the crises which inevitably result from sudden and substantial readjustments.

Gas production must accordingly be supported by sales prices which are sufficiently attractive for the producer to permit the enormous investments needed for the development of this energy source. Moreover, a sufficiently attractive price will certainly encourage additional exploration and development of this energy source; in Algeria there are still substantial prospects for discovering new gas reserves. Such new reserves would increase the energy potential of Algeria and could also make a substantial contribution to the energy requirements of consumer countries: a contribution from those same consumer countries to exploration efforts in Algeria could only be to the mutual benefit of Algeria and its partners.

In conclusion, all those elements favor the optimum management of our hydrocarbon resources in such a way as to facilitate the achievements of the goals both of producer and consumer countries.

Condensate

On 1 January 1979, proven reserves of condensate were estimated at 450 million tons. For Algeria this represents an increasingly important source of foreign exchange. Production in 1978 rose to 2.4 million tons and by 1985, 21 million tons will be produced at Hassi R'm'el and transported to Arzew.

Liquefied Petroleum Gas

Proven reserves of liquefied petroleum gas reached 200 million tons on 1 January 1979. Production is expected to increase rapidly to almost 4 million tons in 1980.

SATISFACTION OF NATIONAL DEMAND

The Demand for Energy

The industrialization of the country, coupled with improvement in the living standard, means that the country's energy needs will increase rapidly.

National energy consumption will accordingly rise from 14.5 Mtep in 1979 to 31 Mtep in 1983, or an annual growth rate of 21 percent, then from 43 Mtep in 1985 to 64 Mtep in 1990, or an annual growth rate of 8.5 percent. The average annual growth rate from 1979 to 1990 is therefore 14.5 percent.

The rapid growth in consumption and its quantitative long-term importance indicates Algeria's interest in managing its hydrocarbon resources prudently in order to be able to satisfy the country's own long-term energy demand.

Moreover, the choice of an energy form for national consumption must be linked to a number of criteria, including:

- The total cost of the energy product made available for use.
- The comparative impact on development of making different energy forms generally available to users.
- The loss of foreign currency earnings resulting from national consumption of otherwise exportable energy.

Balancing these criteria should make it possible to arrive at an optimum long-term consumption pattern.

Demand by Product

LNG. Algeria had substantial quantities of LNG available at the beginning of 1980; consumption of LNG has also been greatly stimulated by the increased number of processing plants and distribution centers as well as by the relatively low sales price to consumers over the past 10 years. Substitution of LNG for refined products and for firewood has been systematically sought and encouraged.

Consumption increased at an annual rate of 15 percent to 1980. From 1980 to 1983 the rate will be 12 percent and from 1983 to 1985, 10 percent. Thereafter it will stabilize at about 5 percent annually.

Light Petroleum Products. Bearing in mind the increase of the automobile population, demand for gasoline will grow rapidly to an annual rate of 12 percent by 1980 and thereafter will stabilize at 10 percent until 1990.

Likewise, the expansion programs of the air carriers will lead to a strong demand for aviation fuel. This will grow at the annual rate of 14 percent until 1985 and will stabilize around 10 percent thereafter.

Heavy Petroleum Products. The demand for gas oil will grow at an annual rate of 12 percent and for diesel oil (FOD), 10 percent.

"Marine" fuel has escalated at a rate of 50 percent until 1980, 25 percent from 1980 to 1983, and 15 percent after 1985. Such growth is linked to the development of the merchant fleet, LNG carriers, and oil tankers.

After 1985, the consumption growth-pattern for oil products will stabilize at 10 percent for LNG, 20 percent for light products, and 70 percent for heavy products.

Bitumens and Lubricants. As a consequence of the development of the road network and of the substantial increase in building activity in coming years, demand for bitumens will increase at an annual rate of 18 percent until 1981; thereafter the rate will drop to 12 percent. Lubricants are also expected to show a substantial increase in growth, reflecting the nation's automobile population.

Natural Gas. The country's abundant resources of natural gas mean that this energy source is called upon to play an increasingly important role in satisfying final demand. The percentage of final demand met by natural gas will rise from 16 percent in 1977 to 42 percent in 1990. The growing role of natural gas can be observed at three levels:

- It will provide the fuel for the electric power stations which have been planned.
- It will be the main energy source for large industrial projects.
- It will provide the bulk of the energy demand of private homes.

Electricity. Increased consumption of electrical energy will lead to a rise in installed capacity from 1,374 megawatts in 1978 to 3,388 in 1985, a virtual tripling of capacity. The development of electricity production is basically a response to two criteria:

- Satisfaction of industrial demand for electrical energy.
- Total electrification of the country by 1987, regarded as a basic choice for the country. This will make electricity available in every home, with substantial impact on living conditions and an improved cultural level.

Evolution of Energy Consumption Per Capita. Estimates of national energy consumption indicate that energy consumption (excluding self-consumption by liquefaction plants) *per capita* will develop as follows: 0.6 Tep in 1980, 1.30 Tep in 1990, or an annual growth rate of 8.5 percent.

This level will approximate that of the Netherlands or Hungary in 1955 (1.68 and 1.32). In 2005 consumption *per capita* will reach 2.6 Tep, so that, on the basis of the population, total consumption will be of the order of 110 Mtep. This figure demonstrates the importance of Algeria's long-term energy needs and the need for prudent management of hydrocarbon resources.

Sectoral Demand

The Industrial Sector. Consumption by the industrial sector will more than double between 1979 and 1983, in particular because of the consumption of the construction materials and siderurgical branches. During the period, consumption will rise from 2.7 Mtep in 1979 to 6.3 Mtep in 1983, representing an annual growth rate of 22.7 percent. By 1990, such consumption will reach 19 Mtep or a growth rate of 14.4 percent between 1985 and 1990.

Natural gas will make the largest contribution to the pattern of consumption of this sector and its share will rise from 44 percent in 1979 to 51.3 percent in 1983. This share will rise further after 1985 and should reach 64 percent by 1990.

The Transportation Sector. The growth of this sector's consumption (excluding inventory) will be at the annual rate of 11.4 percent from 1979 to 1983 and 6 percent thereafter. This growth will be reflected both in light oil products, because of the rise of the light vehicle population and increased air traffic, and in gas oil.

The "Households and Other" Sector. Consumption by this sector will rise from 2.9 Mtep in 1979 to 4.5 Mtep in 1983, or an annual growth rate of 11.6 percent. It will reach 5.5 Mtep in 1985 and 9 Mtep in 1990. The pattern of consumption will be dominated by heavy petroleum products (FOD) which will supply about 35 percent. LNG's share will be about 25 percent, while electricity and gas will account, respectively, for 22 percent and 18 percent of total consumption.

Evolution of the Pattern of Consumption. The industrial sector consumed 35 percent of total energy in 1979 but this share will rise to 54 percent in 1990.

The transportation sector's share will drop from 27 percent in 1979 to about 20 percent in 1990.

The share of the "households and other" sector will decline from 38 percent to 26 percent. This pattern will be closely similar to that of Spain in 1974, with 54 percent for industry, 26 percent for transportation, and 20 percent for households.

Mechanisms for Implementing Energy Policy

Infrastructure

Distribution of Oil Products. In addition to the need for production, an essential condition for the satisfaction of needs is the existence of an appropriate distribution network and accordingly such a network will be developed through the installation of integrated distribution centers with a total storage capacity of 1.5 Mt. Transportation facilities will also be expanded substantially. As Algeria is a relatively large country, the entire storage and distribution network will need to be increased considerably and such expansion will call for total investment in the order of 4,300 million Algerian dinars for the whole period.

The Country's Gas Policy. Estimates of energy demand have demonstrated the growing role which natural gas will play in the supply of energy to the national economy. The development of a public distribution program will make gas available to almost 95 percent of homes by 1985. This will make it possible by 1985 to supply all population centers which had over 8,500 inhabitants in 1978. The public distribution program will reach 84 such population centers. Installation of 3,160 km of pipeline will be needed, and total investment will come to about 740 million Algerian dinars.

Electrification of the Country. The expansion and strengthening of the production and distribution of electric power will be achieved through:

- The construction of new electric power stations which will cover the entire country.
- The installation of new high-tension transportation lines which will ensure supplies to new industrial centers in the interior of the country.
- The total electrification of the country between now and 1987, regarded as a fundamental goal of the country.

The goal of total electrification implies the installation of an additional 12,100 stations, the link-up with a further 1,200,000 homes representing 9 million inhabitants, and the construction of 58,000 km of powerline and 11,700 transformer stations.

Total credits required for these works will be of the order of 5.5 billion Algerian dinars.

Energy Tariff Policy

A tariff policy can only be applied effectively when there is in existence a distribution infrastructure capable of delivering energy products in sufficient quantities simultaneously throughout the whole country. This obstacle can be removed by the installation of the infrastructure described above. In addition, the question of establishing tariffs for energy products plays an essential part in the definition of energy policy. It should:

- Make it possible to satisfy, at the lowest cost and in the long term, the growing energy needs of the various sectors of the national economy.
- Encourage consumption of the most suitable and plentiful forms of energy available in the country. Thus, the price of natural gas fell by 50 percent in 1968 and has not risen subsequently.
- Make cheap energy available to consumers so as to safeguard their purchasing power and to establish uniform prices throughout the national territory in harmony with the principle of regional equilibrium.
- Fight inflation through stabilization of the cost of the energy element in the production costs of other goods.
- Avoid waste through better energy use.

DEVELOPMENT OF NEW ENERGY AND POLICY IN REGARD TO EXPLORATION FOR AND CONSERVATION OF RESERVES

The Role of New Sources of Energy

With a view to strengthening the country's energy capacity to meet future demand, other sources of energy are under study.

The geographical and climatic situation of the country makes it possible to be optimistic regarding the future development of solar energy. At the present time studies and research are under way throughout the country, together with such pilot operations as solar houses.

Furthermore, the availability of uranium reserves makes it possible to think of the post-hydrocarbon period in terms of the future development of nuclear power plants. In that connection, in addition to an existing center for nuclear sciences and technology, a large-scale study and training program for specialists has been started.

Exploration and Conservation Policy

> State action in this field is essentially designed to increase the rhythm of production and to enhance the size of available known reserves on a continuing basis. In fact, while present production dominates the level of income received, the state of power reserves is a safeguard for the future.... The growth of the nation's mineral reserves requires intensified prospecting for underground wealth, both in the interior and in the maritime zones.
>
> --Extract from the National Charter

Accordingly, exploration represents another important element in energy planning in Algeria. During the 1973-1977 period, an average of 23 exploration wells were drilled annually. Seismic activity reached an annual level of almost 300 crew-months during the same period. Nevertheless, bearing the country's size in mind, an additional effort is required, insofar as something less than 50,000 Algerian dinars per km^2 has been spent. Such an effort cannot fail to achieve positive results because the rate of success reached by Algeria by exploratory drilling during the 1972-1976 period has been at least as favorable as that achieved by other North African producers during the same period. Furthermore, the success rate for gas has been very satisfactory and encourages hopes for substantial results in the long term.

While an intense exploration effort will make it possible to maintain, if not increase, reserves, the use of secondary recovery techniques will make increased recovery from existing fields possible, with a 1 percent average gain representing the equivalent of 2 years' production at current rates.

CONCLUSION

The elements described above, taken as a whole, illustrate the complexity of the task faced by a developing country with hydrocarbon resources, when it has to draft an optimal energy plan for the long term.

First, an accurate evaluation of its energy potential, both by exploration for hydrocarbons and the possible development of other sources of energy, presents difficulties, in particular because of the uncertain character of future discoveries. Such an evaluation must then be equated with national requirements for cheap energy. It is only then that a decision can be made on the proper level for hydrocarbon exports and the extent of their valorization. While adjustable short-term sales contracts represent a flexible framework for liquid hydrocarbons, the same is not true for gas exports, which are made under long-term contracts requiring very heavy investments and the development of a specific infrastructure.

An optimal strategy must therefore be drafted which will take such constraints and rigidities into account and which will lead to a long-term production and valorization policy to safeguard the country's own supply needs on a continuing basis and to satisfy for as long as possible its growing energy requirements; it is also essential to give full weight to the need to guarantee the country's long-term independence in energy matters, in particular once industry has been developed and the country has achieved its economic takeoff.

CHAPTER 15

MOTIVATION AND ATTITUDES IN THE LONG-TERM PLANNING OF ALTERNATIVE ENERGY STRUCTURE

Ioan Ursu and Dan Vamanu

Romania's energy policy has proved perceptive with respect to the role of diversity in supply and technologies.

After World War II, diversity of supply was the obvious strategy to match the tremendous demand for energy in the rapidly growing economy of Romania, in view of the diverse but modest natural endowment of primary energy sources and the desire for economic and energy independence. Traditionally well known as an oil-producing country, Romania valued this richness and resisted the temptation of an oil monoculture.

Encouraging power generation from natural gas, extensive district heating and mass transportation, standard double glazing of windows, brick buildings, and a certain permanent trend to energy austerity, suggests a positive attitude with respect to the energy management and stands as an asset of future energy policy.

Nevertheless, Romania faces today serious energy problems. Approaching the limits of primary recovery, the oil reserves look limited and their prospects modest. A highly efficient chemical industry is ready to absorb and turn to account even large quantities of oil and methane. More and more, hydrocarbons are beneficially diverted to nonenergy uses, which already reach the considerable level of 10.5 percent, as compared to the 11 percent and 12 percent target figures for North America and Western Europe in 1980. A net exporter before the War, Romania now imports a quantity of oil nearly equal to its domestic production. A tense world market and skyrocketting prices exacerbate the oil-supply problem.

With a key role in the formation of the GNP, 57.9 percent of the national income in 1978 [1], the industry is the largest energy consumer in Romania, with more than 60 percent of the total demand. Efficiency and good management of energy play a very important role. Starting from scratch after World War II with its infrastructure, the country has had to rely largely on energy-intensive branches, such as metallurgy and building materials. Thus, structural problems of the economy make the pattern of domestic energy even more complex.

For the next decade, Romania has made a firm commitment to a strategy of intense, multilateral development. Table 15-1 shows some relevant indicators for the first 5 years of the 80's. To reach, in 1990, a national income of $3,000-$3,500 U.S. per capita for a projected 25 million inhabitants, the rate of net accumulation must be about 30 percent a year. Unlike the previous phase of growth, the economy will take the qualitative approach: capital investments will go mainly to modernization, increased productivity, competition, and efficiency. An increased share will go to expanding services and to improving the standard of living and quality of life in a balanced way throughout the country.

A centrally planned economy, Romania will build its development in the next decade according to a master plan [2]. The label "decade of science, technology, quality and efficiency" describes in short the ends and means of this program. To stress high reliance on science and technology for development, two additional master programs have been put forward; one is devoted to scientific research, technological development, and application of technical progress, in the decade 1981-1990 and as guidelines for the year 2000 [3]. The second is the national energy master program, known as "Energy Research and Development Programme for 1981-1990 and the Principal Guidelines until the Year 2000" [4]. These drafts of July 1979 encompass the 5-year period 1981-1985.

The program sets, as basic tasks, geological research, discovery and utilization of new energy sources, expansion and improved organization of the national power grid, efficient management of the energy potential of the country, and rational, constantly lower energy consumption in all fields of activity.

The program goal is unambiguously formulated: "Underlying the energy research and development programme is the idea that in the next decade Romania should achieve self-sufficiency in fuel and energy." To attain that end, two approaches were initiated, diversity of supply and technology substitution.

A central target is the thorough exploration of all zones of the country to discover new coal, crude oil, gas, and radioactive mineral deposits, and the exploitation of the energy of the sun, the wind, thermal waters, and biomass. Why all of them? Because they are there, and to reduce significantly the share of hydrocarbons in the country's energy production.

TABLE 15-1. Basic orientations and goals of Romania's development under the 1981-1985 Five-Year Plan [2]

	Provisions for 1985 as against the 1980 plan	Average annual rate 1981-1985 (%)
1. Social product	133.5-137.5	6.0-6.6
2. National income	138.0-143.0	6.7-7.4
3. Net industrial production	154.0-161.0	9.0-10.0
4. Gross industrial production	147.0-154.0	8.0-9.0
5. Gross agricultural production (annual average for the 5-year period)	124.5-127.5	4.5-5.0
6. Overall investment in economy (in 5 years)	130.0-135.0	5.4-6.2
7. Foreign trade volume (in 5 years)	150.0-157.5	8.5-9.5
8. Number of personnel	110.5-112.5	2.0-2.4
9. Labor productivity (computed on a net production basis):		
- in industry	140.0-143.5	7.0-7.5
- in construction-assembly	130.0-135.0	5.4-6.2
10. Reduction of expenditure for 1,000 lei worth marketable production in the national industry	7-8	
11. Total real incomes of the population	123.0-125.0	4.2-4.6
12. Total remuneration fund	141.0-143.0	7.1-7.4
13. Average real remuneration	116.0-118.0	3.0-3.4
14. Real incomes of peasantry from cooperative farms and private farmsteads per active person	120.0-123.0	3.7-4.2
15. Retail sales	130.0-135.0	5.4-6.2
16. Services for the population	147.0-154.0	8.0-9.0

Already diverse and nonexclusive as regards oil, the energy mix of the country is to undergo further shifts to more abundant primary sources, which are neither low-cost nor free of problems.

Prime competitors are coal and bituminous shales, due to increase their share in the structure of electric power production from 40 percent in 1980 up to 55 percent or more in 1985. To that end, the new thermal power stations will be designed to operate on lignite and bituminous shales, while providing conditions for the present liquid-hydrocarbon-fueled stations to shift to solid fuel.

Coal extraction is planned to rise significantly so as to reach 85 to 88.3 million tons by 1985. Lignite production, assigned mainly to energy use will stand at 75 million tons by 1985, or more than three times the 1978 level, a production mainly contributed by large open-pit mines. The next 5 years will see the utilization of bituminous shales (a production of 15 to 16.5 million tons by 1985) by commissioning two large thermal power stations. Research and development is summoned to find solutions for use of all available reserves, and to improve extraction methods and coal-preparation technologies. Among the tasks are resizing of the protective pillars of the surface objectives, approaching thin and very thin layers of pit and brown coal, and recycling the coal-preparation wastes.

Hydropower will be intensively developed, through both large stations on the Danube and inland rivers, and widespread use of microhydropower stations. A determined commitment to harvest 30 percent of the national hydropower potential in 1980, 45 percent in 1985, and 65 percent in 1990 will ensure full utilization of Romania's hydropower potential by the year 2000.

By then, the hydropower share in the primary mix of electric generation will reach 24 percent, a 4 percent increase with respect to 1985 and more than a 6 percent increase as compared to 1980.

Under a specialized Nuclear Power Plant Building Programme, installed electric power of nuclear origin will be 660 MWe in 1985, 3,960 MWe in 1990, and 10,560 MWe in the year 2000. This will reduce the requirement for coal and shales in power generation from a top of 55 percent in 1985 to 44 percent at the end of the century.

A remarkable emphasis is given in the program to the utilization of new energy sources and technologies. In the Romanian approach, "new" sources are geothermal, solar (radiation), wind and waves, biomass (biogas from farm wastes as a first, practical approach), urban wastes, and "trapped" coal reserves, which are potentially usable for underground gasification, not being amenable to conventional mining methods, owing to difficult geological conditions and prohibitive extraction costs. Among the new energy technologies envisaged one can mention heat pump recovery of waste heat; various fuel saving techniques for thermal engines; MHD generation; surface coal conversion to gases and liquids; hydrogen; electrochemical sources and storage; and cryogenic transportation of energy. Fundamental and applied research is conducted, in various degrees of commitment, in all these promising fields. Speculative approaches, such as CO_2 conversion to fuel and other biological, chemical, and physical methods are also considered, with confidence in future breakthroughs somewhere in the world.

Preliminary evaluations of the chances of the "new energies" to penetrate the energy economy of the country show interesting prospects. Their potential, calculated on the basis of existing first-generation technologies (solar flat-plate collectors) seems well in excess of 20 million tons of coal equivalent per year, and fits, in size and energy quality, a comparable market of low-enthalpy heat.

A 5-year research and development effort, largely sponsored by the National Council for Science and Technology, produced an initial inventory of first-generation technical solutions to ensure a sustained deployment plan for 1981-1985. To guide the implementation action and avoid wasteful use of time and materials, standard solutions were examined and approved at a Government level and appropriate Government decisions were taken to enforce their application. In the next 2 years, more than 400 local consumers are due to adopt standard solutions to use solar heat, geothermal water, and biogas to eliminate their hydrocarbon demand.

Of the total fuel projected to be saved by 1985 by "new energies," around 18 percent is expected to be in industry (drying processes of wood and ceramics, washing, preheating, cooling, space heating, and hot water), 19 percent in agriculture (drying, dehydrating, food storage, washing, and feedstock processing), 22 percent in transportation (diesel retrofitting, fuel saving devices, synthetic fuels), and more than 40 percent in household (mainly hot water and space heating).

Technical progress and mass production of hardware will help capital costs to decrease, while also enlarging the market and, thus, the "potential," conventionally defined, of the new energies. Increasing costs and, presumably, prices of oil will do the rest, in favor of the new-energy market penetration. In the 1990 horizon one expects that the whole 1978-defined potential of geothermal water will be tapped, some 10 to 20 percent of the 1978 solar potential, up to 4 percent of the now-accessible wind and waves, almost all biogas potential, and, necessarily, all the urban waste potential.

So far the capital investment to use the new energies appears recoverable on the account of saved (imported) fuel, in 2 to 15 years. Meanwhile, research and development is aimed at cost reduction and life prolongation of the first-generation equipment, and prepares for a timely development of second-generation, better performing equipment.

Till 1985-1990 the new energies will be assigned missions mainly on the low-temperature heat market. Within the last years of the century one expects to achieve the technical capability to approach higher temperature requirements, especially through massive industrial waste-heat recovery, fuels from biomass production, and by appropriately chosen hard solar technologies. According to the strategy promoted by the master program, in the last decade of the century electric energy generated by coal and tars will stay roughly constant, while that supplied by hydropower and nuclear power plants is due to increase; there should be a more marked growth of the production of electricity generated by solar and other new sources of energy as well as by recovered sources of energy, which have to account for at least 20 percent of the total production by the year 2000.

An electric-energy production of 80 to 90 billion kWh is planned for 1985, to reach 105 to 111 billion kWh in 1990. An installed power of 22,000 MW in 1984 and 28,000 MW in 1990 will sustain this growth. In such circumstances, improving the organization and functioning of the electric power grid is essential. Among the guidelines are: steady expansion of cogeneration of heat and electricity, by developing industrial and city facilities and a faster extension of the already major district heating network (savings on this account in the 1981-1985 period estimated at 3 million tons of coal equivalent); more appropriate siting of power stations; simplification of networks and shortening of transport distances; building the capability for the electric power grid to be operated by autonomous zones; increasing the safe operation of the national grid and equipment through new protection and automation methods and devices, new sizing of the power reserve, higher reliability of coal-fueled power aggregates, lines, and stations, and computerization of dispatcher control. In support of all this, an important research effort is expected in such fields as equipment stress, chemical corrosion, electric insulation breakdown, pollution processes in power plants, and higher yield equipment and technologies.

Diversification of resources is a firm command of Romania's long-term energy plans. It comes together with such associated principles as utilization of low-grade, low-energy density, eventually higher cost energy carriers; recycling and recovery of energy and materials; and a vast, persistent effort at technology substitution. The energy sector is not the only one to be shaped by such conduct. Internal demand will be met, by recovery and reuse of all raw and auxiliary materials, by 1985, in the following proportions: 40 percent iron, 41 percent copper, 40 percent lead, 44 percent textile fiber and yarn, 50 percent processed polyethylene products, 44 percent tires, over 41 percent spent mineral oils, 52 percent pulp and waste paper, 60 percent chip board and fiberboard, almost 90 percent of the needs of glass, and about 95 percent of wood package.

This makes evident another important coordinate of Romania's development programs, which is to reduce consumption and more efficiently utilize all materials and energy.

A crucial assertion of the energy master program is that "total energy consumption (in Romania) can rise in the decades to come only if new sources of energy are identified and technological measures applied to economically utilize them. The chief source and decisive conditions of the further steady development of national energy and of the people's welfare is a substantial increase in the economic efficiency of utilization of fuel and electric energy." Achieving and exceeding in the shortest possible time best world figures for energy consumption per national income unit is a basic target in Romania's further development. Within the next decade "an at least 40 percent lower index-average energy consumption per 1,000 lei worth of industrial production is to be achieved...., by the year 2000 the index is to be at least 2.6 times smaller than in 1980" which, certainly, takes into account the expected rapid growth of the value of the industrial production too. The chief lines of action to achieve these conservation goals are "to lower end-use energy consumption by a rigorous reappraisal of the consumers and allocation of energy in the strictly necessary amount, form and quality; modernization of technologies and rationalization of flow charts by restricting the use of large energy-consuming raw and auxiliary materials; steady rise in the conversion yields of energy from primary to end-use forms so as to maximize economic effects with as low as possible primary energy consumption; recovery of energy, heat, and combustible resources resulting from technological processes; substitution of higher quality, depletable resources, mainly hydrocarbons for other, low-grade ones. All fuel-consuming plants, especially industrial furnaces, will be equipped with heat recovery units; appropriate solutions will be applied and generalized to utilize the heat contained in cooling waters, exhaust gas, hot products, and wastes, including heat pumps, thermal pipes, etc." On the other hand, to fully understand the country's conservation policy another crucial statement is to be singled out: "a greater stress will be laid upon low fuel- and energy-consuming industries, while in the processing industries raw and auxiliary materials will have to be utilized better, to a higher extent." And this means structural change. Indeed, reshaping the industrial structure of the Romanian economy is an explicit, central coordinate of the country's new way to grow. Taking into account the international economic pattern, the impact of the world energy and raw materials crisis, a central task of the next 5-Year Plan is the radical improvement of the structure of our industry, the firm direction of the efforts toward the development of the low-energy consuming branches and sectors [5]. In pursuing this task, it is only obvious that we will have to mainly develop those low-energy consuming industries yielding high-value products, embodying complex labor, superior technical skill, and to continuously diminish production involving large consumption of raw materials and energy, as well as to eliminate outdated technologies requiring large consumption of fuel and energy [6]. Thus, conservation in the broadest sense is a strictly enforced rule of conduct in Romania's development.

Specific conservation goals are pursued throughout all of Romania's long-term energy program. But where is the oil and gas in this strategy? The essential aim, inasmuch as the hydrocarbons are concerned, is their utilization to the greatest extent as chemical raw materials. That is why special attention will be paid to continually restricting the use of oil and gas for burning purposes. Likewise, research is to be intensified to develop new types of fuels, synthetic included, for transportation.

Meanwhile, oil and gas production will be limited to maintain geological reserves at an adequate level. Accordingly, extraction in 1985 is planned at 12.5 million tons of crude oil and 26.5 thousand million m^3 of methane.

Through in situ combustion, steam and carbon dioxide injection, thermal, chemical, and biological methods to be investigated and tested, the oil recovery factor will have to rise from 31.5 percent in 1979 to 37 percent in 1981 and to about 40 percent in the 1990's. Improved oil management will comprise centralized control of wells in operation; closed system handling to minimize losses; use of hydraulic and submerged pumping equipment and of a wider range of chemicals to boost the inflow; gas underground storage near large consumers to better meet peak demand. Thus, diversification and conservation are accompanied by implementation of appropriate technology.

The implementation of the Programme provides the prerequisites for an ever more intensive promotion of Romania's cooperation and collaboration with all countries and international organizations interested in the equitable and efficient solution of the problems posed to mankind by needs for energy and its rational utilization.

Cooperation with the developing countries will be enlarged to the end of capitalizing, on a mutually advantageous basis, the energy resources of various categories existing in these countries, by undertaking prospecting and geological exploration, construction of power plants, and joint development of new energy sources and technologies.

Cooperation extends also to other countries, including the developed, industrialized ones, in the construction of nuclear power plants, capitalization of new sources of energy, and promotion of advanced energy technologies.

Romania also is committed to take active part in the implementation of joint research programs sponsored by the United Nations Organization, as well as in regional cooperation aimed at solving energy problems.

REFERENCES

[1] The Statistical Yearbook of the Socialist Republic of Romania, 1979.

[2] "Draft Directives of the Twelfth Congress of the Romanian Communist Party for Romania's Economic and Social Development in the 1981-1985 Five-Year Plan Period and the Guidelines until 1990," Romania--Documents, Events, 9th Year, No. 34, July 1979, Agerpres--The Romanian News Agency, Bucharest.

[3] "Draft Programme of Scientific Research, Technological Development and Application of Technical Progress in the 1981-1990 Period and the Guidelines until 2000," Romania--Documents, Events, 9th Year, No. 36, July 1979, Agerpres--The Romanian News Agency, Bucharest.

[4] "Energy Research and Development Programme for 1981-1990 and the Principal Guidelines until the Year 2000," Romania--Documents, Events, 9th Year, No. 37, July 1979, Agerpres--The Romanian News Agency, Bucharest.

[5] Nicolae Ceaușescu. Address to the Meeting of the Senior Staff in Industry, Civil Engineering, Transportations and Agriculture, March 6, 1979, București, Ed.Politică, 1979, p. 35-36.

[6] Nicolae Ceaușescu. Address to the Joint Plenary Session of the Central Committee of the Romanian Communist Party and the Supreme Council of Economic and Social Development of Romania, October 28, 1973, in "Romania on the Road of Building up the Manysidedly Developed Socialist Society," vol. 9, Ed.Politică, București, 1974; a-623; b-619.

CHAPTER 16

PAST TRENDS OF CAPITAL-OUTPUT AND INVESTMENT RATIOS IN THE FIELD OF ENERGY IN THE ECE REGION

United Nations Economic Commission for Europe[1]

This paper[2] argues that there are a number of developments affecting the traditional energy supply sectors which place an upward pressure on capital-output ratios (the capital required to produce a unit of output). Rising capital-output ratios in the energy supply sectors taken as a whole would be a reversal of the downward trends experienced by most countries of the ECE region[3] during the 1960's. These declining trends, along with increased reliance upon imported energy, were two factors which tended to reduce the investment requirements of the rapidly growing energy sector.

A rise in the capital-output ratio (COR), reinforced to some extent by a stabilization of energy import dependence, does not necessarily imply rising or unsupportable shares of national output having to be devoted to energy investments in the long term. Factors such as the reduced growth of energy demand through higher prices, technological innovation, energy conservation, and slower economic growth may offset rising COR's in an important, but as yet unquantifiable, manner.

CAPITAL-OUTPUT RATIOS

Capital-output ratio trends, or modifications thereof, have often been used to estimate future capital requirements. In the past few years there has been considerable interest in the magnitude of the projected capital needs of the energy sectors. By presenting historical trends in the following section, it is not implied that they will necessarily continue into the future. They

---

[1]Note prepared by Joseph E. Smolik, Projections and Programming Division, United Nations Economic Commission for Europe, Geneva (Switzerland).

[2]For details see a recent study prepared by the ECE secretariat entitled Consideration of Past Trends for Investment in the Field of Energy, EC.AD.(XVI)/AC.1/R.4, 20 July 1979, and Statistical and Methodological Appendix, forthcoming.

[3]North America and Europe, including the USSR.

may not, since trends do not reflect the influence of the new highly capital-intensive renewable and unconventional energy technologies for which the economic relationships between inputs and outputs have not yet been well established. These and other underlying factors which may influence future changes in energy sectors COR's will be examined in the second section.

Past and Recent Capital-Output Ratios

Coal Mining

There exists a wide variation in the quality of coal deposits and coal production policies in the ECE region. On one hand there are those countries with large, good quality coal reserves (Canada, Federal Republic of Germany, Poland, United Kingdom, USA, USSR) which, with the exception of the Federal Republic of Germany and the United Kingdom, have increased or maintained production. The other countries can be divided into two groups: some other centrally planned economies which have reserves of declining quality (mostly brown coal), but which have more or less maintained production; the other market economies which have small reserves of hard coal and which have allowed production to diminish. The capital-output ratios were found to be rising in all the countries shown in Fig. 16-1.[4] In a number of countries the upward trends might be linked to the declining quality of deposits, to the concentration, mechanization, and automation of underground mining, and to the shift to open-cast mining. Also, the recent institution of new mine safety and environmental legislation can be cited as contributing factors.

Oil and Natural Gas Extraction

In Norway, the USA, and the Federal Republic of Germany (Fig. 16-1) COR's were found to have been decreasing at a rapid rate. In Norway, the COR trend is that of a relatively new industry and a few additional years will be required to confirm the downward trend. The United States experience is of particular interest as that of a major producer reflecting a wide variety of extraction conditions. Unfortunately, this series stops before the full development of the recent Alaskan North Slope projects. The general downward

---

[4] Data for France shows similar tendencies and some indirect evidence of rising COR's exists for a number of centrally planned economies.

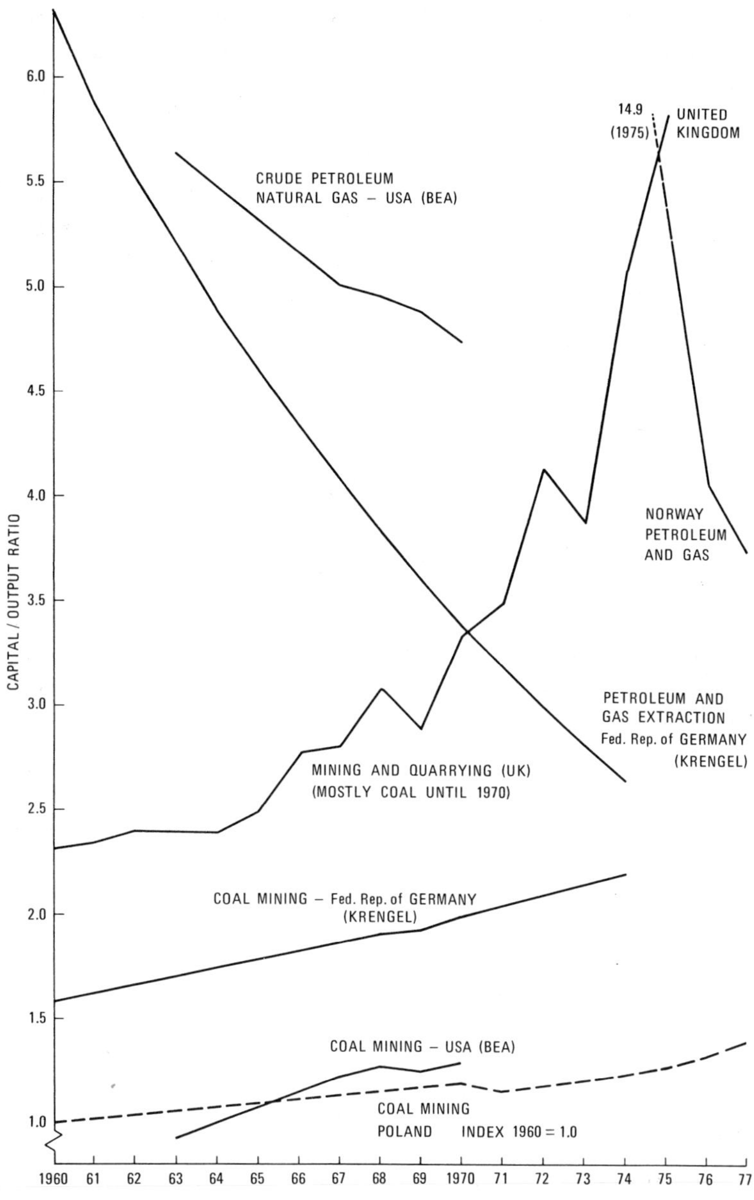

Figure 16-1. Capital-output ratios in the energy extraction sectors

trend is surprising because of the move toward deeper drilling and offshore installations, and might be due to very rapid technological progress.

COR's in the USSR were increasing. Investments for the increase of oil production (rubles/ton) have increased 7.5 percent in the periods 1966-1970 and 1971-1975, while for natural gas production they have gone up 23.9 percent over the same period.[5] This is not surprising given the difficulties of extraction from the new areas in Western Siberia which are being counted on for both compensating and incremental increases in production. The differences in the decreasing trends experienced by the Federal Republic of Germany and the United States (and perhaps by Norway), and the upward trend of the Union of Soviet Socialist Republics might be taken as a rough measure of the capital penalty imposed by geography and climate in the latter country.

Petroleum Refining

In general, the COR's of the petroleum refining sector were declining slowly (Fig. 16-2).[6] This observation is clearest for the two countries, the Federal Republic of Germany and the United States, for which adjustments were made for capacity utilization changes. If the Canadian figures are adjusted for capacity utilization, it appears that the COR fell slightly during the early part of the period and then leveled off. The last few years, 1974-1976, were also strongly affected by the recession. The Polish data show a drop until 1973 after which the trend reverses sharply.

To summarize, it appears that the COR trends in both petroleum refining and chemicals have been downward during the period in question and in some instances have leveled off. This result seems reasonable considering that similar technology had become available internationally during this period and that there was also a common tendency toward larger plants allowing greater economies of scale.

---

[5] A. Lalayanz, "Two Forms of Society--Two Forms of Development of the Fuel and Energy Industry," Planovoe Khozyaistvo, No. 4/1975, p. 28.

[6] The chemicals sector was also included in this chart since there are production similarities between petroleum refining and some chemical products. One objective of the analysis was to determine if the COR trends of the two sectors have been similar, which was, in general, the case.

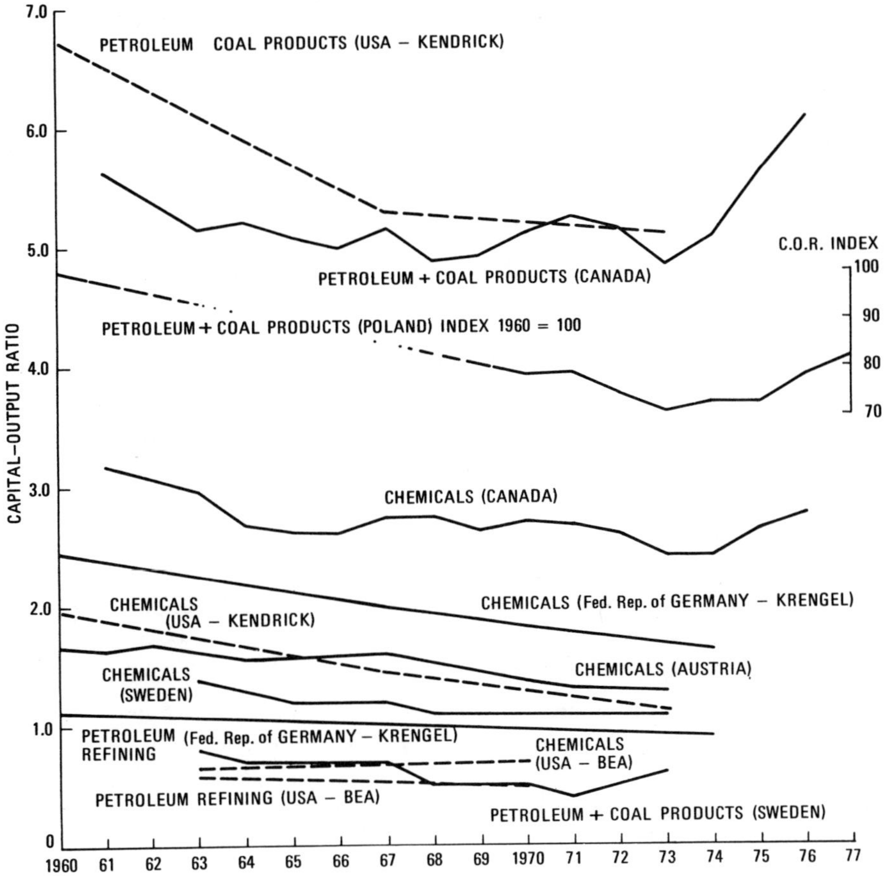

Figure 16-2. Capital-output ratios in the petroleum refining and chemical industries

Electricity Sector

Although the product of this sector is homogeneous, trends may nevertheless be difficult to interpret because of structural changes in production technology and changes in the relative importance of production, transmission, and distribution facilities. Among the centrally planned economies of the ECE region the COR trends have generally been downward in Czechoslovakia, Hungary, and Poland (Fig. 16-3).[7] However, a reversal of the trend has occurred in Czechoslovakia and possibly in Hungary and Poland, where at least a leveling has occurred in the past few years. The continuous increase in the COR of this sector in the USSR may be due to the development of large hydroelectric complexes in Siberia and to the large transmission network required to link producing areas with consumers.

In the market economies of the ECE region, data are rarely available to construct COR's separately for each of the three sectors comprising ISIC 400: electricity generation, transmission, and distribution; gasworks and distribution; and water utilities.[8] In general, the results in Fig. 16-4 show that there has been a long-term decline, the exception being in the Federal Republic of Germany. In a number of countries there were periods when the COR's increased for some time before beginning to decrease again (Federal Republic of Germany, Sweden, and the United Kingdom), which suggests that there have been other determinants, such as structural and load factor changes, at work besides scale economies and technological progress.[9]

In three other countries--Italy, Japan, and the United States--the downward trend appears to have leveled off and then COR's started to increase. In the United States the leveling off and the reversal of the downward trend of electricity, water, and gas in 1969 is also reflected in the results for the electricity sector alone and may be tentatively attributed to the exhaustion of scale economies and technical efficiencies, and to nuclear power

---

[7] In the centrally planned economies the electricity sector also includes the production of heat in the form of district heating and cogeneration.

[8] Limited detailed sectoral data suggest that gas works and distribution and water utilities have tended to raise the COR's in sector ISIC 400 as a whole.

[9] In the United Kingdom, the bulge in the COR starting in 1965 may have been due to a sharp rise in investments in gasworks and distribution.

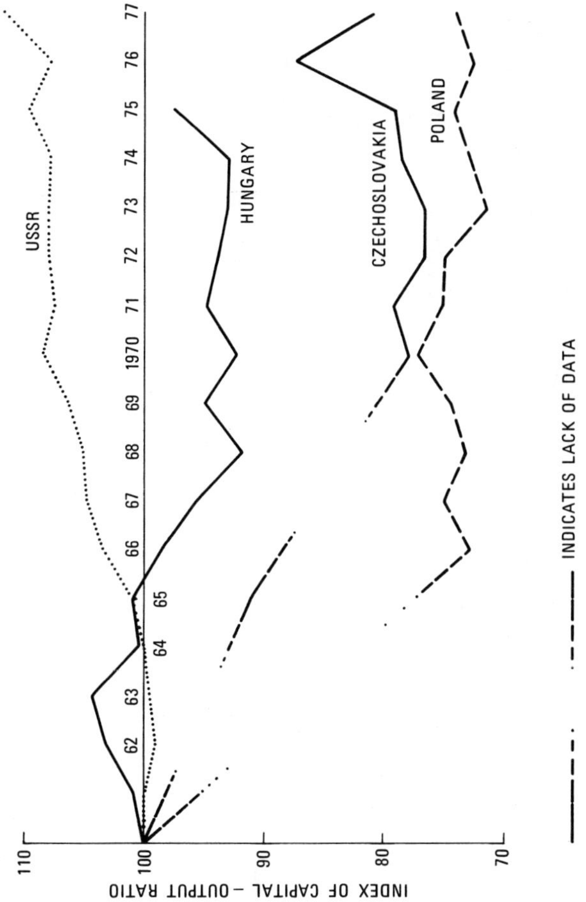

Figure 16-3. Capital-output indices of the electricity-heat sector (centrally planned economies) 1960 = 100

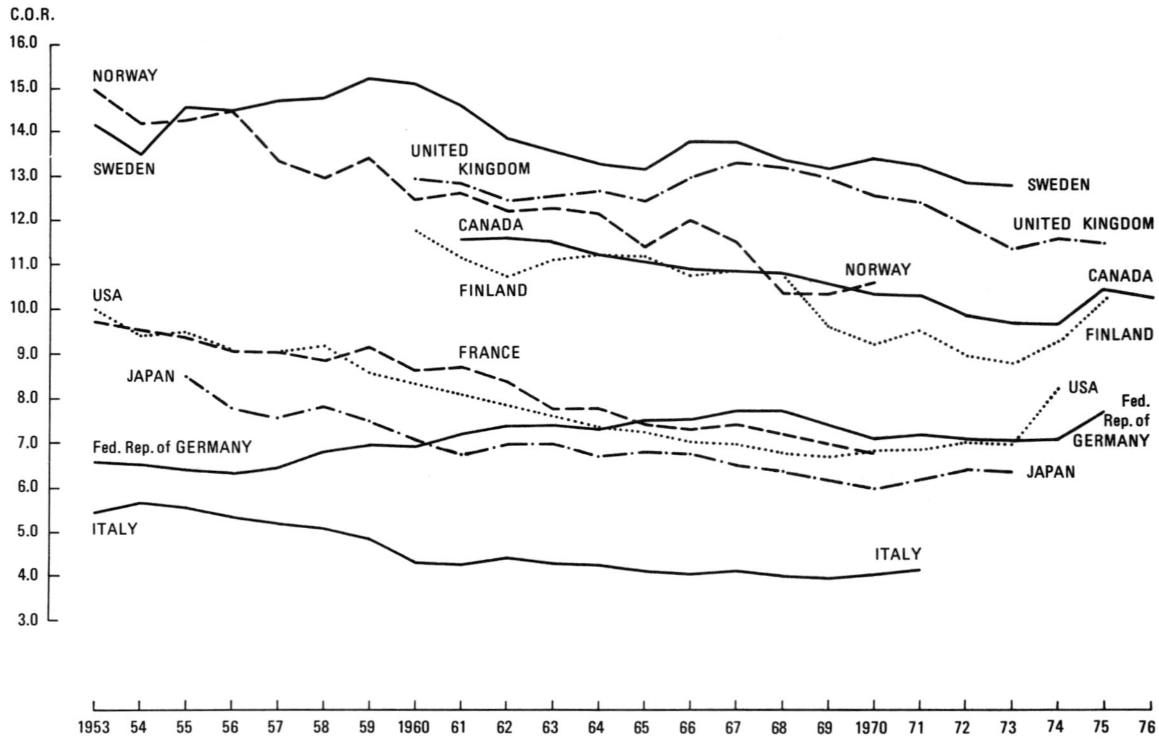

Figure 16-4. Capital-output ratios in electricity, water, and gas (ISIC 400)

investments. Environmental investments were not significant at that time. The scale economy hypothesis is consistent with the leveling out of the installed cost per kilowatt of production capacity which appears to have occurred around 1970.

The Norwegian experience of decreasing COR's presents a case of special interest because almost all electricity generation is by hydroelectric means.

Oil and Gas Pipeline Transport

Trends in these activities might be of interest because of the relatively rapid expansion recently of the pipeline system in much of the ECE region. Pipeline systems offer possibilities of scale economies through the use of large diameter pipes. In Canada, the COR fell 40 percent between 1961 and 1973 after which it rose sharply, seemingly due to a reduction in capacity utilization. In the United States there was also a decline of about 16 percent between 1963 and 1970, but the Alaskan Oil Pipeline project probably reversed or at least slowed this trend.

On the other hand, the COR of pipeline transport in the Union of Soviet Socialist Republics has more than doubled (133 percent for oil and 123 percent for natural gas[10]) between the periods of the last two completed 5-year plans (1966-1970 and 1971-1975) reflecting the increasing difficulty of construction in Siberia.

The Energy Industries as a Whole

It would appear that during the 1960's most countries that relied heavily on oil and gas enjoyed falling COR's in the energy sectors as a whole (Figs. 16-5 and 16-6). Improvement in capital productivity was brought about by falling capital-output ratios in oil and gas extraction and transport (pipelines), petroleum refining, and electricity production, sectors that accounted for a majority of energy investment in most economies. Partially offsetting this decline have been the increasing COR's in coal mining and, to a lesser extent, those in gasworks and gas distribution. Those countries which have relied most heavily upon coal, especially on brown coal, have experienced the most upward pressure on their total energy COR's.

---

[10]Lalayanz, op. cit., p. 28.

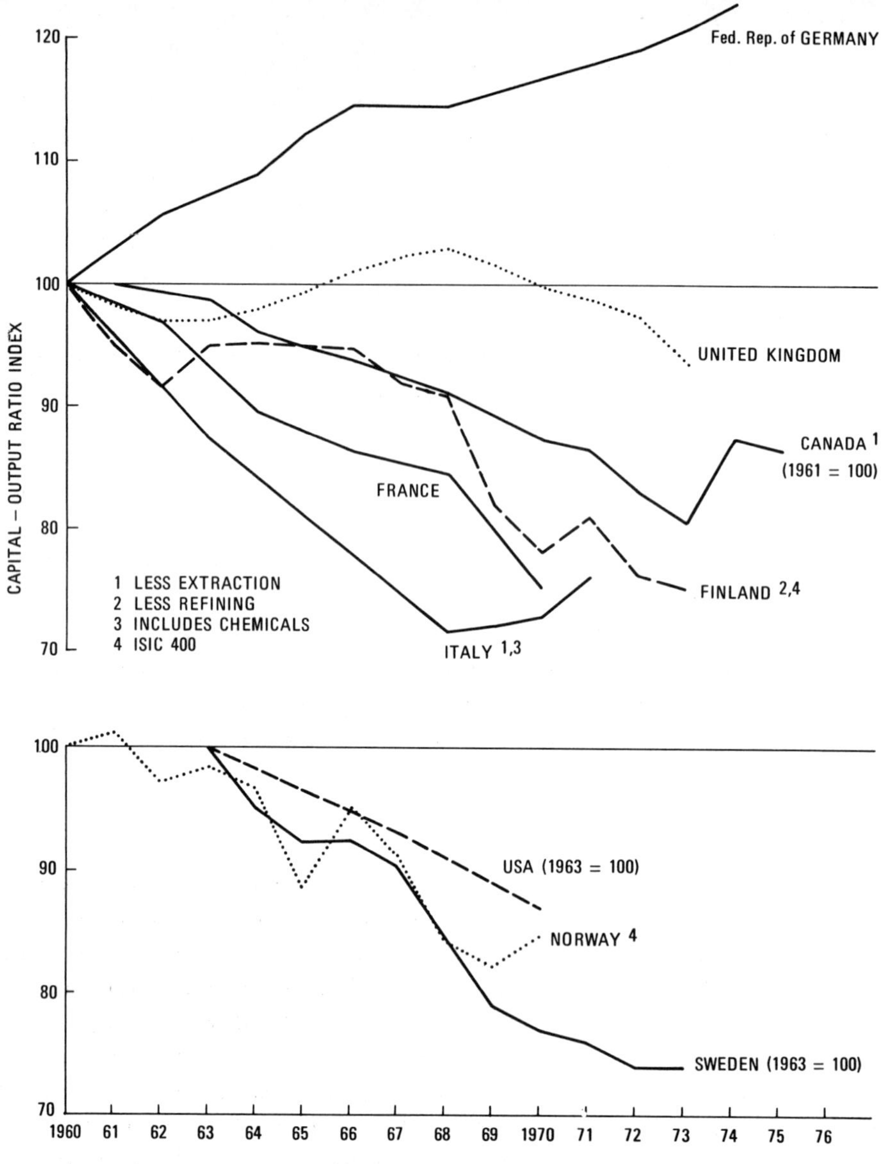

Figure 16-5. Capital-output ratio indices of the energy sectors taken as a whole (market economies) 1960 = 100

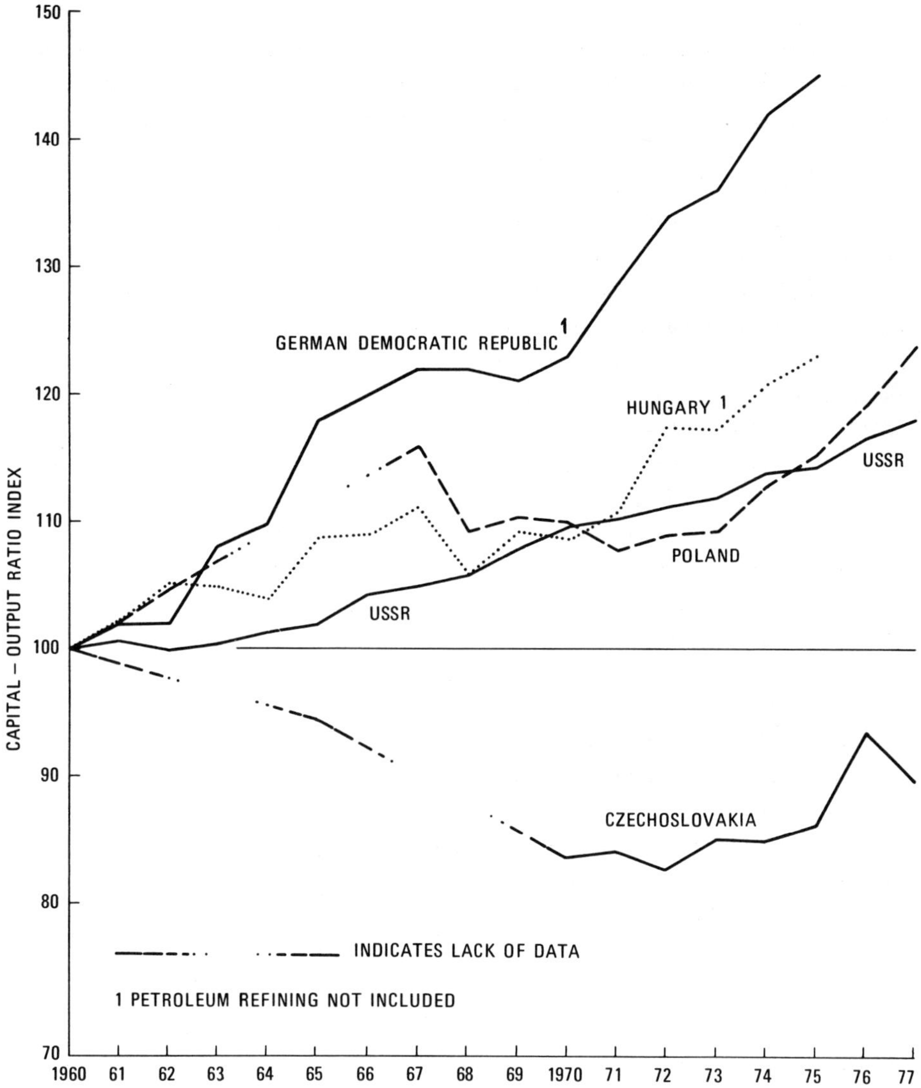

Figure 16-6. Capital-output ratio indices of the energy sectors taken as a whole (centrally planned economies) 1960 = 100

Some evidence has also been presented which shows that there have been decelerations and even reversals in the historical declines of sectoral COR's which might be explained by the exhaustion of certain technical or scale economies, new environmental and safety regulations, lower quality deposits, capital-labor substitution or structural changes within the sector, such as growing role of capital-intensive nuclear power. Countries in which reversals were observed in electricity generation (or ISIC 400) included Czechoslovakia, Italy, Japan, Poland, and the United States. Reversals may also have taken place in petroleum refining in Canada, Italy, and Poland.

Factors Affecting Future Capital-Output Ratios in Energy Supply

In contrast to the generally declining capital needs per unit of output in the energy sector experienced in the past by many countries, a number of relatively recent developments would suggest a reversal of these trends:

- Rising relative price of energy, which would permit the economical application of more capital-intensive supply technologies
- Exhaustion of scale economies
- Implementation of environmental protection legislation
- Work safety legislation
- Growing role of nuclear power
- Stabilization/reversal of the historic trend toward greater import dependence
- A continuing exploitation of progressively lower quality deposits.

On a sectoral basis the following factors would act on balance to place an upward pressure on the capital-output ratios.

Coal Mining

Capital-output ratios can be expected to continue rising due to the exploitation of poorer deposits, environmental regulations, and other institutional factors. To some extent, the COR's may be held down by the application of such new technology as microprocessors in underground mining, which offer large potential savings in both labor and capital.

Oil and Natural Gas Extraction

Although technological progress will probably continue to reduce specific capital requirements, COR trends will be determined largely by the location of production, relative importance of offshore operations, and the intensity of application of secondary and tertiary extraction techniques. Marginal output in the ECE regions is coming from sources (Alaskan North Slope, North Sea, Siberia) which require a greater investment per unit of output than do traditional locations. New sources of oil, such as tar sands, heavy crude, and oil shale, are also highly capital-intensive but their impact on the capital needs of the extraction sector is difficult to estimate, given the uncertainties about their future rates of development. Scale and technical economies realized in new pipeline systems will be lessened or offset in many cases by the increased distance of transportation.

Petroleum Refining

Given the evidence that declining capital-output ratio trends have leveled out in some countries, a full reversal may occur as increased emphasis is placed upon catalytic cracking capacity.

Electricity Sector

Conventional power plants require additional investment for environmental protection or for converting from oil to coal at a time when the offsetting effects of scale economies appear exhausted. A large percentage of marginal electricity capacity is to be provided by nuclear power, which is relatively more capital-intensive than conventional electric power. Furthermore, in some countries nuclear plants have, for various reasons, been run at lower load factors, which effectively increase their COR's.

New Energy Technologies

Many of the renewable (wind, solar, tidal) and nonconventional synthetic fuel sources are very capital-intensive. On the whole, there would appear to be a general trend toward raising capital-output ratios in the energy supply sectors of the ECE region.

INVESTMENT RATIOS

The trends in the ratios of energy investment to Gross Domestic Product (GDP) during the post-war period can be viewed as the net result of the interplay of the COR trends in the energy sector and the changing share of the energy sector in national output. Declining (increasing) COR's and energy sector shares would place a downward (upward) pressure on the ratio of energy investments[11] to GDP.

In the following section, energy investment ratio trends and levels are presented for various countries and preliminary attempts are made in the last section to assess the importance of some possible determinants.

Past and Recent Investment Ratios

Levels and Trends

In the centrally planned economies of the ECE region, energy investment ratios were between 3 and 5 percent; long-term changes were not important in the period 1960-1977, although there were a number of significant short-term fluctuations (Fig. 16-7).

Among the market economies, the ratios were between 1 and 3 percent, remained roughly constant in Belgium, the Federal Republic of Germany, Italy, and the United Kingdom, decreased in France and the Netherlands, and increased in Canada, Finland, Norway, Sweden, and the United States. More recently, the trends have also been increasing in the Federal Republic of Germany and sharply so in the United Kingdom (Fig. 16-8).

The trends in country COR's and investment ratios are not always in the same direction because the share of energy in national product has not remained constant. It would appear that in most countries the energy sector grew more quickly than total GDP since petroleum refining and electricity generation have

---

[11]Energy investments were defined as the total of investments in the energy supply industries to the extent that this was possible. The industries included were: coal mining, crude oil and natural gas extraction, petroleum refining, electricity (production, transmission, distribution), and gasworks and gas distribution. For most industries investments in energy transportation were not separately available and thus would not be included. Petroleum refining investments were not included for those countries in which they are combined along with those of chemicals. National differences also exist with respect to the definition of energy investments which can make international comparisions difficult.

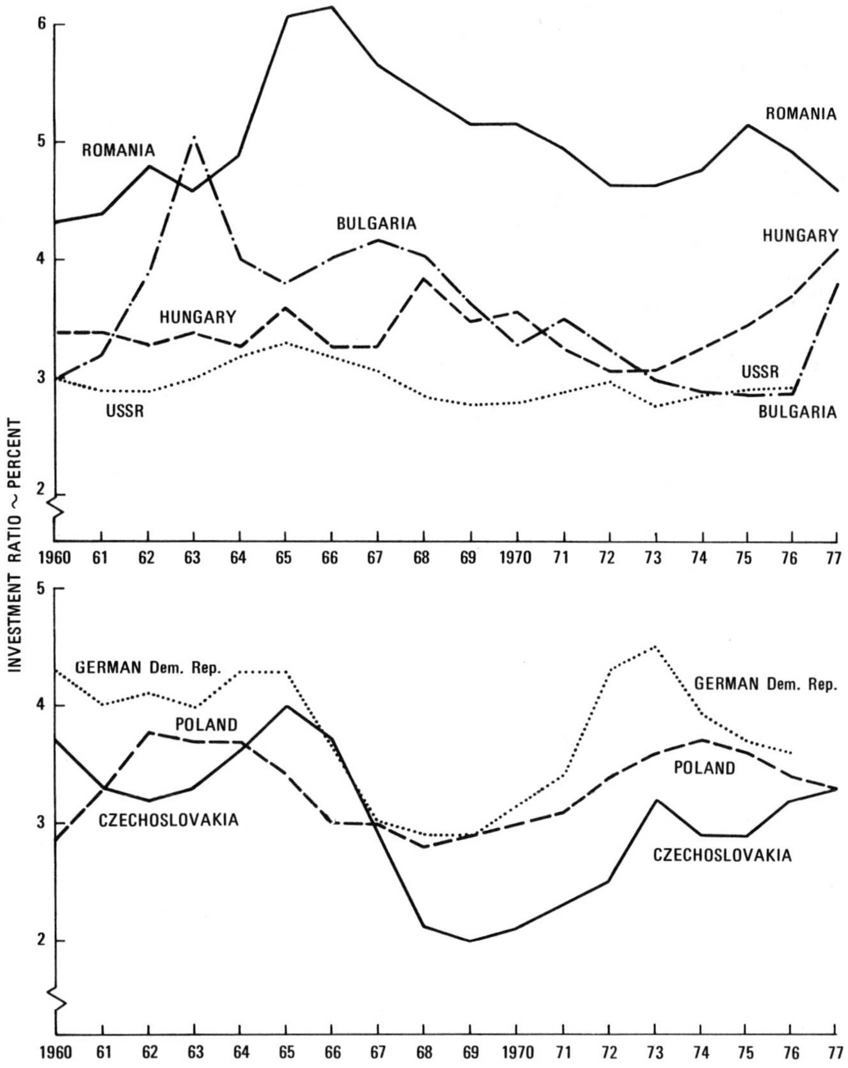

Figure 16-7. Energy sector investments as a percentage of net material product (constant prices)

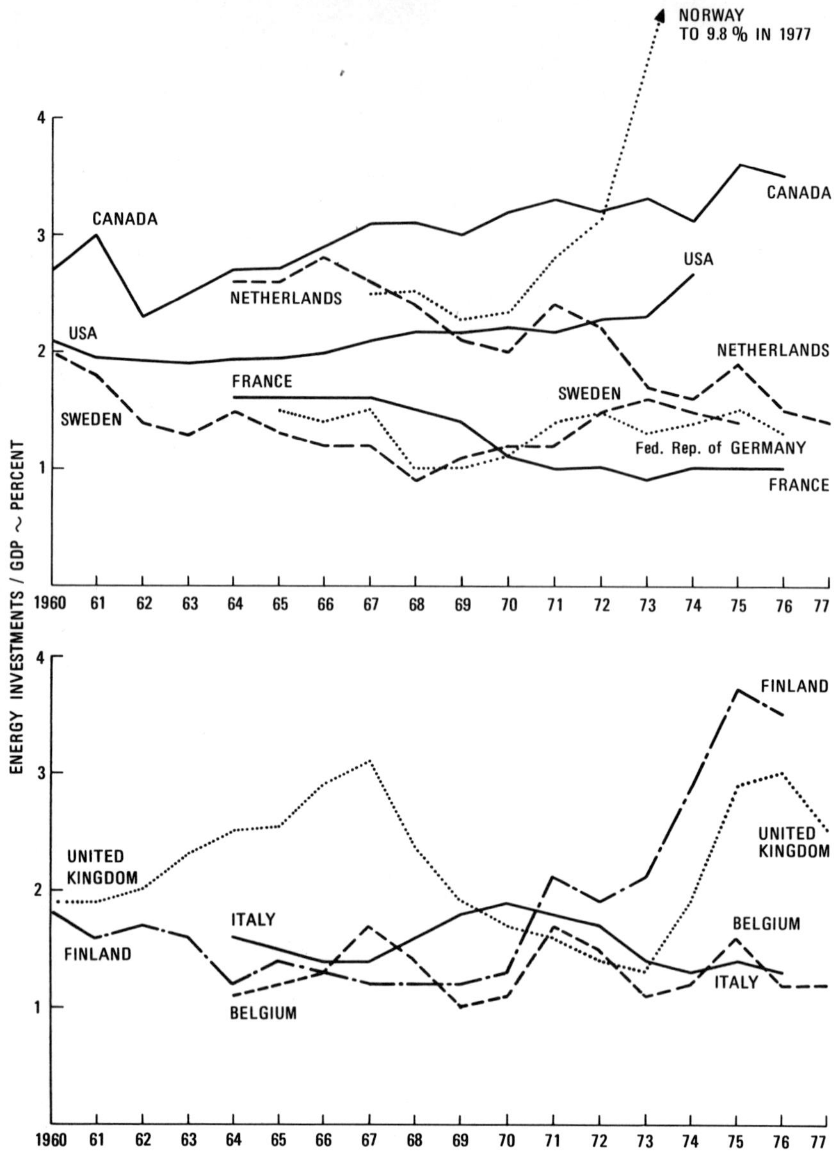

Figure 16-8. Energy sector investments as a percentage of gross national product (current prices)

been among the fastest growing sectors. The rapid growth of these two sectors has been offset to a certain extent by slowly growing or declining output in extraction.

Determining Factors

The shares of energy investments in GDP for the period 1964-1975 are shown in Table 16-1. There are significant differences in the energy investments made by the countries of the ECE region. This section attempts at explaining these intercountry differences. It is hypothesized that the energy investment effort made by a country is directly proportional to the size of its energy deposits. Subsequently, the role of the energy investment ratio as determined by, or as a determinant of the energy-import dependence is examined.

Energy Resources

The magnitude of domestic energy reserves might be expected to act as a major determinant of the investment effort made in the energy supply sector. A preliminary examination of this potential relationship resulted in little statistical evidence to support this hypothesis.[12]

Three indicators of energy reserves were used. The first was simply the economically recoverable fossil and hydrocarbon reserves (i.e., the sum of coal, crude oil, and natural gas reserves) in tons of coal equivalent in 1976. Nine countries in the sample had investment ratios between 2 and 3 percent of GDP, but their reserves varied between 10^9 and 1.1×10^{12} tce (Table 16-1).

Another measure of reserves was taken as the ratio of economically recoverable reserves to GDP (in $1,000 U.S.) in 1973. This measure appeared more appropriate to use than absolute reserves since it also reflects the size of

---

[12]A more vigorous examination is impeded by the lack of data. The investment effort was measured by total investments in the energy supply sectors as a share of GDP during 1964-1971. Accordingly, this indicator includes investments in energy transformation (i.e., petroleum refining and electricity generation), which account for a large percentage of all energy investments, but which are not necessarily linked to the extraction of energy. Redefinition of the investment ratio to include only energy extraction investments reduced the sample of countries significantly because of insufficient disaggregation in national investment statistics. It was observed, however, from the limited sample of countries available, that there was a much greater variation in the proportion of GDP allocated to energy extraction than there is in total energy investments, but that the conclusions would not change.

Country	Coal 10^6 tce	Oil 10^6 tce	Natural gas 10^6 tce	Total reserves 10^6 tce	Reserves $\frac{\text{tce}}{\$1{,}000\ \text{GDP}}$	Reserves $\frac{\text{tce}}{\text{capita}}$	Energy investment ratio[1]	Net import dependence[2]
Market economies								
Belgium	127	-	-	127	3	13	1.4	84
Canada	9,381	1,226	1,800	12,407	107	537	3.0	-14
France	438	11	97	546	3	10	1.4	75
Germany, Fed. Rep. of	34,419	63	228	34,710	143	564	1.4	42
Italy	-	64	216	280	2	5	1.6	86
Netherlands	1,430	16	1,917	3,362	61	244	2.1	49
Norway	-	978	736	1,714	110	429	2.6	49
Sweden	-	-	12	12	0.3	2	1.3	74
United Kingdom	45,000	1,940	889	47,828	222	855	2.4	46
USA	177,588	6,064	6,676	190,328	161	884	2.2	4
Centrally planned economies								
Bulgaria	2,203	3	15	2,221	115	252	3.2	52
Czechoslovakia	4,815	3	17	4,835	106	324	2.4	22
German Democratic Republic	7,660	4	97	7,761	145	462	2.7	23
Hungary	950	40	130	1,120	45	106	2.8	32
Poland	21,790	7	151	21,948	272	638	2.6	-14
Romania	413	235	178	826	21	39	4.8	9
USSR	109,900	11,740	24,132	145,772	226	567	2.6	-9

[1]Average of years 1964-1971.

[2]Primary energy only, 1971 (a minus indicates net exports).

tce = ton of coal equivalent

TABLE 16-1. Energy reserves, investment ratios, and import dependency in some countries of the ECE region

the economy drawing upon the available resources. The use of this indicator resulted in two major groupings:

1. Energy "poor" countries, with reserves averaging 1.5 tce per $1,000 of GDP (Austria, Belgium, Finland, France, Ireland, Italy, and Sweden);
2. Countries with average energy reserves of about 133 tce per $1,000 of GDP. Even among the energy "rich" countries differences in reserves are still quite large, ranging from 21 tce/$1,000 GDP for Romania to 270 tce/$1,000 for Poland.

The energy "poor" group does in fact have a lower investment ratio (about 1.4 percent of GDP) vis-a-vis that of the relatively "energy-rich" countries of the region, which invest at about 2.5 percent of GDP (Table 16-1). An analysis of the data for 13 countries within the energy "rich" group indicates, however, that there is no clear relationship between energy investment ratios and resources.[13]

Energy Import Dependence

One factor which reduced the domestic investment needs of the energy sector had been the gradually growing dependence on imported energy. This trend had been strong in Western Europe until 1973, around which time it was reversed by production from the North Sea. In Eastern Europe the trend was much weaker, with most of the economies (except Romania and the USSR) remaining almost totally coal-based until the 1970's when increased imports of crude oil and natural gas started from the Union of Soviet Socialist Republics.

The inverse relationship between the energy import dependence of a country and its investment effort in the field of energy can be seen in Fig. 16-9. This relationship would suggest that there would be an upward pressure on energy investment ratios as countries attempt to stabilize the trend toward greater energy import dependence. The plans for the growth of nuclear power would work to strengthen this relationship.

---

[13] The third indicator of energy reserves, economically recoverable reserves per capita, provided similar negative results. Regress analysis was used to examine these relationships.

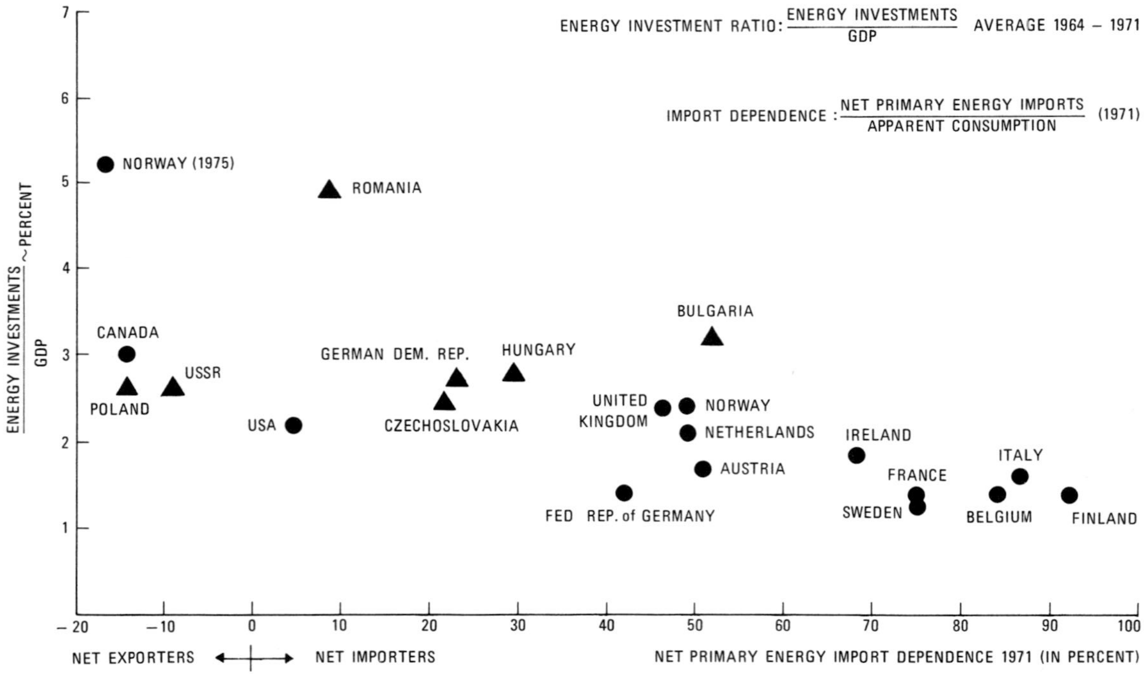

Figure 16-9. Energy investment ratio versus primary energy import dependence

Factors Affecting Future Energy Investment Ratios

It is recalled that the share of national output required for the investments of the energy supply sector depends on

- The capital-output ratios of this sector, and
- The relative importance and growth of the energy sector in total national output.

The evidence presented in the first section suggests that the capital requirements per unit of output in the supply sectors would increase in the future. For most countries, this development would be a reversal of past trends. This would imply, *ceteris paribus*, that energy investments would absorb a progressively larger portion of national output until corrective adjustments in the economy prevailed. Although the relative change in the importance of various sectors in the economy should be looked upon as a normal phenomenon, a period of relatively increasing capital needs by the energy sector might, in some cases, be accompanied by financing problems if energy pricing policies were not sufficiently accommodating. The fairly large differences in the energy investment ratios found between the countries of the ECE region suggest, in fact, that there is some scope for those countries with historically low ratios to accommodate to the kinds of structural changes which relatively higher levels of energy investments might involve.

A host of factors affect the relative importance and growth of the energy sector.

Contributing to a *relative expansion* and, hence, to a higher energy investment demand would be the slowing, or perhaps even the reversal in some countries, of the historical trend toward greater energy import dependence.

Energy conservation, on the other hand, *reduces the relative importance* of energy-related investments because the investment expenditure required to save a unit of energy is in many cases smaller than that required for a marginal unit of output from traditional energy sources.[14] Continuing energy savings can be anticipated in response to the large energy price increases which have been experienced and which should further improve the already superior rate of return on energy conservation investments vis-a-vis those in the

---

[14] Many examples of the higher rate of return on energy conservation investments have been cited. For example, *New Issues Affecting the Energy Economy of the ECE Region in the Medium and Long Term*, United Nations, E/ECE/961, p. 19.

energy supply industries. Also preliminary steps are being taken in many of the countries of the ECE region toward the restructuring of industry away from energy-intensive activities. An unknown factor at this time is the speed and magnitude of the technological response to the present energy situation, as well as the relative capital intensity of new sources of energy.

Some of the most recent estimates, at least for the medium term, already reflect the importance of lower economic growth and energy conservation in reducing the magnitude of capital requirements in spite of increasing COR's in several key sectors. For example, the investment requirements consistent with the national energy program for the countries of the European Communities for 1978 envisage an investment ratio of 1.6 percent of GDP for the period 1976-1985 compared with 1.5 percent in 1977 and 1.8 percent in 1976.[15]

SUMMARY

It is not at all certain at this time whether future energy-related investments will increase or reduce this component of national output. Factors working in one or the other direction are strong and might, on balance, prevent a significant change of historical relationships. However, in the event that relative requirements are greater in the future than in the past, experience would suggest that a certain scope for structural accommodation does exist.

[15] Energy Objectives for 1990 and Programmes of the Member States, Commission of the European Communities, COM/78/613 final, November 1978.

CHAPTER 17

SOME LIMITATIONS ON NEW ENERGY SOURCES DEVELOPMENT

Alexander Arbatov[1]

It is not original to declare that mankind has reached the turning point in its energy policy, based on traditional fuel resources. It also is not new to say that it is high time to reconstruct the contemporary energy balance of our society. Indeed, the well-known bottlenecks in oil supply stipulate the necessity to shift from the conventional energy economy to a new, more reliable, and diversified base.

In this case we face the problem of finding an optimal solution. Which to choose? We can observe different points of view with respect to long-run energy policy. I am not going to compare advantages and disadvantages of each alternative. It is quite another question.

I would like to remind you of some limitations which to my mind play a very important role in development of future energy strategies. These constraints are of different origins and influences. So let us consider the most typical and common roadblocks in the way of the breeder reactor, fusion, solar energy development, and large-scale utilization of coal and gas.

It is quite obvious that environmental, political, economic, and technical constraints will emerge most clearly in the case of the breeder alternative. A very serious threat to our society is connected with the problem of worldwide proliferation of nuclear technology among the countries which may then possess atomic weapons. In addition we have to deal with many problems during the entire economical cycle from producing nuclear materials to their use and final waste disposal. Securing reactor safety also has become a subject of considerable public concern. Great care must be exercised in the handling and containment of large quantities of radioactive materials. The contemporary economic situation in the western countries stipulates high costs, long periods of plant construction, and lack of the capital and uranium enrichment capacities. The shortage of water which is necessary for reactor cooling is another serious bottleneck.

---

[1]State Committee for Science and Technology, Moscow, USSR.

As far as fusion is concerned we face first of all technological constraints. The feasibility of controlled thermonuclear reaction has not yet been demonstrated. A roadblock to achieving fusion has been the engineering problem of magnetic containment in case of the Tokamak reactor--and the problem of developing a very durable laser of high capacity. Besides we have to cope with the challenge of availability of construction materials.

The problems of energy supply and materials availability are closely interconnected. The solution of one of the problems to a large extent depends on the solution of the other. This interdependence is constantly growing. During the entire history of mankind one can observe how the discovery of energy sources cleared the way to new materials and involvement of new materials made it possible to raise energy intensiveness to a qualitatively higher level.

Development of poor ore deposits, super-deep drilling and mining, and extraction of elements from sea water and the earth's crust promises to supply mankind with a vast amount of material needed to sustain optimal economic growth and acceptable quality of life.

But at this point we face the so-called "vicious circle," because the development of new energy sources stipulates a high demand for necessary materials which can't be met easily by the existing supply of minerals, based upon the present energy economy.

Even while solving the problem of nuclear fusion and the possibility of industrial energy plant construction, from a purely physical point of view difficulties may appear in the replacement of present energy generators by different kinds for nuclear reactors due to scarcity of such materials as beryllium, helium, lithium, molybdenum, columbium, lead, tin, titanium, and vanadium.

A scarcity of some materials is typical of solar energy exploitation also. A scarcity of aluminium, silver, and gallium restrains solar energy development on a large scale. Along with this, speeding up the rates of technological development reduces the waiting period for exploitation of new technologies on a large scale. Such speeding up of development of new industries is often stopped by limitations of the mineral raw materials base, which may not be able to be reconstructed in due time and in due direction. Therefore, in order to avoid undesirable consequences it is necessary to form a long-term and universal raw materials strategy, taking into account the most cardinal modifications not only of the energy-fuel balance but the whole industry structure of the economy.

The other restrictions to solar energy exploitation are the limits of the geographic area for solar thermal power-stations, high costs of photoelements for photogalvanic stations, insufficient level of scientific research and financing, as well as unfavorable meteorological conditions, the atmosphere's cover of dust, and high costs of equipment transportation and power transfer in case of earth-orbiting solar power stations.

With respect to the large-scale extraction and exploitation of coal directed to forcing oil out of energy-fuel balance, the most essential restrictions are: environmental pollution resulting from coal extraction and exploitation, high costs, insufficient level of development of coal liquefaction and gasification technology, scarcity of modern high-production equipment for extraction in many countries, a scarcity of labor forces, and unfavorable transportation conditions.

Though natural gas resources are limited, their availability is better than in the case of oil. Therefore the use of gas may soften the tension connected with oil supplies. At the same time a large-scale use of existing gas resources entails a number of restrictions. Gas industry is a very capital-intensive one and a scarcity of investments can be an essential restriction for exploitation of some regions. Gas transportation is a very complicated problem and large investments are needed for its solution. Pipeline transportation demands an enormous quantity of pipe made of the specialized steel, in the production of which iron with low grad of sulphur concentration, vanadium, columbium, and molybdenum is used. The difficulties of materials availability with these appear to be quite a serious restriction.

Powerful gas-pumping plants and special equipment for production of pipe of large diameter are essential and this production will demand the development of a number of industries. Building of large plants for gas liquefaction and special tankers for transportation of the liquefied gas presents a serious problem.

As said above, the mentioned restrictions are not necessary for all countries in the complex. Each country is characterized by its own restrictions which are of great importance for it. This should be taken into account while elaborating large international projects that can become an effective way to avoid these restrictions.

SECTION III. COAL

CHAPTER 18

WORLD COAL RESOURCES AND RESERVES

Harry Perry[1]

ABSTRACT

World coal resources and reserves are much larger than world oil and gas resources and reserves but, like oil and gas, are unevenly distributed geographically, with over 85 percent of the geological coal resources concentrated in the USSR, China, and the USA. However, the coal resources in a number of other countries are still sufficiently large that they could make an important contribution to domestic energy demands and, in some instances, as export to other countries. Despite some potential constraints, international coal trade could help provide a transition fuel from a mineral fuel economy to a world economy dependent on renewable resources. No entirely new technologies would need to be developed, as coal was at one time the dominant commercial fuel worldwide, but improved technology would be desirable to reduce costs and adverse environmental impacts.

INTRODUCTION

In the first quarter of the 20th century coal was the dominant commercial fuel and in 1925 supplied about 83 percent of the world's energy demand. Since that time coal's share of the energy market has declined steadily. Absolute world coal production levels have increased slowly because of the rapid growth in total energy demand. By 1976 coal supplied only about 33 percent of the world's energy demand.

The world is now heavily dependent on petroleum for its energy supplies and much of these resources is concentrated in a very limited geographic area. Estimates of when world oil production will peak vary widely--from the mid- to late-1980's to after the turn of the century. When peak production levels are approached, world oil prices will rise and the world may turn again to coal to provide the energy it needs for economic growth and to sustain industrial production.

---

[1]Consultant, Resources for the Future.

The future potential importance of coal among the conventional mineral fuel resources is shown in Table 18-1. Coal resources (even allowing for only a 50 percent recovery) represent about 80 percent of total conventional fuel resources[2] and about 55 percent of conventional fuel reserves[3] when nuclear fuels are used in conventional nuclear reactors and 12 percent of total world resources and 4 percent of world reserves when nuclear fuels are used in breeder reactors.

Coal reserves alone would be sufficient to supply all of the energy required at current rates of world energy consumption for over 60 years; coal resources would last for nearly 500 years.

ESTIMATION CRITERIA

A large number of estimates have been made of the world's coal resources and reserves and the reported amounts vary due to a variety of causes. Different definitions have been used in different countries for the terms "resources" and "reserves," and estimates made at different times reflect actual changes in resource estimates as a result of new geological information. Even when thickness of seam and maximum depth for hard coal and brown coal are specified there are considerable deviations in a number of countries from these specified values because of the way in which the data were collected. Moreover it is difficult to establish a firm differentiation between hard and lignite coal that can be used by all countries.

In those countries with coal resources sufficient to supply their demand for decades, there is little incentive to explore further to identify the amount of the resources with greater precision; the already identified reserves are sufficient to supply all the coal needed for many years. In that case, resource information is of interest mainly to policy makers, not to coal producers, and most governments are unwilling to expend funds for information needed largely for long-range planning. In many countries that report small amounts of coal there have been only minor amounts of exploratory

---

[2]The total natural stock of a mineral is the "resource base." That portion of the resource base that has been estimated to exist and that might be recovered in the future is termed "geological resources."

[3]That portion of the geological resources which has been located with considerable certainty and that can be recovered with present technology at competitive costs is termed reserves.

TABLE 18-1. World recoverable reserves and resources of conventional mineral fuels

Region or nation	Coal (10^9 metric tons coal equivalent)		Oil (10^9 barrels)		Gas (10^{12} cubic feet)		Uranium (10^3 metric tons U)	
	Reserves	Resources	Reserves	Resources	Reserves	Resources	Reserves	Resources
United States	178	1,285	29	110-185	205	730-1,070	643	1,696
Canada	9	57	6	25-40	59	230-380	182	838
Mexico	1	3	16	145-215	32	350-480	5	7
South and Central America	10	14	26	80-120	81	800-900	60	74
Western Europe	91	215	24	50-70	143	500	87	487
Africa	34	87	58	100-150	186	1,000	572	772
Middle East	--	--	370	710-1,000	731	1,750	--	--
Asia and Pacific	40	41	18	90-140	89	a	45	69
Australia	27	132	2	b	31	500	296	345
Soviet Union	110	2,430	71	140-200	910	2,850	N.A.	N.A.
China	99	719	20	a	25	a	N.A.	N.A.
Other Communist areas	37	80	3	a	10	a	N.A.	N.A.
Total	626	5,063	642	1,450-2,120	2,502	8,710-9,430	1,894	4,288
Quintillion (10^{18}) Btu[c]	17.7	140.6	3.7	8.4-12.3	2.6	8.9-9.7	7.4 (LWR) 443.2 (FBR)	16.7 (LWR)[d] 1,003.4 (FBR)

[a]Included in Soviet Union.

[b]Included in Asia and Pacific.

[c]Heat contents: coal, 27.78 million Btu per metric ton; oil, 5.8 million Btu per barrel; gas, 1,025 Btu per cubic foot; uranium, 390 billion Btu per metric ton U (without breeder) or 23.4 trillion Btu per metric ton U (with breeder).

[d]LWR--light water reactor; FBR--fast breeder reactor.

NOTE: All resource figures are cumulative. They include reserves. The figures for international coal reserves and resources in this table are given in metric tons of fixed heat content rather than actual metric tons.

TABLE 18-1. World recoverable reserves and resources of conventional mineral fuels--Continued

Sources: Coal--World Energy Conference, Coal Resources (Guildford, England, IPC Science and Technology Press, 1978). A recoverability factor of 50 percent is assumed for resources.

Oil and gas--Reserves: Oil and Gas Journal, vol. 76, no. 52 (25 December 1978), p. 102-103; U.S. resources: U.S. Geological Survey; Canadian resources: Department of Energy, Mines, and Resources as reported in International Petroleum Encyclopedia 1978 (Tulsa, Okla., Petroleum Publishing Co., 1978), p. 39; Mexican resources: PEMEX data as reported in Mexico's Oil and Gas Policy: An Analysis, Committee Print, Joint Economic Committee, 95 Cong. 2 sess. (Washington, D.C., December 1978), p. 13. An uncertainty margin of 20 percent for possible resources is introduced; other oil resources, Richard Nehring, Giant Oilfields and World Oil Resources, Report R-2284-CIA (Santa Monica, Calif., Rand Corporation, 1978), p. 88; other gas resources, Joseph D. Parent and Henry R. Linden, A Survey of U.S. and Total World Production, Proved Reserves, and Remaining Recoverable Resources of Fossil Fuel and Uranium (Chicago, Ill., Institute of Gas Technology, January 1977), p. 15.

Uranium--International Atomic Energy Agency/OECD Nuclear Energy Agency, Uranium: Resources Production and Demand (Paris, Organization for Economic Co-operation and Development, 1978), p. 20-21. The 300,000 metric tons of Swedish uranium have been excluded from the reserves.

TABLE 18-2. World resources of bituminous coal and anthracite (10^6 tons of coal equivalent)[a]

Continent (countries)	Geological resources	Reserves
Africa	172,714	34,033
Botswana	(100,000)[b]	(3,500)
South Africa	(57,366)	(26,903)
North America	1,308,541	126,839
United States	(1,190,000)	(113,230)
Canada	(96,225)	(8,706)
Asia	5,494,025	219,226
China	(1,424,680)	(98,883)
USSR	(3,993,000)	(82,900)
Australia and South Seas	213,890	18,164
Australia	(213,760)	(18,128)
Europe	535,664	94,210
Federal Republic of Germany	(230,300)	(23,919)
Poland	(121,000)	(20,000)
United Kingdom	(163,576)	(45,000)
	7,724,834	492,472

[a]"Equivalent Coal" has a heating value of 25.2 million Btu per short ton.

[b]Those numbers shown in parentheses reflect only the resources of the country with the largest reserves within that continent and are not addable to the totals.

Source: World Energy Resources 1985-2000. World Energy Conference, IPC Science and Technology Press, 1978, p. 66-67.

TABLE 18-3. World resources of subbituminous coal and lignite (10^6 tons of coal equivalent)[a]

Continent	Geological resources	Reserves
Africa	190	90
Nigeria	(190)	(90)
America	1,408,838	71,081
United States	(1,380,000)	(64,358)
Canada	(19,127)	(673)
Asia	387,127	29,626
USSR	(867,000)	(27,000)
China	(13,465)	(N.A.)
Australia and South Seas	49,034	9,333
Australia	(48,374)	(9,225)
Europe		
Federal Republic of Germany	(16,500)	(10,500)
Yugoslavia	(10,823)	(8,430)
	2,400,430	143,992

[a]"Equivalent Coal" has a heating value of 25.2 million Btu per ton.

[b]Those numbers shown in parentheses reflect only the resources of the country with the largest reserves within that continent and are not addable to the totals.

Source: Same as Table 18-2.

activity so that little is known about the actual size of the resource that may be eventually discovered. This is particularly true in less-developed countries where funds for exploration activities are lacking.

ESTIMATION OF COAL RESOURCES AND RESERVES

Table 18-2 shows estimates of the world resources and reserves of bituminous coal and anthracite in tons of coal equivalent by continent and by countries with the largest resources within each continent, as reported by the World Energy Conference (WEC).[4] Total geological resources total 7.7 x 10^{12} tons of coal equivalent while reserves total 4.9 x 10^{9} tons of coal equivalent.

Table 18-3 gives the same information for brown coal and indicates a total of 2.4 x 10^{12} tons of coal equivalent of geological resources and 143.9 x 10^{9} tons of coal equivalent of brown coal reserves. Total geological resources of coal of all ranks are 10.1 x 10^{12} tons of coal equivalent and 635 x 10^{9} tons of coal equivalent for reserves.

As shown in Table 18-4, estimates of geological resources of coal have increased between 1974 and 1977 by over 2 trillion tons, while reserves estimates have increased over 160 billion tons. The increase in reserves since 1974 is to be expected because the quadrupling of world oil prices following the 1973/74 embargo led many countries to reevaluate their coal resources.

TABLE 18-4. Comparison of the 1977 World Energy Conference (WEC) estimates with earlier WEC estimates

	(10^{9} tons of coal equivalent)		
	Geological resources	Reserves	Reserves as a percentage of resources
World Energy Conference 1977	10,125	636	6.3
World Energy Conference 1976	9,045	560	6.2
World Energy Conference 1974	8,063	473	5.5

[4] An Appraisal of World Coal Resources and Their Future Availability, W. Peters and H. D. Schilling; World Energy Resources 1985-2020, World Energy Conference, IPC Science and Technology Press, 1978.

The trend toward higher values for coal resources and reserves can be expected to continue. Estimation of coal resources is much simpler and less expensive than for oil and gas resources. Coal deposits tend to be much more uniform in many locations and often extend for hundreds of miles. In addition, many coal deposits crop out at the surface and are generally at shallow depths compared to oil and gas, so that drilling costs for estimating reserves or finding new deposits are comparatively low. In spite of this, wide variations in estimates appear in the literature and these have been discussed by D. C. Ion.[5]

Table 18-2 indicates that for hard coal and anthracite three countries dominate both the resource and reserve base--the USSR, China, and the United States--all countries in the Northern Hemisphere. Three countries of the Southern Hemisphere have relatively large coal deposits--Australia, South Africa, and Botswana. The three countries in the Northern Hemisphere together have 85 percent of the geological coal resources and 59 percent of the reserves. Table 18-3 indicates that the USSR and the United States dominate both the geological resources and reserve base of subbituminous coal and lignite. Together the USSR and the United States have 93.6 percent of the geological resources and 40 percent of the reserves. For coal resources of both ranks the three countries have 87 percent of the geological resources and 60 percent of the reserves.

It should be noted that of the ten major oil producing countries, only the USSR and the United States also have major resources of coal. This is fortunate for those countries which import energy, since it permits a greater diversity of both source and type of supply; this should lead to greater competition among suppliers, which would be in the best interest of the importing countries.

Because of the vast size of world coal resources, countries that appear to have only very modest-sized coal deposits can have very significant amounts of energy supply. For example, Czechoslovakia, whose share of world hard coal is so small that it does not even appear in Table 18-2, has coal reserves containing the same amount of energy as the oil reserves of Asia.

---

[5] Availability of World Energy Resources, D.C. Ion, Graham and Trotman Limited, United Kingdom, 1975.

RESOURCE ESTIMATE LIMITATIONS

The wide disparity in coal resources among the continents may be a true indication of how resources are distributed geographically, but it is probably a reflection, at least in part, of large differences in the intensity of exploration efforts. Africa and South America have been only sporadically examined for coal and other mineral deposits. Even in countries where large coal deposits are known to exist, for example, Colombia, their size and nature have only been superficially examined. Moreover, some countries know so little about their coal resources that they do not report at all.

While the more detailed country reports on which the totals in Tables 18-2 and 18-3 are derived contain some additional information on the rank of the coal, whether it can be recovered by surface or underground mining methods, coking quality, ranges of sulfur content, and heating value, and some information on their chemical analysis, the data are too spotty to classify most coal resources by any of these important parameters.

The value to the user of the reported resources would be greatly improved if information on coal resources could be reported in at least the detail shown for the United States in Table 18-5. This table shows by general geographic area in the United States the amount of coal by rank, whether it can be mined by strip or underground methods, and the amount of each type that is low in sulfur (<1 percent).

The value of total resources that can be classified as reserves is now reported to be 6.3 percent. However, this percentage can be expected to increase as new mining technologies are developed that either permit extraction of coal at lower cost or a higher percentage of the coal in place at the same cost. Increasing the price of competitive fuels (oil and gas) will tend to increase coal prices and lead to additional coal being classified as reserves rather than resources.

AVAILABILITY OF COAL RESERVES

The major producers of coal in 1975 are shown in Table 18-6. The three largest producers are the same three countries with the largest coal deposits--the USSR, the USA, and China. However, three other countries produced over 100 million tons in 1975 and five countries produced approximately 70 million tons. Thus, a large production base exists which could be expanded if worldwide markets for coal developed.

TABLE 18-5. U.S. coal reserve base by area, type, mining method, and sulfur content (billion short tons)

Type	East of the Mississippi River			West of the Mississippi River			Total		
	Surface	Underground	Total	Surface	Underground	Total	Surface	Underground	Total
Anthracite	--	7	7	--	--	--	--	7	7
Bituminous	39	155	195	8	27	34	47	182	229
Subbituminous	--	--	--	61	108	168	61	108	168
Lignite	1	--	1	33	--	33	34	--	34
All types	41	162	203	101	135	235	141	297	438
Percentage of low sulfur coal[a]	19	20	20	69	83	77	55	49	51

[a]Represents all coal with a sulfur content by weight of less than 1 percent as a percentage of all coal for which sulfur content is known. Sulfur content is known for at least 84 percent of all coal in each category presented.

NOTE: Figures may not add to totals due to rounding. The U.S. Bureau of Mines terminology defines reserves as recoverable reserves and reserve base as in-place reserves. Thus, the figures presented in this table refer to in-place quantities.

Sources: U.S. Bureau of Mines, "Demonstrated Coal Reserve Base of the United States on January 1, 1976," in Mineral Industry Surveys (Washington, U.S. Bureau of Mines, August 1977); U.S. Bureau of Mines, The Reserve Base of U.S. Coals by Sulfur Content, U.S. Bureau of Mines Information Circulars 8680 and 8693 (Washington, GPO, 1975).

TABLE 18-6. Estimated coal production of major producing countries in 1975
(millions of tons of coal equivalent)

Australia	69
China	349
Czechoslovakia	80
Federal Republic of Germany	126
German Democratic Republic	75
India	73
North Korea	34
Poland	181
Republic of South Africa	69
United Kingdom	129
United States	581
USSR	614
Total	2,593

Source: Coal Resources--The Full Reports to the Conservation Commission of the World Energy Conference, IPC Science and Technology Press, 1978, p. 106.

A substantial world coal trade has existed for many years. For the past 20 years, shipments have ranged from 160 to 220 million short tons per year, with 210 million short tons being traded in 1977. Most of the coal exports were from the United States, Australia, South Africa, Federal Republic of Germany, and Poland. In 1977, the United States was the largest exporter (24.8 percent) and Poland was second (19.7 percent). The major importers in 1977 were Japan, France, Canada, and Italy, which together received over one-half of total world imports. Seven other countries, of the 26 countries importing coal, imported more than 5 million tons in 1977.

Most of the coal moving in international trade has been metallurgical coal used to produce the coke needed in steel making. Some steam coal has been traded, mainly when it was the lowest cost coal that could be obtained from a neighboring country by users near an international boundary; this was largely in Western European markets.

The oil embargo of 1973-1974 resulted in a reexamination by many countries of the desirability of the continued shift toward oil and away from coal. The increase in the real price of oil made coal competitive in many areas where it had not been before. This, together with a desire by most of the oil-importing countries to reduce their dependence on oil imports, so as to limit their vulnerability to disruptions in oil supply and their negative balance of payments, led to expectations that the historic decline in importance of coal worldwide would be reversed.

A sudden and very large demand for coal would place some strain on the production, transportation, distribution, and utilization infrastructure, even though there are vast quantities of coal resources that could be developed. In the production phase, long lead times are required to bring new mines up to planned capacity, environmental regulations must be met, long-term and high-cost financial commitments must be made, and underground miners must be recruited. The health and safety records of the underground coal mining industry have been very poor, which represent a deterrent to recruiting new underground miners. Labor-management relations in the coal industry have been stormy in a number of major coal-producing countries, so that potential consumers are wary about the security of coal-based energy supplies.

In some coal-producing countries, transportation could be a bottleneck to increased movement of coal in world trade, because of the poor physical and financial condition of the railroads. Maritime movements of coal to countries where coal has not been traditionally imported could be constrained by inadequate port facilities--lack of berths for large coal-carrying ships and limited unloading facilities. In these countries, a coal distribution infrastructure would have to be created and manufacturers found for producing equipment to utilize the coal efficiently. Environmental regulations could limit coal use by increasing the cost of production and transportation and by requiring the use of control technologies during coal combustion.

In spite of these potential constraints to increased utilization of coal worldwide, the large relative size of the coal resource base compared to oil and gas suggests that world oil and gas production will peak soon, and that oil and gas prices will rise much more rapidly than those of coal. Since none of the constraints on world coal trade appear to be limiting, increased coal use worldwide should occur until new, economically-competitive renewable resources can be developed by large energy markets.

CHAPTER 19

WORLD COAL PROSPECTS TO 2000

James R. Reddington[1]

This chapter deals with the prospects for coal in future world energy markets. The International Energy Agency (IEA) welcomes the initiative shown by the United Nations in focusing world attention on the improved prospects for coal usage and world trade.

At the end of last year the IEA published a study--Steam Coal Prospects to 2000--that examined the substitution for oil by coal in both industrialized countries and developing countries. This chapter discusses the findings of that study and how they have been modified in recent months during the preparation of the IEA's upcoming World Energy Outlook to 2000. Events this year have required us to revise upward our projections of oil and energy prices for the next two decades, as well as to revise downward our estimate of economic growth, at least in the near term. These reassessments, however, have had only a modest impact upon our outlook for coal; they do not change the thrust of our conclusions in the Coal Study. If anything, higher oil prices appear to increase the opportunities for substitution of oil by coal.

STEAM COAL

The Steam Coal study contains projections to the year 2000 based on two cases. The first is the reference case, embracing alternative assumptions of lower and higher nuclear capacity expansion that capture the extent of uncertainty about the future of nuclear development. The second case--the enlarged coal case--is characterized by a more intense policy commitment to faster coal development and more extensive substitution of oil by coal.

---

[1]International Energy Agency, Paris, France.

COAL EXPANSION IN THE BASE CASE

The overall results of the reference case projections are that, even with the successful implementation of nuclear power programs implied in the high nuclear subcase, and in spite of the more than two-fold increase in OECD coal consumption expected between 1976 and the latter part of this century, projected energy requirements are not fully met by the sum of the expected supplies of indigenous and imported fuels. The quantity of oil that oil-exporting countries may make available for export is insufficient to satisfy expected demand, and the result of both nuclear subcases is a potential excess demand for energy that rises steadily throughout the period after 1985.

Energy requirements in the OECD are projected to grow by 2.7 percent per year through the year 2000, while economic activity expands at an average annual rate of 3.4 percent over the same period. In the OECD's energy supply sectors, indigenous oil production increases until 1985, then declines. Natural gas production declines steadily throughout the period. Therefore, substitution of other forms of energy becomes increasingly necessary. Nuclear energy grows at average annual rates of 7.9 percent and 9.0 percent in the low nuclear subcase and the high nuclear subcase, respectively, after 1985. Coal demand grows 3.2 and 2.7 percent annually in the low and high nuclear subcases, respectively, from 1976 through 2000.

In the reference case, between 1976 and 2000, OECD coal utilization in the low nuclear subcase rises by 1,096 million tons of coal equivalent (Mtce) or 767 million tons of oil equivalent (Mtoe), the equivalent of almost 15 million barrels of oil per day (Mb/d). As Table 19-1 shows, substantial increases in thermal coal utilization occur in every region of the OECD and in several industrial sectors as well as in power generation. The more than doubling of thermal coal use within 24 years foreseen in the reference case is, itself, an ambitious task that will require substantial new investments in coal production, transportation, and utilization projects.

The reference case shows that--in spite of a very significant increase in coal utilization--substantial potential excess energy demand exists in both nuclear subcases. An enlarged coal case, with even more extensive substitution of coal for oil, was constructed to show how much of the potential excess energy demand of the reference case could be met by the adoption of more intensive coal policies and programs favoring an even faster increase in coal utilization.

TABLE 19-1. OECD thermal coal utilization
[reference case (low nuclear)]

OECD region	Total uses 1976 Mtoe	Total uses 2000 Mtoe	Total uses Percent change	Power generation only 1976 Mtoe	Power generation only 2000 Mtoe	Power generation only Percent change
North America	308	705	129	269	592	120
Europe	159	371	133	126	297	136
Pacific	32	116	263	30	118	293
OECD Total	498	1,193	140	425	1,007	137
IEA Total	455	1,087	139	391	922	136

ENLARGED COAL CASE

If industrial countries implemented coal substitution policies in the years immediately ahead, such as those recommended in the Study, a massive substitution of coal for oil could be realized toward the end of this century. With due consideration for the technical and economic limits to coal substitution determined by process requirements, lead times, environmental restrictions, and the oil price assumption of this study, realization of the substitution potential shown in the enlarged coal case would increase OECD thermal coal utilization over the reference case (low nuclear subcase) levels by 100 Mtce (70 Mtoe), in 1985, 219 Mtce (153 Mtoe) in 1990, and 504 Mtce (353 Mtoe) in 2000.

The additional coal utilization shown in the enlarged coal case would permit corresponding reductions in demand for oil and natural gas. If the freed-up natural gas were also substituted for oil, OECD oil demand and net oil import requirements could be reduced by 1.4 Mb/d in 1985, 3.0 Mb/d in 1990, and 7.0 Mb/d in the year 2000.

During the period in question, coal cannot satisfy the entire potential excess demand shown in the reference case. The remaining potential imbalance between energy supply and demand not filled by coal will require more vigorous policies to promote additional conservation and expansion of other energy supplies, but the burden placed on these alternatives can be alleviated considerably by concerted efforts to promote the expansion of coal utilization and trade.

Thus, coal emerges as an attractive substitute that has the potential, with the proper policy support, to make a significant contribution by 2000. Furthermore, increased utilization of coal in the near future may speed the development of new technologies that would allow even greater substitution for oil by coal-based liquid and gaseous fuels in the latter part of this century.

OTHER FINDINGS: DEVELOPMENT AND TRADE

Faster development of mining must await development of demand but once demand begins to grow strong, present leasing schedules, resolution of environmental problems, fiscal regimes, and, initially, rail and slurry pipeline infrastructure could become constraints upon expansion unless corrective policies are implemented. As international trade grows, port and inland waterways will need expansion. But the precedent expansion of demand also requires a boost in incentives for the needed capital investment, amended ordinances to allow creation of cogeneration and energy parks, and, above all, workable environmental rules on fuel choice, emissions control, and land usage. An agreement among IEA Member governments was announced in April 1979 to ensure freer trade in coal to assist in promoting the confidence of traders and investors in an unobstructed international trade.

PATTERNS OF FUTURE COAL TRADE

The Suppliers

Within the OECD, Australia is thought likely to be the largest exporter, increasing its present exports more than six-fold by the year 2000. Australia's exports of hard coal have been rising since the early 1950's, spurred by the growth of the Japanese iron and steel industry (Japan takes more than 75 percent of all Australian hard coal exports). However, in the recent past, the policy has been to diversify exports, and new demand has made this possible.

By 1985, hard coal exports could be almost double those of 1975, more than a quarter going to Japan for electricity production but also to Western Europe and other Asian countries (e.g., Taiwan and South Korea). In time it should be possible for Australian coal to maintain a favorable competitive position in the Japanese and other fast-growing Asian markets against oil and even other sources of coal.

The Australian coal industry is one of the world's most efficient, measured in terms of labor productivity; in late 1976, hard-coal mining costs for new underground operations were estimated at \$18 per ton, of which 50 percent represents labor costs, and \$8 to \$15 per ton for new surface mines (largely in Queensland).

Estimates of capital costs, on the other hand, are quite high in some areas because of the need for transport infrastructure and new towns for workers and their families in the sparsely populated coalfield area.

The second largest supplier within OECD is projected to be the United States, which also has by far the largest reserves in the OECD area--65 percent of the total and almost 30 percent of those estimated as technically and economically recoverable--but much of it is coking coal. Use of coal is encouraged by the recently enacted energy law which, with some exceptions, outlaws new oil- and gas-fired power plants. But the law tilts toward the Eastern deep mines which are high in sulfur content and high in cost, because they are labor intensive and because of safety, health, and work rules. The Western Plains have more accessible, low-sulfur surface mines, which could be more labor productive and less costly to operate; however, despite the low-sulfur content of the coal, its use will require desulfurization equipment and hence, a cost penalty; transport costs are also higher--up to 75 percent of the cost of delivered coal.

A competitive transportation network is required to realize coal export potential and domestic use of coal. (A controversy waxes as to whether Western coal should be transported by rail or pipeline.)

Most (80 percent) of the potential Western coal fields are owned by the Federal Government (Eastern deposits are almost exclusively privately owned) and subject to a 1971 moratorium on leasing, pending a land-use plan. After 1990, this moratorium could serve as a constraint on production.

The key to greater coal usage in power generation is the cost of controlling sulfur oxide emissions. Coal-burning power plants currently account for one-half of all sulfur oxide emissions, and emissions from nearly half the coal now consumed in power plants fail to meet existing clean air standards.

Still another potential source of coal within the OECD area is Canada, projected to have a small net export by 1985 which could double by 2000, again mainly coking coal. Canada already supplies coking coal to Japan from the Canadian West, where 99 percent of Canadian coal resources are located. Canadian coal could also be exported to Japan for use in power plants at a cost competitive with that of the U.S. and not much higher than Australian coal.

However, three Canadian provinces plan to permit exports only if it can be shown that a surplus exists after taking account of domestic requirements, present and future. Two provinces have introduced environmental policies which prohibit coal projects in environmentally sensitive areas, and limit development in certain others (by prohibiting strip mining, for example). Just how restrictive these policies will be is not yet clear. A mine/power-plant complex proposed in Calgary was rejected on the grounds that it would disturb a prime wheat growing area and that alternative sites for power generation could be developed on less productive farmlands.

Of the sources outside of OECD, South Africa is by far the most promising. It is the only developed market economy primarily fueled by coal (almost 75 percent of total primary energy requirements in 1976 while oil supplied only 25 percent) and is the only country where commercial coal liquefaction is carried out.

Operations are mostly underground, but surface mining has been increasing steadily as labor costs mount in this labor-intensive industry. Costs are low despite the predominance of underground and labor-intensive operations.

South Africa has traditionally exported small quantities of coal, but the amount has increased vastly since 1973, making South Africa the fifth largest coal exporter in the world. This expansion was made possible by the opening of a new coal terminal at Richards Bay. It is one of the world's most modern and can now service 150,000 dwt vessels; ultimately 250,000 dwt vessels could be handled. Richards Bay is connected to the Transvaal coal-producing region by a railway line built exclusively to service the port.

The centrally planned economies could be the next largest group of exporters. Poland has exported roughly a fifth of its coal, mostly coking, to other centrally planned economies and to Western European countries. Because of difficulties in meeting its own demand, Poland cannot be looked to for significant new volumes of coal, at least before 1990. The USSR and China have huge reserves but have exported little, although both economies have been turning to oil and natural gas. In the wake of high energy prices, Soviet planners are trying to reverse the long-term trend away from coal use, and the long-range target for coal production appears to be a level of 1 billion tons by 1990, produced mainly east of the Urals, far from consumption centers. This will certainly pose serious problems of infrastructure.

However, some limited expansion of present exports can be expected from the Soviet Union. A new deep-water port, under construction at Vostochniy on the Siberian Pacific coast, will handle coal from the Southern Yakutian basin, 1,200 km to the West. Much of the coking coal will be exported to Japan, starting in the early 1980's, under a long-term supply contract.

Prospects for coal in China are difficult to establish. The current drive to modernize the economy will intensify future energy needs. But inadequate allocation of resources to coal mining and the favoring of oil development will tax coal authorities in meeting domestic demand and might preclude any important growth of exports. The new coal port in Lienyunkiang seems to imply, however, a moderate increase of coal exports in the future.

India has by far the largest present production of hard coal in the developing countries, using much of it herself. South Korea also produces and consumes coal. Aside from these two countries, there has not been much exploration for coal in the developing world; but, from available surveys, IEA tentatively concludes that countries such as South Korea, Taiwan, and the Philippines may import coal for coastal power plants while Colombia, Venezuela, Mozambique, India, and, to a lesser extent, Botswana, could export coal.

ENVIRONMENT

Before concluding, mention should be made of the Study's conclusions about environmental concerns. Environmental problems stem from coal mining, transportation, utilization, degradation of land, air, and water, as well as increased water consumption; these have been addressed in the light of what has been done, what is being done, and the costs involved.

The known environmental impacts of the entire coal cycle may be divided into health and safety risks (those of combustion to the general public and

Projected world coal trade
(present policies lower nuclear estimate--Mtoe)

	1976	1985	1990	2000
Canada	+2.1	-4.2	-6.3	-9.8
United States	-37.8	-47.6	-55.2	-90.2
OECD Europe	+38.5	+65.7	+107	+217.5
Japan	+42	+67.1	+86	+126.6
Australia	-21.7	-39.9	-62.9	-136.4
Total	+23.1	+41.1	+68.6	+107.7
Centrally planned economies	-26.6	-30.1	-34.3	-46.2
Developing countries	+4.9	+7.7	+1.4	-7
South Africa	-4.2	-23.8	-42	-62.9
Other	+2.1	+4.9	+6.3	+8.4
Total	-23.8	-41.3	-68.6	-107.7

Composition of coal trade in 2000
(Mtoe)

	Metallurgical	Steam
Canada	-16.0	+6.3
United States	-48.9	-41.3
OECD Europe	+39.9	+177.6
Japan	+72.7	+53.8
Australia	-52.0	-84.0
Centrally planned economies	-17.5	-28.7
Developing countries	+18.2	-25.2
South Africa	--	-62.9
Other	+4.2	+4.2

+ = Imports

- = Exports

of mining accidents to miners) and physical risks (those of damage to land, water, plants, and the ecosystem). The second class of risks is better understood. For most of these, manageable control technology exists, although institutional arrangements for assessing and managing these risks are often inefficient and needlessly time-consuming. The first class of risks, health and safety, may not be as amenable to further substantial reduction by existing technology. Furthermore, other possible risks in the coal cycle, of unknown character or magnitude, are the subject of research; included are those linked to carbon dioxide, heavy metals, and synergistic effects, but as yet the evidence is not such as to offer any policy guidance for coal development. For example, a public concern has risen over trace elements in coal, such as lead, cadmium, and uranium; however, it is not yet established whether these elements are released into the atmosphere during combustion or retained in the ash.

Overall, the expanded use of coal will require some compromise or harmonization between environmental and energy objectives, with careful selection of control measures that are most cost-effective, and with the higher costs borne by energy consumers. The costs of meeting current environmental standards are given in the Study and are taken into account in the comparative costing of utility fuels.

CHAPTER 20

BROWN COAL--AN IMPORTANT ENERGY RESOURCE

Hans Weise[1]

Often neglected as an energy source because of its low calorific value, brown coal has seen an upsurge in development in recent years, primarily as a result of oil price increases. Countries, however, with limited alternative resources have emphasized brown coal recovery for many decades. These include some Eastern European and Asian countries, as well as West Germany and Australia.

As an energy source with very low heat value, brown coal is sensitive to transportation. Consequently, regional importance prevails, with oil, gas, and hard coal dominating the markets distant from the deposits. Mine-mouth use is preferred with--at present--emphasis on power generation. Upgraded brown coal in the form of briquettes is sometimes transported in barges or railroad cars as far as 400 miles, competing successfully with oil and gas.

Beyond doubt, recent developments in the oil-producing countries will stimulate conversion of brown coal into more convenient energy forms by making use of its high reactivity, thus widening its market. However, gasification and liquefaction of brown coal have already experienced more financial and environmental setbacks than expected and the road to production-size plants may be long.

Very recently, tremendous planning activities have started to make lignite available in the Gulf Coast States of the United States of America, primarily for use in mine-mouth power stations but also for gasification or other industrial use. It is expected that production of lignite in the United States of America will reach more than 200 million tons in the year 2000, up from approximately 60 million tons in 1978. This compares to 1978's world production of brown coal of more than 900 million tons.

---

[1]Rheinbraun-Consulting U.S.A., Inc., Englewood, Colorado.

Among fossil energy resources coal ranks first in total reserves, with brown coal contributing 2,600 billion tons or somewhat more than 10 percent in terms of total heat value. One hundred sixty billion tons are considered recoverable with present day technology.

Table 20-1 shows world production subdivided by countries. A production increase of barely 1 percent in the last 3 years has to be compared with more ambitious targets that sum up to 1.2 billion tons of world production of brown coal in 1985. This includes, for instance, a doubling of production in countries like Poland, Turkey, Spain, and India, each of which has started tremendous investment programs.

Though amounting to only 5 percent of world energy consumption, brown coal provides 50 percent and more of the energy needs of the GDR, USSR, and Yugoslavia.

In West Germany brown coal's share of primary energy production is 21.6 percent. Since 35 percent of all power produced in public power plants requires lignite, a secure long-term market exists. Production will hover around 120 million tons with additional mining capacity making up for a deterioration of the brown coal:overburden ratio.

The USSR, with 1.7 billion tons of resources--making it by far the richest brown coal nation--finds most of its deposits that can be economically surface mined in the Asian part of the country. Thus brown coal mining will be in step with the overall development of this part of the continent. At present, the USSR still relies heavily on underground mining.

Brown coal deposits are quite often significant in three ways:

- They are close to the surface.
- The thickness of brown coal seams may reach 100 m or more.
- Overburden and interburden are often unconsolidated.

All three factors favor surface mining, which is also reflected in Table 20-2.

Surface mining allows a high degree of mechanization and the use of the heaviest earthmoving equipment known today. Mining operations often are aided by the unconsolidated nature of overburden and interburden.

Both Germanys depend heavily on brown coal as a prime energy resource. Naturally both countries have developed mining technologies that are outstanding for their efficiency. Long before the Second World War continuous mining machines were manufactured and in use to allow reduction of

TABLE 20-1. World brown coal production
(1,000 tons)

Year	1974	1975	1976	1977	1978
France	2.760	3.185	3.189	3.081	2.733
Italy	1.960	2.040	2.028	1.871	1.890
Germany (FRG)	126.044	123.377	134.536	122.948	123.587
Albania	850	850	890	1.000	1.000
Bulgaria	23.998	27.515	25.188	24.864	24.960
CSSR	82.788	86.268	89.472	93.240	95.292
Germany (GDR)	243.468	246.706	246.888	253.704	253.272
Greece	14.268	18.408	22.572	23.400	22.368
Yugoslavia	32.976	34.944	36.264	38.592	39.012
Austria	3.629	3.400	3.216	3.132	3.072
Poland	39.828	39.864	39.300	40.776	41.004
Romania	19.788	19.776	18.732	19.416	19.416
Spain	2.880	3.384	4.140	5.784	8.256
Hungary	22.553	21.864	22.320	22.524	22.716
USSR	157.716	160.212	160.032	160.032	159.600
Canada	3.486	3.549	4.680	5.484	5.040
USA	14.058	17.976	23.256	23.796	26.232
Australia	27.300	28.176	30.936	29.268	30.396
New Zealand	2.148	1.956	1.992	2.000	2.000
India	3.012	2.784	3.900	3.642	3.612
Japan	80	64	64	60	36
Korea, North	7.890	9.000	10.000	11.000	11.000
Mongolia	2.337	2.549	2.650	3.078	3.078
Thailand	480	456	684	444	526
Turkey	6.336	7.116	8.256	8.880	8.520

Source: Jahrbuch fuer Kohle 78/79, p. 805.

manual labor and concentration of high-capacity production. Continuing development led to large bucketwheel excavators (BWE), which may reach an output of up to 240,000 m^3 per day.

TABLE 20-2. Open-pit production as percent of total brown coal production (1972)

Area	% open-pit
World	84.3
GDR	100.0
USSR	67.6
CSSR	79.0
Poland	99.2
Bulgaria	88.9
Hungary	18.9
Romania	42.0
North America	100.0
Australia	100.0
FRG	99.6
Yugoslavia	44.6
Greece	97.4
Others	62.2
Estimate 1979:	90.0

Source: Gaertner "Die Bedeutung der Braunkohle als Energietraeger und Rohstoff." Braunkohle, Waerme und Energie, Sonderdruck S.6, Heft 12, 1973.

Though these units of 13,000 tons service weight are impressive in all respects, the principle of bucketwheel excavator mining is equally applicable at much smaller outputs as a result of its economics. Hence it is not surprising that brown coal mining in Australia, Greece, Poland, Romania, and India relies heavily on these machines, with Turkey and Spain soon joining this group of countries.

As a supplement to BWE mining, belt conveyors may take care of transportation. Two reasons why neither conventional truck haulage nor train haulage can be the ultimate choice are:

- The discontinuity of the aforementioned transportation system.
- The high and ever-rising cost of diesel fuel and rubber tires.

Also with the aforementioned giant-sized bucketwheel excavators, capacities are reached that simply cannot be matched by these transportation systems.

Mining depth is not critical. BWE's advance on a number of benches that are developed to recover the deepest-lying brown coal, if this is economically feasible.

A depth of 350 meters is presently being reached in West Germany with 600 meters soon to come. Obviously to open a deep-surface mine means the removal of a large volume of overburden first. A feature of a deep BWE mine therefore is the outside-overburden dump, which has to be integrated into the landscape as well as possible.

Spreaders are employed that dump overburden and interburden into the mined-out area. They have the capability to reshape the landscape within the limits dictated by the volume removed.

In Eastern European countries development of brown coal has followed along these lines, though not with equally rapid progress. Older mines in Russia are known to employ draglines for overburden removal, a mining method known the world over whenever a shallow-enough depth of deposit is given. Draglines typically expose brown coal by removing the overburden in one step. Consequently, the depth of the overburden determines the viability of the operation:

- The one-step operation is limited to approximately 180-ft depth of consolidated overburden and approximately 150-ft depth in the case of unconsolidated overburden.

There are mining operations known which exceed this depth with draglines. These, however, usually involve a considerable amount of rehandled overburden which rapidly decreases the economics of the mining venture.

Other conventional mining methods, but for lower capacities, involve shovels and trucks which in principle operate the same way in mining operations all over the world. Scraper operations gain importance under certain circumstances:

- In selective mining of top soil.
- In case of multiple-seam deposits.

Though fleets of more than 100 scrapers are known to be in operation in one mine, their application is also limited to lower production figures.

As mentioned earlier, brown coal deposits often stand out because of their nearness to the surface. Hence, surface mining is the preferred exploitation method. In surface mining one of the decisive economic factors is the stripping ratio, which determines the amount of overburden and interburden to be moved in order to expose 1 ton of lignite. Usually the overburden is measured in cubic meters. For simplicity's sake in many European countries brown coal is also measured in cubic meters. Since in situ brown coal weighs between 1.10 and 1.18 tons, the ratio already implies a mining loss of 10 to 18 percent.

Due to the very low calorific value of brown coal, the limit of economical mining is quickly reached if the volume of overburden to be removed is excessive. Stripping ratios in lignite mines are known to range from 1:1 to 1:20, the latter applying to better quality lignites found, for example, in the Gulf States of the United States of America. Whether lignite affected by a high stripping ratio is still economically mineable is very much dependent upon where it will be utilized. At present most lignite power plants are located near the mines to avoid long-distance transportation. This holds true for brown coal-fired power plants in the United States and Western Europe and for most of the future power stations as well. In the USA, though coal of higher quality from Wyoming and Montana is available south of the sunbelt, tremendous price increases for railroad transportation have stimulated lignite development in the Gulf States, despite the approximately three times higher mining cost per million Btu.

Australia in particular is blessed with lignite deposits of great extent, enormous thickness, and minimum overburden. The mining technique applied in Australia is comparable to the one used in Europe. Recent acquisition of bucketwheel excavator-spreader equipment also assures a future parallel development.

India's only major lignite deposit, in Tamil Nadu, was explored in the late 1950's and opened for mining in the early 1960's. Here also the same technology was applied.

The United States of America traditionally emphasizes draglines and heavy trucks in surface mining. But Gulf Coast lignite does not generally lend itself to the application of this type of equipment. It does not fall into the category of large-sized seams and occurs mostly in multiple seam deposits with single seam thickness seldom exceeding 10 feet. Furthermore, dragline operations prefer a more consolidated overburden than occurs in this region for slope stability reasons.

Mining methods planned for these deposits aim to achieve the following goals:

- Recovery of even the smallest seams, as thin as 1½ to 2 ft.
- Improvement of the stripping ratio by following the lignite to depths of more than 150 ft. (The generally accepted cut-off line for a one-bench dragline operation.)

Gulf Coast lignite offers an outstanding opportunity for the marriage of the well-known dragline (shovel) mining technique with the continuous mining technique of a bucketwheel excavator. The BWE would "pre-bench" the dragline operation, thus confining the latter to its optimum digging depth. Also, the unique capability of the BWE to mine topsoil, coal, and any useful overburden material selectively, enables it to recover local and lenticularly developed stray seams. It is not surprising that almost every Gulf Coast mine which is now in the planning stage studies the above-mentioned application.

Closely linked with brown coal surface mining is the dewatering of the mining area. Extensive dewatering is done in the Germanys and also in Poland. Wells equipped with submersible pumps are installed well in front of the mining operation to assure a drawdown of the water table once the mining operation approaches. In some cases (Australia) the brown coal itself is the aquifer. In the USA (Gulf Coast) a water table above the lignite is not usually encountered. Locally in Louisiana, dewatering may play a very important role. Besides the aforementioned pump-well method, cut-off by means of a slurry trench is also considered. It can be said that the occurrence of water does not impede the mining operation if an agreement is reached with regulatory agencies to discharge the water without harm to the environment.

A particular dewatering problem occurs in Southern India. Here the ground water table underneath the brown coal is under artesian pressure.

For years it was felt that this would effectively stop any lignite development. The pump-well method, appropriately applied, made it possible to exploit this deposit on a large scale.

Underground mining of brown coal has been restricted to areas in which surface mining is not possible, primarily because of the depth of the deposit. Much more labor intensive than surface mining, underground mining is less economical. A few countries, such as Russia and Yugoslavia, still rely heavily on it because local substitution of cheaper energy is not possible.

In West Germany attempts were made after the Second World War to mine Rhenish brown coal underground. Due to the thickness of the seams, mining losses in the order of 50 percent to 60 percent, however, could not be avoided. This terminated underground brown coal mining and consequently the characteristically deep and large surface mines were developed.

In the United States underground mining of lignite occurs in Wyoming and Montana. These lignites, however, run very high in heat value and top the scale of low Btu coal.

Large-scale brown coal mining calls for major investments. During the last 25 years the Rhenish brown coal mining industry has spent more than US $3.3 billion in the development of mines and an additional $3.9 billion for power stations. Since the environment is tremendously impacted by Germany's brown coal mining activity, approximately 9 percent to 10 percent of the brown coal production cost goes into reclamation (including relocation of roads, rivers, and villages). As a result of large-scale mining, the production cost per ton of brown coal still ranges in the vicinity of U.S. $1.00 to $1.20 per million Btu.

In the Gulf Coast area, economics are somewhat worse. Due to the aforementioned extremely high cost of railroad transportation, lignite produced at U.S. $1.30 per million Btu can presently be sold to mine-mouth consumers.

Australian coal costs are not known but can be expected to be much less than those quoted for West Germany. Unfortunately no costs are published for East European brown coal operations. South India's brown coal production costs are also unknown but are said to be hovering near the break-even point.

In the USA two factors might slow down lignite (brown coal) development in the South:

- The authorization of a coal slurry pipeline from Western States (Montana, Wyoming).
- The permission to use natural gas in newly built industrial boiler units.

Surface mining of brown coal cannot be adequately described without mentioning the environmental impact of the mining operation. Outstanding examples exist in Western Germany of the compatibility of modern technology and nature. West Germany's laws have lived up to the situation. A way was found to create, in cooperation with communities and industrial interests, a landscape in the mined-out areas that now provides recreation for a growing population.

The kind of reclamation to be carried out is determined by a committee in which the mutual interests of communities and industrial groups are represented. The mining industry has only two votes. Meticulous planning, sometimes decades in advance, has to be done by the mining industry and accepted by the committee. If these plans are followed carefully, no further interference by government agencies will occur.

In comparison hereto, the laws governing reclamation in the United States appear to be much less mature. A number of agencies, both federal and state, sometimes issue contradictory regulations, making it very difficult to comply. Also, whereas the German mine inspector has to be a graduate mining engineer, U.S. agencies often employ personnel with a lack of understanding of mining operations. This unfortunately leads to mutual distrust rather than cooperation. To my knowledge the excellent opportunity of land-value enhancement through reclamation has not yet been recognized. Restoration alone misses this opportunity. Again BWE conveyor-spreader technology facilitates land-value enhancement because the shaping of the landscape can be affected, within certain limits, without additional costs.

Reclamation in Eastern European countries follows very much along the lines of German reclamation. In Australia, the stripping ratio is so favorable that reestablishment of contours will never be achieved. In India not too much emphasis is given to reclamation; however, the mine drainage water is used in agricultural projects, converting a nearby semi-desert into much needed farmland.

With today's technology brown coal can be competitive if converted into forms of more valuable energy in the vicinity of the mines. The following uses are feasible in many countries:

- Direct reduction of iron ore.
- Gasification and liquefaction.
- Production of special coke.

With direct reduction the use of furnace coke can be avoided and energy saved.

Coke derived from brown coal has metallurgical importance, but it could also be used as a means to clean water and gas.

Liquefaction of brown coal, i.e., the production of gasoline, is a technology known for decades. Modernized methods may gain importance after the recent oil price hikes. But it looks as if gasification is the most promising method of coal conversion for the near future.

To maintain brown coal production or to increase it is in the interest of all the countries blessed with vast brown coal deposits.

REFERENCES

Jahrbuch fuer Bergbau, Energie, Mineraloel, Chemie 77/78, Verlag Glueckauf, Essen, p. 805.

Gaertner, "Die Bedeutung der Braunkohle als Energietraeger und Rohstoff," Braunkohle, Waerme und Energie, Sonderdruck Heft 12, 1973, p. 6.

UNO, Bulletin of Statistics.

Rheinische Braunkohlenwerke, unpublished documents.

CHAPTER 21

THE MEXICAN COAL INDUSTRY

M. Castañeda Perez[1]

THE ENERGY SECTOR

The Mexican energy sector has historically grown approximately 6.3 percent every year, the total energy demand in 1978 being 713 x 10^{12} kcal. Of primary energy, 87 percent of total demand is satisfied by oil and gas and 7.20 percent by geothermal and hydroelectric power, with coal accounting for only 5.20 percent, utilized in the steel industry.

This investment structure has been relatively unimportant for the generation of electricity, actually supplying only one small coal-fired plant with an installed capacity of 37.5 MW that represented only 0.31 percent of the total Mexican installed capacity of 12,109 MW in 1977, the last year the plant was in operation.

THE INDUSTRIAL STRUCTURE

Various entrepreneurial groups set up to produce coal are in existence at the present time:

- Siderurgica Mexicana (SIDERMEX), a state group composed of several iron-manufacturing enterprises, Altos Hornos de México, Fundidora de Monterrey y Siderúrgica Lázaro Cárdenas, and Los Truchas, of which only the first two have the capability, through their subsidiaries, of satisfying their own demands. Las Truchas must import its coal.
- The private sector. Under this category fall two enterprises with national majority capital--Industrial Minera México and Carbonífera San Patricio--that produce coal, the first to satisfy its own requirements and the second for sale to the subsidiary steel sector. There is also a group of small- and medium-scale producers that together account for approximately 18 percent of Mexican output.

---

[1]Ingeniero de Minas y Metalurgista, Mexico City, Mexico.

This share is totally committed for sale to the steel sector, either directly or through the state enterprise, Comercial Carbonera.

- The electrical sector. The Federal Electricity Commission, an agency of the public energy sector, began to produce coal in 1966 with an output of limited significance that has ranged from 130,000 to 180,000 tons per year.

  As a result of the expansion plans the electrical industry is making to generate power from coal, the enterprise Minera Carbonífera de Río Escondido was set up in 1978. Most of its capital stock derives from the government, with only a minimum of private initiative involved. This enterprise will exploit mines with a combined capacity of approximately 16,500 tons per day, supplying four thermoelectric units with a total installed capacity of 1,200 MW. The first of these plants will go into operation in 1981 and the last at the end of 1983.

EXPLORATION

Exploration for coal in Mexico is conducted both in the public and in the private sector, mainly by drilling.

The Council on Mineral Resources, a government agency, is now completing a program of exploration on the national level. The Federal Electricity Program is developing other projects in regions where the right kind of coal exists for its particular purposes.

The private sector is also conducting exploration to meet its own future needs.

COAL PROVINCES

At the present time, although precedents exist for black coal and lignite in different parts of the Mexican Republic, it is only the one coal province in the northern state of Coahuila that is of any significance (Figs. 21-1 and 21-2). It is there that the widest explorations have been carried out and hence where the major Mexican coal reserves have been defined.

On the other hand, in the states of Sonora in the northeast and Oaxaca in the south, deposits of anthracite have been explored in different areas, although their exact extent is not known.

Figure 21-1. Localities of carbonaceous deposits in Mexico

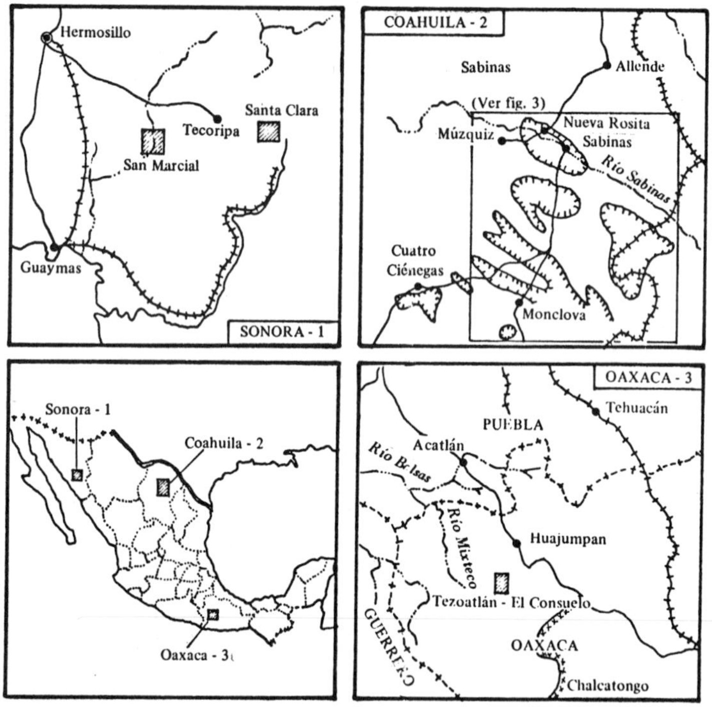

Figure 21-2. Locations of the most important carbonaceous provinces in Mexico

The Coahuila province covers a large part of the state. It consists of two subprovinces, Sabinas and Fuente-Río Escondido, the first with an area of 10,000 km^2 and the second with 2,000 km^2.

Geologically, the coals of these subprovinces are of the same Upper Cretaceous age. Nevertheless, they are of different grades. In the Sabinas subprovince, most are bituminous and subbituminous, with low or medium volatile matter and with ash content ranging from 20 percent to 45 percent. The Fuente-Río Escondido province, on the other hand, which is delimited by the border with the United States, contains nonbituminous and bituminous coals with a high content of volatile matter and high-flaming characteristics. In their physicochemical properties the coals from the Sabinas subprovince are coking types and those from Fuente-Río Escondido are by themselves noncoking.

In the Oaxaca province the deposits make up part of the Middle Jurassic, with semianthracite and subbituminous coals. The most important Upper Triassic and Lower Jurassic sites, where anthracite and metaanthracite coals are found, are in the area around San Marcial, Sonora State.

COAL RESERVES AND RESOURCES

Research and exploration conducted so far has shown that, as of December 1978, there were 1,916 million tons of measured coal reserves in situ in the state of Coahuila, 1,436 million in the Sabinas subprovince, and 480 million in Fuente-Río Escondido. Another 58 million tons are estimated to exist in Sonora and Oaxaca (Figs. 21-3 and 21-4).

The amount of the country's resources varies in accordance with various sources of information. Nevertheless, the geological data now available indicate a potential for increasing reserves by 1,300 million tons in Sabinas and by 1,000 million in Fuente-Río Escondido. The resources of the states of Sonora and Oaxaca are very difficult to estimate because of the nature of the geological formations and the scant data available up to now.

Thus we may say that estimated reserves and resources are of the order of 4,274 million tons in situ. Their grades and properties are sufficient to supply the needs of the steel (2,736 million) and electrical (1,480 million) industries.

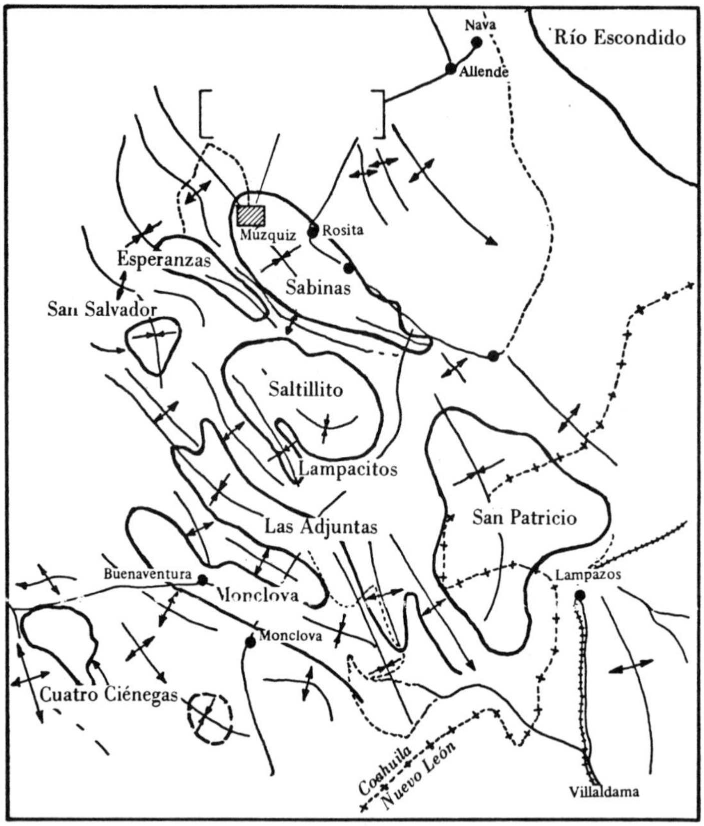

Figure 21-3. Location of carbonaceous subbasins in Sabinas subprovince, Coahuila

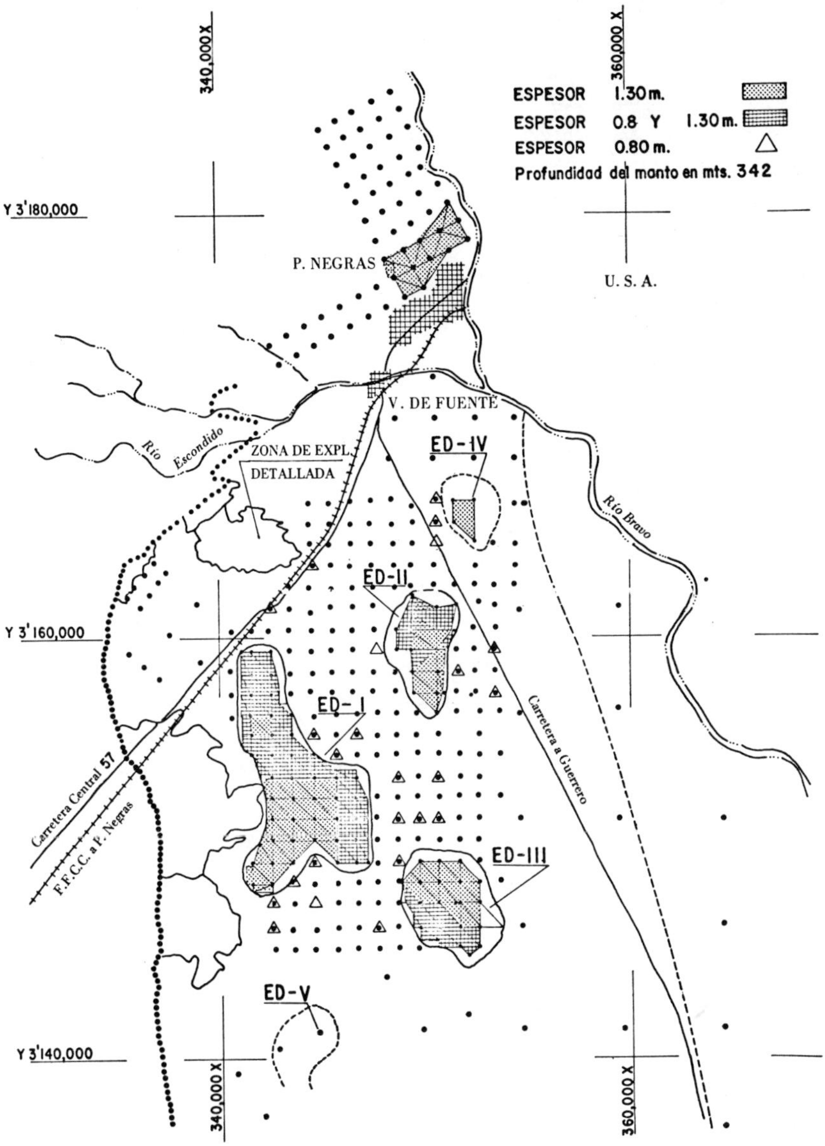

Figure 21-4. Location of coal deposits in Fuente-Río Escondido subprovince, Coahuila

PRODUCTION

The national production of coal has all-in-all paralleled the development of the steel and nonferrous metals industries. The 2.96 million tons produced in 1970 increased to 6.76 million in 1978 (Fig. 21-5), and about 8 million are expected for 1979.

In spite of these increases, the coal sector must supplement its deliveries with imports. About 660,000 tons of dressed coal were imported in 1977 and 490,000 in 1978.

DEMAND FORECASTS

From 1980 to 2000 it is predicted that the demand for coal in Mexico will be on the order of 750 million tons for the steel industry, while nonferrous metals will account for another 96 million. Electrification programs envisage an accumulated demand for 250 million tons.

These forecasts imply that a productive capacity of 66 million tons per year will have to be attained in the year 2000 through mining projects deriving from the steel industry, while the electrical power industry will put mining units into operation that will have to attain a production capacity of no less than 30 million tons per year to satisfy its coal-fired plants.

The former means at the very least that by 2000 the country will have to attain an annual production capacity of 100 million tons of coal. This will be a challenge, considering that our present productive capacity is only about 8 million tons per year.

EXPANSION OF THE ELECTRICAL SECTOR

Of special significance are plans for expanding the use of coal in the electrical sector, including the construction of 28 turbine-generator units of 300 MW each. This means that 1.4 units will have to be built every year during the period in question, divided into 7 projects, each with an installed capacity of 1,200 MW.

Given these prospects, which will permit coal to participate with approximately 26 percent of the total anticipated power for the year 2000 in Mexico, we shall be obliged not only to develop mines to satisfy demands

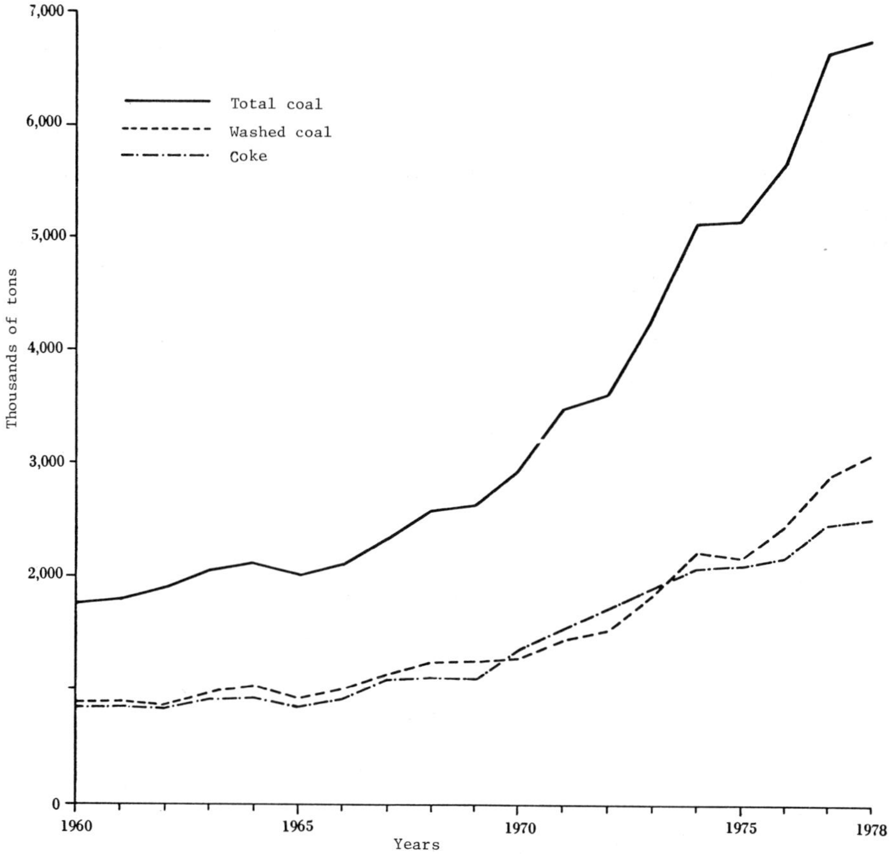

Figure 21-5. Production of total coal, washed coal, and coke

for 30 million tons of coal per year but also to initiate exploration programs allowing the present reserves of coal to be tripled.

The first group of coal mines devoted to the electrical power sector is now being developed. It is estimated that they will go into production in 1981 with a capacity of only 250,000 tons per year, which will grow little by little until stabilization in 1985 at an annual output of 4 million tons of coal overall. In that same year, a second group will go into operation, beginning at 585,000 tons and attaining the same capacity as the first in 1988. The remaining mining projects, according to the Federal Electricity Commission's coal-firing plans, will go into operation in 1988, 1991, 1994, 1997, and 1999, while the annual demand for the period 1981-2000 will increase at a mean annual rate of 28 percent.

Joint expansion of the electrical, steel, and nonferrous metal sectors with respect to coal will generate about 50,000 direct jobs. These will have a multiplying effect resulting in about 100,000 indirect positions.

RESEARCH

Finally, we will refer in general terms to the research being conducted on coal in Mexico as a scientific basis for the consumer sector. This research includes:

- The classification of coals, with the goal of precisely determining the best way to use the different types,
- Methods for ensuring the total utilization of coal products in order to reduce losses to a minimum,
- New mixtures of coking and noncoking coals,
- Methods for utilizing the fly ash produced in coal-fired power plants, and
- Environmental pollution.

CHAPTER 22

POLAND'S ENERGY RESOURCES WITH SPECIAL REFERENCE TO BLACK COAL

Dr. Ing. Zygmunt Falecki[1]

SURVEY OF DEVELOPMENTS IN THE ENERGY ECONOMY: THE PART PLAYED BY COAL

One of the basic factors in the very great economic and social impetus that has occurred in Poland in the last few decades has been the design and execution by political decision makers of a strategy of maximization, where economically justifiable, of the utilization of national economic resources, especially black coal and, more recently, lignite (Falecki, 1974, 1978). Table 22-1 shows the dynamics of this development in the past as well as one projection to the year 1990 (Sikora et al., 1979). Tables 22-2 through 22-5 show some of the major elements that have characterized the use of primary energy sources for 20 years.

The indices relative to the development of the national economy (Table 22-6) reveal a dynamically increasing national income. This increase has been the major stimulus for the augmentation of energy consumption in the past.

It should also be stressed that during this period the rate of increase in the consumption of primary energy has not kept pace with that of the national income. The average energy rate has increased to 4.4 percent in the last 20 years, for example, while that of the national income has climbed to 7.1 percent, so that specific energy consumption itself has actually been seen to decrease relative to the national income as a whole.

During the period under consideration here, except for significant changes in the specific consumption of primary energy in relation to the total national income (expressed in millions of zlotys), we have also witnessed profound alterations both in the structure of primary energy utilization and in that of the ultimate utilization of fuels and energy.

The abundant reserves of black coal and lignite, as well as what the present level of geological exploration indicates to be only modest natural gas and low oil and water-power resources, have been dominant factors in the

[1]Scientific and Development Centre for Energy Problems, Katowice, Poland.

TABLE 22-1. History and structure of primary-energy consumption in Poland from 1956 to 1976 and until 1990

Year	Total consumption of primary energy (million mcte)	Consumption dynamics in % (1950 = 100%)	Share of individual energy sources in the total consumption of primary energy											
			Coal		Lignite		Other solid fuels		Petroleum		Natural gas		Water power and nuclear	
			million mcte	%	million mcte	%	million mcte	%	million mcte	%	million mcte	%	million mcte	%
1950	51.9	100.0	48.7	93.8	0.6	1.1	1.6	3.1	0.6	1.2	0.3	0.6	0.1	0.2
1960	77.0	148.4	69.2	89.8	1.2	1.6	2.2	2.9	3.2	4.2	1.0	1.3	0.2	0.2
1970	121.5	234.1	91.8	75.5	7.4	6.1	2.3	1.9	12.2	10.1	7.4	6.1	0.4	0.3
1976	171.7	330.3	124.8	72.7	11.2	6.5	1.8	1.0	22.1	12.9	11.3	6.6	0.5	0.3
1980	185.0	356.4	125.2	67.7	12.4	7.7	1.5	0.6	31.5	17.0	13.9	7.5	0.5	0.3
1990	292.0	562.6	165.6	56.7	27.7	9.5	1.5	0.5	63.8	23.9	20.9	7.9	7.3	2.5

mcte = 29.31 GJ

Source: Sikora, W., 1979.

TABLE 22-2. Consumption of primary energy (10^6 cte)

	1955	1960	1965	1970	1975	1976	1977
Black coal	54.8	69.1	80.4	90.3	104.4	112.2	115.0
Lignite	0.6	1.2	5.4	8.8	10.6	10.6	10.1
Peat	0.2	0.2	0.2	0.2	0.1	0.1	0.1
Wood	2.0	2.0	2.0	2.2	1.6	1.6	1.6
Petroleum	2.0	3.2	6.1	11.7	19.0	29.6	24.1
Natural gas	0.7	1.0	2.1	7.6	9.7	1.1	12.2
Water power	0.3	0.3	0.4	0.7	0.9	0.9	1.0
Total	60.6	77.0	96.6	121.5	146.3	158.1	165.0

TABLE 22-3. Index of unit consumption of energy in cte per million zlotys of the national income

Year	Tonnes of coal equivalent (cte)
1955	190
1960	195
1965	163
1970	154
1975	118
1976	119
1977	116

TABLE 22-4. Indicators of the energy economy from 1955 to 1977

	Units	1955	1960	1965	1970	1975	1976	1977
Per capita energy consumption	cte	2,200	2,570	3,070	3,730	4,280	4,576	4,735
Rate of increase in per capita energy consumption (average value per 15-year period)	%		3.09	2.83	3.89	2.80	6.90	3.50
Elasticity: energy/national income (average value per 15-year period)			0.72	0.76	0.79	0.57	0.50	0.57

TABLE 22-5. Structure of the consumption of primary energy from 1955 to 1977
(percent)

	1955	1960	1965	1970	1975	1976	1977
Black coal	90.4	89.6	82.7	74.1	71.5	70.9	70.2
Lignite	1.0	1.6	5.8	7.3	7.2	6.7	6.1
Peat	0.3	0.3	0.2	0.2	0.1	0.1	0.1
Wood	3.3	2.6	2.1	1.8	1.1	1.0	1.0
Petroleum	3.3	4.2	6.5	9.7	12.9	13.7	14.6
Natural gas	1.2	1.3	2.0	6.3	6.6	7.0	7.4
Water power	1.5	0.4	0.4	0.6	0.6	0.6	0.6
Total	100.0	100.0	100.0	100.0	100.0	100.0	100.0

TABLE 22-6. Indicators of the development of the national economy

	Units	1955	1960	1965	1970	1975	1976	1977
Population	millions	27.55	29.60	31.55	32.66	34.18	34.55	34.85
National income (1971 base)	million zlotys	319	439	593	791	1,238	1,326	1,424
Per capita national income (1971 base)	mill. zlot./year	11.58	14.73	18.80	24.22	36.22	38.48	40.86
Rate of increase in national income per 15-year period	%	4.90	5.06	5.22	9.3		7.1	7.4

way primary energy demands have been satisfied in recent years. Nevertheless, the requirements that result from the development of the country, and especially from transportation, the mechanization of agriculture, and chemistry, as well as from production, have led to structural alterations in the direction of increased consumption of petroleum and gas. We have also noted during this period a greater role for lignite, which is now being completely exploited as the result of new discoveries intended to supply electric power plants in the neighborhood of open-pit mines. The consumption of primary energy increased almost three times between 1955 and 1957, from 60.6 to 165 million metric tonnes of coal equivalent (cte), with an accompanying decrease in the black-coal share from 90.4 to 70.2 percent and an increase in that of lignite from 1.0 to 6.1 percent. The hydrocarbon share rose from 4.5 percent in 1955 to 22 percent in 1977. The increased utilization of hydrocarbons, taking into account their greater efficiency, is one of the reasons for the decreased specific energy consumption in relation to the national income. The ultimate utilization of fuels and energy increased almost three times from 1955 to 1977, from 44.5 to 125.3 cte, although it should be remarked that the black-coal share dropped from 53.6 to 16.0 percent, while that of the hydrocarbons increased from 10 to 14 percent. All of these indices are shown in Tables 22-2 through 22-5. It is evident nevertheless that the secondary-energy share has increased at a regular rate in terms of total ultimate utilization in the last few years.

The dynamic social and economic development of Poland will be maintained over both the medium and the long term. One of the main bases for the continuation of this impetus is the guarantee of sufficient energy supplies, to ensure which we are undertaking very intensive and systematic anticipatory efforts. Look, for example, at one of the projections that have been made in various publications, both Polish and foreign (Frisch, 1977) with respect to the year 2000 (million cte):

Coal	Petroleum	Natural gas	Water power	Nuclear energy	New forms of energy	Total consumption
280	110	45	5	45	5	470

It must be emphasized that the prediction shown above may differ significantly from prevailing opinion and, on the bottom line, from the values actually attained. Nevertheless, it is now possible to establish the boundaries of two large areas, defined in a certain sense by different probabilities.

The High-Probability Balance-Sheet Area

Here, the order of magnitude of energy demand may be established at 400 million cte.

That the energy economy in the context of its mutual relations with the total economic potential of the country will have the capacity to ensure this order of magnitude of energy requirements is due above all to the potential for a perceptible increase in the production of black coal and lignite. Any deviations from this order of magnitude will be above all a function of two basic parameters:

a. the dynamics of any alterations in the industrial structure (depending on whether the specific energy consumption demanded by the national income will be heavy or light) and
b. the actual and economically justifiable level of rationalization in the utilization of energy.

The high probability of this area is due primarily to a high potential for the continuation of a perceptible growth in the coal industry, for both black coal and lignite.

Present plans provide that the production of black coal will attain a level of 300 million tonnes in the year 2000 and that of lignite a level of 200 to 250 million tonnes. This scenario is technically and economically attainable.

Very significant efforts are now being undertaken to provide investments in this area. A new coal basin in the Lublin region and an energy complex in the Belchatów region, based on a deposit that is very rich in lignite, are now being opened.

What is significant in this respect is that the national coal resources are sufficient to ensure approximately 300 million cte within the framework of the country's energy requirements for the year 2000, more than 70 percent of the total energy balance sheet. The role of coal, and especially of black coal, is therefore of overwhelming importance to the future development of the energy economy and hence to the social and economic growth of the nation.

The Medium-Probability Balance-Sheet Area

This area is characterized by a certain number of problems and will not be developed here.

The Share of Petroleum and Natural and Synthesis Gas

The gas resources that we know today are modest and those of petroleum meager. The prices of hydrocarbons in the last few years have risen strongly, with the price of petroleum exceeding $25 a barrel recently. On the other hand, the technologies for converting coal into liquid and gaseous fuels are beginning to mature (considering that their costs may well become competitive in view of the trend just mentioned, which is in fact why these technologies are being developed in Poland). As an illustration of this potential, there are certain projections for the year 2000 showing outputs of 5 to 10 million mcte of liquid fuels and 10 to 15 million m^3 of synthesis gas with a high methane content.

The Share of Nuclear Energy

In spite of the fact that Poland has not yet developed any energy from nuclear reactors, the nation still has an ample scientific and industrial infrastructure in the area (being in fact a member of the Club of London). Plans for the construction of nuclear reactors are a function first of all of the macroeconomic factors and world trends (of varying intensity) in this area. The 1980's will witness the first Polish nuclear plant, with an output of 880 MW, going into operation as part of a program of construction of this type, with 10 to more than 20,000 MW planned for the year 2000.

The Rationalization of Energy: The Share of the New Technologies

We are now witnessing vast socio-economic alterations on a worldwide scale. Without going into the complexities of the problem we may note the birth of a new energy geopolitics. Of the fundamental characteristics of this new approach, one of the most important is undoubtedly the necessity of rationalizing energy policies on a worldwide basis as well as within groups now in existence or in the process of formation, and of course on a nationwide scale. The rationalization of energy must provide for the inclusion within the energy chain of all the necessary links from extraction to final processing.

In this respect it is indispensable to concentrate on overall efficiency in the utilization of particular energy technologies, especially those of coal. This overall efficiency, especially with regard to the steering of specific resources toward precise socio-economic goals, will therefore be a function of the selection of an appropriate technology, and particularly of its level of industrial maturity. The problem of developing lignite production to a level of 250 million tonnes will be tied to the solution of various problems, among which are:

(1) the effect of large, open-pit mines with an annual output of 50 to 100 million tonnes and that of their associated electric-power plants with outputs of 4,000 to 8,000 MW on the social and economic milieu (ecology, industrial concentration, etc.),

(2) the maturity of the technology for converting lignite into synthesis fuels,

(3) the overall economics of accelerating the dynamics of the development of the lignite industry with respect to that of black coal and nuclear energy,

(4) accelerating the maturity of the technology of fluidized beds and its influence on the management of solid-fuel resources,

(5) the rate of development of new technologies within the framework of local energy systems (small mines, solar energy, geothermal, etc.), and

(6) the development of combined heat and electricity production.

All these considerations show, although in a very general way, some of the fundamental problems in elaborating an energy policy and plainly demonstrate the policy, which is thoroughly familiar at the present time, of using as many solid fuels as possible. It is evident nevertheless that we cannot at the present time limit the concept of energy sources only to conventional and national sources of primary energy or to their standard combinations. Energy sources must on the contrary include the whole spectrum of potential energy bases that can be utilized as the result of the most efficient action of the energy system as conceived in all its complexities (energy technologies, refuse energy, and local energy systems).

RESOURCES FOR PRIMARY ENERGY MATERIALS

Poland's geological structure is fairly complicated, which explains the variety of mineral resources (copper, sulfur, salt, silver, coal, tin, etc.) included in numerous geological structures. It is therefore natural for the art of mining to have developed in the land several thousand years before our era (in the Góry Swietokrzyskie region, for example). One of many remains from the history of Polish mining, known throughout the world, is the salt mine at Wieliczka.

In the seventeenth century, when coal deposits were discovered in Upper and Lower Silesia, we were direct witnesses to the growth of both coal production and its associated science. In the nineteenth century, the discovery of petroleum deposits in the Carpathians was one of the starting points for the world history of petroleum as a result of the efforts of Polish scientists (the separation of kerosine by Łukasiewicz).

The historical avatars through which the nation progressed almost totally inhibited the growth of Polish geology. Still, Polish geologists have contributed to motivating geological studies throughout the world. It was not until after the Second World War, and in spite of considerable destruction, that the Poles were able to get under way an intensive geological reconstruction not only on the level of the central government, but also on that of science and industry (Dembowski, 1975), with a resulting series of explorations and documentation of numerous mineral resources, especially energy resources. The national economy is now profiting from a production of mineral substances amounting to 500 mcte per year.

Within the complex geological structure of Poland the Tertiary and Quaternary eras are especially rich in black coal, lignite, and peat. Deposits of natural gas are also found, as well as petroleum (although much more meager according to our present state of knowledge). The geographical situation of these reserves is shown in the map of Poland (Fig. 22-1). Of all these energy resources the richest at the present time are those of black coal.

Figure 22-1. Map of coal, petroleum, and natural-gas beds in Poland

Black coal
Lignite
Natural gas
Petroleum

BLACK COAL RESOURCES

The discovery and documentation of rich deposits of black coal in Poland is primarily the result of developments in the geological sciences, especially in the area of Carboniferous formations (paleontology, biostratigraphy, lithostratigraphy, etc.), and of intensive activities in the drilling industry. During the last 30 years almost 1.5 million m have been bored in an attempt to identify and augment black coal resources (Dembowski, 1975). These activities have enabled the drawing of geological maps of both the Carboniferous system and of known coal deposits. As a result of this scientific and industrial infrastructure, the Central Bureau of Geology is in a position to proceed to a systematic realization of what is known about black coal resources.

Three main coal basins have now been discovered: the Upper Silesian (Namurian and Westphalian), the Lower Silesian (Namurian and Westphalian), with productive series in the Upper Carboniferous, and the Lublin (Visean-Namurian and Westphalian), with productive series in the Upper Carboniferous and to a lesser extent in the Lower.

According to the most recent publications (Sikora et al., 1979), these resources are estimated at nearly 140 billion mcte, of which 43 percent belongs to ranks A, B, C_1, and C_2. Table 22-7 shows the structure of these resources.

TABLE 22-7. Structure of black-coal resources in Poland (in billion mcte)

Ranks of coal	Exploitable resources			Potential resources down to 1,500 m			
	ABC_1	C_2	Total	down to 1,000 m	1,000-1,200	1,200-1,500	Total
Coal resources, of which:	19.82	40.10	59.92	42.02	21.00	21.00	84.0
Rank 31-33	13.92	21.70	35.63				
Rank 34	3.12	7.90	11.02				

Source: Sikora, W., 1979.

Table 22-8 shows the resource balance of black coal. The documented resources amount to 60 billion mcte, of which 36 billion are high-quality steam coals (types 31-33) and 11 billion gas and coking coals (types 34-38).

TABLE 22-8. Resources of black coal in Poland

	Total		Type 35-38	
	10^9 mcte	%	10^9 mcte	%
All of Poland	57.5	100	13.0	100
Mines under exploitation	25.1	43.6	5.5	42.3
Mines under construction	1.7	3.0	0.7	5.4
Beds destined for exploitation	30.7	53.4	6.8	52.3

Source: Ministry of Mines, 1978.

Systematic and extensive research has produced an accurate evaluation of the utility of Polish coal. One result has been the establishment of two systems of classification, one with respect to the qualities of the coals and the other to the evaluation of the resources from the standpoint of their potential exploitation.

The Qualitative Classification of the Coals

The Polish coal classification (Krupiński, 1967) is based on the content of volatile matter, the caking properties (indexed in accordance with what is called Roga's number, LR), on the plastometric properties (contraction and dilatation), and, for coals with a cinder content of no more than 10 percent, on heat of combustion (Table 22-9). The main types of coal in accordance with this classification are shown in Fig. 22-2 (Mielecki, 1972) in comparison with foreign systems.

The Classification of Resources

The Polish classification of resources (Fig. 22-3) is based on two types of data (Krupiński, 1967): (1) geological data, in accordance with which the resources are classified in a range from C_2 (least known) to B and A, depending on the amount of geological information we have about

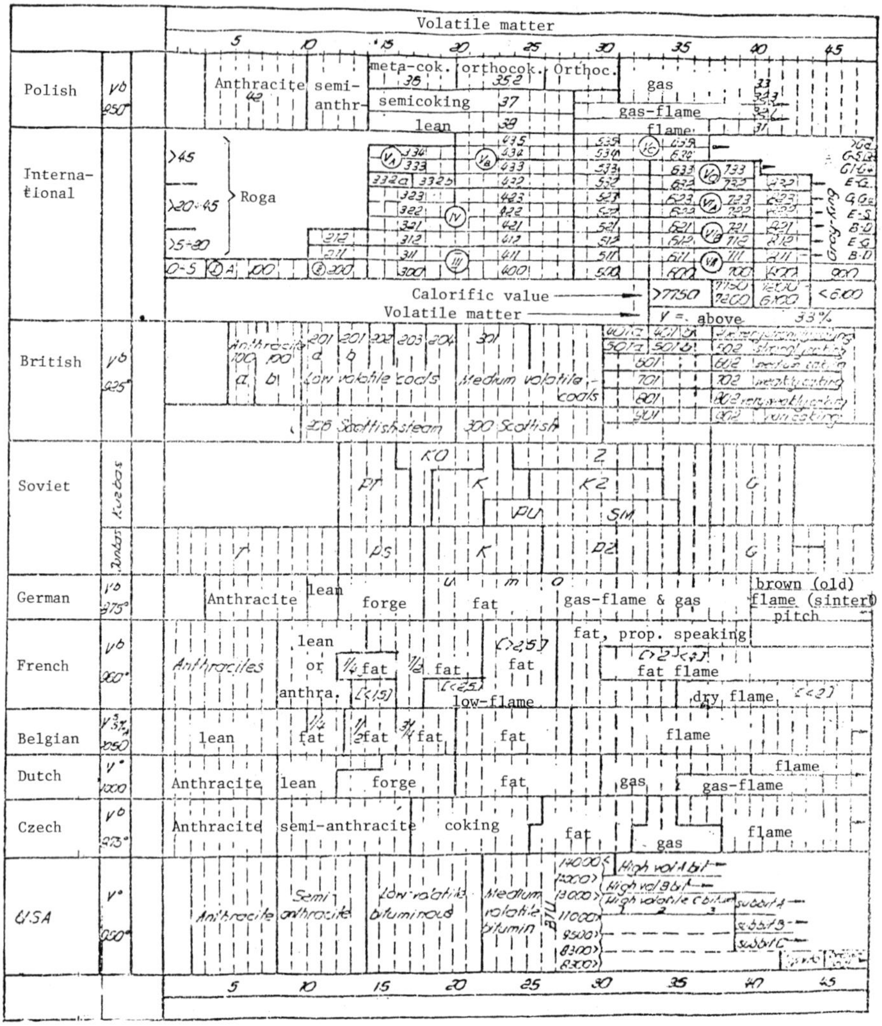

Source: Mielecki, T., 1972.

Figure 22-2. Classification of coal

them; and (2) project data, used for the adequate treatment of resources that have been classified as balance resources in accordance with the demands for projects for mines to be constructed. Figure 22-3 (Krupiński, 1967) provides an outline of the systematics as applied to Poland for the management of coal deposits.

TABLE 22-9. Coal classified in accordance with rank

Grade of coal	Rank	Volatile matter (%)	Roga number (LR)
Gas-flame	31	36-48	0-10
	32	32-43	10-35
	33	30-40	30-55
Fat and lean (coking)	34	28-39	50-85
	35	22-30	50-85
	36	17-22	25-55
	37	14-18	10-30
	38	10-16	0-10
Semi-anthracite	41	7-10	0
	42	3-7	0

Source: Krupiński, B.

The Upper Silesian Coal Basin

The total area of this basin in Poland (Fig. 22-4) is 5,400 km^2. It projects below the Carpathians into Czechoslovakia. The productive Carboniferous, which is as thick as 6,900 m in the west and 2,700 m in the east (Passendorfer, 1965) is of limnic and paralic origin (lower part of the productive series). This basin is shaped like a tub, with the Lower Carboniferous tangent to its rim. The productive series is composed of three basic groups: (1) an edge group (Lower Carboniferous); (2) an anticlinal group (characterized by the presence of thick beds of coal); and (3) a synclinal group, in the central part of the basin. The coal resources of this basin are presently estimated at up to 1,500 m in thickness, more than 120 billion mcte (Sikora, 1979), distributed among nearly 400 beds, of which about 100 are of industrial value. The quality of the coal varies

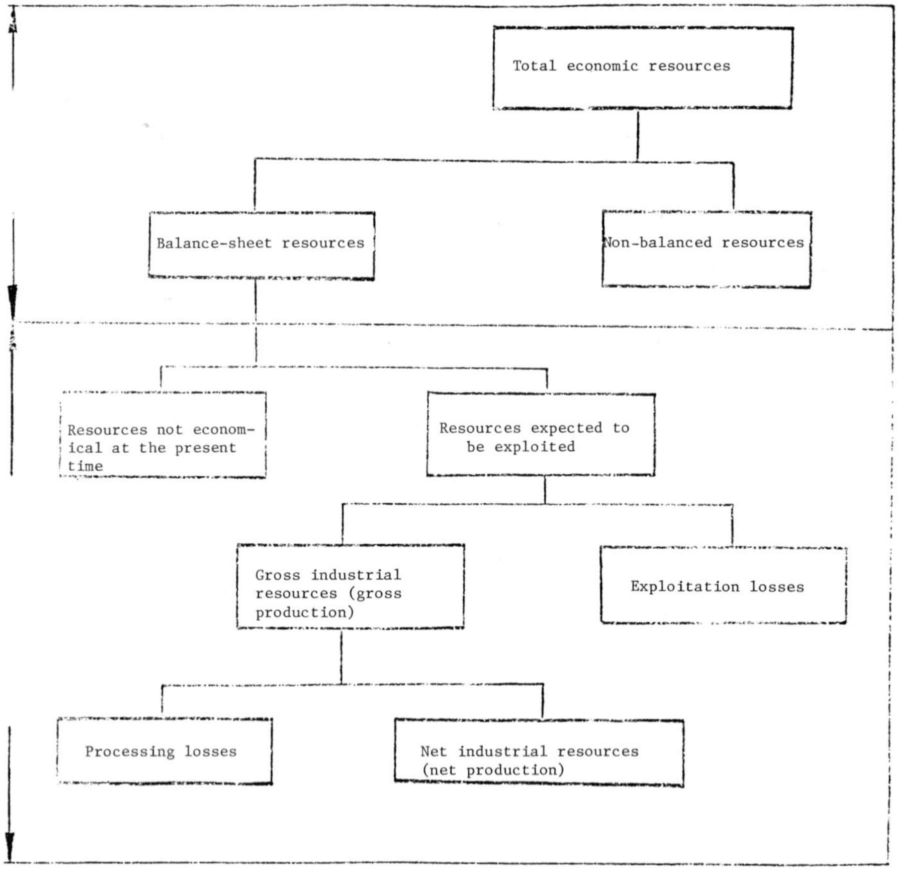

Source: Krupiński, 1967.

Figure 22-3. Classification of resources

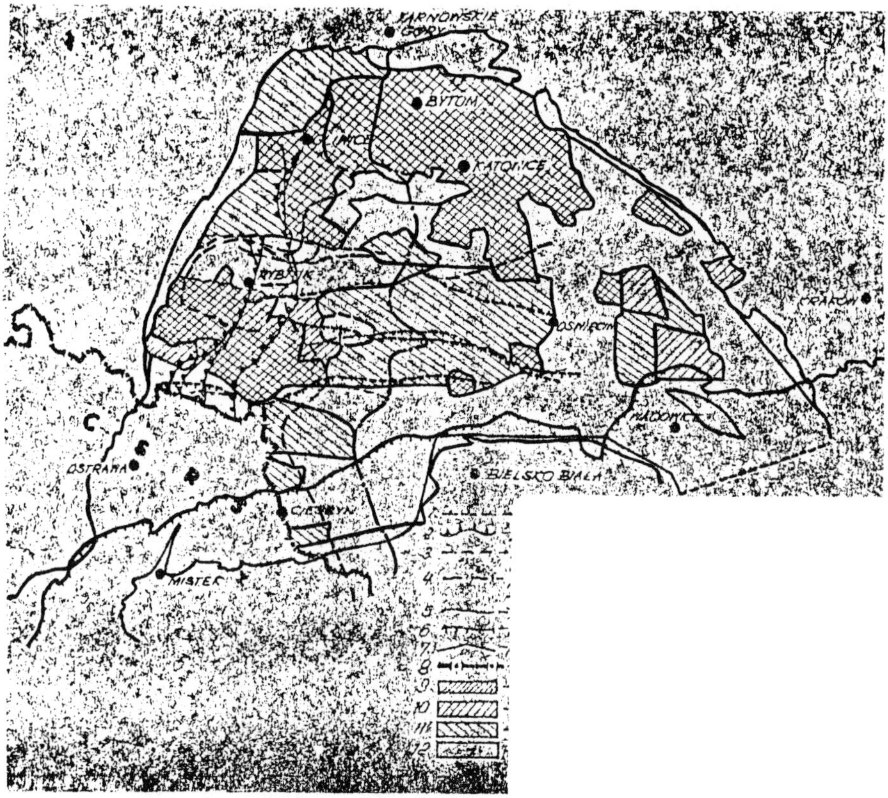

1. Faults delimiting zones of different grades of coals.
2. The Michalkawicé and Orlaw faults.
3. Borders of Rank 35-coal beds extending down to 1000 m (eastern part).
4. Eastern border of Rank 37-coal beds down to 1000 m.
5. Border of coal basin.
6. Border of Carpathian
7. Mining-concession border.
8. National border.
9. Areas with mines under exploitation.
10. Category C_1 + B regions prospected.
11. Category C_2 regions prospected.
12. Potential regions (and regions in reserve).

Source: Z. Dembowski, Polish Coal Review, 1975.

Figure 22-4. Map illustrating present state of geological knowledge of Upper Silesian coal basin

in accordance with the individual deposit, ranging from steam coals to coking coals (in the edge group), with an average thickness of 1 to 4 - 15 m (in the anticlinal group). It should also be stressed that 20 percent of the exploited resources lie within protective pentices that are the result of the development of an appropriate technology of exploitation.

Quite a lot is known about these resources. There is still a potential for research and coring, especially with respect to coking-coal reserves in the southwest of the basin and to steam coals in the central and eastern part. The present importance of this basin is particularly evident from the fact that it contributes nearly 98 percent of all the black coal produced in Poland, the other 2 percent coming from the Lower Silesian Basin.

The Lower Silesian Coal Basin

The surface area of this basin is as much as 440 km^2. Its geological structure is very complex. The Upper Carboniferous can be as thick as 2,500 m. Approximately 40 beds have been identified in the prospected regions (Sikora, 1979). The coal beds are rather thin, usually no more than 2 m. The quality of the coal depends to a great extent on the depth of the bed. Coking coals of a high quality are being exploited. Known resources amount to about 0.7 billion mcte. There are probably potential reserves at depths of more than 1,000 m.

The Lublin Coal Basin

This basin was only recently discovered and is a good example of the results of the type of systematic research conducted by Polish geologists (Samsonowicz, 1932). Samsonowicz had predicted the existence of the Carboniferous (Passendorfer, 1956) in the neighborhood of the Bug River. In the 1950's, the Geological Institute conducted intensive studies of the region (Dembowski, 1975) which resulted in the localization of rich resources of black coal at depths of 350 to 750 m over an area of 5,000 km^2 (Fig. 22-5). Various preliminary estimates place these resources, down to 1,000 m, at from 20 billion mcte (Sikora, 1979) to 40 billion (Ministry of Mines, 1979). The most productive series are in the Westphalian, especially in the Upper. Known resources now amount to 4 billion mcte. More than half of these resources are steam coals, the remainder coking (Ministry

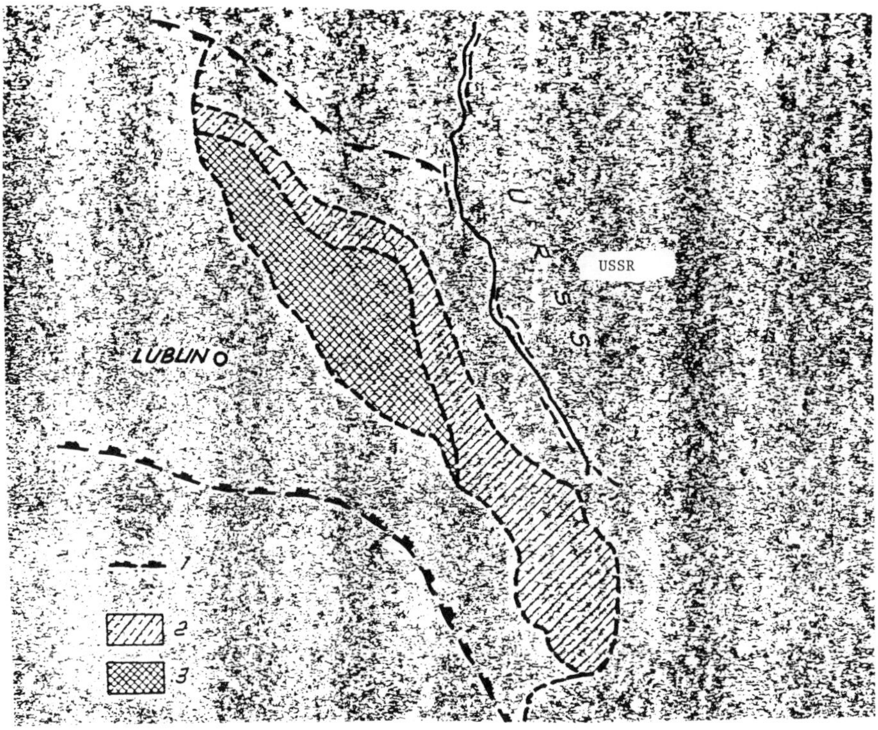

Source: Sikora et al., 1979.

Figure 22-5. Outline of the Lublin black-coal basin

1. Probable limits of the Lublin black-coal basin.
2. Area containing exploitable coal beds.
3. The richest coal region.

of Mines, 1979). Known resources are located in the central coal district. The first mines are now starting to be dug over an area of 120 km^2. The known coal beds suggest that the thickness varies, attaining 3 m in some cases. The qualities of the coals are no different from those of the Upper Silesian Basin.

Problems in Optimizing the Exploitation of the Black-Coal Beds

The richness of Poland's black-coal reserves has been the basis of the country's whole energy economy. Production developments over the last few decades and up to the end of this century (Ciśzak, 1979) are shown in Table 22-10. The dynamics of investments intended for this purpose in the coal industry reflect the history of the industry itself. Based on constant 1976 prices, the costs are as follows:

Period	Costs (in billions of zlotys)
1951-1960	51
1961-1970	96
1971-1980	272
1981-1985	219

TABLE 22-10. Developments in black-coal production and annual increase for the period 1946-2000

Year	Production (1,000 tonnes)	Mean annual increase in production (1,000 tonnes)
1946	47,288	7,678
1950	78,001	2,644
1960	104,438	3,566
1970	140,101	6,562
1978	192,600	
1979 (anticipated)	200,000	7,300
1980	207,200	6,560
1985	240,000	
2000	300,000	4,000

Source: Ciśzak, E., 1979.

Seventeen mines have been dug since 1945, and six others are in process. More than 200 million tonnes per year are now being produced, present plans calling for 300 million tonnes in 2000.

Tables 22-11 through 22-14 show certain base indices for the coal industry as well as for the conditions under which it can be exploited (Ministry of Mines, 1978-79). The present structure of the black coal industry has a three-level system of management. The industry as a whole is administered by the Ministry of Mines. The managerial infrastructure is composed of hundreds of institutions. There are now three main branches (Ministry of Mines, 1978): a scientific, research, and project-planning base; associations of production enterprises; and cooperative associations, federations, and enterprises. One of the main missions of the scientific and research base is to optimize the exploitation of coal mines, in other words to maximize recovery (minimize loss) while taking into account the whole complex of existing local constraints. In this respect it is possible to note three main types of exploitation (among many others): the longwall system with caving; the room-and-pillar system with hydraulic filling; and coal extraction with protective pentices. The first two methods provide for maximum coal extraction with a potential for slice exploitation of thick beds because of the hydraulic filling. Nearly 76 percent of present output is obtained by slicing and caving and 24 percent by slicing and hydraulic filling.

The extraction of black coal from protective pentices is undoubtedly one of the most successful results of Polish mining science, enabling the exploitation of beds lying under cities and large industrial complexes (Borecki et al., 1975). The product obtained by using these pentices now amounts to almost 39.2 percent of the total domestic output of black coal. Finally, developments in science, engineering, and mining design, especially in the area of the complex management of deposits and mining districts (Krupiński, 1967) and based on the principle that the number and sequence of mine workings must be viewed within the context of the deposit as a whole, should also be mentioned. Two examples of solutions to the complex problems in the Polish coal industry are the Rybnik Coal District ("ROW") and the Central District of the Lublin Coal Basin (Paździora, 1979).

TABLE 22-11. Indicators characterizing the development of black-coal production in Poland

	1978	1979	1980	1985
Output	192.6	200	207.5	240.0
Yield (kg/year)	3,577	3,610	3,655	
Number of workers (1,000's)	334	350	368	

Source: Ministry of Mines, 1978.

TABLE 22-12. Changes in exploitation depth from 1950 to 1977

Year	Mean depth (m)	Output share as function of depth (percent)		
		Down to 300 m	300 to 600 m	Below 600
1960	363	46.6	45.7	7.7
1965	383	43.1	49.2	7.7
1970	413	35.0	54.1	10.0
1975	472	24.7	56.9	18.4
1977	498	25.4	53.1	21.5

Source: Ministry of Mines, 1978.

TABLE 22-13. Black-coal output share in accordance with bed thickness (percent)

Year	Bed thickness		
	1.5 m	1.5 to 3.5 m	3.5 m
1968	21.1	44.8	34.1
1969	20.1	46.5	33.4
1970	18.4	49.6	32.0
1971	19.5	50.3	30.2
1972	15.5	51.9	31.6
1973	16.3	53.1	30.6
1974	16.8	53.3	29.9
1975	15.6	57.4	27.0
1976	14.6	57.8	27.0
1977	13.8	56.9	29.3

TABLE 22-14. Black-coal output share in accordance with bed dip (percent)

Year	Bed dip		
	15°	15° 30°	30°
1968	76.9	19.3	3.8
1969	78.3	18.0	3.7
1970	78.3	17.9	3.4
1971	80.4	16.2	3.4
1972	79.4	17.7	2.9
1973	80.2	15.6	3.2
1974	80.0	17.9	2.1
1975	79.9	18.4	1.7
1976	82.3	17.4	2.3
1977	80.1	17.0	2.9

LIGNITE RESOURCES

The formations that include deposits of lignite extend over a large part of Polish territory, forming an almost unbroken area in the center and north of the country (Falecki and Matl, 1978). Lignite beds are generally of the Miocene. According to present knowledge it is estimated that the reserves amount to 36 billion mcte, of which 15 billion are fairly well documented (Kozłowski, 1976). On the whole, several dozen lignite deposits have been identified up to the present time. They are distributed among four main basins: Wrocław, Zielona Góra, Poznań, and Łódź (Fig. 22-6). The calorific value of lignite varies from 1,600 to 2,400 kilocalories. The ratio of dead-ground coverage (overburden) to bed thickness is not expected to exceed 6 during the coming decades.

Known deposits of lignite of industrial quality are situated at a depth of 200 to 300 m (Bednarczyk, 1979). The richness of the lignite reserves is the reason that present plans for production development for the year 2000 are on the order of 250 million tonnes per year. Lignite is mainly used and will be used in the future for burning in large electric power plants. Other uses however are also planned, like gasification and as a fuel for local demands.

OTHER PRIMARY-ENERGY SOURCES

Natural Gas and Petroleum

Known resources of natural gas are on the order of several hundred billion m^3, whereas those of petroleum are much more modest.

Water Power

Water-power resources are very limited in Poland (12.1 TWh/year). About 1.7 terawatt hours per year of these resources are now being exploited. A very significant plan for managing the Vistula is being prepared at this moment. It extends to the end of the century and will enable the utilization of a large percentage of Polish water resources. We are also looking into the possibility of extending our small-scale water-power systems to contribute practically to local requirements within the framework of a total program of rationalization of the energy economy.

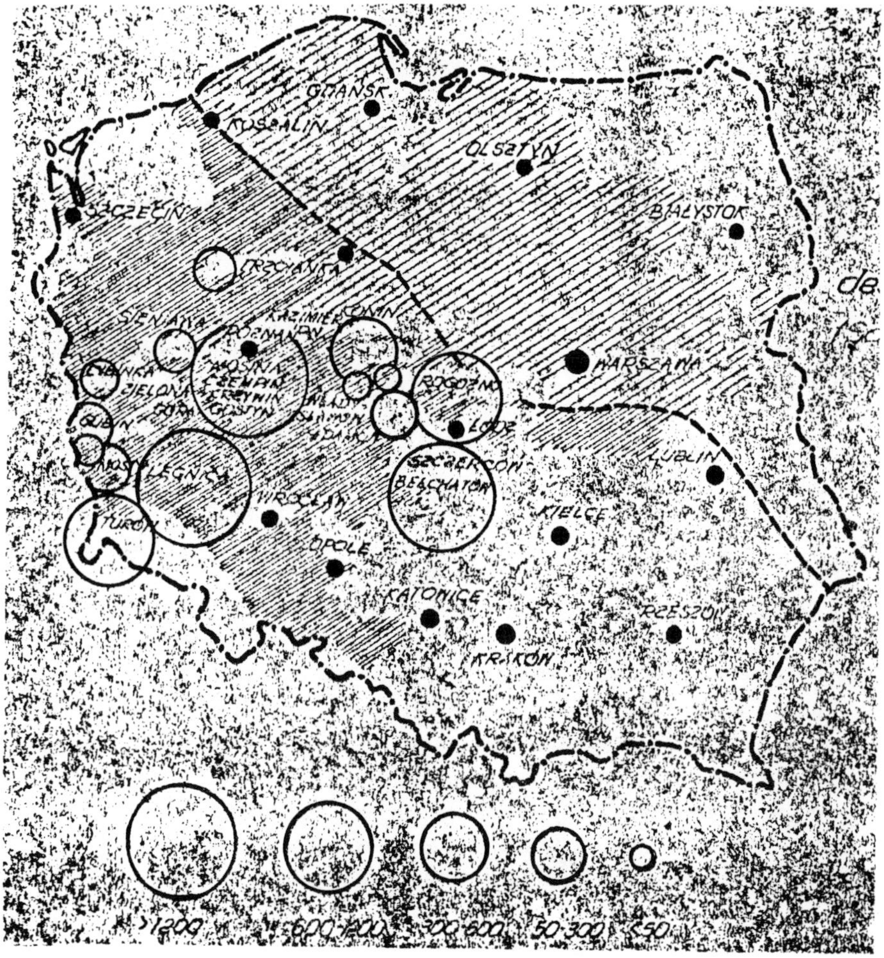

Source: Kozłowski, 1976.

Figure 22-6. Distribution of lignite resources

Circles represent amount of reserves in 10^3 tonnes.

Nuclear Energy

The dominant and economically justified part played by the solid fuels has made it unnecessary as yet to develop nuclear energy. Nevertheless, the country has the research and development infrastructure for presently contemplated plans for the construction of nuclear power plants with an output that will be as high as 15 to 20 gigawatts in the year 2000.

Other Energy Sources

Research and development is being carried out in Poland in the field of new sources of energy (solar, biomass, etc.). We do not however foresee that these forms of energy will play a significant part in coming decades. Finally, it should be mentioned that Poland has significant reserves of peat, and that we are now looking into its potential for local energy requirements.

PROBLEMS IN RATIONALIZING THE ENERGY ECONOMY

The maximum economically justifiable and efficient utilization of energy resources is one of the basic factors of the Polish energy policy. The realization of a policy of this type requires important changes in the scientific application of many technologies, for example. The various links in the overall industrial structure must not be treated independently, but within the context of that structure. Only the total effect can at the present time serve national decision makers as a basic indicator of the economic efficiency of a technological and, on the bottom line, social enterprise. In many countries, the systems approach worked out in particular at the International Institute for Applied Systems Analysis (IIASA) in Laxenburg, Austria, is being used more and more. In Poland, this method is now being developed and is starting to be applied among other uses to evaluating and selecting different large-scale technologies, especially in the field of energy economy.

The increasingly dominant part played by the new technologies in the overall efficiency of energy-resource utilization should also be stressed. Certain aspects of these problems that directly affect overall energy-resource conservation are mentioned below.

The Development of Centralized Heating Systems and the Combined Production of Heat and Electricity

The specific nature of the Polish energy balance sheet has led to large-scale development of technologies that allow overall maximization of the energy efficiency of the total heat production and distribution systems for heating and engineering applications (Table 22-15). Within the framework of a predominantly coal-utilization policy, it is only centralized heat distribution that will allow rationalization of the energy policy as a result of, among other factors: the utilization of the combined production of heat and electricity for an efficiency of as much as 80 percent, and the burning of lower-quality coals (fines, for instance).

TABLE 22-15. Different methods of centralized heat production

Methods	Output (Tcal) 1975	1980
Combined heat and electricity output plants:		
Public service	30,000	55,000
Industrial	58,000	75,000
Municipal heating	29,400	37,800
Industrial heating	61,200	77,700

Source: Wójcicki, J., 1978.

The economic efficiency of the combined production of electricity and heat is especially evident from, among other factors, the economy of coal in comparison to the production of heat by itself: from 1976 to 1990, for example, we will save nearly 10 million cte and will double the present output.

Coal Gasification and Liquefaction

The richness of coal resources all over the world has been a strong stimulus for research into and the development of methods for gasification and liquefaction. In Poland, predictions have been made (Sikora, 1979) that 3 million cte will be devoted to this purpose in 1990. One of the main reasons for using coal is undoubtedly the potential of lower-grade coals. The methods selected will depend to a certain extent on this possibility. The problem however must be viewed as part of a total policy of coal-reserve conservation, so that the possibility of utilizing lignite should also be considered. This technology forms part of a strategy for the physical and chemical conversion of coal.

Fluidized-Bed Combustion

This method is becoming increasingly familiar throughout the world. The basic technique has been frequently described. Still, it is important to particularly emphasize its value as one component in the total energy chain. Fluidized-bed combustion allows, among other factors, the exploitation of lower-grade coal beds, the fully mechanized exploitation of trivial deposits (neglected up to now because they have been too expensive to extract) of coals that are not easy to burn by conventional methods, and simpler processing plants. These factors contribute appreciably to lower production costs.

Poland has devoted a great deal of research to this area. Results so far are allowing us to proceed now to the industrial phase of building fluidized-bed boilers.

Heat Recovery

This very brief review of the various technologies discussed is obviously not exhaustive, bearing in mind all the methods directly involved in increasing the efficiency with which existing energy sources are being employed. Another area that needs to be investigated is undoubtedly that of the recovery of lost heat. Analyses conducted in Poland (Wójcicki, 1979) have resulted in estimates of about 7 million cte per year of heat-loss resources from the major branches of the national economy. We are now saving 7 million cte per year as a result of the modernization of exploitation plants. The General Inspectorship for the Energy Economy is responsible for accomplishing objectives in this field.

FINAL REMARKS

The energy reserves of Poland are both responsible for and dependent on the dominant future part to be played by black coal, 300 million tonnes of which is expected to be produced in the year 2000. A coal-utilization strategy of such dimensions implies the necessity of solving a large number of complex problems connected with the maximum utilization of this raw material at optimum efficiency. This means that research and development efforts in the field of new technologies must be carried out throughout the energy chain from extraction to consumption. It is only the systems approach that will provide any hope of appreciably improving the management of black-coal resources. This methodology obviously applies to the whole range of the energy economy.

The other energy source that is becoming increasingly important is lignite. Present plans call for 250 million tonnes of lignite to be produced in the year 2000. To satisfy energy requirements over the next decades it will be necessary to develop nuclear energy to an in-place output of more than 20 GW in the year 2000. Petroleum and natural gas will play a passive part in the national energy balance, in spite of the fact that they will be used more and more on a quantitative basis.

The main element of the Polish energy policy is the maximum possible utilization of national energy resources. This corresponds to a new energy strategy throughout the world. It must be stressed that the policy of rationalizing the energy economy is beginning to be a dominant factor in the nation's energy policy. The goal is the maximum economically justifiable utilization of all existing energy resources, including local ones (small mines, biogas, solar, heat recovery, etc.).

REFERENCES

Bednarczyk, J., 1979, Problemy rozwoju wydobycia i racjonalnego zagospodarowania krajowych zasobów węgla brunatnego: Koks Smoła Gaz, "Śląsk," 1979.

Borecki, M., et al., 1970, Doświadczenia z eksploatacji w filarze ochronnym szybu: Przegląd Górniczy, 1970, no. 3.

Ciśzak, E., 1979, Design and Construction of Hard Coal Mines: United Nations Symposium on World Coal Prospects, Katowice, 1979.

Dembowski, Z., 1975, Polish Coal Review--Węglokoks, Katowice.

Falecki, Z., 1978, Wstęp do wytycznych energetyzacji Kraju: GIG, 1974; OBR-GPE, 1978.

Falecki, Z., and Matl, K., 1978, Zagadnienia związane z oceną gospodarczą zasobów energetycznych Kraju w perspektywie do 2000 r.: OBR-GPE, Katowice, 1978.

Frisch, J., 1977, L'équilibre mondial entre besoins et ressources d'énergie à l'horizon 2000: Revue de l'Energie, 1977 Novembre.

Kozłowski, Z., 1976, Brown Coal in Poland: Polish Coal Review, 1976.

Krupiński, B., 1967, Wykorzystanie zasobów węgla: "Śląsk," Katowice, 1967.

Mielecki, T., 1972, Wiadomości o badaniu i własnościach węgla: "Śląsk," 1972.

Ministerstwo Górnictwa, 1978, Polskie Górnictwo Węgla Kamiennego: Katowice, 1978.

Passendorfer, E., and Maurice Gignoux, 1955, Geologia Stratygraficz na: Warszawa, 1955.

Paździora, J., 1979, Koncepcja Modelu Górniczego zagospodarowania i eksploatacji złoża LZW: Górnictwo, Gliwice, 1979.

Sikora, W., et al., 1979, Estimation des réserves de charbon en Pologne par rapport aux besoins futurs en énergie: Nations Unies Colloque sur la gazéification et la liquéfaction du charbon, Katowice, Pologne, 1979.

Wójcicki, J., 1978, Oszczędność paliw w ciepłownictwie mieszkaniowym i przemysłowym: Gospodarka Paliwami i Energią, Warszawa, 1978.

Wójcicki, J., 1978, Problemy zmniejszenia energochłonności: Gospodarka Paliwami i Energią, Warszawa, 1978.

SECTION IV. PETROLEUM

CHAPTER 23

THE OUTLOOK FOR CONVENTIONAL PETROLEUM RESOURCES

Richard Nehring[1]

Conventionally produced petroleum is not a long-term energy resource for the world. The known and prospective conventional petroleum accumulations of the world are geologically limited. Petroleum is a replenishable resource only in terms of the geologic time scale. Current rates of consumption of petroleum liquids and natural gas can only continue for several decades, given what we know about the size of the recoverable resource base.

Even though conventionally produced petroleum liquids and natural gas are not a long-term energy resource for the world, considering the prospects for them is both an appropriate and crucial task for a conference organized to explore long-term energy possibilities. Conventionally produced petroleum liquids and natural gas provide more than half of the energy used in the world today. The extent to which they will be available in the decades ahead is the most important factor determining the timing and scope of the inescapable transformation which will have to occur in world energy production and consumption.

This chapter addresses the prospects for conventionally produced petroleum. This includes crude oil, natural gas liquids, and natural gas produced from accumulations where some commercial production can occur without injection or fracturing. It excludes petroleum liquids produced from tar sands, oil shale, coal, and biomass. It excludes natural gas produced from coal, shales, tight (microdarcy permeability) sandstones, and landfills. In some cases, the dividing line between the conventional and nonconventional is fuzzy, for example, between conventionally produced heavy crude oil and some tar sands or between natural gas production from low and very low permeability sandstones. These, however, are rare enough to make the distinction a workable one in practice.

A survey of the outlook for conventional petroleum resources could take any one of several directions. For example, it could summarize an individual

[1]The Rand Corporation, Santa Monica, California, USA.

world petroleum resource assessment or it could critically survey existing assessments, in both cases focusing on the results of one or more resource assessments. This paper will focus primarily on general approaches to the problem of world petroleum resource assessment. Because differences in how people think about petroleum resources are the main factor in explaining differences in what they think, addressing general approaches promises to be more conducive to clear thinking than merely discussing results. After a brief overview of known conventional world petroleum resources, I will discuss three key points to consider in evaluating the outlook for world petroleum resources. Using that discussion as a point of reference, three common misconceptions in petroleum resource assessment will then be described and critiqued. I will conclude with a brief summary of the outlook for conventional petroleum resources.

As of the end of 1978, approximately 1.1 trillion barrels ($175 \times 10^9 m^3$) of petroleum liquids and 3.7 quadrillion cubic feet ($105 \times 10^{12} m^3$) of natural gas had been discovered and made recoverable in the world (Table 23-1). Approximately 424 billion barrels ($67.4 \times 10^9 m^3$) and 1,030 trillion cubic feet ($29.2 \times 10^{12} m^3$) of these amounts have been produced, leaving proved and probable reserves of 676 billion barrels ($107.4 \times 10^9 m^3$) and 2,670 trillion cubic feet ($75.6 \times 10^{12} m^3$). The known conventional recoverable resources of petroleum liquids are highly concentrated in three regions of the world: the Middle East, North America, and the Soviet Union. Together, these three regions contain slightly more than 80 percent of total known conventional resources. The Middle East contains nearly half of the world's known recoverable resources of petroleum liquids. The Soviet Union contains slightly more than one-third of the world's known recoverable resources of natural gas. This concentration by region provides an appropriate introduction to the first and most important point about world petroleum resources, namely, petroleum resources are highly concentrated in a small number of sedimentary provinces.

The existence of hydrocarbons in sedimentary areas is a common phenomenon. There are approximately 600 sedimentary provinces in the world, the exact number depending on how one chooses to define an individual province. Exploratory drilling has occurred in more than 400, resulting in discoveries of producible accumulations at current world prices in nearly 240 (60 percent). Indications of hydrocarbons have been encountered in most of the remaining 160. The near ubiquity of the existence of hydrocarbons in the sedimentary areas of the world should not, however, blind us to the more important fact that accumulations of conventional petroleum on a significant scale are a statistically rare event.

TABLE 23-1. Known conventional recoverable resources[a] of petroleum liquids and natural gas in the world by region as of December 31, 1978

Region	Petroleum liquids		Natural gas	
	10^9 bbls	($10^9 m^3$)	10^{12} ft^3	($10^{12} m^3$)
North America	242.2	(38.5)	930	(26.3)
South America	69.9	(11.1)	126	(3.6)
Western Europe	24.4	(3.9)	203	(5.8)
Soviet Union/Eastern Europe	115.5	(18.4)	1,275	(36.1)
Africa	74.8	(11.9)	204	(5.8)
Middle East	523.3	(83.2)	788	(22.3)
Asia/Oceania	50.3	(8.0)	174	(4.9)
World total	1,100.4	(174.9)	3,700	(104.8)

[a]Known recoverable resources are defined as cumulative production plus proved and probable reserves.

Measures of significance are of course subjective, depending on the perspective that one chooses. Because this conference addresses long-term energy resources for the world, an appropriate measure of significance is current world consumption of petroleum liquids and natural gas. In 1978, world consumption of petroleum liquids was slightly more than 23 billion barrels ($3.7x10^9 m^3$). World consumption of natural gas was slightly less than 54 trillion cubic feet ($1.5x10^{12} m^3$). Using the standard equivalency conversion of 6,000 cubic feet per barrel, total world consumption of conventional petroleum was approximately 32 billion barrels ($5.1x10^9 m^3$) of liquid and liquid-equivalent petroleum resources. The latter number provides a useful standard of significance for world energy resources.

After more than a century of petroleum exploration covering nearly all of the prospective areas of the world, only ten provinces containing oil and gas equal to a year or more of current world consumption have been discovered (Table 23-2). Those ten provinces contain 72 percent of all the conventional petroleum liquids and 61 percent of all the natural gas known to be recoverable. Only two provinces--the Arabian-Iranian and the West Siberian--are known to contain more petroleum resources than are currently consumed in 2 years. The Arabian-Iranian province clearly dominates, with recoverable resources equivalent to more than 20 years of current world consumption. The West Siberian province has recoverable resources equivalent to slightly more than 5 years current world consumption. Together the two contain 51 percent of known recoverable resources of petroleum liquids and 42 percent of the natural gas resources. (Combining the Mississippi Delta province with the Texas Gulf Coast-Burgos province, as some do, would create a Gulf Coast province with resources equivalent to nearly 3 years current world consumption.) Because of their significance, these ten provinces clearly merit the designation as <u>super-provinces</u>.

TABLE 23-2. The major petroleum provinces of the world

Province	Known recoverable resources as of 12-31-78			
	Petroleum liquids		Natural gas	
	10^9 bbls.	($10^9 m^3$)	10^{12} ft^3	($10^{12} m^3$)
1. Arabian-Iranian	523.0	(83.1)	c.768.0	(21.7)
2. West Siberian (USSR)	c. 37.0	(5.9)	c.780.0	(22.1)
3. Volga-Ural (USSR)	c. 40.0	(6.4)	c.108.0	(3.1)
4. Maracaibo (Venezuela)	41.2	(6.5)	54.0	(1.5)
5. Mississippi Delta (USA)	21.5	(3.4)	169.4	(4.8)
6. Permian (USA)	30.0	(4.8)	73.9	(2.1)
7. Texas Gulf Coast-Burgos (USA-Mexico)	18.4	(2.9)	138.0	(3.9)
8. Reforma-Campeche (Mexico)	36.4	(5.8)	25.2	(0.7)
9. Sirte (Libya)	30.0	(4.8)	31.8	(0.9)
10. Alberta (Canada)	16.3	(2.6)	94.2	(2.7)
(Super-province subtotals)	793.8	(126.2)	2,242.5	(63.5)
11. Amarillo Anadarko Ardmore (USA)	9.8	(1.6)	116.4	(3.3)
12. Niger Delta (Nigeria-Cameroon)	20.5	(3.3)	51.0	(1.4)
13. Northern North Sea	20.0	(3.2)	40.2	(1.1)
14. Triassic (Algeria)	10.8	(1.7)	90.0	(2.5)
15. E. Texas-Arkla (USA)	15.2	(2.4)	50.1	(1.4)
16. N. Caucasus-Mangyshlak (USSR)	c. 11.8	(1.9)	c. 54.0	(1.5)
17. Netherlands-NW Germany	2.0	(0.3)	93.0	(2.6)

TABLE 23-2. The major petroleum provinces of the world--Continued

Province	Known recoverable resources as of 12-31-78			
	Petroleum liquids		Natural gas	
	10^9 bbls.	($10^9 m^3$)	10^{12} ft^3	($10^{12} m^3$)
18. Eastern Venezuela (Venezuela-Trinidad)	13.1	(2.1)	21.0	(0.6)
19. South Caspian (USSR)	c. 12.0	(1.9)	c. 15.0	(0.4)
20. North Slope (USA)	9.9	(1.6)	26.0	(0.7)
21. San Joaquin (USA)	11.6	(1.8)	12.0	(0.3)
22. Tampico-Misantla (Mexico)	10.7	(1.7)	9.0	(0.3)
23. Amu Daryu (USSR)	0.5	(0.1)	c. 60.0	(1.7)
24. Appalachian (USA)	4.0	(0.6)	38.4	(1.1)
25. Los Angeles (USA)	8.6	(1.4)	7.2	(0.2)
26. Murgab (USSR)	0.0	(0.0)	c. 57.0	(1.6)
27. Dnepr-Pripyat (USSR)	2.0	(0.3)	c. 45.0	(1.3)
28. Sung-liao (China)	c. 8.5	(1.4)	c. 6.0	(0.2)
29. Central Sumatra (Indonesia)	9.0	(1.4)	c. 1.2	(*)
30. Timan-Pechora (USSR)	c. 4.0	(0.6)	c. 27.0	(0.8)
(Major province subtotals)	184.0	(29.3)	819.5	(23.2)
Super and major provinces	977.8	(155.5)	3,062.0	(86.7)
World total	1,100.4	(174.9)	3,700.0	(104.8)

A relaxation of the standard of significance further confirms the point. There are only 20 provinces with conventional petroleum resources equivalent to 3 to 12 months of current world petroleum consumption ($8\text{-}32\times10^9$ barrels). These 20 provinces contain another 17 percent of known conventional petroleum liquids resources and 22 percent of known natural gas resources. The 30 largest provinces contain 89 percent of the known recoverable petroleum liquids and 83 percent of the known recoverable natural gas provinces. The other 210 provinces with producible accumulations contain only 11 percent of the petroleum liquids and 17 percent of the natural gas, over half of which is in 35 or so provinces with 2 to 8 billion barrels liquid and liquid-equivalent resources each. Thus, nearly all of the world's conventional petroleum resources are concentrated in only 15 percent of the explored sedimentary provinces.

Two interrelated aspects of the geographical distribution of the most significant provinces are noteworthy. First, they are concentrated in two arcs, one in each hemisphere. In the Eastern Hemisphere, the most significant provinces are concentrated in an arc traversing North Africa, the Middle East, and the central Soviet Union. In the Western Hemisphere, the most significant provinces are concentrated in an arc traversing Venezuela, the western Gulf of Mexico, the central United States, and western Canada (Fig. 23-1). All of the super-provinces and half of the major provinces are found within these two arcs. Secondly, not one of the 30 most significant provinces lies south of the equator. Central Sumatra straddles it. The Niger Delta, the Maracaibo Basin, and the Eastern Venezuela Basin are between 0° and 10° N latitude. Twenty-three of the 30 are between 20° and 60° N latitude.

The great variation in the distribution of petroleum resources by province and region should not be surprising. The sedimentary provinces of the world differ widely in sedimentary volume. More importantly, they differ even more in their geologic history and characteristics. For large accumulations of petroleum to exist, several conditions must have existed in an appropriate temporal and spatial relationship with each other. Sufficient organic material must have been produced, accumulated, and preserved. This material must be buried to a depth where sufficient temperature and pressure exist to generate petroleum. After the petroleum is generated, it must be expelled from the source rock into permeable carrier or reservoir rocks. Traps with porous and permeable reservoir rocks covered by impermeable sealing rocks must exist where the migrating petroleum can accumulate. These traps must not be breached after accumulation occurs (Meyerhoff, 1976; Tissot and Welte, 1978).

Where these conditions are met to a high degree, we encounter the major accumulations of the Middle East, West Siberia, or the provinces circling the western Gulf of Mexico. Where production of organic material was insufficient, where the organic material that was produced was oxidized, where unoxidized organic material was buried to an insufficient depth, where migration did not occur or occurred prior to the formation of traps, where reservoir rocks are poor or nonexistent, where sealing rocks did not exist, or where traps with petroleum accumulations were destroyed by subsequent erosion or tectonic activity, only negligible amounts of petroleum, if any,

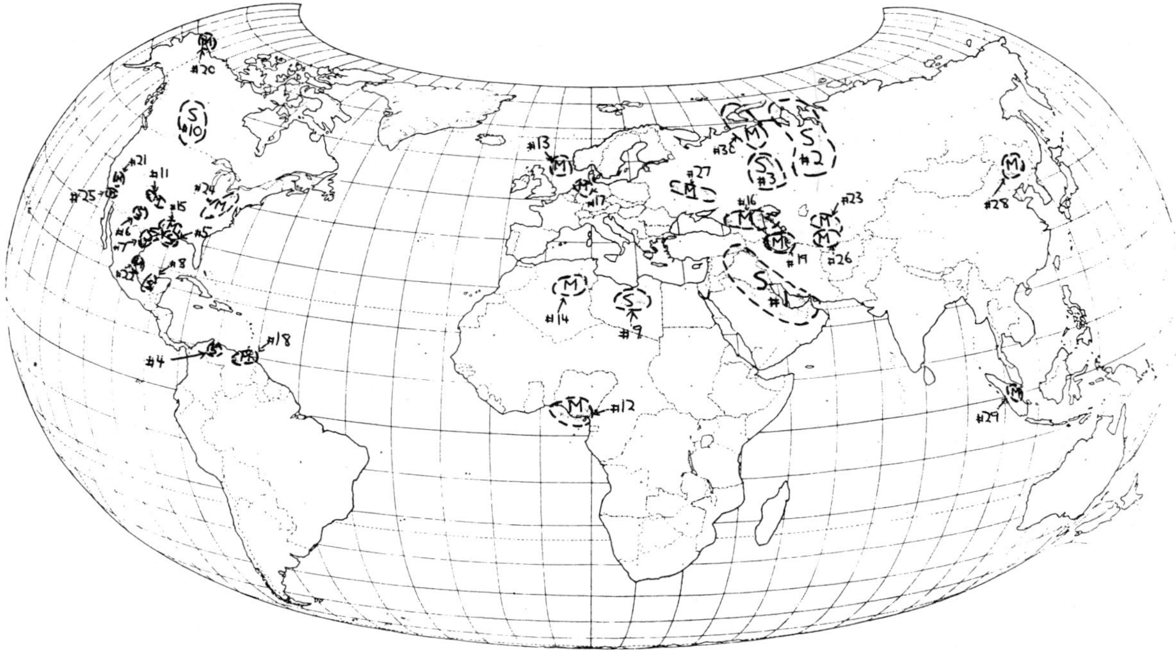

S - Super Province

M - Major Province

Figure 23-1. The major petroleum provinces of the world

are likely to be found. Most provinces lie between the extremes of the optimal and the deficient. But significant concentrations of petroleum are found only where optimal or near-optimal conditions are found as well.

The second point to consider in evaluating conventional world petroleum resources is that the world's oil and gas resources are highly concentrated in a small number of large fields. Since petroleum exploration began, approximately 30,000 fields have been discovered. But more than 90 percent of these fields are insignificant as far as world petroleum resources are concerned. The 30 largest oil and gas fields discovered to date--those with 10 billion barrels ($1.59x10^9m^3$) or more of liquid or liquid-equivalent recoverable resources--contain 475 billion barrels ($75.5x10^9m^3$) of petroleum liquids and 1,124 trillion cubic feet ($31.8x10^{12}m^3$) of natural gas. In other words, just 0.1 percent of the known fields contain 43 percent of the petroleum liquids and 30 percent of the natural gas discovered to date. The 52 super-giant fields discovered to date--those with 5 billion barrels ($0.79x10^9m^3$) or more of liquid or liquid-equivalent recoverable resources--contain 569 billion barrels ($90.5x10^9m^3$) of petroleum liquids and 1,500 trillion cubic feet ($42.5x10^{12}m^3$) of natural gas, 52 percent and 41 percent, respectively, of the world's conventional oil and gas resources. The approximately 400 known giant fields--those with 500 million barrels ($79x10^6m^3$) or more of liquid or liquid-equivalent recoverable resources--contain approximately 850 billion barrels ($135.1x10^9m^3$) of petroleum liquids and 2,600 trillion cubic feet ($73.6x10^{12}m^3$) of natural gas, 77 percent and 70 percent, respectively, of the world's conventional oil and gas resources. The approximately 1,400 known large fields--those with 50 to 500 million barrels ($8-79x10^6m^3$) of liquid or liquid-equivalent resources contain at least another 175 billion barrels ($27.8x10^9m^3$) of petroleum liquids and 750 trillion cubic feet ($21.2x10^{12}m^3$) of natural gas. Thus, only 6 percent of the known fields contain over 93 percent of the known conventional petroleum liquids and over 90 percent of the known conventional natural gas resources of the world. (For the most recent treatment of the importance of giant oil fields, see Nehring, 1978.)

Significant petroleum fields and significant petroleum provinces are closely associated. The 10 super-provinces contain 44 of the 52 super-giant fields. Six of the remaining eight are found in the 10 largest major provinces. The importance of the Arabian-Iranian and West Siberian provinces stems from the fact that they contain 30 and 7 super-giant fields, respectively. The Mississippi Delta and Texas Gulf Coast-Burgos provinces are unusual in that they are the only super-provinces with no super-giant fields, nothing resembling a super-giant field, and relatively few giant fields. But the former has more than 250 large and giant fields while the latter has 140 large and giant fields. The Niger Delta is the only other major province in which the known oil and gas resources are concentrated in large rather than giant fields.

The third key point is that modern petroleum exploration is an efficient process. Operationally, this means that if giant and large accumulations of petroleum exist in a sedimentary province, most will be found by the time 25 to 200 new field exploratory wells have been drilled in the province. This number is likely to be exceeded only when there is a very large number of potential large accumulations within the province or if the pattern of drilling within the province is shaped heavily by political or technological constraints. Exploration efficiency is the result of the geologic fact that most giant and large fields occur in obvious traps, that is, traps that are detectable prior to drilling (Klemme, 1973/1974); that improvements in geologic knowledge and exploratory technology are enabling the petroleum industry to locate these traps with increasing effectiveness; and that any rational exploration philosophy emphasizes drilling the largest prospects first in exploring a virgin province.

Intra-province exploration efficiency is particularly evident in those provinces in which exploration began since World War II. Nearly all of the giant fields discovered in major provinces were found in less than 10 years after the initial giant field discovery (Klemme, 1971; Nehring, 1978, p. 37). In the provinces opened up since 1965, the process may even have accelerated, given the high success rates and rapid rates of discovery witnessed in the Reforma-Campeche, Oriente, Southern North Sea, Northern North Sea, Cambay, Gippsland, Dampier, and Mahakam Delta provinces and subprovinces.

A recent development of the past 10 to 15 years appears to be a marked increase in inter-province exploration efficiency. Geological investigations prior to exploratory drilling are becoming more and more capable of distinguishing between geologically favorable and unfavorable conditions. Geologic knowledge and exploratory technology have progressed to the point where only a handful of exploratory wells may be necessary to indicate whether a province is likely to contain significant amounts of petroleum. If a few wells encounter unfavorable geologic conditions of regional significance, as has occurred in the Gulf of Alaska and most of the East African coastal provinces, those wells are sufficient to indicate that the entire province will not be a major source of petroleum.

These three points considered together provide a powerful means of understanding the present status of world petroleum resources and for predicting the ultimate conventional petroleum resources of the world. Because only major and super-provinces make any appreciable difference, the task of world petroleum resource assessment is predominantly one of estimating future discoveries and reserve growth in known major provinces, of determining which other producing provinces have a chance of becoming major provinces with further discoveries and development, and of assessing which unexplored provinces have a meaningful probability of becoming major provinces. Because most of the oil and gas is found in giant and large fields, the assessment of unexplored provinces can focus on whether and how many large traps exist in a province if other conditions are favorable for petroleum accumulation. Because modern petroleum exploration is an efficient process, resource assessment in known producing provinces can focus on known field-size distributions, those areas of a province in which past exploration has been constrained for technological or political reasons, and the possibilities of less obvious traps within the province.

The implications of these three facts for conventional world petroleum resource assessment are not favorable. The Arabian-Iranian, West Siberian, and Reforma-Campeche provinces are the only super-provinces that are likely to have several giant field discoveries in the future. The Amarillo-Anadarko-Ardmore, Niger Delta, Northern North Sea, and Triassic provinces are the only major provinces with a good probability of becoming super-provinces. The South Caspian, North Slope, Tampico-Misantla, and Timan-Pechora provinces have a low probability of becoming super-provinces. None of the eight appears to have a potential for becoming more than just a minimal super-province. Of the nonmajor producing provinces, only six--Lena-Vilyuy (USSR), Illizi-Ghadames (Algeria-Libya), Suez (Egypt), North China and Szechwan (China), and Kutei (Indonesia)--appear to have the potential to become major provinces. Of the provinces that are unexplored or are only lightly explored there are probably no more than a dozen with the potential to become major provinces, including the Labrador Shelf, Beaufort Sea/ Mackenzie Delta, and Sverdrup Basin (Canada), the Sabinas Basin (Mexico), the Gulf of Venezuela (Venezuela), the Malvinas Basin (Argentina-UK), the Norwegian Coastal Basin, the Lena-Anabar and East Siberian Sea provinces (USSR), and the East China Sea province (China). Currently the Beaufort Sea/Mackenzie Delta province is the only one of these that has a reasonable probability of becoming a super-province. Ultimately, of the world's 600 sedimentary provinces, no more than 40 to 45 will be major provinces (including 12 to 18 super-provinces).

During the past decade, several attempts have been made to evade the implications of these facts. These attempts have had enough surface plausibility to mislead both the political leadership and the public of many countries as to the true petroleum supply situation. But they can readily be shown to lack scientific support.

The first of these misconceptions is that sedimentary volume (or area) is a useful indicator of petroleum potential (Grossling, 1976). The basic argument from this premise is that a substantial proportion of the world's sedimentary volume (or area) is still unexplored or lightly explored, particularly in the less-developed countries, and that with extensive drilling its full promise will be revealed. The argument gains plausibility from the fact that some of the world's most significant provinces, specifically the Arabian-Iranian and West Siberian, are also among the largest in sedimentary volume. However, sedimentary volume is only a minimally necessary condition for petroleum accumulation. The characteristics of the sediments are vastly more important than mere volume for explaining petroleum accumulation, as indicated by the wide variations in known recovery per unit of sedimentary volume among basins (Klemme, 1977). Moreover, when applied to provinces that have been at least lightly explored, this argument ignores the extensive amount of information that can now be obtained about the petroleum prospects of a province from a small number of exploratory wells.

A second major misconception is that significant amounts of oil remain to be discovered in small fields (Wood, 1979). The argument is that because the most intensely drilled area of the world, the United States, has a significantly greater proportion of its known recoverable petroleum resources in nongiant fields, an intensive drilling effort in the rest of the world will result in the discovery of hundreds of billions of barrels in large and small fields. The argument is woefully ignorant of the important geologic differences between most of the U.S. petroleum provinces and most of the provinces of the rest of the world. First, two of the three largest provinces in the U.S.--the Mississippi Delta and the Texas Gulf Coast--have atypical field size distributions. They and the Niger Delta are the only three major provinces in which the majority of the petroleum resources are found in large rather than giant fields. Secondly, most of the oil in small fields (less than 50 million barrels liquid or liquid-equivalents) in the United States is found in wholly Paleozoic provinces. Provinces of this age contain a large proportion of the petroleum found in the United States and a small proportion of that found in other countries of the world. (The most prominent examples of Paleozoic provinces elsewhere in the world--Volga-Ural, Alberta, and Timan-Pechora--appear to have roughly similar field size distributions as the Mid-Continent/Permian Basin area of the United States.) The distribution of petroleum resources by field size in the Mesozoic and Tertiary provinces of the United States (excluding the Gulf Coast) does not differ significantly from that of Mesozoic and Tertiary provinces elsewhere in the world. (For a detailed survey of the distribution of U.S. petroleum resources by field size, see Nehring, 1980a.) Third, on the other extreme of field sizes, the concentration of the petroleum resources of the Arabian-Iranian and West Siberian provinces in super-giant and giant fields with large drainage areas makes it extremely unlikely that significant amounts of petroleum will be found in small fields in either of these two provinces. The argument is plausible that discoveries of small and large fields will be relatively more important in the future than they have been in the past. But there is no reason to expect major percentage increases in world petroleum reserves from this source.

The third major misconception is that higher prices will result in substantial increases in world petroleum reserves. The argument is that high prices will induce drilling in expensive environments and encourage the development of known but previously sub-economic resources. This argument is qualitatively correct, but quantitatively inaccurate. Because most oil and gas is found in large and giant fields and because nearly all large and giant fields have good to excellent reservoir characteristics, the costs of finding and producing most of the world's conventional oil and gas are less than 10 percent of the current world oil price (or less than U.S. $2.50 per barrel or barrel-equivalent). Higher prices only make a difference for recovery from fields with poor reservoir characteristics such as the Chicontepec area in Mexico, for enhanced oil recovery from some fields, for the discovery of very small fields, and for exploration and production in expensive environments such as the high latitudes or deep-water provinces (Nehring, 1980b). None of these categories appears to contain more than 10 percent of the world's ultimate petroleum potential. Most contain substantially less.

Considering the prospects for future discoveries and additional recovery in the known producing provinces and the prospects of unexplored or lightly explored provinces, I estimate that ultimate conventional production of petroleum liquids will be between 1.6 and 2.0 trillion barrels (254-318x $10^9 m^3$). Ultimate conventional production of natural gas will be between 5,000 and 6,500 trillion cubic feet (142-184x$10^{12} m^3$). The majority of the increase from known recoverable resources of petroleum liquids is likely to come from additional recovery from known fields, not future discoveries. Although the increase from known to ultimate recoverable resources of natural gas is slightly less than the increase for petroleum liquids, the prospects for future natural gas discoveries, particularly in the Middle East and the Asiatic and Siberian provinces of the Soviet Union, are more favorable than the prospects for future oil discoveries. Future reserve growth from known natural gas fields will be only a minor proportion of future reserve additions. A major breakthrough in oil recovery technology could add as much as 0.5 trillion barrels (80x$10^9 m^3$) to this prognosis. No other possibilities for larger conventional resources appear to have any plausibility.

REFERENCES

Grossling, B. F., 1976, Window on Oil: Financial Times, London.

Klemme, H. D., 1971, Trends in Basin Development: Possible Economic Implications: World Petroleum, October.

Klemme, H. D., 1973, 1974, Structure-Related Traps Expected to Dominate World Reserve Statistics: Oil and Gas Journal, December 31, 1973 and January 7, 1974.

Klemme, H. D., 1977, World Oil and Gas Reserves from Analysis of Giant Fields and Petroleum Basins (Provinces), in R. F. Meyer, ed., The Future Supply of Nature-Made Petroleum and Gas: Pergamon Press, New York, p. 217-260.

Meyerhoff, A. A., 1976, Economic Impact and Geopolitical Implications of Giant Petroleum Fields: American Scientist, September-October 1976, p. 536-541.

Nehring, R., 1978, Giant Oil Fields and World Oil Resources: R-2284-CIA, The Rand Corporation, Santa Monica.

Nehring, R., 1980a, The Discovery of Significant Oil and Gas Fields in the United States: The Rand Corporation, Santa Monica, in press.

Nehring, R., 1980b, The Cost of Oil: The Rand Corporation, Santa Monica, in press.

Tissot, B. P., and Welte, D. H., 1978, Petroleum Formation and Occurrence: Springer-Verlag, Berlin.

Wood, P. W. J., 1979, There's a Trillion Barrels of Oil Awaiting Discovery: World Oil, June, p. 141-148.

CHAPTER 24

PRESENT STATUS OF OIL AND GAS FROM THE OCEAN FLOOR[1]

K. O. Emery[2]

ABSTRACT

About 19 percent of the world's oil came from the continental shelves in 1978, a percentage that is slowly increasing; a smaller percentage of the gas is from offshore owing to handling problems. In deeper water beyond the shelf the best prospective provinces appear to be the inner marginal basins, the lower parts of mature continental slopes, and the inner continental rises. All three provinces probably contain petroleum source beds, reservoir beds, impervious caps, stratigraphic and structural traps, and temperatures at depth sufficient for maturation of their marine organic matter. These inferences are based upon geophysical studies and shallow drill-hole samples, but to date not a single deep exploratory hole in them has been drilled and published. Although prospects for the presence of deep-ocean petroleum are believed to be good, its presence in exploitable quantities is not proven nor is the energy cost of its production known.

INTRODUCTION

Since the July 1976 petroleum conference in Laxenburg, Austria, sponsored by the United Nations Institute for Training and Research (UNITAR) and the International Institute for Applied Systems Analysis (IIASA), there has been little addition to the store of hard facts about the petroleum (oil and natural gas) potential of the deep-ocean floor. However, some progress has been made on theoretical concepts, perhaps justifying an attempt to bring up to date a previous summary (Emery, 1977). Very active investigation of deep-ocean, as well as continental-shelf, sources of petroleum might have been expected to result from the escalating prices of oil sold by the Organization of Petroleum Exporting Countries (OPEC): now, as much as $23/barrel contract

---

[1]Contribution No. 4484 of the Woods Hole Oceanographic Institution.

[2]Woods Hole Oceanographic Institution, Woods Hole, MA 02543, USA.

and $45/barrel spot. However, a broad array of factors contributed to indecision and inactivity in exploration of world offshore petroleum resources. These negative factors include a severe inflation followed by economic recession in the industrialized nations (partly brought on by high prices of imported oil), high government taxes (partly to fund social transfer payments), demands by the United Nations for broad sovereignty over ocean-floor mineral resources, strong government bias against big business (especially in the United States), and effective political activities by zealous environmentalists. Typically, such negative attitudes undergo major reversals with time and mature consideration. Summaries of existing knowledge, such as are included in the present volume, may help to avoid some of the waste inherent in crash programs when the trend eventually reverses.

CONTINENTAL SHELVES

The continental shelves of the Earth have a total area of about 23×10^6 km^2, which corresponds to 6.4 percent of the area of the entire ocean and 15 percent of the area of the land. The shelves have been termed extensions of the land, because they are shallow and are underlain mainly by continental crust. However, the older and broader shelves, most of which are due to plate divergence, consist of wedges of sediments that thicken oceanward atop the crust. Most of their sediments are debris shed by weathering and erosion of the land and then deposited on the adjacent shallow ocean floor. These shelves are depositional features, unlike the land and many shelves within belts of plate convergence. The latter are erosional or volcanic in main. Great thicknesses of shallow-water sediments beneath the shelves of divergent plates result from continuous downwarping of the continental margins due to cooling and sinking of crust after rifting (Montadert et al., 1977; Pitman, 1978) and to loading by sediments. An estimate of the average thickness of compacted sediments deposited on all continental shelves since Paleozoic time (after major rifting) is 3 km (allowing for the minor thicknesses on inner parts of nearly all shelves and on the full width of many shelves of convergence). This thickness times the area of the shelves yields a total volume for their post-rift sediments of about 75×10^6 km^3 (Table 24-1), about double that of the sediments deposited on the land during the same period of time.

TABLE 24-1. Estimated volumes of post-Paleozoic sediments of the Earth

Region	Area ($\times 10^6$ km^2)	Average thickness (km)	Volume ($\times 10^6$ km^3)
Land	149	0.3	45
Continental shelves	23	3	75
Continental slopes	32	6	200
Marginal basins	12	3	35
Continental rises	50	3	150
Beyond continental margins	244	0.1	25
Totals	510		530

The thick wedges of sediment under shelves of divergent margins contain marine source beds rich in organic matter and both sandy and reefal reservoir beds; thus the shelves inherently are more likely to be petroliferous than the land, where only remnants of a former cover of marine sediments remain uneroded in filled structural basins and in coastal plains having gentle slopes. The search for petroleum on the continental shelves has been most successful through the tracing beneath the ocean floor of extensions of fields originally found and produced on land. Closely following is the success in discovery of salt diapirs as parts of general distribution patterns known on the adjacent land. Much less success has attended the more difficult search for oil fields in strata and structures unrelated to those of the adjacent land; productive examples that have been found are the Bass Strait of Australia, the North Sea off Norway, and the northern North Sea. Many other searches have been unproductive owing to absence of source beds (as for most carbonate platforms), of good reservoir beds, and of impervious caps. Drilling costs alone amount to as much as $15 million per well, aside from much larger costs of geophysical exploration, nonproductive bonus bids required by governments, and protracted delays in the name of environmental preservation.

General geological considerations indicate that the most prospective parts of the continental shelves may be those shown in black on Fig. 24-1. Only about 15 percent of the shelf area is considered to be well explored by geophysics and test drilling. Much of the remainder is essentially

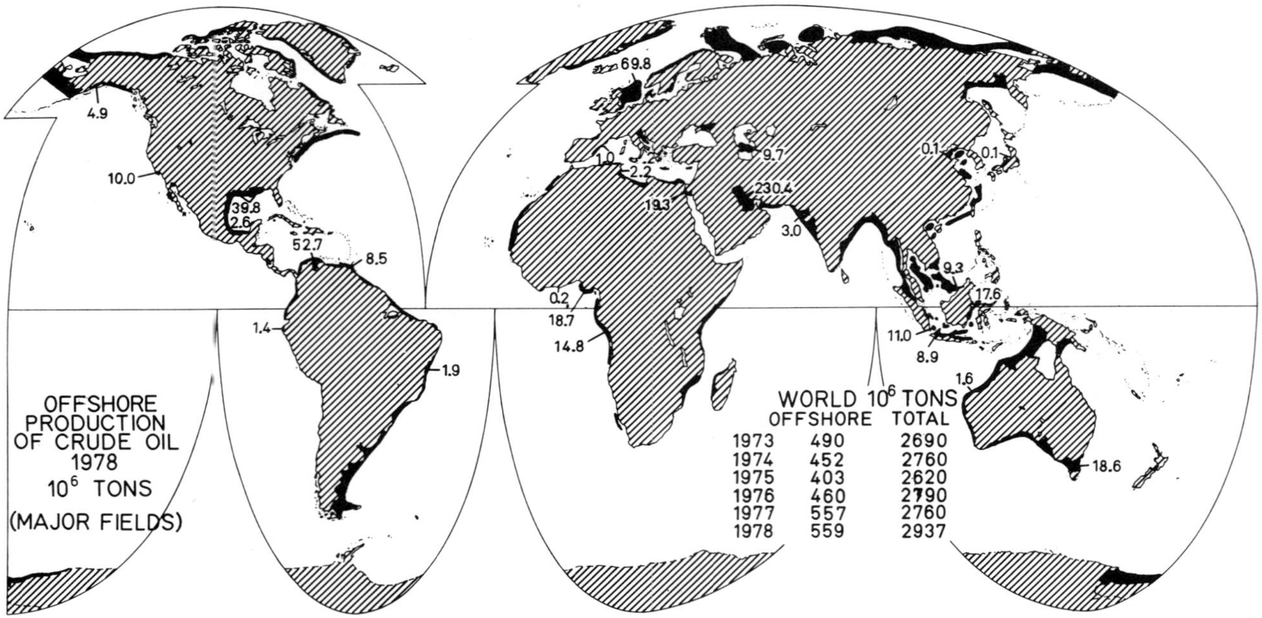

Figure 24-1. Most prospective parts of the continental shelf for petroleum (black), and actual production of oil in different provinces of the shelf during 1978

uninvestigated owing to both climatic and political inhospitality. Production of oil during 1978 in the chief regions that are being exploited also is indicated in Fig. 24-1; note that the Persian Gulf, the North Sea, Lake Maracaibo, and the Gulf of Mexico produced 70 percent of the total from the shelves. Production from the North Sea and the southern Gulf of Mexico is expected to increase, and that from the Persian Gulf and Lake Maracaibo to decrease in the near future. Total production from the continental shelves during the years 1973 through 1978 (tabulated in the figure) is somewhat erratic owing to new discoveries, local interruptions to production, and changes in demand caused by price rises. However, the trend for the shelf exhibits an increase averaging 17×10^6 tons per year, corresponding to 3.5 percent per year. In contrast, production from the land (a much larger area that is more intensively investigated) increased only 0.7 percent per year. This difference may reflect the greater oil prospects for offshore and the lesser ones for land.

Many attempts have been made to predict in advance the recoverable oil and gas in unexplored regions (Hubbert, 1974; McKelvey, 1974; Grossling, 1975; Haun, 1975; Miller et al., 1975; Skinner, 1975) in order to aid decisions regarding investment and tax potential. Comparison of predictions and results reveals a poor correlation (Uman et al., 1979) and particularly bad ones for the Atlantic continental shelf of the United States and the Destin Dome in the Gulf of Mexico. The lesson to be learned is that predictions for small regions cannot be realistic in the absence of drill-hole data, and they should be avoided in spite of strong pressure from congressmen and news media.

CONTINENTAL SLOPES

Continental slopes have a larger total area than do continental shelves (32 versus 23×10^6 km^2--Emery, 1979). Their origin and development is directly associated with plate movements and with post-rift sedimentation. A classification by Emery (in press) begins with an initial stage that is produced by rifting of a protocontinent. Resulting continental terraces (shelves and slopes) consist of pre-rift (generally pre-Mesozoic) rocks whose original content, if any, of petroleum was lost by fracturing associated with the rifting, and whose post-rift sediments are too thin and too young for their organic matter to have matured into petroleum. By youth, sediments derived mainly from the land have filled to shelf levels many of the basins (as in and off southern California and eastern Asia), and the basins may contain both source beds and reservoir beds of petroleum. Some youthful continental terraces are thicker than others because of additional damming effects of coralgal reefs atop the ridges produced by tectonism. By late youth, sediments have mantled the continental slope especially between outcrops of tectonic ridges. By maturity (physiographic--not temporal), sediments have kept pace with the subsiding shelf and have prograded far beyond the shelves onto continental slopes and still farther as continental rises. Sediments beneath the outer shelf, the slope, and the inner rise may exceed even 20 km thickness. Eventually, and not necessarily only after maturity is reached, plate movements can change from divergence (or translation) to convergence. The resulting change of the continental terrace to old age (which equals destruction) causes extreme folding, shearing, intrusion, and metamorphism of the sediments in the former continental rise. Most or all of the petroleum that may have been present is lost by this tectonic activity.

Thickness of post-rift (generally post-Paleozoic) sediments on continental slopes increases from essentially zero on those of initial and early youth stages to a maximum of possibly 20 km for some in maturity, followed by losses during subduction processes of old age. A rough estimate of average thickness is 6 km, yielding a total volume of about 200 x 10^6 km^3 (Table 24-1).

Clearly, the best prospects for petroleum on the continental slope are to be found in the 25 percent of it that belongs to mature continental terraces, because the continental slopes of initial and youthful stages have only a little post-rift sediment, and those of old age are so disrupted by collision tectonics that most petroleum is able to escape. These mature stages border much of the Atlantic coasts of North America, South America, and Africa, the Arctic Sea, and India (Emery, in press), and they are especially well developed in large deltas. Best prospects for petroleum may be in regressive beds of large deltas and on mature slopes that have been truncated by repeated mass movements and turbidity currents so that the distal ends of the strata have been removed, and drilling can reach older strata without having to pass through younger ones (as off Cape Hatteras). Fortunately, truncated continental slopes appear to be more common than prograded ones. However, the same mass movements (Moore, 1977; Embley and

Jacobi, 1978) that have truncated the strata pose a distinct hazard to the safety of oil wells on truncated continental slopes.

MARGINAL BASINS

Along most continental margins of convergence and some of translation are basins adjacent to continental slopes with floors that are deeper than those of the shelves and shallower than those of the deep ocean floor. Most of these marginal basins were mapped by Menard (1967), who believed them to be underlain by crusts that are transitional between typical continental and oceanic crusts, possibly being in the process of transformation from one kind of crust to the other. Later studies by Karig (1971) indicate that at least some of the marginal basins along eastern Asia owe their origin to secondary sea-floor spreading associated with plate convergence. Support is provided by median ridges, patterns of magnetic reversal, and high heat flow.

The basins that were studied by Menard have an area of 7.5×10^6 km^2; to them must be added others in the Scotian Sea, the Mediterranean Sea, and the East Indies (Emery, 1977, Fig. 5-2), yielding a total area of 12×10^6 km^2. Many of these basins are well placed to intercept the sediment discharged from large river systems, and they have received thick fills of sediments. Basins that have become filled flush with the surface of the adjacent continental shelf are included in the previous discussion of the continental shelf. The total volume of post-Paleozoic sediments in all of the incompletely filled marginal basins is estimated at very roughly 35×10^6 km^3.

One of the first drill holes of the Deep Sea Drilling Project encountered crude oil in the Sigsbee Knoll, a salt diapir of the Gulf of Mexico Basin (Davis and Bray, 1969). Another D.S.D.P. hole, number 88, drilled atop a different salt diapir in the Gulf included samples having gas that was chemically and isotopically identical to natural gas associated with petroleum (Claypool, 1975). Additional oil shows and gas having a methane/ethane ratio of 10 (comparable to natural gas) were found on the floor of the Balearic Basin in the western Mediterranean Sea (McIver, 1974). Gases sampled in cores from other marginal basins, however, have compositions that are ascribed to bacterial origin.

The thick fill, the high content of marine organic matter from nutrients discharged in river waters, and the maturing effect of high heat flow may warrant serious consideration of incompletely filled marginal basins as potential provinces for future petroleum. A major limitation appears to be the uncertain presence of thick reservoir beds, particularly in outermost basins that are separated from turbidite sources by intervening submerged ridges. Completely filled inner basins that are near land sources of sands for turbidity currents are known to contain adequate reservoir beds, as in southern California. The various depths of the floors of the marginal basins below the depths of continental shelves provide a sort of ladder to permit gradual exploitation into deeper and deeper water as the technology of drilling improves with experience, and so these basins may be the first of the ocean-floor provinces beyond the shelf and the upper slope to be exploited. In fact, the Santa Barbara Basin off southern California already is on the verge of such exploitation.

CONTINENTAL RISES

Continental rises lie at the foot of most mature and some late youth continental terraces, covering about 50×10^6 km^2 area, according to the distribution mapped by Emery (1977, Fig. 5-2). The total volume is poorly known, but clearly enormous. A world estimate based upon seismic reflection and refraction mapping in the Atlantic Ocean yields an average thickness of about 3 km for all sediments beneath the surface of continental rises; thus the volume is about 150×10^6 km^3. This is about 40 percent of the total volume of all post-rift ocean-floor sediments. Note that some continental rises may be underlain at depth by abyssal-plain sediments onto which the rises prograded. This subtlety is beyond the ability of present knowledge to evaluate quantitatively. All estimates of Table 24-1 are intended to give only a general idea of the relative volumes, but hopefully they will be much improved within a decade.

Detailed seismic profiles show that continental rises contain many good acoustic reflecting horizons that depict stratigraphic traps, folds, and faults. The reflecting horizons themselves mark possible reservoir beds--mainly sandy turbidites deposited in deep water. The turbidites are not restricted to Pleistocene glacial epochs of low sea level. They also were deposited during late Miocene, Oligocene, and late Early Cretaceous times of glaciation and during Middle Jurassic because of tectonically lowered sea level, as suggested by the discovery of repeated episodes of submarine-canyon cutting (Emery and E. Uchupi, in preparation) and by unconformities revealed by multichannel seismic reflection profiles (Vail et al., 1977). Turbidite sands of any age that are adequately coarse-grained, permeable, and thick are most likely to occur on the inner continental rises near the mouths of submarine canyons.

Source beds are not reliably indicated by geophysical studies. Perhaps the best kinds of potential source beds are the massive piles of debris brought by landslides down the continental slopes to many inner continental rises. These are the same slides that truncated beds and caused them to crop out on many continental slopes. In this way, organic-rich sediments from intermediate depths of oxygen minimum in the open ocean are incorporated into the otherwise organic-poor sediments of continental rises. Maturation of the organic matter into petroleum may be followed by movements of the petroleum into adjacent sandy beds deposited by turbidity currents. Another kind of source bed may be the Middle Cretaceous black shales that were encountered in the Atlantic Ocean during the Deep Sea Drilling Project (Lancelot and Seibold, 1977; Jansa and Macqueen, 1978; Tucholke, 1979). Organic matter in the shales of the western Atlantic, however, is reported to be largely terrestrial in origin, a composition that generally leads to natural gas rather than to oil. Analyses of gases in some of the core samples reveal a downward decrease in the methane/ethane ratio from about 10^5 by volume near the sediment surface to 10^3 at about 1 km (Claypool, 1975; Whelen and Sato, in press). The normal ratio in natural gas fields is about 10^1, which would be reached if the rate of change from 0 to 1 km could safely be extrapolated to 2 km depth. The high ratios in the bottom parts of the D.S.D.P. cores and their reasonably uniform change with depth implies that these and other less abundant gases in the samples result from degradation of organic matter by anaerobic bacteria followed and accompanied by low temperature chemical reactions.

Geothermal gradients are poorly known beyond the continental shelf, but Emery and Skinner (1977) compiled about 300 heat-flow measurements from continental rises, applied thermal conductivities believed typical of sediments at 1 km depths, and concluded that the temperature gradient must average about $3.5^{\circ}C/100$ m, somewhat steeper than the average $2.3^{\circ}C/100$ m gradient in oil-field wells. If these data are correct, the temperature of maturation ($60^{\circ}C$ for Cretaceous or older shales--Hunt, 1979) can be reached in only 1.6 km of drilling (plus water)--less than the average of about 2.2 km of drilling on land. Somewhat similar but generally lower thermal gradients were computed by Jansa and Macqueen (1978) for the North Atlantic Ocean. In either set of calculations oil maturation can be expected beneath much of the continental rise, although adequate reservoir beds may be restricted to the inner rise.

BEYOND CONTINENTAL MARGINS

The ocean floor beyond the outer limits of the continental margins (shelves, slopes, and rises) consists of a variety of physiographic units, some of which are diastrophic in origin (mid-ocean and other ridges and associated abyssal hills, plateaus, seamounts, and trenches) and others that are sedimentary in origin (abyssal plains). Sediments are mainly less than 50 m thick in the diastrophic units and less than 500 m thick in the abyssal plains. A reasonable average thickness for the entire region is about 100 m, yielding a total volume of only about 25×10^6 km^3 (Table 24-1). The thicknesses and their derived volumes of sediments for all regions given in Table 24-1 are based entirely upon seismic and drill-hole data. The supply and the geochemical-ratio methods that were much used before the discovery of sea-floor spreading (Kuenen, 1950, p. 376-403) are not valid, because pre-Mesozoic marine sediments have been subducted during convergence of crustal plates.

Gases collected from Deep Sea Drilling Project drill samples have the same composition as those from continental rises. A core of D.S.D.P. site 49 on the Shatsky Rise in the Pacific Ocean contained somewhat more mature hydrocarbons, probably indicating some movement from deeper and older strata (McIver, 1974). In any event, the sediments are too thin for the geothermal gradient to have matured the organic matter into petroleum, the original organic content is very low (generally less than 0.5 percent--Hedberg et al., 1979), because slow deposition allowed nearly complete oxidation before burial, and sufficiently coarse and permeable reservoir sands are scarce. For these reasons, the ocean floor beyond the ends of the continental margin is considered to have extremely poor prospects for sizable accumulations of petroleum.

GENERAL OBSERVATIONS

Millions of line-km of single-channel seismic profiles accompanied by measurements of seismic refraction, geomagnetics, and gravity have been run in the ocean. The results have provided much information about the general internal structure and underlying crust of the ocean floor, but their usefulness for thick sediments of continental margins is limited by low power of sound sources and low resolution of sensors, processing, and recording. Far more detail has been provided by multichannel seismic systems operated chiefly by industry on the continental shelves and by government agencies on some continental slopes and rises. Results of the geophysical studies on deep-ocean floor have been supported and extended by rock dredgings and especially by drill-hole samples obtained by the Deep Sea Drilling Project. On the continental shelves oil companies have secured much information from stratigraphic and production wells, and on the continental slopes both oil companies and geological surveys (Hathaway et al., 1979) have drilled shallow-penetration test wells. Much less information has come from drill holes on the deeper parts of continental margins owing to the limitation on number and depth of holes permitted in order to avoid oil pollution that could be caused by inadequate capability of the drilling ship for the D.S.D.P., the *Glomar Challenger*.

The time has now arrived for a near order-of-magnitude expansion of effort on the deeper parts of continental margins (lower continental slopes and continental rises) in order to assess the nature of the contact between continental and oceanic crusts, and the composition, origin, and history of the overlying sedimentary prism. An obvious fall-out will be a stratigraphic evaluation of the petroleum prospects of the deep-ocean floor of the world. A report by an ad hoc committee of the National Research Council chaired by Bally (1979) recommended the funding of an intensive upgrading of survey ships and geophysical equipment for better delineation of the major problems to be solved by a new drilling program for the 1980's, probably through the drill ship *Glomar Explorer* (Anonymous, 1979). Additional efforts are being bent toward deep seismic exploration by multi-ship arrays, and toward other new approaches. These efforts can mean that knowledge of the ocean to be acquired during the coming decade can eclipse that accumulated during the entire past. The subject of today's debates may become obsolete, and new questions not now visualized can be asked.

Three major obstacles exist before the deep-ocean floor (probably only the inner marginal basins, lower continental slopes, and inner continental rises) can be exploited for possible petroleum contents. One obstacle is the question of sovereignty, related mainly to demands for new sources of government or United Nations taxes. A second obstacle relates to environmental safeguards, for prevention of oil spills and to satisfy a vociferous and influential body of environmentalists. A third obstacle is the matter of production costs, which are likely to be very high because of the hostile deep-marine environment. These costs can be expressed in terms of energy, such that more energy must be produced than is spent on the search and production activities plus that required by governmental and environmental demands and regulations. The very long-delayed timetable for petroleum production on the more accessible continental shelves suggests that deep-ocean production is not as near as many had hoped and expected several years ago. Moreover, we have yet to cross the very first major threshold--the drilling to discover whether or not any petroleum is present. I repeat the last sentence of my previous article for UNITAR-IIASA (Emery, 1977): "Until this drilling is done, let us not make absurd speculations about how much oil and gas may or may not be present on the deep-ocean floor."

REFERENCES

Anonymous, 1979, Post-IPOD science: EOS, v. 60, p. 777-778.

Bally, A. W., chairman, 1979, Continental margins, geological and geophysical research needs and problems: Natl. Acad. Sciences, 302 p.

Claypool, G. R., 1975, Manual on Pollution-Prevention and Safety, 3rd ed.: JOIDES Journal, p. C1-C11.

Davis, J. B., and E. E. Bray, 1969, Analysis of oil and cap rock from Challenger (Sigsbee) Knoll; in M. Ewing, J. L. Worzel, et al., Initial Reports of the Deep Sea Drilling Project, v. 1: Washington (U.S. Govt. Printing Office), p. 415-500.

Embley, R. W., and R. D. Jacobi, 1978, Distribution and morphology of large submarine sediment slides and slumps on Atlantic continental margins: Mar. Geotechnology, v. 2, p. 205-228.

Emery, K. O., 1977, The potential and time frame for the development of petroleum in the deep ocean; in R. F. Meyer, ed., An International Conference Sponsored by UNITAR and IIASA, 5-16 July 1976, Laxenburg, Austria: New York, Pergamon Press, p. 101-111.

Emery, K. O., 1979, Potential for deep-ocean petroleum: Ambio, no. 6, p. 87-92.

Emery, K. O., in press, Continental margins--Their classification and petroleum prospects: Bull Amer. Assoc. Petroleum Geologists.

Emery, K. O., and B. J. Skinner, 1977, Mineral deposits of the deep-ocean floor: Mar. Mining, v. 1, p. 1-71.

Grossling, B. F., 1975, In search of a statistical probability model for petroleum-resource assessment: U.S. Geol. Survey Circ. 724, 18 p.

Hathaway, J. D., and others, 1979, U.S. Geological Survey core drilling on the Atlantic shelf: Science, v. 206, p. 515-527.

Haun, D. C., 1975, Methods of estimating the volume of undiscovered oil and gas resources: Amer. Assoc. Petroleum Geologists, Studies in Geology No. 1, 206 p.

Hedberg, H. D., J. D. Moody, and R. M. Hedberg, 1979, Petroleum prospects of the deep offshore: Bull. Amer. Assoc. Petroleum Geologists, v. 63, p. 286-300.

Hubbert, M. K., 1974, U.S. Energy Resources, a review as of 1972: A background paper prepared at the request of Henry M. Jackson, chairman, Committee on Interior and Insular Affairs, United States Senate, 93rd Congress, 2nd session, 201 p.

Hunt, J. M., 1979, Petroleum Geochemistry and Geology: San Francisco, W. H. Freeman and Co., 617 p.

Jansa, L. T., and R. W. Macqueen, 1978, Stratigraphy and hydrocarbon potential of the central North Atlantic Basin: Geoscience Canada, v. 5, p. 176-183.

Karig, D. E., 1971, Origin and development of marginal basins in the western Pacific: Jour. Geophys. Research, v. 76, p. 2542-2561.

Kuenen, Ph. H., 1950, Marine Geology: New York, John Wiley and Sons, 568 p.

Lancelot, Y., and E. Seibold, 1977, The evolution of the central northeastern Atlantic--summary of results of DSDP Leg 41; in Y. Lancelot, E. Seibold, and others, Initial Reports of the Deep Sea Drilling Project, v. 41: Washington (U.S. Govt. Printing Office), p. 1215-1245.

McIver, R. D., 1974, Evidence of migrating liquid hydrocarbons in Deep Sea Drilling Project cores: Bull. Amer. Assoc. Petroleum Geologists, v. 58, p. 1263-1271.

McKelvey, V. E., 1974, Revised U.S. oil and gas resources estimates: U.S. Geol. Survey news release, March 26.

Menard, H. W., 1967, Transitional types of crust under small ocean basins: Jour. Geophys. Research, v. 72, p. 3061-3073.

Miller, B. M., and others, 1975, Geological estimates of undiscovered recoverable oil and gas resources in the United States: U.S. Geol. Survey Circ. 725, 78 p.

Montadert, L., and others, 1977, Rifting and subsidence on passive continental margins in the north east Atlantic: Nature, v. 268, p. 305-309.

Moore, D. G., 1977, Submarine slides; in B. Voigt, ed., Rockslides and Avalanches, Vol. 1, Natural Phenomena: Developments in Geotechnical Engineering, v. 14A, p. 563-604.

Pitman, W. C., Jr., 1978, Relationship between eustacy and stratigraphic sequences of passive margins: Bull. Geol. Soc. America, v. 89, p. 1389-1403.

Skinner, B. J., chairman, 1975, Mineral Resources and the Environment; Chap. V, Resources of the Fossil Fuels: National Acad. Sciences, p. 79-126.

Tucholke, B. E., 1979, Relationships between acoustic stratigraphy and lithostratigraphy in the western North Atlantic Basin; in B. E. Tucholke, P. R. Vogt, and others, Initial Reports of the Deep Sea Drilling Project, v. 43: Washington (U.S. Govt. Printing Office), p. 827-846.

Uman, M. F., W. R. James, and H. R. Tomlinson, 1979, Oil and gas in offshore tracts--Estimates before and after drilling: Science, v. 205, p. 489-490.

Vail, P. R., R. M. Mitchum, Jr., and S. Thompson, III, 1977, Seismic stratigraphy and global changes of sea level; Part 4, Global cycles of relative changes of sea level; in C. E. Payton, ed., Seismic Stratigraphy--Applications to Hydrocarbon Exploration: Amer. Assoc. Petroleum Geologists Mem. 26, p. 83-97.

Whelen, J. K., and S. Sato, in press, C_1-C_5 hydrocarbons from core gas pockets, Deep Sea Drilling Project Legs 56 and 57, Japan Trench transect; in Initial Reports of the Deep Sea Drilling Project: Washington (U.S. Govt. Printing Office).

CHAPTER 25

THE FUTURE OF DEEP CONVENTIONAL GAS[1]

John M. Hunt[2]

INTRODUCTION

In this paper, the term deep conventional gas refers to accumulations found at depths greater than 4,600 m (15,000 ft) in which gas is the continuous phase. The potential for such gas accumulations worldwide may be much greater than has been assumed in the past. Considering that structural and trapping conditions can be just as favorable in this deep range as at shallower depths, the critical factors remaining are (1) an adequate source for the gas, (2) sufficient porosity and permeability for economic production, and (3) conditions for preservation of the hydrocarbons since their accumulation.

Today drilling companies talk of building rigs capable of drilling to 15,000 m (about 50,000 ft). The deep Tuscaloosa gas discoveries in south Louisiana have sparked new interest in deep drilling and raised the question as to the potential of deep gas accumulations worldwide. This paper discusses the geochemical aspects of deep gas, namely, the sources of these hydrocarbons, the depth of peak gas generation, the factors causing a gas floor, and the prospects for finding large accumulations in the future.

GENERATION PEAKS FOR OIL AND GAS

It is well established that oil and gas accumulations in the earth's crust originate from the organic material deposited with fine-grained sediments. The volume of oil and gas generated depends upon the quantity, type, and maturation state of the organic matter. Sapropelic organic matter such as fats, oils, resins, waxes, and pigments, which have hydrogen to carbon ratios greater than one, yield primarily oil during early stages of maturation and gas in the later stages. Humic organic matter such as the components

---

[1]Contribution No. 4474 of the Woods Hole Oceanographic Institution.

[2]Woods Hole Oceanographic Institution, Woods Hole, MA 02543, USA.

of peat and humus with hydrogen to carbon ratios of less than one yield primarily gas. The thermal history of the source rock, that is, the time at which the organic matter has been subjected to various temperatures, determines the varying amounts of oil or gas that are formed. As a rough generalization, most oil forms in the subsurface temperature range from 60° to 150°C (140° to 300°F), with the peak around 100°C (212°F). Some gas forms throughout the oil range, but it generally peaks around 150°C (300°F) and begins to phase out around 200°C (about 400°F). Condensate formation peaks between the oil and gas peaks. The initiation and termination of oil and gas generation does not occur suddenly, but rather follows an exponential increase and decrease which may occur over a broad subsurface temperature interval (Hunt, 1979, p. 344).

For example, the generation peaks for oil and methane gas in the Cretaceous and Jurassic formation of the Aquitaine Basin in France are shown in Fig. 25-1. Gas yields are in milliliters per gram of organic carbon. Le Tran et al. (1974) analyzed the fine-grained rocks in this basin to a depth of 5,500 m (18,000 ft) for their C_{15+} hydrocarbons (oil), methane, and hydrogen sulfide gas. The yield of oil in both formations peaked at a subsurface temperature in the 90° to 100°C (194° to 212°F) range while the methane peaked in the range from 140° to 150°C (285° to 300°F). After peaking, the methane gas, like the oil, showed a clearly-defined decrease at higher temperatures (greater depths), indicating that the organic matter was rapidly losing its capability to generate methane gas.

The chemical process that occurs in an oil source rock as oil and gas are formed is shown schematically in Fig. 25-2. At the top is a molecular unit of the organic matter (kerogen) with a hydrogen to carbon ratio of 1:43. As the temperature increases, chains and rings in the oil range are cracked from the kerogen which then condenses somewhat in order to release the hydrogen required to form the oil molecules. In effect, this is an internal hydrogen-exchange process in which the kerogen loses hydrogen to form small molecules with a higher proportion of hydrogen. The hydrogen to carbon ratio of the kerogen decreases as shown in the column on the right. Eventually this ratio drops below one. Beyond this point methane is the dominant hydrocarbon formed. Methane continues to be formed until the H/C ratio drops in the range of 0:3, after which generation essentially is

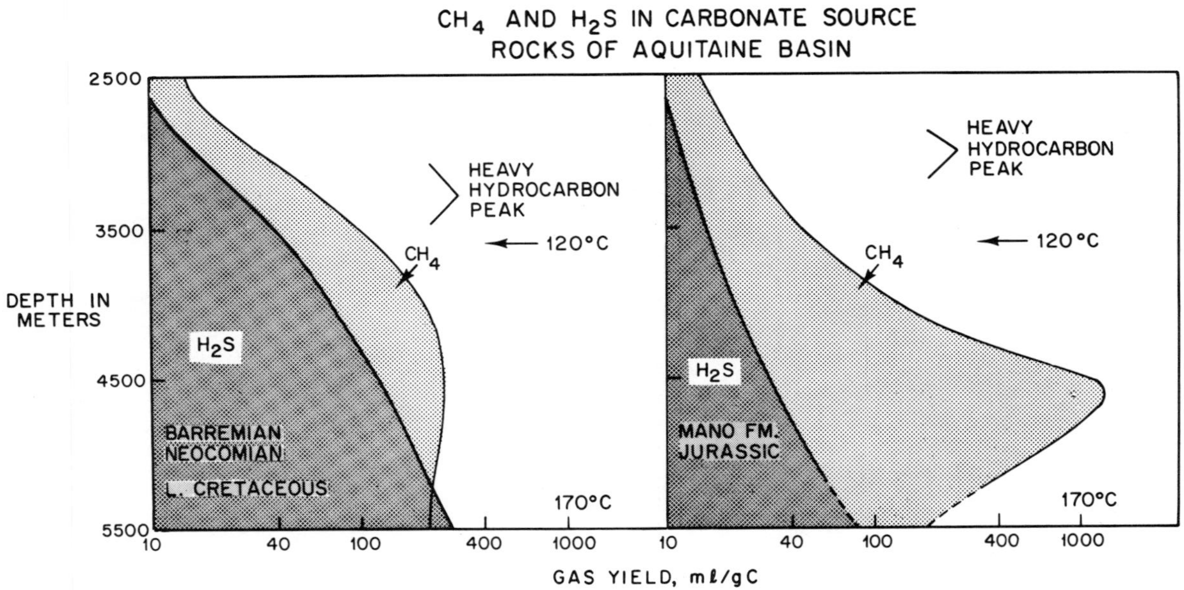

Figure 25-1. Generation of heavy hydrocarbons (oil), methane, and H_2S in carbonate source rocks of Aquitaine Basin, France. Data from Le Tran et al. (1974)

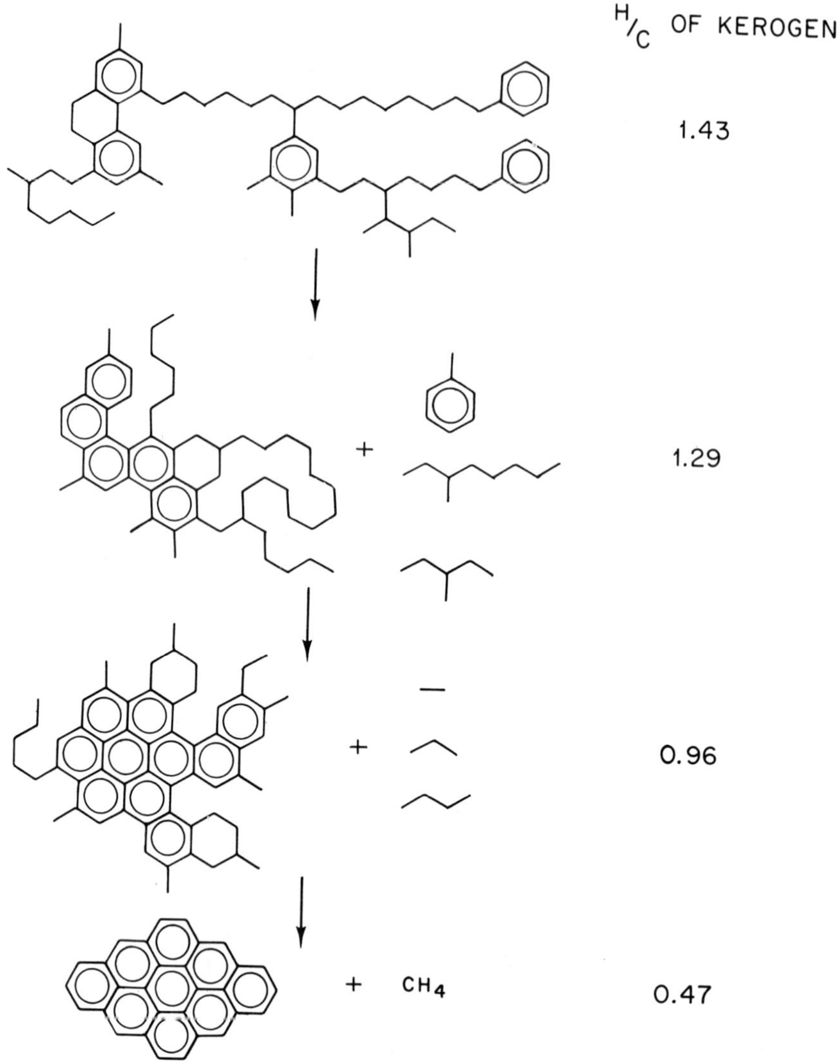

Figure 25-2. Molecular changes in source rock kerogen during thermal alteration to form oil, condensate, and gas (Hunt, 1979, p. 316-317)

terminated. Again, this is an exponential decrease rather than a sudden termination at a particular ratio.

This process, which goes on in sedimentary rocks throughout the world, means that wherever well-defined oil source rocks are found, they will have the potential for gas if buried deeper.

The depth of the generation peaks for oil and gas will vary widely depending on the geothermal gradient and time of burial. It would be around 3,300 m (10,000 ft) and 4,900 m (16,000 ft), respectively, for an Eocene source rock with a gradient of about 2.7^{o}C/100 m (1.5^{o}F/100 ft). The depths of peak oil and gas generation are not necessarily the depths at which most oil and gas accumulations will be found, since these fluids will readily migrate up or down from the source beds depending on the fluid pressure gradients. Nevertheless, the fact that the gas generation peak is always deeper than the oil generation peak means that, as a rule, gas will be found deeper than oil in all petroliferous basins where the source beds are buried to gas-generating temperatures. A typical example is the Upper Cretaceous Woodbine and Eagleford shales of the U.S. Gulf Coast. In southern Mississippi, central Louisiana, and east-central Texas, these source beds have been buried to oil-generating depths, resulting in prolific oil accumulations in the associated Tuscaloosa sands. South of this trend, the Woodbine and Eagleford source beds have been buried to depths up to 8,000 m (about 26,000 ft), equivalent to formation temperatures above 200°C (400°F). These temperatures for Cretaceous time indicate that the source beds have passed through the stage of maximum gas generation. Consequently, a trend of condensate and gas reservoirs should lie south of the Tuscaloosa oil accumulations in proximity to these more deeply buried source beds. Such a trend is now being exploited as a 20- to 30-mile-wide belt extending 200 miles northeast of New Orleans to the Louisiana-Texas border. This trend should extent farther west into Texas if adequate sand porosity exists.

Similar trends of gas accumulations deeper than oil should be observed in other deltaic deposits around the world where oil generating source beds have been buried deep enough to reach peak gas generation.

In some sedimentary basins, the organic matter may be so predominantly humic that little or no oil is formed, even though temperatures exceed those of maximum oil generation. In such cases, there is still the possibility of deep gas, but the drilling will have to extend into this higher temperature regime from 150° to 200°C (300° to 400°F) before the gas potential of a basin can be verified. The fact that no oil is found in drilling through the zone of oil formation does not condemn the potential for deeper gas.

THERMAL STABILITY OF HYDROCARBONS IN SOURCE AND RESERVOIR ROCK

Hydrocarbons in both source beds and reservoir rocks are continuously undergoing changes in molecular structure toward a state of equilibrium. The equilibrium mixture of hydrocarbons that finally forms depends on the free energy of the system at the reservoir temperature and pressure. Reactions usually proceed in the direction of lower free energy and the most stable hydrocarbons in a mixture are those with the lowest free energy. Within the temperature range of sedimentary basins, paraffins are the most stable hydrocarbons and their stability increases (free energy decreases) as the number of carbon atoms in the molecule decreases (Andreev et al., 1968; Hunt, 1975). Thermally, methane is the most stable of all hydrocarbons. In the absence of other compounds, it is unchanged at temperatures up to 550°C (1,022°F) which is much higher than any sedimentary rocks. This means that in both reservoir beds and source rocks, all hydrocarbon mixtures will gradually become paraffinic and of higher API gravity until they are eventually converted to condensates and ultimately to methane.

Free energy data indicate the direction that a reaction will go to reach equilibrium, but not how long it will take. Nevertheless, over geologic time, we would expect to find only methane in the deepest and oldest sediments.

This process is similar to the generation of gas from organic matter shown in Fig. 25-2, but in a reservoir, the source of hydrogen is not the organic matter but the heavier fractions of crude oil. Figure 25-3 shows this process schematically (Connan et al., 1975). The increasing temperatures at depth cause cracking of the 17 carbon n-paraffin into a 5 and 12 carbon chain. The hydrogen for this reaction comes from the linking of two phenanthrene molecules. Further cracking of the chains results in smaller and smaller molecules eventually leading to condensate and finally only methane. The hydrogen for these reactions comes from further condensation and aromatization of the heavy ends of crude oil eventually leading to a black pyrobitumen in the reservoir. Such bitumens have been identified in

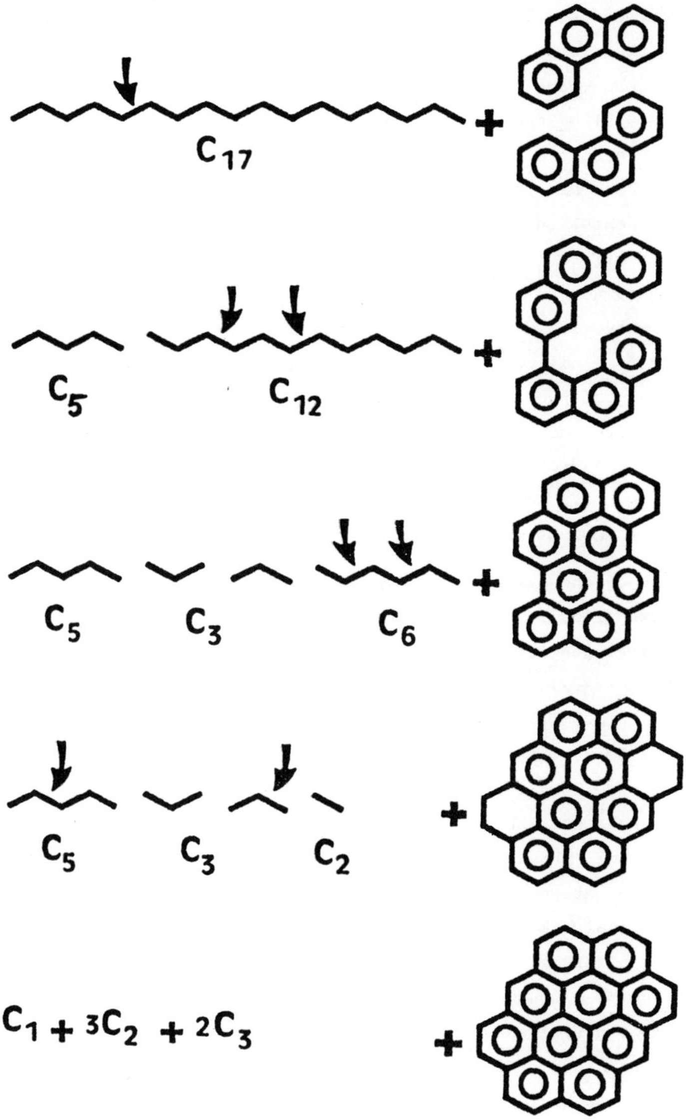

Figure 25-3. Molecular changes in reservoir crude oil during thermal alteration to form condensate and gas (Connan et al., 1975)

many reservoir cores. They are not always observed because the gases can migrate from their area of origin in the reservoir rock.

This process means that oil accumulations which previously have formed in reservoirs will be converted in the reservoir to gas with deeper burial. Consequently, there are two major processes contributing to gas in deep reservoirs. These are the peaking of gas generation at a higher temperature (greater depth) than oil in the source rock and the conversion of previously formed oil to gas with deeper burial. The combination of these processes causes gas to be found in virtually all reservoirs buried at temperatures above 150°C (about 300°F). Exceptions to the reservoir maturation change from oil to gas may occur if the original oil was heavy due to bacterial degradation or washing by meteoric waters. A low API gravity oil might not have enough available hydrogen to convert an appreciable percentage of the reservoir hydrocarbons to distillate or gas.

OIL AND GAS PHASEOUT DEPTHS

The depth of the oil phaseout zone was estimated from well data in a paper by Landes (1967). He developed a chart which showed that the commercial oil floor ranged from about 4,300 m (14,000 ft) at a geothermal gradient of 3.65°C/100 m (2°F/100 ft) to 8,400 m (27,000 ft) where the gradient is 1.8°C/100 m (1°F/100 ft). In order to keep his chart simple, he placed the oil phaseout zone at the maximum depth where oil occurred in the youngest sediments. Figure 25-4 is a modification of Landes' chart in which the approximate oil phaseout zone is shown for the Cenozoic, Mesozoic, and Paleozoic eras. The oil phaseout temperatures were estimated from published time-temperature data on oil and gas accumulations. Figure 25-4 is useful only for gross evaluations because it shows the present day temperature and age of the reservoir rather than the geothermal history. Nevertheless, it can be used for a rough estimate of the depths at which gas will be found below oil. In old sediments in areas of high geothermal gradients this may be as shallow as 2,000 m (6,600 ft). In young segiments with a low geothermal gradient it could be as deep as 8,000 m (26,000 ft).

A more critical question for deep drilling is, at what depth will gas accumulations phase out? Most geologists have assumed that this is a porosity problem and Landes' (1967) chart simply states that porosity and gas volume decrease downward. Methane, however, can be destroyed chemically

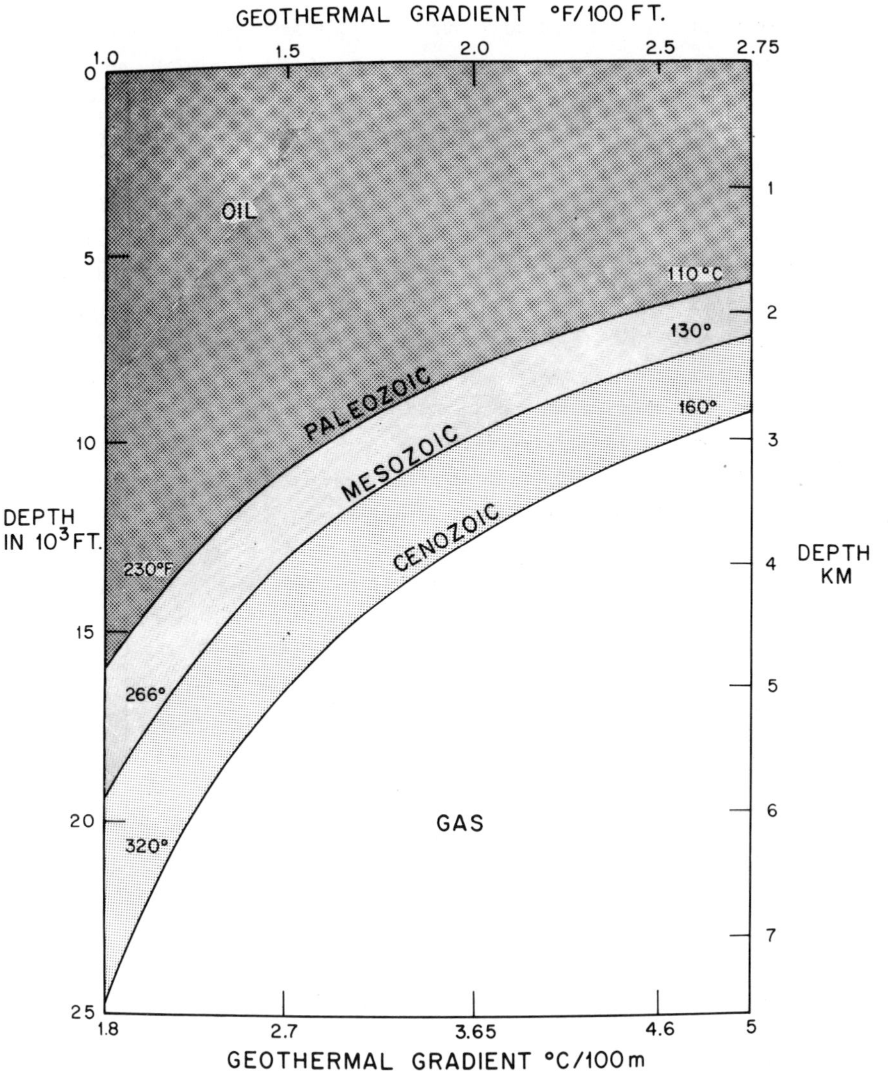

Figure 25-4. Estimated depth of oil phaseout zones for different geothermal gradients and time periods (Hunt, 1979, p. 373)

and it also would have a high diffusivity under the temperature-pressure conditions of deep burial. In some cases these factors would be equally important to porosity in determining a gas floor.

Methane can be destroyed either by sulfur or oxygen, both of which are present in the molecules of the organic matter and fluids of deeply buried sediments. Hydrogen sulfide increases with depth in both source (Fig. 25-1) and reservoir rock due to the high-temperature cracking of organic sulfur compounds. Sulfur compounds in both crude oil and kerogen first begin to break down around $100^{\circ}C$ ($212^{\circ}F$), after which there is a gradual increase that shows no evidence of peaking at temperatures as high as $170^{\circ}C$ ($338^{\circ}F$). Elemental sulfur also has been identified in many crude oils and is encountered in reservoirs at depths to 9,200 m (30,000 ft). Hydrogen sulfide and sulfur can react to form polysulfides which are powerful oxidizing agents. In the presence of water, they can convert methane completely to carbon dioxide at elevated temperatures. If sulfates are present, as in some carbonate-evaporite reservoirs, the oxidation of methane will proceed rapidly at moderate temperatures. Toland (1960) has demonstrated that methane is converted 90 percent to carbon dioxide in 70 minutes at $325^{\circ}C$ ($617^{\circ}F$) in the presence of sulfate ion. Although this temperature is higher than a typical reservoir, the reaction rate would be high enough at $150^{\circ}C$ ($300^{\circ}F$) to destroy most reservoir methane in a few hundred years. Other chemical reactions that can destroy methane have been discussed by Hunt (1975; 1979, p. 392).

Hydrogen sulfide and other sulfur compounds are more apt to be in carbonate rather than sand-shale reservoirs due to the presence of iron in the latter (Hunt, 1979, p. 170-171). Iron and other heavy metals would remove the H_2S as metallic sulfides. In sand-shale reservoirs with little or no H_2S, methane is less likely to be destroyed chemically.

As previously mentioned, a gas floor in terms of generation will occur when the hydrogen to carbon ratio of the kerogen is around 0:3 or the vitrinite reflectance value is around 3.5 percent. As the methane cannot be thermally destroyed at reservoir temperatures, the reservoir methane floor is determined by the porosity decrease, chemical reactions of the methane, and its upward migration.

Porosity loss in sandstone reservoirs has been discussed by Klemme (1972) and Funkhouser (1978). Klemme pointed out that in areas of high geothermal gradients the zones of oil and gas generation are reached at shallower depths than in areas of low gradients. Consequently, the porosity will be higher at shallower depths and the volume of hydrocarbons correspondingly greater. For example at a gradient of 5.5°C/100 m (3°F/100 ft), a sandstone at 121°C (250°F) will have a porosity of 38 percent. This same temperature is reached at a depth of about 4,000 m (13,000 ft) where sandstone porosity is only 26 percent in an area with a gradient of 2.5°C/100 m (1.4°F/100 ft). The shallower field would contain about 50 percent more oil and gas than the deeper field.

Compaction and intergranular cementation by silica and carbonate often reduces sandstone porosity to values below 10 percent at depths greater than 6,100 m (20,000 ft). Porosities greater than predicted frequently occur, however, due to the high pressures present in most deep basins, the development of secondary porosity, or the prevention of intergranular deposits. For example, the Tuscaloosa sand grains apparently are coated with some type of mineral that prevented intergranular growth of quartz. They still have a porosity of 25 percent at 18,000 ft whereas the predicted porosity would be about 15 percent (Funkhouser, 1978).

Porosities also are frequently higher than predicted in carbonates and calcareous shales which fracture on structure. Although porosity loss can eventually define a floor for economic accumulations of methane, there are so many exceptions to predicted porosity trends that no area can be condemned without testing with the drill.

This entire discussion refers to accumulations where gas is the continuous phase. Geopressured gas, that is, high pressure gas dissolved in water, has intrigued the media as a major energy source of the future. Unfortunately, it has severe problems in recovery. As soon as dissolved gas is produced from a reservoir, gas bubbles form in the sand and reduce the permeability for water flow to an uneconomic level. In contrast, where gas is a continuous phase, even tight sands which can be hydrofractured offer a much greater potential than more permeable beds where gas is in a discontinuous phase.

Another concept which has been played up by the media is mantle methane from the outgassing of the earth. As yet, there is no factual basis for this concept. Every phenomenon attributed to the presumed mantle gas is readily explained as a manifestation of conventional deep gas.

THE PROBLEMS AND POTENTIAL OF DEEP GAS RESOURCES

It is futile to estimate the value of deep gas resources in the world if politics and economics prevent their recovery. Environmentally, gas is the cleanest and least toxic of all fossil fuels and yet some countries such as the USA have blocked off huge areas of potential gas resources as unavailable for exploration. Gas also has an advantage over oil in that primary production techniques recover 70 percent of the hydrocarbons in the gas reservoir compared to 25 to 30 percent in an oil reservoir.

A major problem with deep gas is the increased cost of seismic surveys and drilling. There have been remarkable advances in recent years in 3-dimensional recording and the interpretation of data from depths beyond 6,000 m (20,000 ft), but these more sophisticated techniques are more costly. Also, the cost of drilling a well below this depth increases exponentially with a linear increase in depth.

In the South Louisiana Tuscaloosa play, dry holes to 6,000 m (20,000 ft) are costing up to $4 million while completed wells cost up to $7 million. Governments worldwide need to recognize that these large reserves of deep gas will need free market prices and tax incentives to be economic.

In some countries where oil reserves are plentiful there is little incentive to drill where the source rocks go deeper especially if there is no ready market for large gas discoveries. The Persian Gulf area should have some giant gas accumulations associated with the more deeply buried Jurassic source beds to the east but it is doubtful if these horizons will be tested until the oil reserves begin to reach exhaustion in the next century.

Worldwide, the addition of new reserves of gas currently exceeds consumption (Klemme, 1980) and with the addition of future deep gas this comparison will probably continue long after oil reserves have topped out. These future reserves of deep gas will be found in basins with higher than normal temperatures and greater than average depth. Using Klemme's (1980) basin classification, this would include borderland basins of the downwarp, pull-apart, subduction, median, and deltaic types. Most of these basins worldwide have not been drilled through the peak of gas generation so it is impossible to make realistic estimates of gas resources. In a specific basin, an educated guess could be made of the deep gas waiting to be discovered from data on the geothermal gradient, the volume of source rock buried through the gas-generating range and the yield of oil from equivalent formations in the oil generating interval.

The Upper Cretaceous Gulf series of the U.S. Gulf Coast, in which the Tuscaloosa is the major producing formation, has potential reserves in excess of 2 billion barrels of oil. The deeper gas trend to the south is estimated to have potential reserves of as much as 60 trillion ft^3. This is a ratio of 30,000 ft^3 of gas per barrel of oil. As drilling goes deeper in major deltas of the world and other deep borderland basins, the Tuscaloosa trend will be recognized not as an anomaly but as a commonplace situation wherever good oil source beds are buried to the gas-generating range. Deep conventional gas accumulations may well be our major fossil fuel world energy source when the oil begins to run out sometime in the next century.

REFERENCES

Andreev, P. F., A. I. Bogomolov, A. F. Dobryanskii, and A. A. Kartsev, 1968, Transformation of petroleum in nature: London, Pergamon Press, 466 p.

Connan, J., K. Le Tran, and B. Van Der Weide, 1975, Alteration of petroleum in reservoirs, in 9th World Petroleum Congress, Proc. Vol. 2: London, Applied Science Publishers, p. 171-178.

Funkhouser, L. W., 1978, Ultra deep gas exploration--an expanding frontier, Institute on Petroleum Exploration and Economics Symposium: The Southwestern Legal Foundation, Dallas, Texas, March 8-9.

Hunt, J. M., 1975, Is there a geochemical depth limit for hydrocarbons?: Petrol. Eng., v. 47, no. 3, p. 112-127.

Hunt, J. M., 1979, Petroleum geochemistry and geology: San Francisco, W. H. Freeman and Co., 617 p.

Klemme, H. D., 1972, Heat influences size of oil giants, Parts 1 and 2: Oil and Gas J., v. 70, nos. 29, 30, p. 76-78, 134-144.

Klemme, H. D., 1980, The geology of future petroleum resources, Proceedings of 26th International Geological Congress, Paris, July 7-17.

Landes, K. K., 1967, Eometamorphism and oil and gas in time and space, Part 1: AAPG Bull., v. 51, no. 6, p. 828-841.

Le Tran, K., J. Connan, and B. Van Der Weide, 1974, Problems relatifs à la formation d'hydrocarbures et d'hydrogene sulfuré dans Le Bassin Sud-ouest Aquitain, in B. Tissot and F. Bienner (eds.), Advances in organic geochemistry, 1973: Paris, Editions Technip, p. 761-789.

CHAPTER 26

OIL AND GAS FROM THE NORWEGIAN CONTINENTAL SHELF

K. Am[1] and J. A. Ludvigsen

This year marks the 10th anniversary of the first commercial find of hydrocarbons in the Norwegian sector of the North Sea. It was late in 1969 that the Ekofisk field was found by the Phillips Group. The decade has seen a development of great importance to Norwegian society, the installation of approximately 30 platforms of different kinds and 4 major pipelines to bring the oil and gas ashore. This illustrates the extent of development during these first 10 years.

In 1978 the production of oil and gas from the Norwegian Continental Shelf (NCS) was about 30 million tonnes of oil equivalent (toe). Figure 26-1 shows estimated production from existing fields reaching a level of 60 million toe/year during the 1980's and then declining in the early 1990's. If existing marginal fields are developed, the maximum production level will be about 80 million toe in the late 1980's and the early 1990's, with decline during 1990's. If the production level is to be increased or maintained at a constant level beyond the 1980's, new reserves must be found and developed.

WATER DEPTHS AND PROBABLE OIL AND GAS RESERVES

Figures 26-2 through 26-4 show the distribution of water depths on the NCS. Drilling activity has so far been confined to the area south of 62°N. The present trend in exploration drilling is to deeper water, presently in water depths up to 400 m. So far discoveries and development have been in less than 200 m water depth but several "deep-water" blocks have been allocated in the recent fourth round of Licenses, as shown in the table on p. 362. Results of exploration drilling in some of these deep-water blocks look promising, which means that there soon may be concrete incentives to develop production systems for deep water in the North Sea.

---

[1]Manager, Research and Development Department, Statoil, Stavanger, Norway.

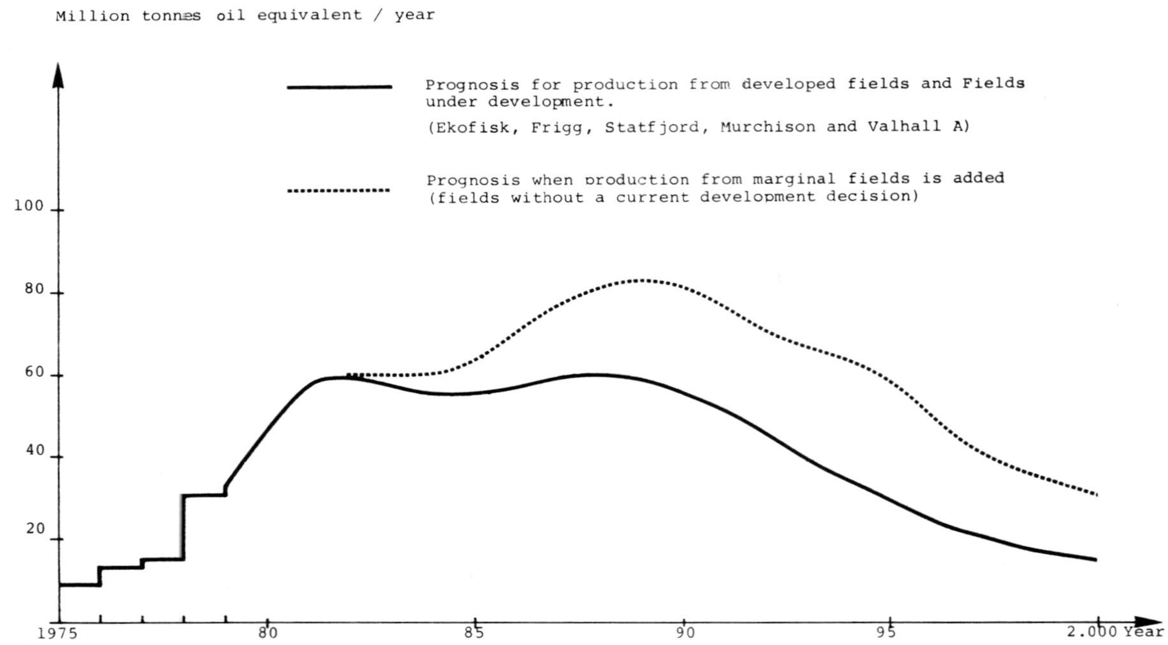

Figure 26-1. Prognosis for production of petroleum from Norwegian shelf

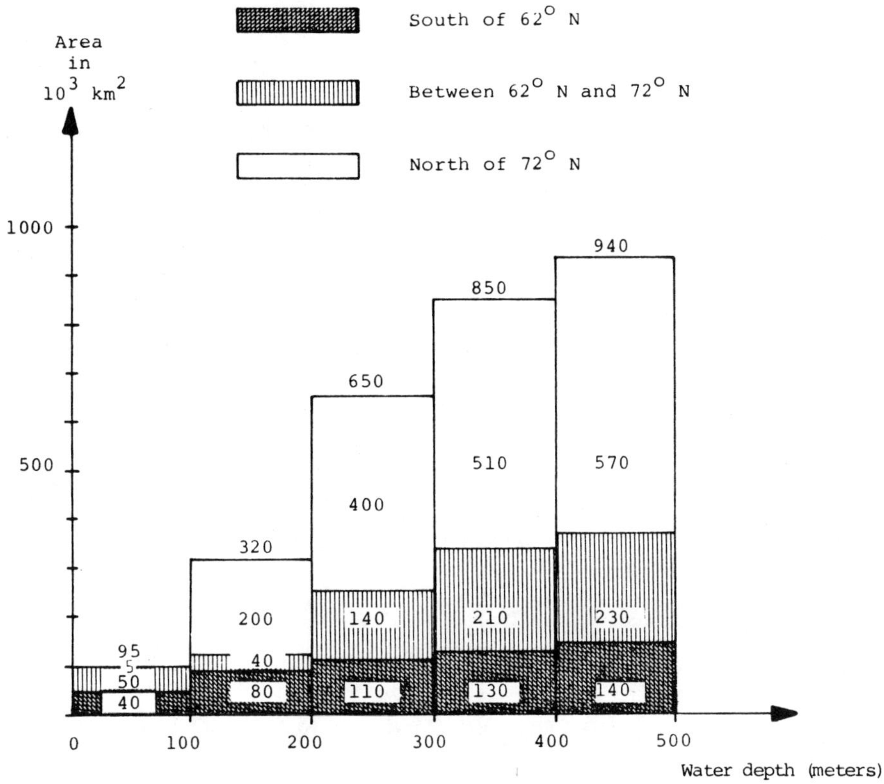

Figure 26-2. Distribution of water depths on Norwegian shelf

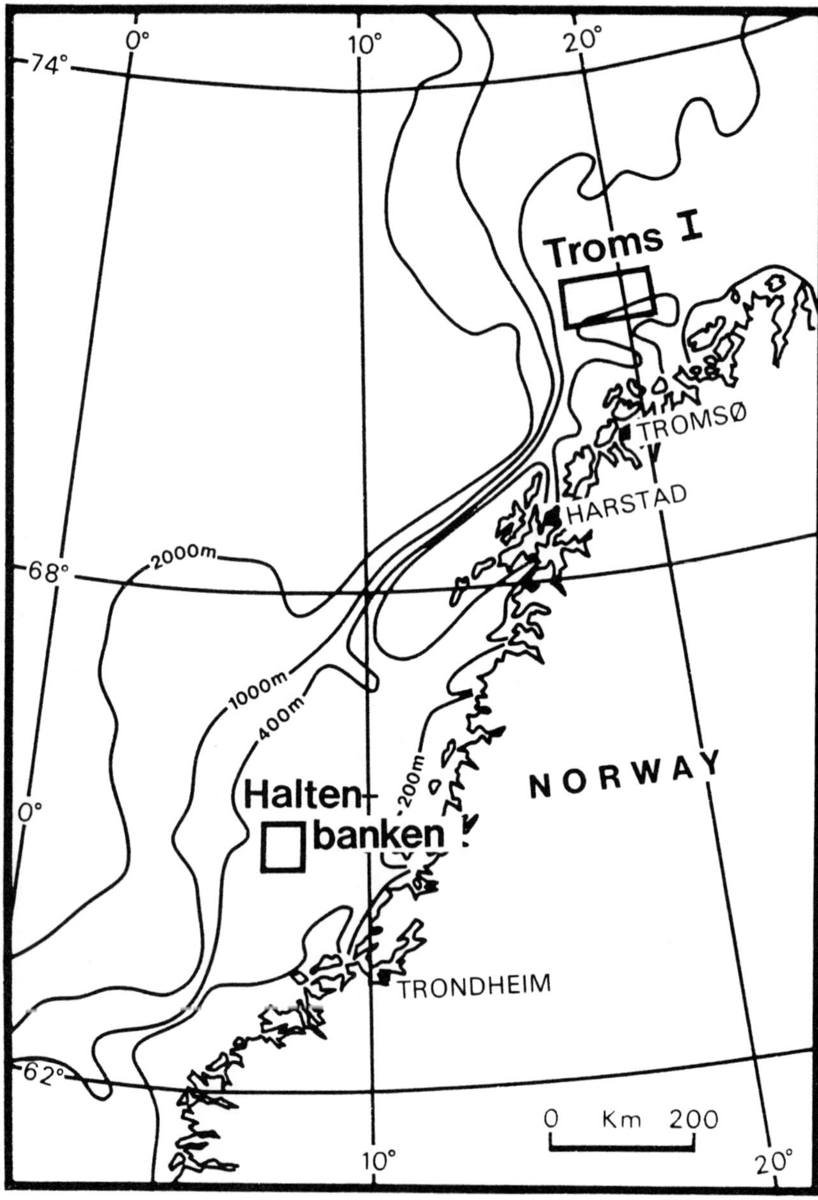

Figure 26-3. Water depths and blocks north of 62° N

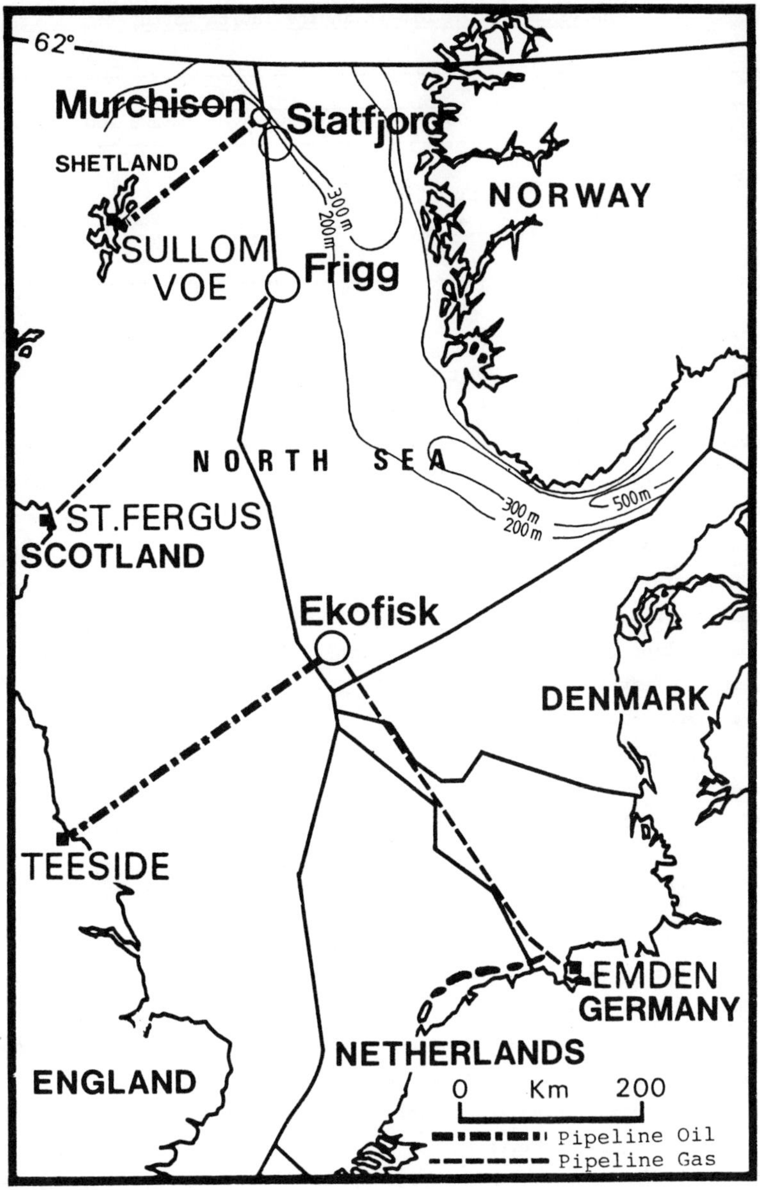

Figure 26-4. Pipelines from the Norwegian continental shelf

Water depths of blocks in the fourth round

Blocks	Depth (meters)			
	<100	100-200	200-300	>300
Already awarded		5		4
Remaining	1	5	1	
Total	1	10	1	4

Proven recoverable reserves of oil and gas on the NCS are estimated to be approximately 1,600 million toe. Of these, about 1,100 million toe are located in producing fields or fields under development. Adding the yet undiscovered reserves, the total recoverable reserves south of 62°N are expected to be above 4,000 million toe. Figure 26-5 shows that the main part of the reserves yet to be discovered south of 62°N are expected to be found in water depths of from 300 m to 350 m. For the vast shelf areas north of 62°N, data are still too scarce to justify quantitative estimates of undiscovered reserves.

Exploration drilling on the NCS north of 62°N will start, probably next year, in two restricted areas:

1. Tromsø I, at 71°N
2. Haltenbanken, at 65°N

These areas are shown in Fig. 26-3. Haltenbanken consists of 6 blocks and Tromsø I of 20 blocks.

FIELDS ON THE NCS

Producing fields and fields under development are located in three separate and geologically different areas: Ekofisk, Frigg, and Statfjord.

Three fields (Statfjord, Frigg, and Murchison) straddle the boundary between the British and the Norwegian sectors. Statfjord and most of the platforms on Frigg are concrete, while only one platform in the Ekofisk area is of that type. The water depths at these fields do not exceed 160 m.

In the Appendix, a brief account of platforms installed and under construction is given for the following: Frigg, Statfjord, Ekofisk, Valhall/Hod, and Murchison.

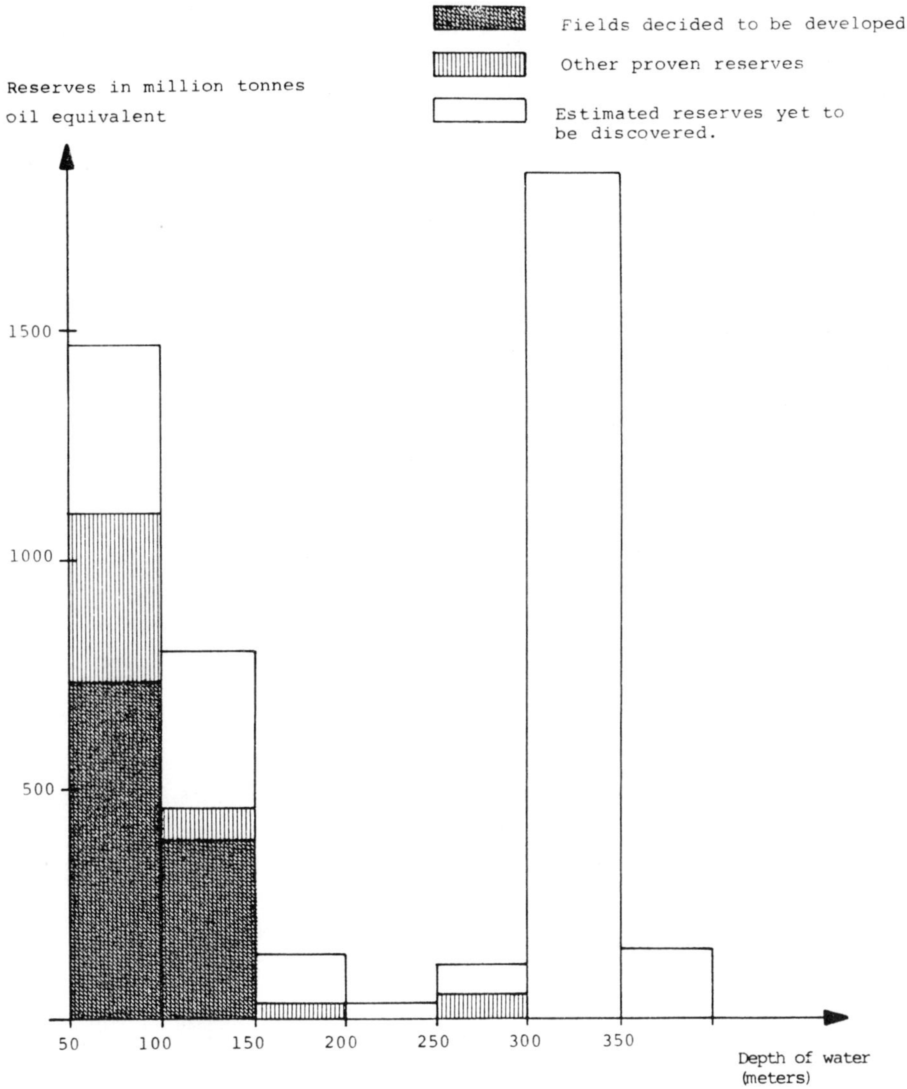

Figure 26-5. Water depths and recoverable reserves of oil and gas on Norwegian shelf south of $62^{O}N$ (Date 1/6/79)

NEW TYPES OF PRODUCTION PLATFORMS

A major technical problem facing the offshore oil industry is the development of oil and gas fields in deeper and more hostile waters. This technical challenge may be met in a number of ways. There is some room for stretching the present fixed platform technology into even deeper water, but the limit, likely to be more economic than technical, has nearly been reached. Figure 26-6 shows the variation in platform jacket weight with water depth while Fig. 26-7 gives costs for vertically moored and fixed platforms as a function of water depth. It is evident that new platform types will have to take over. In the following, three different types of such new platform concepts are described.

Vertically Tethered Production Platform

The vertically tethered platform is, to some degree, similar to the well-known semi-submersible drilling platform. It has a number of large-diameter buoyancy columns and, in some instances, rather small pontoons. The deck structure is supported by the columns, and horizontal and vertical trusses give the platform the required structural rigidity. The platform is moored to one or more sea-bed foundations by vertical cables from the columns. This type of mooring system reduces the heave, roll, and pitch motions to virtually zero, but allows small horizontal motions. Figure 26-8 shows the Tension Leg Production Platform (TPP), developed by the Aker Group in 1974 and based on experience gained from their semi-submersible drilling platform, the H3.

About 20 years ago, Exxon described such a concept in a U.S. patent. However, it is only during the last few years that the concept has aroused any serious interest from potential platform operators. The main reasons for this delayed interest have been:

- further development of the fixed platform
- lack of technology for subsea field development and production risers
- doubts as to the reliability and the economics of the concepts.

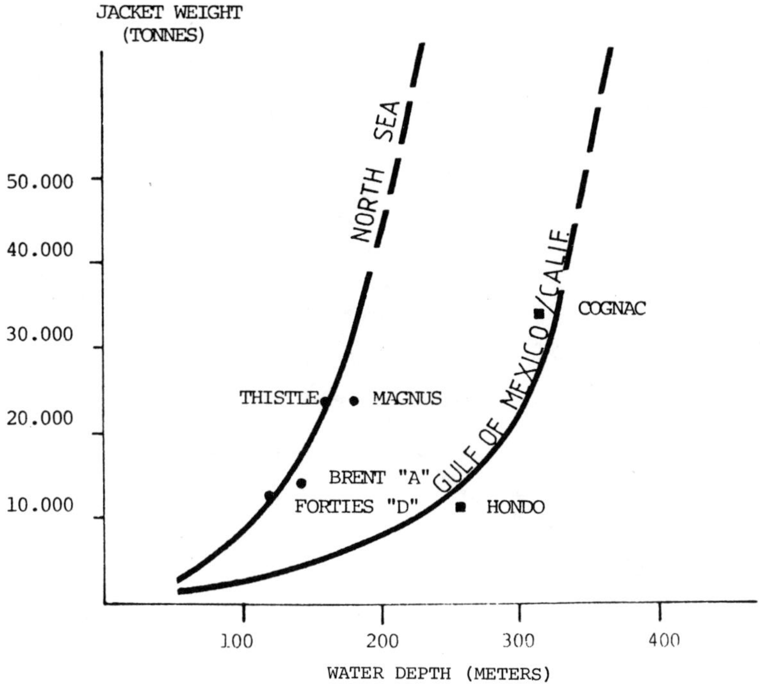

Figure 26-6. Variation of platform jacket weight versus water depth

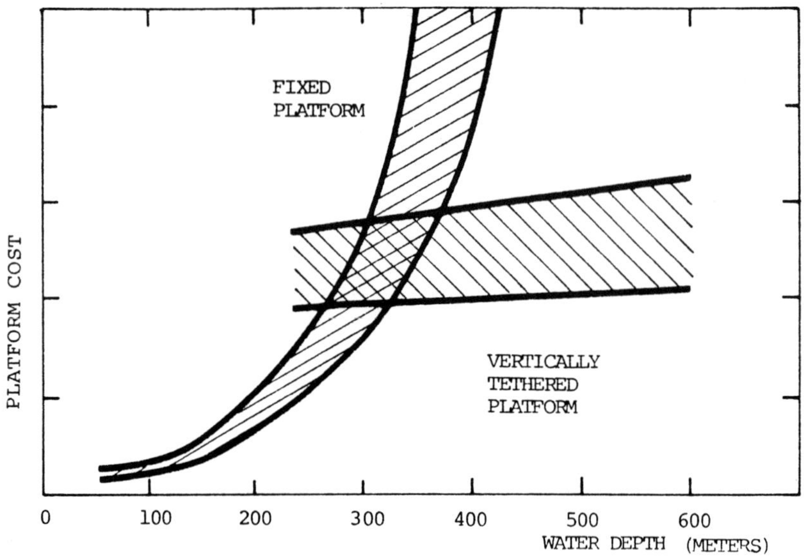

Figure 26-7. Platforms cost comparison

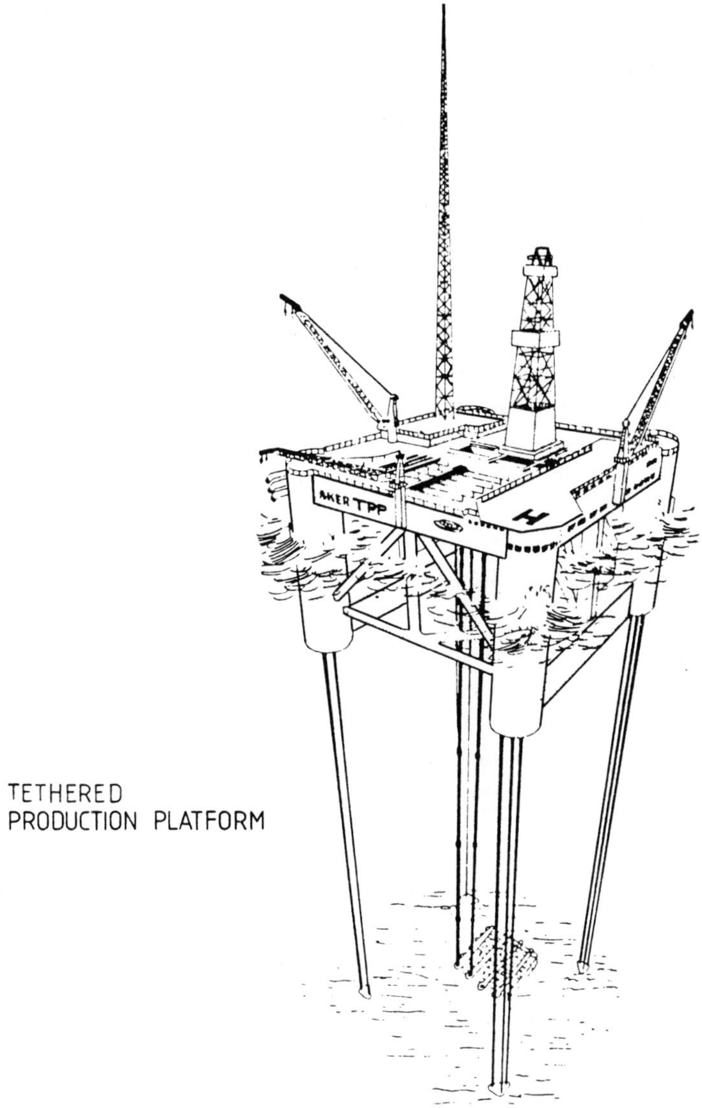

Figure 26-8. Tethered production platform

Guyed Tower Platform

A new, compliant offshore structure, the guyed tower, has been proposed as a deep-water production and drilling platform for use in the North Sea.

The guyed tower is a trussed structure that rests on the ocean floor, extends upward to a deck supported above the waves, and is held upright by multiple guide lines. The base of the tower is supported on a truss-reinforced shell foundation called a spud can.

The concept has two major advantages over other proposed schemes:

- The structure can be manufactured and installed using available equipment and technology.
- The platform is anticipated to be less expensive to build and maintain than present alternatives, such as conventional platforms and subsea production systems, for water depths from 200 m to 600 m.

Figure 26-9 shows a guyed tower designed for 450 m water depth developed by EXXON.

The Semi-Submersible Production Platform

A large number of semi-submersible platforms are in use all over the world for exploration drilling. Some have been converted to serve other functions, such as living quarters or workshops. The first semi-submersible to serve as a production platform was the Transworld 58, a converted drilling rig installed at the Argyll field in the British sector in 1975.

An example of a semi-submersible production platform system appears in Fig. 26-10. This production system consists of the following main components:

1. anchored floating production platform
2. riser
3. multiple production line-riser base
4. subsea wells
5. articulated tower for loading.

The existing North Sea semi-submersibles have limitations with respect to deck load and mooring system capacity.

Another possibility is to replace the anchored floating production platform by a tanker moored to a Single Anchor-Leg Mooring (SALM). The process equipment is placed on the tanker and the hull of the tanker serves as oil

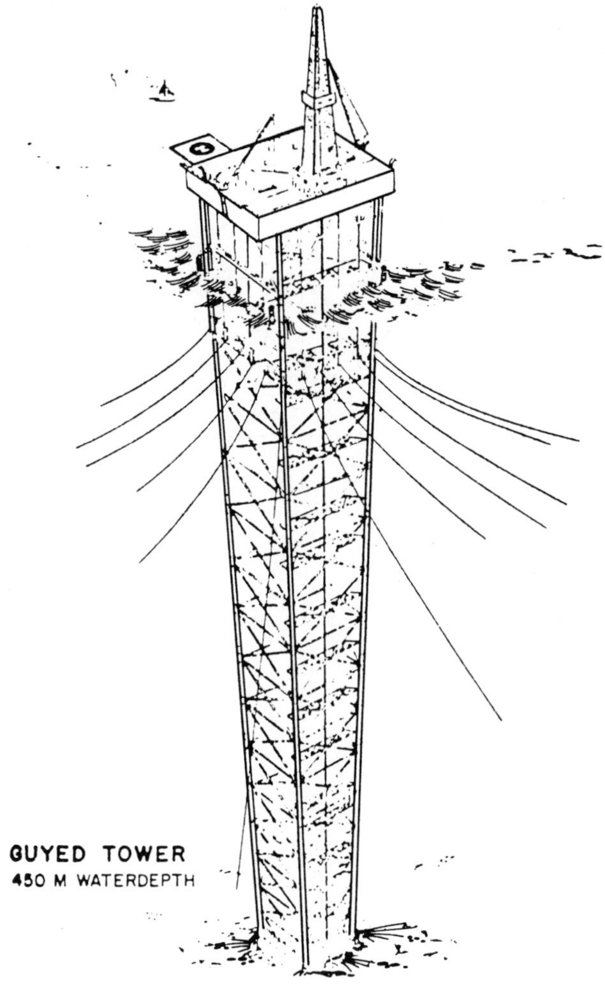

Figure 26-9. Guyed tower

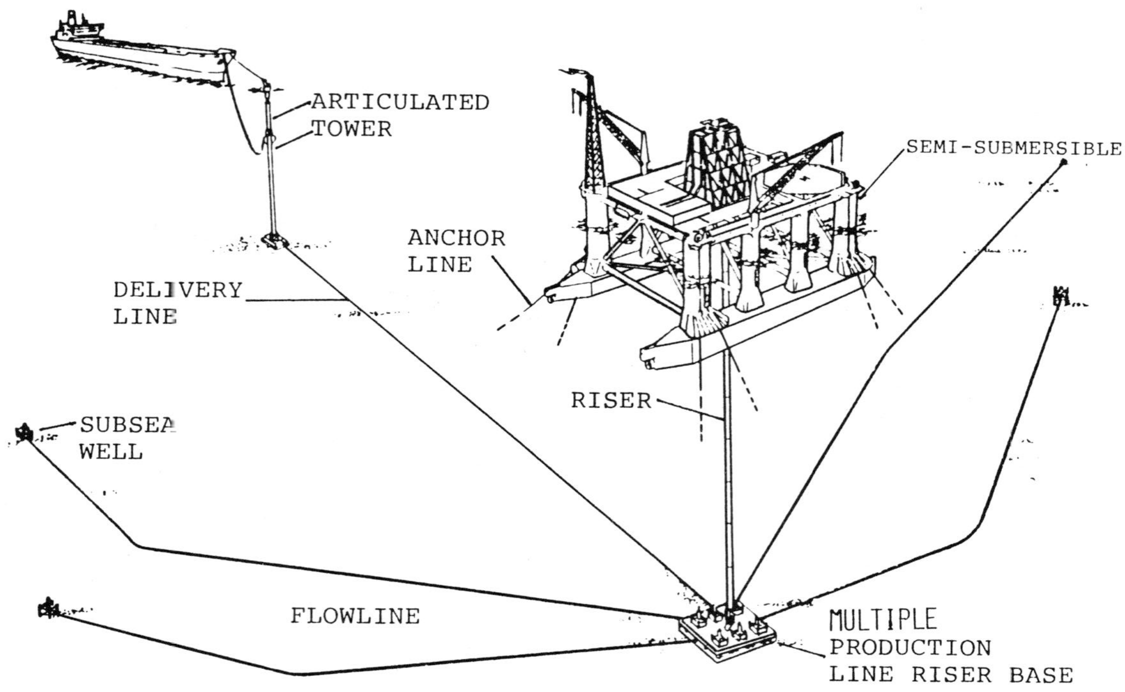

Figure 26-10. Semi-submersible production platform system

storage space. This tanker production system has limitations with respect to environmental conditions.

TRANSPORT OF OIL AND GAS FROM THE NCS

A few years ago offshore loading in the North Sea was thought of as only a temporary measure to allow some movement of crude oil from a field while a pipeline was being laid. The environmental conditions were considered too hostile, the ships too difficult to maneuver, and Single Point Moorings (SPM) too vulnerable, to permit regular transfer of oil. Today, experience from the Beryl and Brent fields on the U.K. shelf has shown that offshore loading can be a safe transport system for any North Sea field discovered thus far.

Gas from NCS is transported by pipeline, as at Ekofisk and Frigg. In an oil field, as Statfjord, the gas will be reinjected into the reservoir in the first production phase.

Ekofisk

Loading oil into moderately sized tankers at the Ekofisk production facility was used temporarily at an early stage. The landing of pipelines in Norway is economically impractical because of the presence of a deep trench along the Norwegian coast. Onshore terminal sites had to be found elsewhere. In northeast England, the estuary of River Tees, with its deep-water access for tankers, was a suitable place to land crude oil. Ekofisk natural gas had been sold on a long-term contract to customers on the European continent, and the city of Emden on the West German coast was found to be a suitable site.

Laying pipelines to England and West Germany constituted another major feat of offshore engineering. Because of the size of the jobs, contracts were given to three separate pipeline construction companies who worked simultaneously on different parts of the lines. The 34-inch line, designed to transport crude oil over the 354 km distance from the Ekofisk Complex to Teesside, was laid first and became operational in 1975. The 36-inch natural-gas pipeline, stretching 441 km to Emden, was laid soon afterward and became operational in September 1977. Prior to the completion on this line, associated natural gas produced at Ekofisk was reinjected into the reservoir as a conservation measure. Oil and gas from Valhall A is proposed to be connected to Ekofisk by pipelines and landed together with Ekofisk oil and gas.

Frigg

The gas from Frigg is transported 360 km to St. Fergus in Scotland by two 32-inch pipelines, one British and one Norwegian. A compression platform of the Doris type is installed midway between Frigg and St. Fergus to step up the pressure and hence increase the transport capacity.

Statfjord

The decision has not been taken yet for a permanent transportation system from the Statfjord field but the technical feasibility of laying and operating a pipeline to shore in Norway has been demonstrated. In any case, offshore loading by means of SPMs will be used as a temporary system.

An Articulated Loading Platform (ALP) has been installed in the Statfjord field and another will be installed in 1981. In the beginning of the production phase, three dedicated tankers of 125,000 dead-weight tons (tdw) each will be in operation.

Murchison

Oil from Murchison will be transported in the Brent pipeline to Sullom Voe, Shetland. This field is scheduled to go on production in 1980/1981.

SUBSEA COMPLETION SYSTEM

A subsea completion system is an offshore well or group of wells installed on the ocean floor. The subsea production system can be the only realistic method for development of deep-water fields and some marginal fields. Development costs can be lower and production can start earlier, compared with conventional platform installations. The total development of a field in some cases may be solved more economically (increased production) by adding subsea completions to a fixed platform. The subsea wells would be installed in the peripheral parts of the reservoir which cannot be reached by directional drilling.

From a safety point of view, subsea production systems are less sensitive to such severe accidents as explosions of process equipment, collisions between ship and platform, or fires on the platform. Future technical development of subsea completion is necessary, with special reference to control systems, electrical connectors, flowline connectors, service, and maintenance systems.

Subsea Completion Installed on the NCS

Four completed subsea wells were put on production in January 1972 by Phillips Petroleum Company in the Ekofisk area. This was an early production system in which the jack up drilling rig "Gulftide" was converted to a production platform. The oil was loaded into tankers through SPMs. This early production system was abandoned in the spring of 1974, after having produced 28 million barrels of oil.

Subsea Completion Planned for Installation on the NCS

A marginal gas field is planned to be developed with 6 subsea completed wells directionally drilled from a template. This will be the first gas field to be produced entirely by use of subsea completed wells. Several subsea completions are planned for water injection and oil production in connection with a fixed production platform in one major oil field.

Experience

Where subsea completions of simple design have been installed in shallow water, experience has been promising. Disappointments, however, have appeared when more sophisticated systems, for instance complicated multiplex control systems, have been introduced. From an operational point of view, problems have been experienced due to low accessibility, hydrate formations, and the fact that flow lines, service lines, and control lines are vulnerable. On the other hand, reliable sophisticated systems are needed for development in deep water and in the North Sea. Such systems will also be beneficial in relatively shallow water.

CONCLUDING REMARKS

Major recoverable oil and gas reserves on the Norwegian Continental Shelf lie in water depths from 200 m to 500 m. Exploration drilling north of 62°N will start probably next year.

Production from existing fields is expected to reach a level of 60 million toe/year during the 1980's and then decline in the early 1990's. If existing marginal fields are developed, the maximum production level will be about 80 million toe.

Achievements in the first decade could be summed up as development of five oil and gas fields through installation of about 30 fixed platforms and four major oil and gas pipelines.

The trend toward field development in deeper and more hostile water requires new deep-water technology that may include platforms other than fixed, subsea completion systems, and permanent means of transport.

APPENDIX

Platforms Installed and Under Construction

Ekofisk Area (Phillips)

This area consists of 7 fields and 18 platforms in the southern part of the NCS. The fields are developed in four steps.

	Field	Start of production
Step I:	Ekofisk	1971
Step II:	West Ekofisk/Cod	1977
Step III:	Tor	1978
Step IV:	Edda, Eldfisk, Albuskjell	1979

The Ekofisk development is now virtually completed. Albuskjell went on stream in June and Edda and Eldfisk should start production later this year.

Frigg (ELF)

The Frigg gas field, discovered in 1971, is among the largest offshore gas fields in the world. The annual production is estimated at approximately

15×10^9 Nm^3 gas over a period of at least 15 years. The Frigg field has been developed jointly by Britain and Norway and when fully completed will consist of seven platforms.

Statfjord (Mobil)

The Statfjord field, discovered in 1974, is the largest oil field in the North Sea.

The development of the field has been divided into two steps:

Step I: Statfjord "A" (PDQ--Production, Drilling, and Living Quarter platform).
Concrete platform of the Condeep type built by Norwegian contractors.
Installed: May 1978
Depth of water: 145 m
Start-up date: 1979

Step II: Statfjord "B" (PDQ--Platform) of the Condeep type with four shafts. Statfjord "B" is scheduled to be installed in 1981.

A decision has not yet been taken for a step III, the so-called Statfjord "C." Further development of the field will depend on:

a. The crude oil transportation system

b. New finds of oil and gas near the Statfjord field.

Valhall/Hod (Amoco)

Valhall/Hod, south of Ekofisk, is planned to be developed in four steps.

Step I: Development of Valhall center by 3 steel platforms (Drilling platform, production-/compression platform, and living quarters).

Step II: Hod

Step III: Valhall "B"

Step IV: Valhall "C"

A firm development decision has been taken for step I only.

Murchison (Conoco)

Murchison was discovered in 1975. The field is planned to be developed by one steel platform for production, drilling, and living quarters, and three subsea completed wells (two production wells and one water injection well). Start-up date is planned to be 1980/81.

CHAPTER 27

CURRENT AND FUTURE HYDROCARBON EXPLOITATION IN MEXICO

Eng. Abundio Juarez Mendez[1]

SUMMARY

To meet the country's hydrocarbon demand, by rationally exploiting the oil and gas fields of the Mexican Republic, is the basic objective of Petróleos Mexicanos.

This work presents a brief historical synthesis of hydrocarbon exploitation in Mexico, the basic development program for the 1977-1982 term, a panorama of the major producing areas, and their long-term perspectives.

Moreover, it provides general data about the secondary recovery systems, current hydrocarbon production, the new production program, and the transportation and distribution system for commercializing the gas produced.

Finally, it states the policies which have been determined by the current management for conducting hydrocarbon exploitation in our country.

BACKGROUND

Historical Synthesis

At the turn of the century, with the discovery of the Ebano and Pánuco fields, located in the neighborhood of the Port of Tampico, Mexico started its oil exploitation operations; important discoveries, among which the Golden Lane area stands out, were made in this early phase. As a result of this finding, Mexico reached an annual production of 530,000 BD in 1921, which made it the world's second largest oil producing country at the time.

In 1938, the country's crude oil production was 115,000 BD, with 60 percent coming from the Tampico-Tuxpan area, where the Poza Rica field stood out as the main producer.

In the period 1938-1972, a number of discoveries of important hydrocarbon fields were made in: Reynosa, Poza Rica, Cerro Azul, Agua Dulce, and the New Offshore Golden Lane located off the Tuxpan seashore in the Gulf of

[1]Manager for Exploration Planning, Petróleos Mexicanos, Mexico City, Mexico.

Mexico; however, in spite of all of these new reservoirs, production reached only 450,000 BD by the end of 1971.

In 1972, the Chiapas-Tabasco Basin Mesozoic fields were discovered, which made possible to increase production to such an extent that by 1977 it reached about 1.25 million barrels per day.

Wells Drilled

In 1938, Petróleos Mexicanos had only 16 drilling rigs, all of outmoded design and in the worst mechanical condition; our oil industry was initiated with this equipment, and it has gradually grown to its current 214 drilling rigs, of which there are 191 for onshore, 19 for marine offshore, and 4 for swamp applications.

Approximately 16,000 wells have been drilled to date and the rate of success has improved with time, as shown by the fact that from 1954 onward rate of success has been maintained at 25 percent for exploratory and 77 percent for development wells.

In 1978, the rate of success was 42 percent for exploratory and 85 percent for development wells.

CURRENT EXPLOITATION PROGRAM

Program for the Six-Year Term 1976-1982

In December 1976, Petróleos Mexicanos' current management presented the program of activities to be carried out in the period 1976-1982. This program includes doubling oil production and refining capacity of the country, as well as tripling the capacity of the basic petrochemical industry.

Plans include drilling of 3,476 wells, with 1,324 of them for exploration purposes and the other 2,152 for developing fields that have been discovered.

Oil production will increase from 953,000 BD in 1977, to 2,242,000 BD in 1982, and natural gas production will increase from 2,186 million CFD in 1977 to 3,630 million CFD in 1982. Export surplus will increase from 150,000 BD in 1977 to 1,105,000 BD of refined products and oil by 1982.

The main sources for these increases are expected to be the results of intensified drilling in the Chiapas-Tabasco Mesozoic area, of applying secondary recovery methods, and of developing the Campeche continental shelf.

Hydrocarbon-Producing Areas

Hydrocarbon-producing areas are mainly located in the coastal plains of the Gulf of Mexico; additionally, gas production has for the first time been developed in other than traditional sites, so that, in general terms, our country's hydrocarbon production comes from the Sabinas area in the Coahuila and Nuevo León States, Reynosa, Tampico, Ebano, Arenque, Cerro Azul, Poza Rica, the Offshore Golden Lane, Chicontepec, the Papaloapan Basin, the Tertiary producing fields of the southern part of the Republic, the Chiapas-Tabasco Mesozoic Basin, and the Campeche Sound.

Details of the major geological provinces, whose hydrocarbon production potential has been confirmed, are given below.

Chiapas-Tabasco Mesozoic Province

The Chiapas-Tabasco Mesozoic province, commonly called Reforma area, covers around 7,000 km^2 and is located in the southeastern part of the country. It occupies the northern part of the State of Chiapas and a large portion of the State of Tabasco.

The hydrocarbon-producing formations of this area are Cretaceous-Jurassic limestones found at depths of about 4,500 m. The pay zone ranges from 240 to 1,080 m in thickness. The oil varies from 27° to 52° gravity API.

Current State. The Chiapas-Tabasco Mesozoic Basin has more than 100 potential hydrocarbon-producing structures, 27 of which already are productive, with the remainder still under exploration, according to programs prepared for that purpose.

This area has the largest concentration of drilling rigs, with a total of 96, 21 of which are engaged in exploratory work and the other 75 are being used for developing known fields.

There are 170 producing wells, with an approximate output of 1,100,000 BD, which amounts to 73 percent of the country's total production, and an average output per well is 6,500 BD.

Based on the plentiful data available from this area, its development prospects are excellent, for which reason long-term plans contemplate drilling over 300 wells by 1985, including exploratory, development, and waterflood injection wells.

It is thought that the above will make it possible to increase current production levels for all of these fields.

Campeche Sound

The Campeche Sound region is located in the western part of the Yucatán Shelf. For this region a high-priority area of 8,000 km^2 has been considered. Its approximate boundaries are as follows: east and south, the shoreline; west, the meridian of Ciudad Frontera; and north, the parallel of the City of Campeche.

Exploratory drilling was begun in 1974 and by the middle of 1975 Chac No. 1 well was completed, producing oil from a Paleocene limestone breccia at a depth of about 3,500 m. The oil has an 18° API gravity, with a dissolved gas-oil ratio of 64 m^3/m^3. Upon developing this area, fields producing 32° API gravity oil, that is, lighter than that initially found, have been discovered.

Current State. Results from the Campeche Sound geophysical-geological prospection indicate the presence of more than 200 structures with oil production possibilities over an area of 50,000 km^2. Their geological evaluation allowed the selection of the aforementioned 8,000 km^2 high-priority area, which has 60 structures with hydrocarbon-bearing possibilities.

Results obtained to date point to the presence of nine oil-producing fields with five additional structures currently under exploration. It is important to mention that the rate of success achieved in the search for new fields is higher than 70 percent.

At present, this area has 5 rigs (4 jack-up and 1 drilling vessel) for exploratory activities, 10 fixed platforms for developing fields, and 3 platforms for handling production.

The first 7 wells on stream are producing 165,000 BD.

Development Perspectives. The significance of discoveries made at the Campeche Sound led to its being regarded as a high-priority area and, hence, it was decided to intensify exploratory and development activities there.

Because of the above, a program involving installation of 30 drilling platforms and about 16 platforms for handling oil and gas production is being contemplated. Additionally, and as drilling advances, platforms for water-flooding and habitational purposes will be installed.

Development of this plan is complex and led to analysis of various alternatives in order to be able to make long-term plans in accord with exploitation policies established by this Corporation.

For this reason, the program contemplates, in the short term, the drilling of 52 wells in the Akal and Nohoch fields and the best alternative for exploitation of Campeche Sound will be defined in accordance to productivity of those wells.

Within the context of the exploitation program it is advisable to make some comments about the risks involved in its development: it is thought that the only way to avoid those risks is to resort to total immobility, which is not possible in the oil industry, in any other area of activity, or in any country of the world.

Risks are consciously accepted and it is imperative that it be so since, on the one hand, there is a serious requirement for energy resources which are basic factors for obtaining the various economic goods demanded by man, and, on the other hand, until there is another form of energy capable of substituting them, these energy resources must be searched for, found, and extracted wherever they might be.

Mesozoic Sabinas Gulf

The Mesozoic Sabinas Gulf Basin is the first province producing non-associated gas outside the coastal plain of the Gulf of Mexico. It is located in the central-eastern part of the State of Coahuila and the western portion of the State of Nuevo León.

Productive formations are Cretaceous fractured limestones and Jurassic sandstones; depths at which production is being obtained range from 2,000 to 3,500 m.

These fields produce dry natural gas without liquids or impurities, which permits its immediate delivery to consuming centers.

Current State. Geologically, this is an ancient gulf filled with Cretaceous and Jurassic marine sediments, with an approximate area of 80,000 km^2 where 65 structures with hydrocarbon-bearing possibilities have been determined, with 9 of them producing gas at present.

For exploring and developing fields that have been discovered there are 16 and 17 drilling rigs, respectively.

Eighteen gas-producing wells have been completed to date, with an average output of 180 million CFD.

Development Perspectives. Because of the high-production characteristics of completed wells and the extension of the area with possibilities, it is thought that these are the largest nonassociated gas accumulations ever discovered in our country.

By reason of the above, it has been planned to drill 170 exploratory and 273 development wells, which are expected to yield a substantial increase in gas production.

Chicontepec

The Chicontepec area is located between the coastal plain of the Gulf of Mexico and the foothills of the Eastern Sierra Madre range; it extends mainly over the State of Veracruz and the eastern portions of the States of Hidalgo and Puebla.

The productive Chicontepec formation lies at an average depth of 1,800 m.

Hydrocarbon accumulation was confirmed with data obtained from more than 1,200 wells drilled in that area; the original objective for most of these wells was exploitation of deeper reservoirs and later, when some of them became depleted, they were completed at the Chicontepec formation, which partially confirmed its productivity.

Current State. This Tertiary geological basin covers an area of 11,300 km^2; study of origin and distribution of sediments within the formation led to detection of a paleochannel 123 km long, 25 km wide on average, and with an area of 3,300 km^2, filled with Eocene sandy-shale sediments whose average thickness is around 800 m.

In order to attain a production of 750,000 BD, about 16,000 wells will be required, as may be inferred from the integral and intensive development program.

However, because of complexity involved in its implementation, industrial development which should accompany it, its socioeconomic repercussions, etc., assistance from various sectors is required for developing said program as the drilling phase alone will take 13 years to be completed.

Petróleos Mexicanos plans for developing this area contemplate solely the use of domestic financial resources. Consequently, work to be performed shall be in agreement with the increase of the Mexican Industry activities.

Other Areas

The Mexican Republic has other, traditional hydrocarbon-producing areas, such as Reynosa, Tampico, Poza Rica, and Tertiary fields of the Southern Zone; in addition, more recently, the Offshore Golden Lane, Arenque, and the Cotaxtla area reservoirs have also been incorporated. Oil production from all these fields is about 350,000 BD and gas production amounts to 1,350 million CFD.

OIL AND GAS DOMESTIC PRODUCTION

Current Oil and Gas Production

Total oil, condensate, and natural gas liquids production amounts to 1,725,000 BD as of November 1, 1979, with 5 percent coming from the northern part of the country, 7.5 percent from the central portion, and 87.5 percent from the southeastern portion of Mexico.

The country's gas production amounts to about 3,000 million CFD as of November 1, 1979, with 80 percent coming from the Chiapas-Tabasco Mesozoic area.

Reserves Production Ratio

The world oil industry uses two parameters for judging the degree of advancement or retrogression of an oil company:

- o hydrocarbon production, and
- o value of hydrocarbon reserves.

In 1938, proven hydrocarbon reserves amounted to 1,240 million barrels.

This figure has gradually increased since then so that, as a result of exploitation and a better knowledge of the Chiapas-Tabasco Mesozoic oil fields, and of data obtained from the Campeche continental shelf and the Chicontepec formation, crude oil, natural gas, and natural gas liquids proven reserves amount to 45,803 million barrels as of September 1, 1979.

Based on size and production potential of the aforementioned reservoirs, the current proven reserves/annual hydrocarbon production ratio is at its highest level in 60 years.

New Production Program

Over many years, Petróleos Mexicanos has been making continued efforts into research using the most advanced technology. Hence, a careful survey of the national territory has led to detection of new hydrocarbon-producing areas, which have widened perspectives and realities relevant to existing areas.

As a result of these discoveries, new forecasts for oil and gas production were made. These forecasts exceeded values initially mentioned for the 6-year program in such a substantial way that it has been estimated that it will be possible to attain the basic program production goal earlier than expected, that is, by 1980. Upon reaching this production plateau for 1980, the country shall be able to decide which shall be the exploitation policy most appropriate for the national interest.

SECONDARY RECOVERY

The Mexican Republic has several reservoirs whose inherent characteristics--drive mechanisms, depletion status, pressure-drawdown rates, or need for pressure maintenance--have made it necessary to apply secondary recovery methods, with water flooding most common.

At present, this system is being applied to 16 fields, which are getting a total injection rate of about 900,000 BD. Among these are Poza Rica and A. J. Bermúdez.

It is estimated that by 1982 water injection rates will amount to 3 million barrels per day.

NATIONAL HYDROCARBON TRANSPORTATION NETWORK

Oil Pipelines

Once it has been separated from gas, oil production is conveyed from separator batteries and storage centers through a pipeline network to refineries for processing and to marine terminals for export and coastwise shipping between domestic ports. For these purposes, a national pipeline network about 3,000 km long and diameters of 8 in to 36 in is available.

Gas Pipelines

Gas produced from the different fields, both associated and nonassociated, is conveyed through a pipe-gathering system to various processing plants for sweetening of gas and recovery of liquids. Residual gas is conveyed to consuming centers located in different parts of the country.

Gas is supplied to processing plants through a gas pipeline network about 1,000 km long and with diameters of 8 in to 36 in.

For distributing the dry gas the following gas lines are used: Ciudad Pemex-Mexico City-Guadalajara for the southern and central parts of the country; the Reynosa-Monterrey-Torreón-Chihuahua for the northern and northeastern areas, and a recently built National Trunk Gasline System with a diameter of 48 in.

For the purpose of supplying gas to those regions where liquid fuels are currently being used, construction of the following gasline branches has been planned:

1. Ciudad Pemex-Campeche-Mérida
2. Cactus-Tuxtla Gutierrez
3. Minatitlán-Salina Cruz
4. Mexico City-Toluca-Lázaro Cárdenas
5. Chihuahua-Ciudad Juárez (construction in progress)
6. Saltillo-San Luis Potosí-León-Salamanca-Guadalajara-Manzanillo.

This network will make it possible to diversify the distribution of available dry gas, thus favoring creation of new industrial centers.

EXPLOITATION POLICIES

In order to carry out exploitation of the country's hydrocarbons, the following policies have been established by the current Petróleos Mexicanos management:

1. All exploratory activities concerning geological, geophysical, and interpretative tasks shall be performed in agreement with the most modern techniques available and conducted in a balanced manner in all of our geological provinces.

2. All reservoirs shall be rationally exploited so that their maximum utilization is achieved.
3. Atmospheric burning of gas shall be reduced to the point deemed advisable both from a technical and economical standpoint.
4. In order to be able to determine the different production options available, there shall be a permanent and strict evaluation of our reserves.
5. Refining capacity increase and diversification of plants shall be conducted in accordance with regional needs; and additional capacity shall be installed in anticipation of future needs.
6. Basic petrochemical industry shall be developed for meeting domestic consumption needs, substituting imports, and for having surpluses available for export.
7. An efficient, nationwide conveyance and distribution network shall be maintained.
8. Domestic consumption shall be rationalized so that energy resources are used in accordance with technically established priorities. Gas consumption, for instance, has been intensified and shall do so to an even higher extent, thus benefiting our country as to availability of a wider range of fuels.
9. A program shall be undertaken for accelerated and careful training of all technical, administrative, and working personnel required for the future development of our industry.
10. Measures shall be taken for achieving an exceptionally independent level in the technological world, to which end the Mexican Petroleum Institute has been and is a significant factor.
11. Our programs shall be financed by means of the solid credit standing consolidated by the Government of the Republic in the international capital market.

This statement has aimed at giving a general outline of oil and gas exploitation being currently carried out by Petróleos Mexicanos as well as its long-term plans for becoming an element decisively contributing to the development of our country.

CHAPTER 28

SOME ASPECTS OF EXPLORATION PLANNING FOR OIL AND NATURAL GAS IN MEXICO

Eng. Jerzayn León Alvarez[1]

GENERALIZATIONS ON THE OIL AND BASIC PETROCHEMICAL INDUSTRY PLANNING IN MEXICO

The groundwork and goals established by Petróleos Mexicanos for exploration during the 1977-1982 term were determined in close coordination with those of all other branches of the oil and basic petrochemical industry of Mexico.

Because the Mexican oil industry is state-owned and its activities range from exploration to marketing, planning is being conducted continuously, setting different goals defined in time, striving for an integrated effort, but retaining sufficient flexibility to analyze various alternatives which might give better direction to the industry's development.

Activities under the development plan are divided into two basic stages. The first pertains to the Diagnosis of the Industry, which allows one to learn the causes and dynamics of its evolution as well as its accompanying problems. The second stage is Prognosis, which includes guidelines and alternate development programs in order to attain the previously set goals and objectives.

In order to maintain congruence between the development plan and the socioeconomic development of our country, macroeconomic indexes established by the different government agencies are taken into account and exhaustive and permanent research is conducted to determine the industry's quantitative objectives.

Once foreseeable objectives and goals have been set in accordance with government policies, planning is commenced in two areas, demand and exploration, in order to obtain preliminary figures on oil products consumption as well as on the oil potential of the subsurface, thus putting the finishing touches to the original objectives and goals; on the other hand, these

---

[1]Assistant Manager for Reserves and Alternate Resources, Petróleos Mexicanos, Mexico.

figures are fed back to the different branches of the industry, in a closed iterative system which ultimately achieves coherent and congruous results defining hierarchies and priorities.

OBJECTIVES AND CHARACTERISTICS OF EXPLORATION PLANNING

Basic objectives of exploration planning are: (1) to determine those activities which may be conducted within a general and integrated framework for the timely location of sufficient oil and natural gas to meet expected goals and (2) to determine the necessary means to implement these activities.

Here one should point out that exploration planning has very distinctive characteristics, because the exploratory process includes factors of chance stemming from geological conditions of the various areas where this activity is carried out, and also because oil exploration results are not immediately known but may take from 5 to 10 years or more to become effective.

DIAGNOSIS

In general terms, oil exploration diagnosis covers two aspects, statistical and geological.

Statistical Diagnosis

Statistical diagnosis is made by systematically introducing growth rates of representative statistical series, as indicators of the true growth of development; structural composition and requirements of each exploratory method are also determined, by calculating the fraction each part represents in the total; thus the variation of relationship between exploration activities and economic and human resources is obtained.

At the turn of the century, Mexico started its commercial oil operations, focusing activity on the Gulf of Mexico's coastal plain, where the first oil fields, enabling the country to become the world's second largest oil producer in 1921, were located. This production rapidly declined, due to the lack of proper exploration, to volumes that only allowed self-sufficiency at the time of nationalization, which took place on March 18, 1938.

It was not until 1974 when there was a significant increase in production, as a result of the development of the new fields discovered in Mesozoic formations in southeastern Mexico. This is considered as the culmination of a systematic and increasing exploratory effort which started in 1944 and which, during the last few years, has received greater impetus, thus resulting in the discovery of new and important productive areas.

In order to achieve the current goals in oil exploration, field crews of all types have accumulated 23,000 party-months of both reconnaissance and highly detailed surveying from 1938 to 1978. This includes surface geological surveys on the majority of accessible areas with oil-bearing possibilities as well as ground, marine, and airborne geophysical work over 1,162,000 linear km from 1955 to 1978.

From 1938 to date more than 1,100 reservoirs have been located in slightly more than 400 fields, by drilling 3,300 exploratory wells with a rate of success of 29.5 percent for the 1938-1976 term and 45.5 percent for 1977-1978.

Investments made from 1938 to 1975 were in the neighborhood of 18,500 million 1978 pesos (U.S. $800 million) for exploration which, added to the expense of exploratory and development drilling, totals 149,000 million pesos (U.S. $6,480 million), an investment which allowed the location and development of sufficient hydrocarbon reserves to meet domestic demand until 1971, when imports began to be necessary. On December 31, 1975, proven reserves were 6,338 million barrels, with an annual production of 452 million barrels, resulting in a reserves/production ratio of 14 years.

By applying new exploration criteria and techniques, 4,700 million pesos (U.S. $200 million) were invested from 1976 to 1978, as well as 26,900 million pesos (U.S. $1,170 million) for drilling; this allowed the development of some known reservoirs as well as the location and development of new reservoirs, both in partially known and new areas. This not only permitted meeting the domestic demand, but allowed the attainment of surpluses for export, which reached a value of 42,000 million pesos for 1978 (U.S. $1,830 million). Proven reserves were increased to 40,194 million barrels of total liquid hydrocarbons by December 31, 1978, with a substantial increase in the reserves/production ratio, from 14 to 60 years.

Data from the last 3 years indicate that the policy to stimulate exploration and drilling has yielded excellent results for the size of the investment made, which represents only 21 percent of the investments made over the 1938-1975 period. This investment has been recovered in just 1 year of exports, and has allowed reserves to increase 6.34 times and production, 1.61 times, in only 3 years.

Geological Diagnosis

The first phase corresponds to an exhaustive analysis of geological and geophysical data gathered in Mexico over many years by geologists of Petróleos Mexicanos. From these data it has been possible to build paleogeographic maps for different periods of geological history in Mexico which, in turn, very clearly synthesize the distribution of sediments in those areas where favorable depositional environments exist. Such areas are the ones in which hydrocarbons are capable of having been generated, as well as sediments which may have served as seals to prevent the leakage of the hydrocarbons migrating from the source rocks to the reservoir rocks.

The paleogeographic map for the Upper Jurassic Kimmeridgian (Fig. 28-1) shows an ancient geography represented by a distribution of continental areas and islands, continental shelves, coastal belts, and, finally, deep sea basins, where sediments with hydrocarbon-generating organic matter were deposited; subsequently, the oil and gas migrated to porous rocks within the shelves and coastal belts where they accumulated into the different types of existing traps.

The following map (Fig. 28-2) depicts the paleogeography of the Lower Cretaceous (Neocomian) in Mexico, using the same nomenclature as Fig. 28-1. During this period continental areas within the present geographical boundaries of the country were reduced and large calcareous (limestone) shelves were developed from shallow seas, the edges of which present the geological characteristics of rocks with hydrocarbon-bearing potential, which have been preserved from erosion; that because of their ample distribution these strata represent first-class objectives in the search for new reserves.

The paleogeography of the Middle Cretaceous (Albian-Cenomanian) in Mexico (Fig. 28-3) represents a new distribution of continents and seas, in which a remarkable reduction of continental areas may be noted together with development of large calcareous shelves, along the edges of which the most important oil reservoirs of Mexico have been located and where large expanses are yet to be evaluated by further exploratory drilling.

It is advisable to remark that at this time only three paleogeographical maps have been presented, which show the near-term objectives of oil

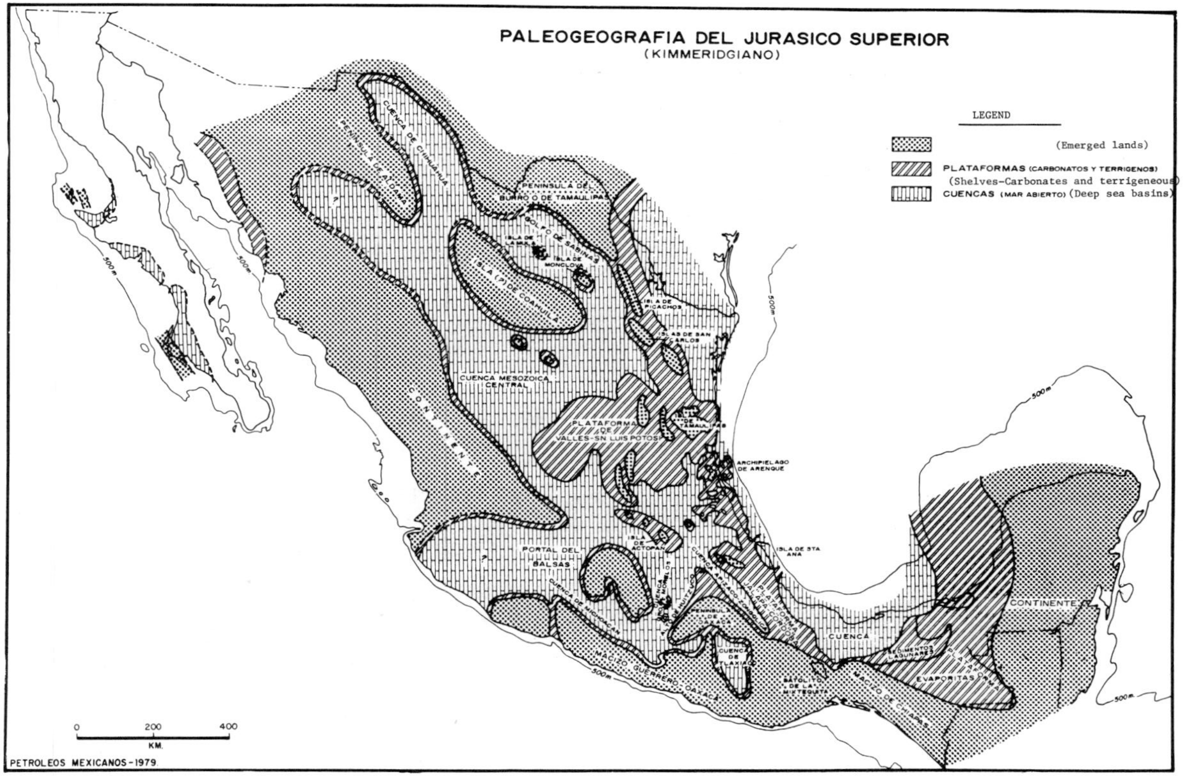

Figure 28-1. Paleogeography of the Upper Jurassic (Kimmeridgian)

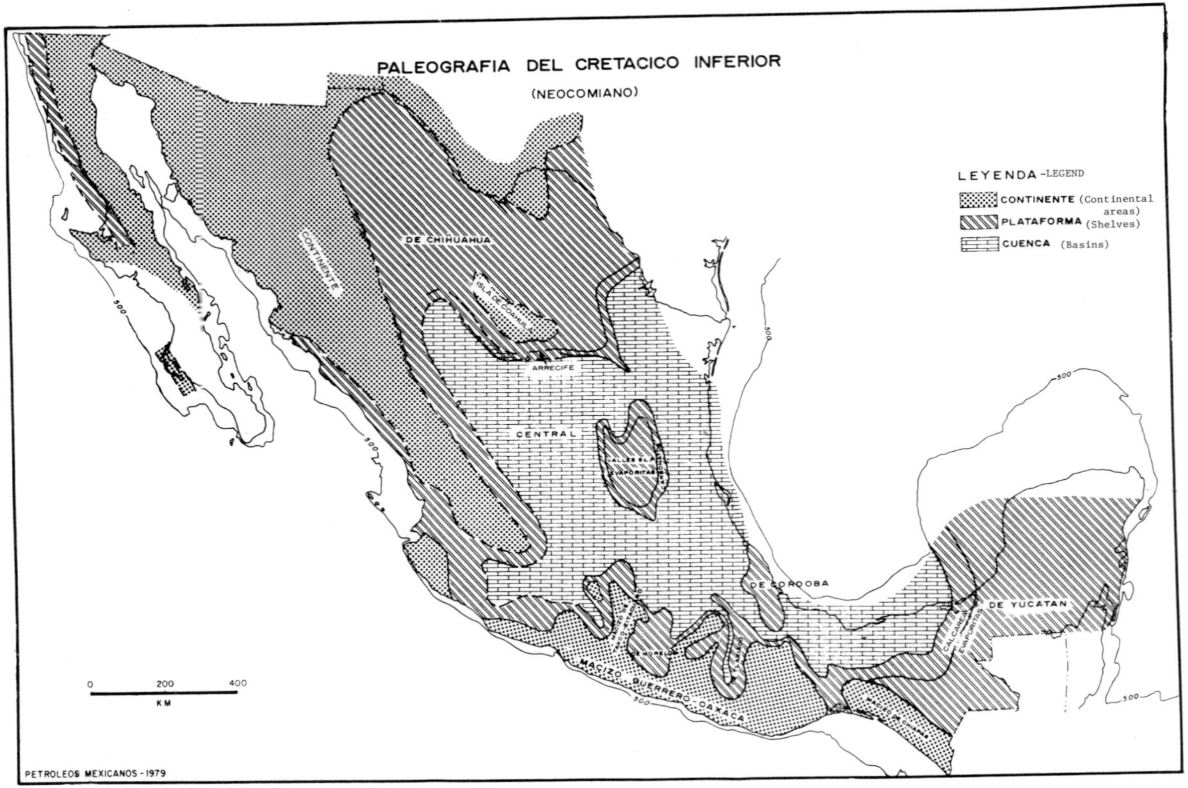

Figure 28-2. Paleogeography of the Lower Cretaceous (Neocomianian)

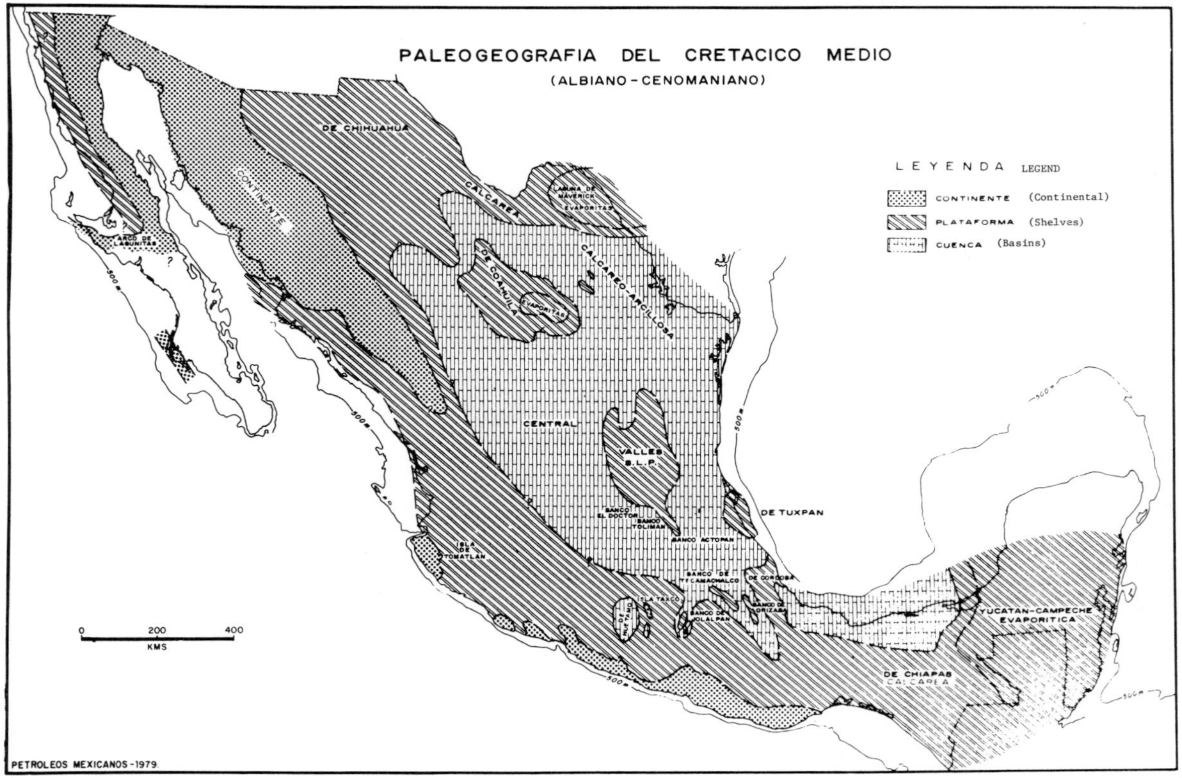

Figure 28-3. Paleogeography of the Middle Cretaceous (Albian-Cenomanian)

exploration in Mexico; however, exploration planning includes Tertiary formations, which are already producing, and Paleozoic formations, which have been only partially explored because these rocks have very few surface outcrops and the depths at which they are found turn the strata into long-term objectives.

The analysis of sites where oil and/or gas seepages have been observed (Fig. 28-4), grouped and presented by areas for greater objectivity, shows their presence even in some areas of the country where no production has yet been established. This, along with the paleogeographical studies, permits the establishment of additional objectives for hydrocarbon exploration in new areas.

The second phase of geological diagnosis has consisted of dividing the Mexican Republic into various provinces in order to reasonably evaluate results of exploratory work and thus hierarchically classify its oil-bearing possibilities in accordance with the potential of their sediments.

With regard to this, an analysis of the characteristic geological conditions of the country's territory was performed, taking into account physiographic and tectonic regional features as well as the sedimentologic and stratigraphic features of basins existing in different geological periods. As a result of this analysis, 36 land and 14 offshore areas on the continental shelves have been identified which have been termed Geological-Oil Provinces (Fig. 28-5).

In order to better evaluate the data, some of the geological provinces were divided into subprovinces, in accordance with their local sedimentologic and structural features, and these subprovinces into even smaller areas, according to the exploratory stage attained in each and their oil-bearing possibilities at the date of evaluation.

As an example of results obtained from exploratory work and of the oil bearing potential of Mexico's subsurface, we shall mention some of the fields located in rocks belonging to the different periods of the geological history of Mexico.

To the Upper Jurassic belong: the reservoirs located in terrigenous sediments in Lampazos in the Burro-Picachos Province; the oolitic limestones of Tamaulipas-Constituciones and Arenque fields in the Provinces of Tampico-Misantla and Tampico, respectively; the shelf edge clastic-calcareous sediments of some fields of the Reforma Area in the Southeastern Basins Province;

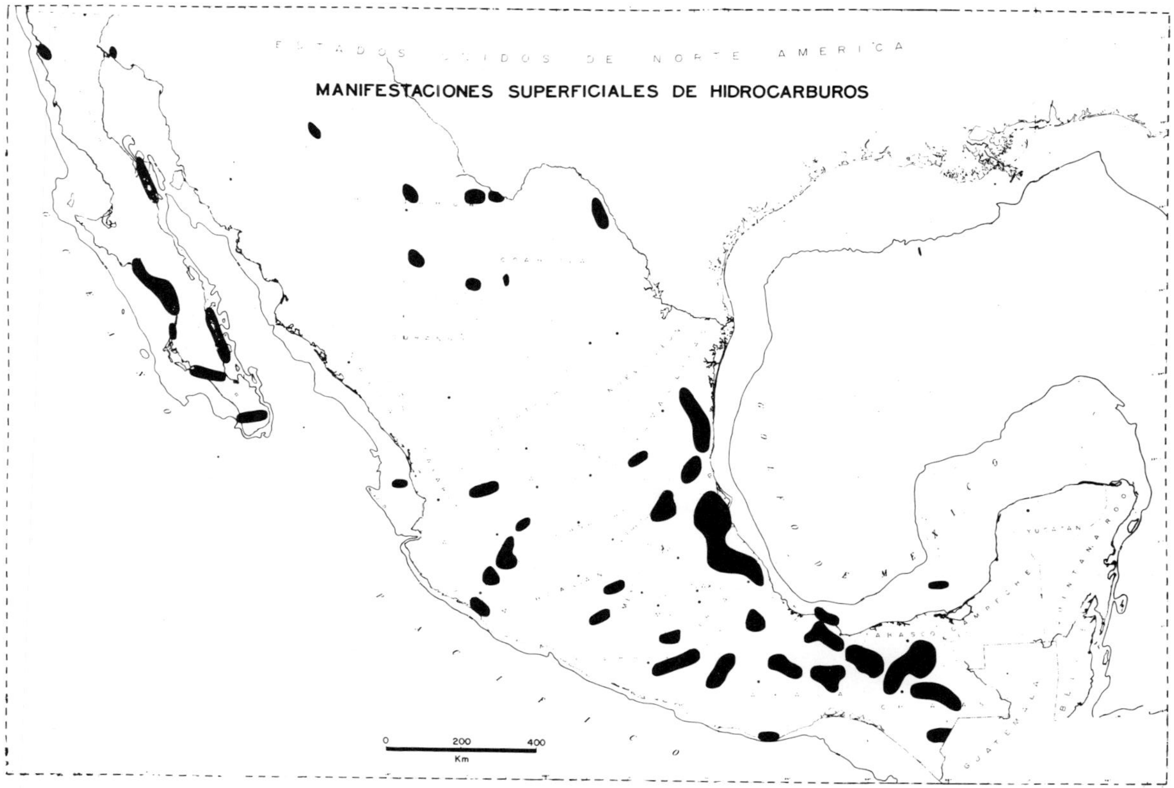

Figure 28-4. Surficial manifestations of hydrocarbons

Figure 28-5. Geological provinces of Mexico

and those of Akal and Nohoch of the Campeche Sound in the Coatzacoalcos Province.

Some examples of the economic significance of the Lower Cretaceous calcareous shelves are the rich gas reservoirs located in the Buena Suerte and Monclova fields in the Sabinas Province; the oil and gas reservoirs of Tamaulipas-Constituciones and Barcodón in the Tampico-Misantla Province; the Arenque field in the Tampico Offshore Province; the fields named Agave, Cactus, and Samaria in the Reforma Area in the Southeastern Basins Province; and the Cantarell Complex in the Campeche Sound oil region in the Coatzacoalcos Province.

The most important oil reservoirs of Mexico are located in Middle Cretaceous carbonate sediments. Some examples of these are the Golden Lane and Poza Rica fields in the Tampico-Misantla Province, those in the Reforma Area in the Southeastern Basins Province, and the important new reservoirs of the Campeche Sound in the Coatzacoalcos Offshore Province.

In short, it is concluded that in the subsurface of the Mexican Republic hydrocarbon reservoirs have been located in rocks ranging in age from Jurassic to Tertiary, both in terrigenous and carbonate sediments, with the latter being the most productive.

Important reserves have been located to date in the onshore provinces of Burgos, Sabinas, Tamaulipas, Tampico-Misantla, Papaloapan, and Southeastern Basins; less important reserves have been located in the Burro-Picachos Shelf as well as in the Chiapas Mountain Range, and gas accumulations, not yet commercial, have been located in the Vizcaino Province. Important oil and gas reserves have also been located in the Tampico and Coatzacoalcos Provinces.

The Mexican Republic total area, including its continental shelves out to the 500 m bathymetric curve, is 2.5 million km^2; of this area, 1,831,100 km^2 correspond to sedimentary marine basins and shelves and constitute the geological-oil framework of Mexico. Only 10 percent of this area has been explored to a reasonable degree, meaning that, besides basic exploration work, most exploratory and development drilling has been concentrated in this portion; another 45 percent has been rated as having very good oil-producing possibilities, and the other 45 percent as having possibilities to be investigated on a longer-term basis. This means that even with intensification of the exploratory activities a considerable expanse will remain to be explored for some years to come and what is even more important, there are very good possibilities for locating new hydrocarbon reserves for the country.

PROGNOSIS 1977-1982

The program for this period includes field operations and interpretation work in high-priority producing provinces as well as in new provinces with good possibilities.

High-Priority Producing Provinces

Southeastern Basins Province (Chiapas-Tabasco Mesozoic Area). This area of 7,000 km^2 was turned into an important oil region beginning in 1972; since then 28 oil and natural gas fields have been discovered in Mesozoic formations. This area still offers excellent possibilities, because there are several structures yet to be drilled, and because the rate of success for exploratory wells has been 59.2 percent. For these reasons, 18 structures are currently being drilled, and there are 30 other structures scheduled to be tested with the drill.

Coatzacoalcos Province (Campeche Sound Offshore Area). Out of an area of more than 50,000 km^2, where 250 structures were located during preliminary exploration work, a high-priority area of 8,000 km^2 was selected through paleosedimentation studies and the application of the structural model previously established. In this area 60 structures have been detailed. These structures are slated to be drilled and have very good possibilities of finding new fields with important reserves. The reasons are: thicknesses of oil-saturated reservoir rocks of up to 1,000 m have been encountered in some of the producing fields, porosity has been found to be greater than that observed in the Reforma Area, and the rate of success for exploratory drilling has been 61 percent.

The above, coupled with the fact that the crests of probable traps lie at depths ranging from 1,250 to 4,000 m, makes this an outstandingly important area; for this reason, in order to evaluate a good part of its reserves, there are 17 high-priority exploratory sites ready for drilling plus another 36 dependent upon results from the former.

Sabinas Province. This province has Jurassic and Cretaceous sediments where large-size structures are found, some of them up to 60 km long, having been determined by surface geology and seismology. Reservoirs which have been located to date contain dry gas (methane) and some wells have absolute potentials of up to 100×10^6 ft^3 per day. The type of gas, the production potential, and the area's geographic location, close to the industrial consuming centers, give this area a first-class rating within the exploration program.

New Provinces with Good Possibilities

Besides searching for new structures within the high-priority producing provinces, which constitutes the bulk of exploration, a program is in progress in the new geological provinces with good possibilities. In these provinces, intensification of field exploratory work, laboratory and interpretation studies will not suffice; a larger number of exploratory wells will have to be drilled to evaluate their oil-bearing possibilities. As examples of exploratory programs which are being conducted and planned for provinces of this type, some aspects are mentioned below.

Vizcaino, Purisima-Iray, and Bahia Magdalena-Cedros Provinces. After several years of exploratory work, these geological provinces have acquired significance because two wells found gas in Cretaceous sand formations. For the country this discovery represents its first significant though not-yet-commercially productive area on the Pacific Ocean slope. In these provinces, surface geological work, land, marine, and airborne geophysical surveys, exploratory drilling, and data processing and integration are being carried out in order to evaluate their medium- to long-term possibilities.

San Felipe-Tiburón Island Province. Within this province, located along the continental shelf of the Gulf of California, geophysical studies predict the existence of a sedimentary column extending down to 6,000 m, but with stratigraphic and stratigraphic-structural characteristics that indicate few anticlines. In order to complement this information and decide on future exploratory well sites, special geophysical studies are being conducted, and operational geochemical surveys for direct hydrocarbon detection on the sea bottom are already being programmed.

Chiapas Mountain Range Province. Surface geological, as well as airborne magnetometric, gravimetric, and seismologic studies, wherever prospection with this method is feasible, have been carried out. These studies have permitted the detection of structures with alignments up to 100 km in length that are likely to contain hydrocarbons in Lower Cretaceous, Upper Jurassic, and, possibly, Paleozoic formations. Some of the wells drilled in this important province had oil and gas shows and encountered sedimentologic conditions that justified continued evaluation of this large area. Work performed in the Lacantun area near the Guatemalan border shows the presence of several structures that will be initially evaluated with the drilling of two test wells. Data from these wells will enable the scheduling of further exploratory work.

In order to evaluate the oil and gas potential of all these new provinces more than 50 drilling sites are already at hand both onshore and offshore, which will certainly provide important data for the location of new producing areas.

Synthesis of Activities and Resources

As material resources and manpower have gradually been incorporated, the volume of exploratory activity initially considered as necessary for the 1977-1982 term has increasingly expanded. In 1977, activities were commenced with 982 group-months and it is expected that during 1982 a total of 1,600 group-months will be carried out. At present Pemex has 214 drilling rigs, 72 of which are assigned to exploratory drilling. The number of rigs shall increase in accordance with development and production requirements.

In order to perform exploratory work scheduled, in the planning, for 1977-1982, the technical staff is being increased in number and is continuously trained. This staff includes 1,000 persons to date and this figure shall be increased in accordance with basic needs of the geological, geophysical, and interpretative operations. Moreover, more than 20,500 million pesos (U.S. $890 million) have been budgeted for exploratory operations.

RESULTS

Goals attained by Pemex through exploration have culminated in the location of significant oil and gas fields in southeastern, central, and northern Mexico, which, upon having been developed, have put Mexico in sixth place in the world among the countries with proven oil reserves, with 45,803 million barrels of total liquid hydrocarbons that, added to the 45,003 million barrels of probable reserves estimated as of July 31, 1979, permit one to forecast that, by the turn of the century, Mexico shall have enough oil energy resources to meet its interests and needs.

Source: Technicians and records of the Oil Exploration Management, Petróleos Mexicanos.

CHAPTER 29

EXPLORATION IN DEVELOPING COUNTRIES

K. Pipes[1]

INTRODUCTION

None of us needs any reminder that there is a continuing energy problem. As we all know, we face the difficult task of finding and producing crude oil in sufficient quantities to meet constantly increasing demand around the world. This is not a problem just for the United States. Europe, Japan, and the rest of the industrialized countries have also had to struggle to balance supply and demand. For the developing countries, particularly those which have to rely extensively on energy imports, energy has become even more of a concern. For these nations, as you know, scarce foreign exchange spent on increasingly costly energy comes at the expense of development projects desperately needed to create employment opportunities, generate income, and meet other basic requirements of their peoples.

Some observers, concerned about the developing world's ability to grow and to compete, have suggested that private companies have not shown sufficient interest in helping non-OPEC developing countries to explore and develop their petroleum potential. The data I will review today show that the oil industry has been working vigorously to discover and develop more oil and gas wherever it can be found.

We believe there is a significant undiscovered petroleum resource base in the non-OPEC developing countries. Projections developed by Exxon suggest that these countries may be producing 10 million barrels of oil a day by 1990, perhaps one-third of all newly discovered oil produced in the non-Communist world.

In this chapter I wish to discuss the extent of exploration activity which has taken place in the non-OPEC developing countries, constraints to exploration, and an estimate of the amount of oil which might be produced in the non-OPEC developing countries by 1990. The discussion is generally

---

[1]Manager, Exploration Division, Producing Department, Exxon Corporation, New York, N.Y.

limited to conventional oil, and mostly excludes heavy crudes and tar sands. For the most part the paper also excludes natural gas. However, because of the nature of the exploration activity, expenditures and wells usually relate to both oil and gas and cannot be distinguished before the fact.

Experts agree that considerable petroleum remains to be found, but estimates cover a wide range. Analyses of appraisals of remaining recoverable oil resources result in a mean estimate of 2.1 trillion barrels, of which 1.0 trillion, or about half, has already been discovered.[2] These resources are located in about 600 sedimentary basins, which have been identified worldwide. About 400 of these basins have been explored and 160 basins in this group are now capable of producing commercial quantities of oil and gas. A World Energy Conference poll indicated 60 percent of remaining discoveries will likely be made in these 400 basins.

The remaining 200 basins have had essentially no drilling as yet--in most cases, because they are in harsh physical environments, or in a few cases because of governmental restraints which have restricted access to them.

NORMAL SEQUENCE OF EXPLORATION

Three conditions are required for exploration to take place: (1) estimate of geologic potential; (2) access to acreage; and (3) acceptability of terms of exploration to all parties involved. If these conditions are met, exploration will almost automatically follow.

Data for an initial appraisal of the geology of a basin frequently comes from surface geology plus gravity and magnetic surveys gathered over a period of as much as several decades of reconnaissance exploration.

Once an area has been targeted as favorable, acreage there must be acquired. For a private company, this involves time-consuming negotiation with governments on many factors ranging from work commitments to taxes. These negotiations are not always successful.

Following an agreement on the acreage involved, detailed geophysical surveys are made to confirm the geology and provide the basis for deciding the location of exploratory wells. If this seismic data does not indicate the presence of traps where oil might be expected to be found, the operator may abandon the area.

---

[2] Poll of 29 "experts," Tenth World Energy Conference.

However, the only way to completely prove or disprove the presence of oil and gas is by actually drilling exploratory wells.

Most of the time, major accumulations of oil or gas are found in the early stages of exploration with the drilling of a few exploratory wells. When a discovery occurs, it sparks a high level of exploration activity; typically, the larger the discovery, the greater such activity. Drilling activity, therefore, follows exploration success rather than leading to it, contrary to the popular impression.

The prospectiveness of an area (part or all of a basin) can be severely downgraded on the basis of a few exploratory wells if these are dry and show unfavorable geology. For example, Mali-Mauritania onshore basin in West Africa contains almost no source rocks and the potential reservoir rocks are extremely hard and tight.

If the geology continues to appear favorable, even though the initial exploratory drilling does not find oil and gas, the exploration process may be reformulated.

Although the larger fields in any given area are searched for first and usually are found early in the exploration process, this does not indicate a lack of interest in small fields. It merely reflects the fact that the economic development of small fields is often dependent on the discovery of the larger fields and the consequent infrastructure created for them. Economics depend on producing rate, and such cost factors as field location, depth, and size.

EXPLORATORY DRILLING (EXCEPT U.S., CANADA, AND COMMUNIST COUNTRIES)

During the 1967-1976 decade a total of some 12,959 exploratory wells were drilled in the world outside of the USA, Canada, and Communist countries.[3] About 24 percent were drilled in the industrial countries of Europe, Australia, and Japan, and 26 percent were drilled in the OPEC nations. The other half were drilled in the developing countries (Table 29-1).

The number of exploratory wells drilled in the United States/Canada substantially exceeds those drilled in the rest of the world, but this is

---

[3] Data Source: American Association of Petroleum Geologists and Oil and Gas Journal.

because of a number of very favorable economic, technical, and geologic factors unique to this location: (1) infrastructure for both drilling and distribution of petroleum is extremely well developed, making costs as much as an order-of-magnitude less; (2) access to onshore acreage is mostly open, with virtually no obligation and full title; (3) in large areas, the presence of petroleum is already established and the geology is conducive to multiple deposits; and (4) financial incentives are superior to many foreign areas. Inclusion of U.S. data tends to distort comparisons with the other OECD nations and the rest of the non-Communist world and has been excluded from this analysis.[4]

TABLE 29-1. Exploration wells drilled ex. U.S., Canada, and Communist areas 1967-1976

	No. of wells	%
Industrial countries	3,145	24
OPEC	3,313	26
Developing countries	6,501	50
Total	12,959	100

Both the level of total exploratory drilling and the shares drilled in the industrial, OPEC, and developing countries remained about the same throughout the decade of 1967-1976. However, there were some important shifts between countries within each group. Many geologically prospective areas were tested, some successfully, many unsuccessfully. For cost and technological reasons, the most readily accessible prospects were tested first, and as a result, exploration was progressively moved into ever-harsher environments--the deeper offshore and more remote continental interiors--where the time to position and drill a well is much longer and the cost much higher than elsewhere. For example, onshore wells in mature producing areas may cost up to $100,000; similar wells in the interior of areas

[4] During the period 1967-1976 approximately 19,000 exploratory wells were drilled in Canada and approximately 84,000 were drilled in the United States.

lacking infrastructure may now cost \$3 to \$5 million; and wells located offshore in deep water can cost \$20 million or more.

Seismic surveys and other exploration techniques, as well as exploratory drilling, are required to give a complete picture of exploration activity.

EXTENT OF EXPLORATION ACTIVITY IN NON-OPEC DEVELOPING COUNTRIES

The extent of all exploration activity in the developing countries for the year 1976 and for the 1967-1976 decade as a whole is shown in Fig. 29-1. Exploration activity is broken down into drilling, seismic surveys, and a catch-all of "other" exploration.[5]

Drilling and seismic activity during any one year is less extensive than for the decade as a whole, because not all areas are worked simultaneously.

By one count, there are 113 developing countries and semi-independent territories in the world, excluding the Communist areas. In 1976, the most recent year for which data are available (Table 29-2), exploratory drilling took place in 37 of these countries, detailed seismic surveys but no drilling in 15, and other exploratory activity, as indicated by acreage holdings, in 30 countries. There was no activity in that year in 31 countries (Table 29-3). This latter group does not represent a very significant amount of surface area and is, to a large extent, made up of countries with poor geologic prospects. However in some, activity was inhibited by governmental policies or uncertainties.

During the full decade, exploratory wells were drilled in 71 countries. Another 19 countries were explored by seismic surveys without drilling, and 3 countries were explored by other means. Only 20 countries were not explored at all. These are mostly small island nations or are countries where geologic prospects are poor.

INTENSITY OF EXPLORATORY DRILLING IN THE NON-OPEC DEVELOPING COUNTRIES

The intensity of exploratory drilling in the non-OPEC developing countries over the 1967-1976 decade is presented in Fig. 29-2 and details given in Annexes I and II. The left hand bar deals with the 71 countries in which

---

[5]Details in Tables 29-2 and 29-3.

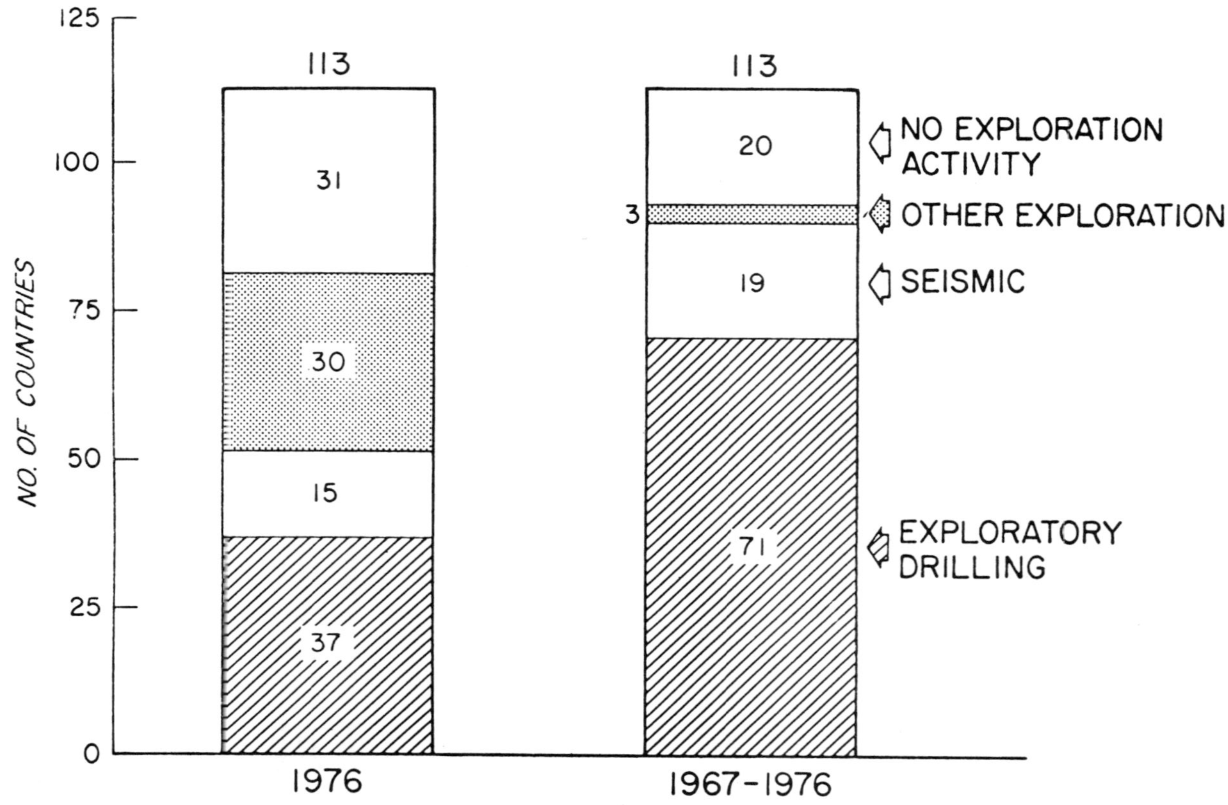

Figure 29-1. Extent of exploration activity in non-OPEC developing countries

TABLE 29-2. Exploration activity in non-OPEC developing countries

KEY: X Seismic took place or acreage in force.
DATA SOURCES: American Association of Petroleum Geologists, Oil and Gas Journal, Petroconsultants and Society of Exploration Geophysicists.

Country	Year 1976			1967-1976			Results of drilling
	Exploratory wells	Seismic	Acreage	Exploratory wells	Seismic	Acreage	
Latin America							
1. Argentina	83	X		1,060	X	X	Discovery since 1907
2. Bahamas	0	0	X	2	X	X	No encouragement
3. Barbados	0	0	X	3	X	X	Small disc.
4. Belize	0	X	X	14	X	X	No encouragement
5. Bolivia	16	X	X	127	X	X	Disc. since 1927
6. Brazil	105	X		934	X		Disc. since 1941
7. Chile	22	X		221	X		Disc. since 1954
8. Colombia	21	X	X	201	X	X	Disc. since 1918
9. Costa Rica	0	0	0	1	X	X	No encouragement
10. Dominican Republic	0	0	0	9	X	X	No encouragement
11. French Guiana	0	X	X	1	X	X	No encouragement
12. Guatemala	5	X	X	13	X	X	Small disc.
13. Guyana	0	0	X	12	X	X	No encouragement
14. Haiti	0	0	X	0	X	X	No encouragement
15. Honduras	0	X	X	9	X	X	No encouragement
16. Jamaica	0	0	0	2	X	X	No encouragement
17. Leeward & Windward Islands	0	0	X	0	X	X	No encouragement
18. Mexico	77	X		1,190	X		Disc. since 1900
19. Netherlands Antilles	0	0	X	0	X	X	No encouragement
20. Nicaragua	0	X	X	24	X	X	No encouragement
21. Panama	0	X	X	2	X	X	No encouragement
22. Paraguay	1	X	X	22	X	X	No encouragement
23. Peru	36	X	X	351	X	X	Disc. since 1869
24. Puerto Rico	0	0	X	0	X	X	No encouragement
25. Surinam	0	X	X	31	X	X	No encouragement
26. Trinidad & Tobago	23	X	X	203	X	X	Disc. since 1902
27. Uruguay	2	X	X	2	X	X	No encouragement

TABLE 29-2. Exploration activity in non-OPEC developing countries--Continued

Country	Year 1976			1967-1976			Results of drilling
	Exploratory wells	Seismic	Acreage	Exploratory wells	Seismic	Acreage	
Africa							
28. Angola	5	X	X	222	X	X	Disc. since 1962
29. Benin (Dahomey)	0	X	X	11	X	X	No disc. (shows)
30. Burundi	0	0	0	0	X	X	No encouragement
31. Cameroon	7	X	X	41	X	X	Disc. 1972
32. Central African Empire	0	0	X	0	0	X	No encouragement
33. Chad	4	X	X	12	X	X	Disc. 1974
34. Congo	0	X	X	26	X	X	Disc. 1969
35. Egypt	34	X	X	213	X	X	Disc. since 1938
36. Equatorial Guinea	0	0	X	7	X	X	No encouragement
37. Ethiopia	0	0	X	14	X	X	No encouragement
38. Gambia	0	0	X	0	X	X	No encouragement
39. Ghana	0	X	X	22	X	X	No disc. (shows)
40. Guinea	0	0	X	0	X	X	No encouragement
41. Guinea-Bissau	0	0	0	6	X	X	No encouragement
42. Ivory Coast	1	X	X	8	X	X	Small disc.
43. Kenya	2	X	X	8	X	X	No encouragement
44. Lesotho	0	0	0	1	0	X	No encouragement
45. Liberia	0	0	X	4	X	X	No encouragement
46. Malagasy	0	0	X	17	X	X	No encouragement
47. Mali	0	0	X	0	X	X	No encouragement
48. Mauritania	0	0	X	12	X	X	No encouragement
49. Mauritius	0	0	0	0	X	X	No encouragement
50. Morocco (incl. Sp. Sahara)	8	X	X	75	X	X	Small disc.
51. Mozambique	0	0	0	36	X	X	No disc. (shows)
52. Niger	0	0	X	6	X	X	No disc. (shows)
53. SaoTome & Principe	0	0	X	0	X	X	No encouragement
54. Senegal	2	0	X	26	X	X	Heavy crude disc *
55. Sierra Leone	0	0	0	0	X	X	No encouragement

*Reflects current data not in preceding texts and charts.

TABLE 29-2. Exploration activity in non-OPEC developing countries--Continued

Country	Year 1976			1967-1976			Results of drilling
	Exploratory wells	Seismic	Acreage	Exploratory wells	Seismic	Acreage	
Africa (continued)							
56. Somalia	1	0	X	7	X	X	No disc. (shows)
57. South Africa	8	X	X	92	X	X	No disc. (shows)
58. Namibia (SW Africa)	0	X	X	4	X	X	No encouragement
59. Sudan	2	X	X	3	X	X	No encouragement (possible disc. 1977)
60. Tanzania	3	0	X	5	X	X	No encouragement
61. Togo	0	0	0	3	X	X	No encouragement
62. Tunisia	10	X	X	93	X	X	Disc. since 1964
63. Uganda	0	0	X	0	0	X	No encouragement
64. Zaire	2	0	X	29	X	X	Disc. 1970
Middle East							
65. Bahrain	0	0	X	0	X	X	Disc. since 1932
66. Cyprus	0	0	X	2	X	X	No encouragement
67. Israel (ex. Sinai)	8	X	X	59	X	X	Small disc.
68. Jordan	0	0	X	6	X	X	No encouragement
69. Lebanon	0	0	X	1	X	X	No encouragement
70. Oman	0	X	X	28	X	X	Disc. since 1962
71. South Yemen	0	X	X	0	X	X	No encouragement
72. Syria	0	X	X	0	X	X	Disc. since 1956
73. Turkey	43	X	X	310	X	X	Disc. since 1940
74. Yemen	0	0	X	0	X	X	No encouragement
Far East							
75. Afghanistan	0	X	X	0	X	X	Gas disc.*
76. Bangladesh	5	X	X	5	X	X	Gas disc. (since 1959)
77. Burma	9	X	X	37	X	X	Small disc. (old)
78. Cambodia	0	0	X	3	X	X	No disc. (shows)
79. Fiji	0	0	X	0	X	X	No encouragement
80. India	0	X		28	X		Disc. since 1889

*Reflects current data not in preceding texts and charts.

TABLE 29-2. Exploration activity in non-OPEC developing countries--Continued

Country	Year 1976			1967-1976			Results of drilling
	Exploratory wells	Seismic	Acreage	Exploratory wells	Seismic	Acreage	
Far East (continued)							
81. Malaysia (incl. Brunei)	21	X	X	185	X	X	Disc. since 1911
82. Maldives	0	0	X	0	X	X	No encouragement
83. New Hebrides	0	0	0	0	0	X	No encouragement
84. New Zealand	7	X	X	54	X	X	Gas disc. since 1959
85. Pakistan	1	X	X	50	X	X	Disc. since 1937
86. Papua-New Guinea	2	X	X	19	X	X	No disc. (shows)
87. Philippines	8	X	X	82	X	X	No disc. (shows)
88. South Korea	0	0	X	7	X	X	No encouragement
89. Sri Lanka	3	X	X	3	X	X	No encouragement
90. Taiwan	10	X	X	120	X	X	Small disc.
91. Thailand	17	X	X	70	X	X	Gas disc. 1973 (minor oil pre-1967)
92. Tonga	0	0	X	0	X	X	No encouragement
93. Vietnam	0	X	0	4	X	X	No disc. (shows)
Total	616			6,501			

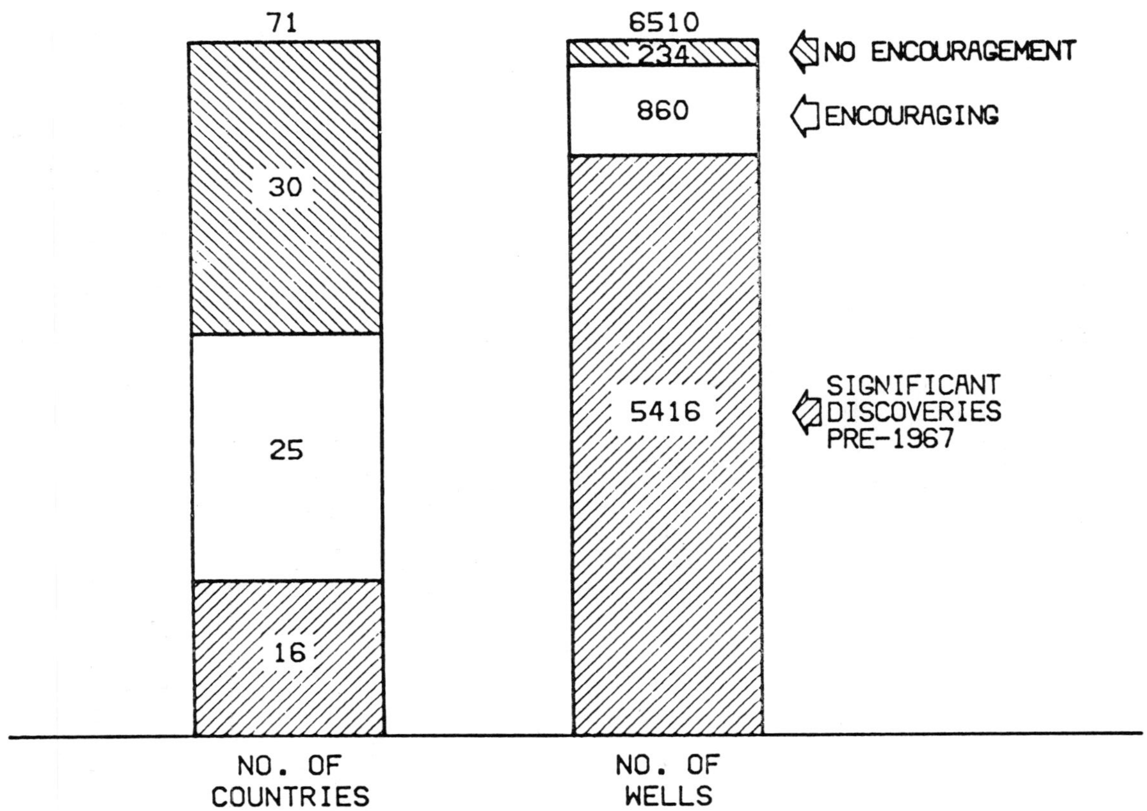

Figure 29-2. Intensity of exploratory drilling in non-OPEC developing countries, years 1967-1976

exploratory drilling took place over the period. "Encouragement" refers to countries where there were small discoveries prior to 1967, discoveries during the decade, or shows of oil which are noncommercial to date.

TABLE 29-3. Non-OPEC developing countries with no exploration activity, 1967-1976

Country
1. El Salvador
2. Granada
3. Botswana
4. Cape Verde Islands
5. Comoros
6. Djibouti
7. Malawi
8. Rhodesia
9. Rwanda
10. Seychelles
11. Swaziland
12. Upper Volta
13. Zambia
14. Bhutan
15. Laos
16. Nauru
17. Nepal
18. Samoa
19. Solomon Islands
20. Singapore

Of these 71 countries, 16 had significant discoveries prior to 1967; encouraging results were obtained during the decade in 25 others where no oil had been found before; and there were 30 countries where 1967-1976 drilling yielded little or no encouragement. However this latter group includes 11 countries which were still under active seismic or drilling exploration in 1976.

The bar on the right deals with the 6,501 exploratory wells which were drilled in the non-OPEC developing countries over the decade. It shows that the vast majority of the wells were in those 16 countries which had significant discoveries prior to 1967. A total of 851 exploratory wells were drilled in the 25 countries considered to be encouraging, and only 234 exploratory holes were drilled in the 30 countries where no positive encouragement was indicated.

These 6,500 wells yielded a wide range of results. Areas which have been disappointing include offshore Burma, the Peruvian Montana, and the shelf of the west coast of Africa. In the latter area, 167 wells were drilled to test many structures, but very little oil was found, and as a consequence exploration activity is now much decreased. On the other hand, important basin discoveries were made in countries such as Chad, Cameroon, Congo, and offshore Vietnam.

Records indicate that over 167 companies, both private and government, were involved in exploratory drilling in the non-OPEC developing countries during 1976.[6] The data in the right hand column confirm that these companies view the most attractive exploration drilling areas as those where the presence of hydrocarbons has already been established. However, a wide diversity of opinion is evident in the diversity of countries selected by various companies, and this in turn enhances the number of prospects tested.

CONSTRAINTS TO DEVELOPING COUNTRY EXPLORATION

There have been and will continue to be restraints to exploration in some countries both developed and developing. A few of these relate to the increasing scarcity of attractive prospects and others are attributable to physical and governmental factors.

Probably the most prevalent physical factor is unsuitable geology. Every country has areas where the occurrence of hydrocarbons is unlikely; in areas such as the southern African shield, where there are large metallic mineral deposits, an entire country may be unattractive for oil and gas. In other countries, the general geology may indicate areas of oil and gas potential, but limited testing has convinced the explorers that the probability of finding oil and gas is very low. Because of the small size of many developing countries, this too may apply to the whole country.

Another physical factor is harsh environments. Much of the remaining geologically prospective area in South America and Africa is either in deep water offshore or in the remote interior and presents a particular problem if infrastructure is lacking.

---

[6] Data Source: Petroconsultants Ltd.

It should be emphasized that the prospect of small discoveries is not a factor which inhibits exploration at the initial stage, because this is still one of the unknowns at that point in time.

GOVERNMENTAL RESTRAINTS TO EXPLORATION

Governmental restraints to exploration--which can apply to both private and state companies--include political uncertainties which increase the explorer's risks, and the willingness or ability of a country to make acreage available.

Exploring companies--whether government or private--will show reluctance to enter a country if there is evidence of political instability, or if there is a government philosophy hostile to foreign enterprise. However, the assessment of and willingness to accept political risk varies among companies, apparently in direct relation to their estimates of geological attractiveness. Thus some companies enter a country when others will not. A survey of current industry exploration activity indicates that there are very few countries where political risk is a totally insurmountable factor; however, during the past 10 years, exploration has been severely restricted at times in such countries as Cambodia, Vietnam, Angola, and Tanzania.

There are some instances where exploration is being restricted by boundary disputes between neighboring countries; this is particularly true in the offshore areas of the East China and South China seas.

Private exploration companies have been excluded from a few countries, where petroleum reserves are sizeable, by national policies which restrict exploration to their own government companies. India and Mexico are good examples. Exploration activity is usually limited when only one company is involved in a large area; when various companies are present, different geologic ideas are tested, and additional technical and managerial capabilities are applied. It should be noted that Argentina, Brazil, and Chile have recently modified their policies to permit entry of foreign operators in order to attract exploration.

Private companies often may have difficulty negotiating terms which they consider to be compatible with the investment risk they are taking; however, while difficult and time-consuming, these negotiating problems normally can be resolved. Malaysia is a recent example where this has occurred.

In general, government constraints can limit the access to acreage in a particular area at a particular time. This is true in industrially developed countries as well as in developing nations. Availability of acreage is a major factor in determining the number of companies exploring as well as the overall level of activity. Over time, however, these constraints tend to be resolved and exploration proceeds.

OIL PRODUCTION IN DEVELOPING COUNTRIES

Figure 29-3 shows the rapid growth in production for the non-OPEC developing countries as a result of past and future exploration, from a level of 4 million B/D in 1976 to 10 million B/D in 1990. Some growth occurs in all geographic regions. The most rapid acceleration is projected to be in Latin America, with Mexico, which has already had significant discoveries in recent years, providing an important component.

Overall, this rapid rate of growth is in contrast with some other areas where production is projected to be more static during this period, in part because production from previously discovered fields will be in decline. The projected 1985/1990 production level of 8 to 10 million B/D in developing countries will be a significant part of total world supply at that time. In 1990, this amounts to about 14 percent of total free world production, and one-third of free world production exclusive of the OPEC nations.

Of total free world oil consumption in 1990, more than one-quarter may come from reserves yet to be discovered. The developing countries are expected to account for about one-third of this new oil. It is not possible to predict where these discoveries will be made because the estimates are based on a probability assessment of the undiscovered resource base. But it is clear that significant quantities of undiscovered oil and gas exist in the developing countries.

CONCLUSIONS

From this discussion, several conclusions may be reached.

First, there is a significant undiscovered resource base in developing and other countries.

Second, only a few wells are needed to make an initial discovery. This initial discovery in a new area usually sparks activity and many subsequent

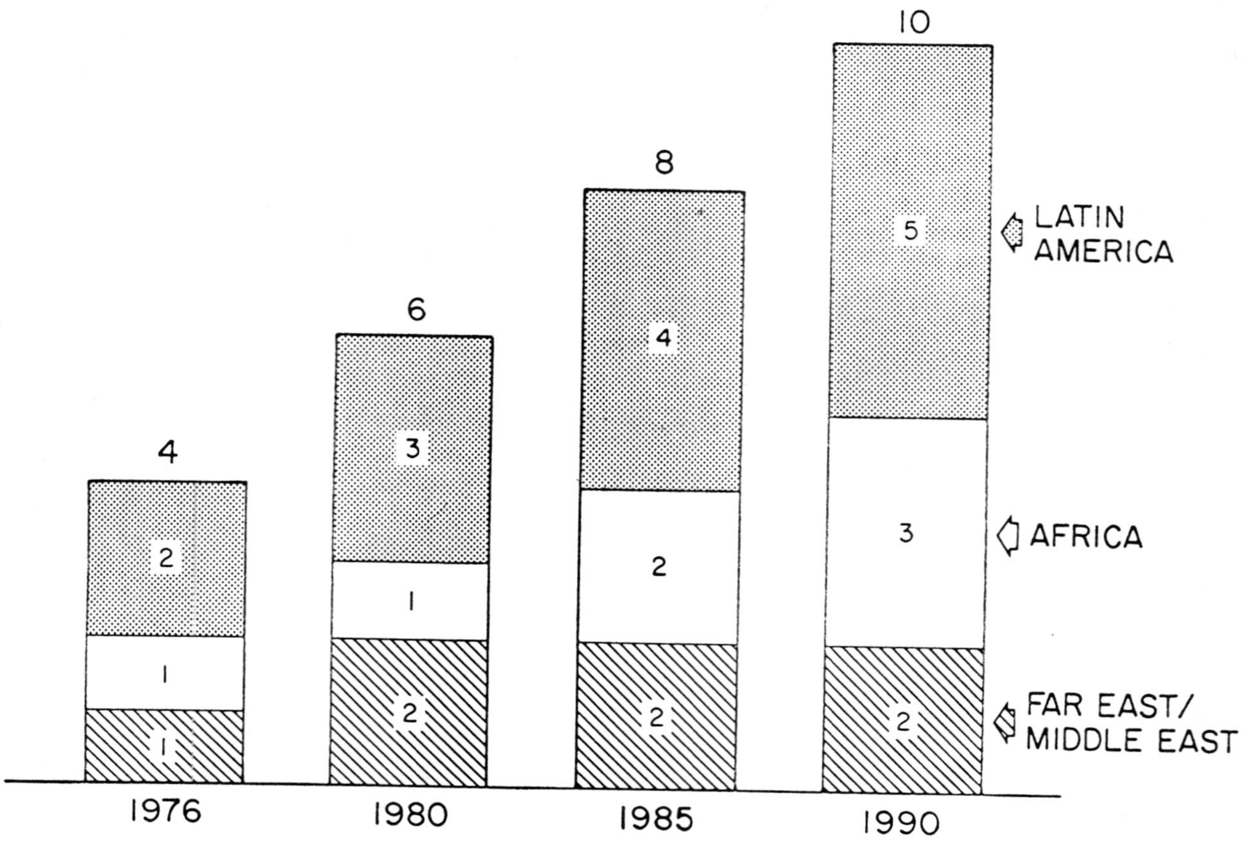

Figure 29-3. Developing countries oil supply, million barrels/day (incl. NGL)

exploratory wells are likely to be drilled to search for additional fields. The largest structures on the available acreage are normally drilled first, and the larger fields are most often discovered early in the exploration cycle. However, the companies will develop whatever sized reserves they find as long as it is economic to do so. Small fields can be important to a developing country and, cumulatively, to the diversity and total size of a company's (and the world's) reserves.

Thirdly, most non-OPEC developing countries have had some exploration (94 out of 113) and 71 have had some exploratory drilling during the last decade. In fact, the total number of exploratory wells drilled in the non-OPEC developing countries over the last decade has been about equal to that in the rest of the non-Communist world, excluding the U.S. and Canada.

Fourth, there are both physical and governmental restraints to exploration in some developing countries. The industry's widespread drilling history suggests that political (or "country") risk is not usually an overriding factor if geologic prospects are good. The situation is dynamic, with conditions in some countries improving and in others becoming more difficult. Government policies affecting access to acreage and the terms of exploration are also important. However, history suggests that, over time, obstacles are likely to be resolved.

Lastly, exploration success has been and will likely continue to be irregular, and results very uncertain. Nevertheless, production in the developing countries is projected to increase in the future, from 4 million B/D in 1976 to 10 million B/D in 1990, assuming governmental restraints do not preclude the activities required.

CHAPTER 30

NATURAL GAS PROSPECTS IN DEVELOPING COUNTRIES AND THE ROLE OF LNG TRANSPORTATION

V. A. Ovcharenko[1]

INTRODUCTION

As the century draws to a close, the thrust of worldwide energy development might have to move increasingly away from oil in the direction of alternative energy supplies. A major contribution may, however, be made by natural gas, conventional and unconventional. Gas expansion in the past has been held back by technological obstacles, which have, however, been greatly reduced, while economic barriers have considerably shrunk. The contrast among gas production rates in the industrialized, centrally planned, and developing countries is striking; 32 percent of commercial primary energy production is accounted for by gas in the industrialized countries, 17 percent in the centrally planned countries, and only 6 percent in the developing countries, with proved reserves of conventional gas constituting nearly 50 percent of the world total.

The lag in natural gas growth in the developing countries results from handicaps that have affected gas on a worldwide basis, but especially in smaller, more remote markets. Unlike oil, gas cannot be widely marketed until a costly pipeline network for local distribution has been built.

The optimism on the future of gas is based partly on technological progress in overcoming the difficulties of transport of natural gas, which is a highly volatile, dispersed fuel; one thousand volume units of natural gas are required to furnish the same calorific value as one volume unit of oil. Improving technology has allowed more than one-half million miles of large diameter gas lines to be laid throughout the major producing areas. Storage, which is likely to emerge as one of the main obstacles to expansion of the gas industries, has been greatly eased by the improvement of refrigeration processes. In the highly developed gas markets, seasonal peaks in demand are now often met by gas drawn from liquefied natural gas (LNG)

[1] Energy Section, Technical Cooperation Department, United Nations.

storage. This technique also makes it possible to transport gas over long distances across the seas. Conversion of gas into methanol and ammonia, which could be transported in conventional tankers, is another promising transportation alternative.

The challenging task of expanding gas production and utilization in the developing countries, especially through an effective use of gas associated with petroleum deposits and modern technological developments, such as the transportation of LNG, will require a substantial reassessment of natural gas prospects and the formulation of long-term technical cooperation policy commitments with regard to investment, manpower, and the transfer of technology. This chapter identifies the basic factors involved in dealing with those problems.

EXPANSION OF TRADE IN NATURAL GAS

Production, Consumption, and Role of International Trade

Past history suggests and latest forecasts confirm that, over the period through 1985, natural gas will be the fastest-growing energy source in the developing countries, as well as in the world. [1-5] The growth of natural gas production over the last two decades has already been impressive in certain developing countries. Gas production in the non-OPEC developing world reached a total of above 71 billion m^3 in 1978 (Table 30-1). This is more than a 20-fold rise compared with the level of production in 1955.

Figures for the latest two years, 1977 and 1978, are less complete than for earlier years and are based partly on estimates but Table 30-1 indicates that in recent years, growth of gas production in the non-OPEC developing countries has slowed down significantly, especially in Latin America and the Caribbean, and only since 1977 is the upward trend gaining momentum. It can be assumed that further investments in transport and distribution network would further increase production at a faster rate.

The possible growth of gas supplies in the non-OPEC developing countries to the year 1985 is outlined in Table 30-2. Within this group, those importing oil consist of some 70 developing nations, most of them with very little known endowment of oil or gas. The gas surplus available for export from the oil-importing developing countries will be smaller than in the case of the oil-exporting non-OPEC developing countries, although it is expected to double by 1985. Taking the two groups together, it is projected that the

TABLE 30-1. Total production of natural gas in non-OPEC developing countries, by region, 1974-78[a]
($10^9 m^3$)

Country/region	1974	1975	1976	1977	1978
Latin America and Caribbean					
Mexico	20.8	16.9	17.2	20.5	26.5
Trinidad and Tobago	1.4	1.5	1.5	1.7	1.8
Argentina	7.2	6.3	7.5	9.7	7.8
Brazil	1.5	0.6	0.6	0.6	0.8
Bolivia	4.1[b]	1.8	1.9	1.8	1.9
Chile	3.6	3.6	3.6	3.7	3.5
Colombia	2.6	2.1	2.2	2.3	2.4
Peru	1.0	0.9	0.9	0.8	0.9
Subtotal	42.2	33.7	35.4	41.1	45.6
Middle East					
Bahrain	1.6	2.1	2.2	2.4	2.7
North Africa					
Egypt	1.4	0.4	0.4	0.5	0.8
Tunisia	0.2	0.2	0.2	0.2	0.3
Morocco	0.1	0.1	0.1	0.1	0.1
Subtotal	1.7	0.7	0.7	0.8	1.2
East Asia					
Brunei/Malaysia	6.0	6.0	7.8	8.9	8.9
Taiwan	1.4	1.9	2.0	2.1	2.3
Subtotal	7.4	7.9	9.8	11.0	11.2
South Asia					
India	0.7	1.0	1.2	2.1	2.1
Pakistan	4.0	5.0	5.2	5.9	5.7
Afghanistan	2.9	3.0	3.0	3.0	2.9
Subtotal	7.6	9.0	9.4	11.0	10.7
Total	60.5	53.4	57.5	66.3	71.4

[a]Excluding flared or reinjected.

[b]Including gas repressured and wasted.

NOTE: Some small gas producers with noticeable proven reserves, such as Bangladesh, Thailand, Burma, Israel, Sudan, and others are not reflected in this table because of space limitation.

Source: Petroleum Economist, July 1976, July 1977, September 1978, and August 1979.

TABLE 30-2. Projected commercial production of natural gas in non-OPEC developing countries by region by the year 1985
($10^9 m^3$)

Country/region	1978*	1978	1982	1985
Latin America and Caribbean				
Mexico	26.5	24.1	42.5	56.6
Trinidad and Tobago	1.8	0.7	1.0	1.4
Argentina	7.8	8.2	12.0	14.0
Brazil	0.8	0.6	0.8	2.1
Bolivia	1.9	1.6	3.5	4.2
Chile	3.5	3.8	5.2	7.1
Colombia	2.4	5.2	7.8	8.5
Peru	0.9	0.6	0.8	1.1
Subtotal	45.6	44.8	73.6	95.0
Middle East				
Bahrain	2.7	2.4	2.1	2.1
North Africa				
Egypt	0.8	0.1	1.4	2.8
Tunisia	0.3	0.3	2.1	3.3
Morocco	0.1	0.1	0.1	0.1
Subtotal	1.2	0.5	3.6	6.2
East Asia				
Brunei	8.8	7.4	8.1	8.5
Malaysia	0.1	-	8.5	9.9
Thailand	-	-	1.4	2.1
Subtotal	8.9	7.4	18.0	20.5
South Asia				
Sri Lanka	-	-	2.1	4.1
India	2.1	1.4	28.3	28.3
Pakistan	5.7	5.7	8.5	10.3
Afghanistan	2.9	3.3	5.4	7.1
Bangladesh	0.6	0.8	2.8	5.7
Viet Nam	-	-	1.4	2.8
Subtotal	11.3	11.2	48.5	58.3
Total: Non-OPEC developing	69.7	66.3	145.8	182.1

*Petroleum Economist, August 1979.

NOTE: Table does not reflect such countries as Oman, Syria, Cuba, Angola, Congo, Burma, and some others, which are making a vigorous development effort to increase or initiate gas production.

Source: Petroleum and Gas in Non-OPEC Developing Countries: 1976-1985; World Bank Staff Working Paper No. 289, Washington, D.C., April 1978.

non-OPEC developing countries will raise their gas demand to 122 billion m^3 by 1985, compared with the projected production level of 182 billion m^3, leaving a balance of above 60 billion m^3 available for export in 1985.

Progress expected in the major oil-exporting regions provides further grounds for optimism on the prospect for gas expansion in the developing countries. An example is provided in Table 30-3, which gives estimates that the local demand for gas in the 13 OPEC member countries might rise more than fourfold during the period through 1985 to reach a total of 220 billion m^3. But the surge in commercial gas production in those 13 countries is expected to be even greater, with a sixfold growth to 340 billion m^3, as indicated in the table. Consequently, the surplus available for export will rise more than twentyfold to reach 120 billion m^3 in 1985. The member countries of OPEC have already embarked on international pipeline and liquefaction projects that would allow this large surplus to be absorbed on the world market, some of it in developing countries that remain unsuccessful in their efforts to expand their indigenous energy supplies.

TABLE 30-3. Estimated natural gas supply and demand in OPEC developing countries, 1974, 1980, 1985
($10^9 m^3$)

	1974	1978*	1980	1985
Indigenous production	56.6	81.4	161.4	339.8
Net exports	5.7	25.2	36.8	118.9
Calculated demand	50.9	56.2	124.6	220.9

*Petroleum Economist, August 1979.

Source: The OECD reference case is used. Computed from World Energy Outlook, Organization for Economic Cooperation and Development, Paris, 1977.

Although world trade in natural gas is as yet not more than a fraction of the trade in oil, some regions have already established an important international trade in gas, which is projected to increase significantly in the future (Table 30-4). The volume of 128 billion m^3 shown in the table is predicted to rise more than 70 percent by the year 1980.[2]

Problems of Transportation in National and International Gas Trade

At present, the international trade in natural gas constitutes about 10 percent of world gas production and is growing noticeably faster than the average growth rate of production. Long-distance pipelines have made possible these increasingly valuable flows. Transportation of natural gas in liquefied form (LNG) had also begun to develop strongly in recent years (Table 30-5). Tables 30-6 and 30-7, which were prepared on the basis of the most recent forecasts, indicate the possible scope for LNG exports over the next few years. Pipeline exports would also rise, with both the Soviet Union and the OPEC countries expected by 1985 to be exporting more than 55 billion m^3. The 1985 total of international gas movements of more than 250 billion m^3 indicates the scope for rapid growth in this new type of traffic. Achievement of such large movements, especially in the form of LNG, will require construction of major new projects to begin in the very near future; a 7-year lead time is often required between the start of construction and full operation of a sizeable LNG venture.

Assuming that the necessary decisions and authorizations do not introduce further costly delays, present indications are that LNG deliveries, and certainly pipeline deliveries, will grow strongly and will remain attractive to major gas consuming countries. Yet major economic and policy unknowns must be removed before the future pace of LNG progress can be assessed with any confidence. Some of these uncertainties will be discussed further.

PROSPECTS AND PROBLEMS OF LNG TRADE

The progress in shipping liquefied natural gas (LNG) from remote producing areas to major consuming centers in specially designed cryogenic tankers might have great importance for the long-term growth of the gas industry in the developing world. The international trade in LNG is

---

[2]The Petroleum Economist estimations of world total natural gas trade for 1977 and 1978 of 147 and 158 billion m^3 a year respectively, published in August 1979, compare rather well with those in Table 30-4.

TABLE 30-4. World natural gas trade, 1976 and 1980
($10^9 m^3$/year)

Producer country/area	Consuming country/area	1976[1]	1980[2]
North America			
Alaska	Japan	1.40	1.38
Canada	USA	25.49	14.56
Western Europe			
Netherlands	W. Europe	49.03	50.95
North Sea	W. Europe	--	15.59
Norway	W. Europe	--	23.38
Latin America			
Bolivia	Argentina	1.43	2.86
	Brazil	--	2.45
Colombia	USA	--	1.53
Africa			
Algeria	W. Europe	4.93	9.72
	USA	0.26	11.45
Libya	Italy	2.60	2.40
	Spain	1.25	1.12
Middle and Far East			
Indonesia	Japan	--	12.78
Brunei	Japan	6.38	7.67
Abu Dhabi (UAE)	Japan	--	5.11
Iran	USSR	9.44	9.91
Afghanistan	USSR	2.52	2.96
Eastern Europe			
U.S.S.R.	E. Europe	12.41	20.24
	W. Europe	11.12	23.81
Grand total		128.26	219.87

[1]Source: World Energy Supplies 1972-1976, United Nations, New York, 1978.

[2]Source: Proes, Nevil, "Is LNG still a viable option?," Hydrocarbon Processing, April 1977.

TABLE 30-5. LNG projects currently in operation

Project	Term	Volume ($10^9 m^3$/year)	Estimated LNG c.i.f.[1] US$/$10^6$ Btu	Effective year	US$/ton	Remarks
Algeria-United Kingdom	1964-1979	1.0	1.00	1977	51.25	Maximum c.i.f. cost $1.15
Algeria-France	1965-1990	0.5	1.50	1977	76.88	Contract extended 10 years in July 1976
Alaska-Japan	1969-1984	1.4	1.70	until end 1976	87.13	C.i.f. price adjusted each year starting 1973 from the initial 52 cents
Libyan Arab Jamshiriya-Italy	1971-1991	3.0	1.52	1977	77.90	C.i.f. price raised periodically starting 1971 from the initial 35.5 cents
Libyan Arab Jamshiriya-Spain	1971-1986	1.4	1.25	1975	64.07	C.i.f. price raised periodically starting 1971 from the initial 37.4 cents
Algeria-France	1972-1997	3.5	1.40	1977	71.76	Original c.i.f. price 41.6 cents
Brunei-Japan	1972-1992	7.4	1.90	1976	97.38	Original c.i.f. price 40.6 cents
Algeria-Spain	1976-1996	4.5	1.47	1977	75.34	Full delivery to start in 1979
Abu Dhabi (United Arab Emirates)-Japan	1977-1997	2.9	2.00	1977	102.51	Agreed c.i.f. price 1978/79-1980 to be based on the average cost of all crude imported into Japan
Indonesia-Japan	1977-1997	10.5	2.37	1977	121.47	In addition, a further volume of 5.0 billion m^3 may be made available
Algeria-United States (Distrigas)	1978-1998	1.2	0.98 2.25	1976 1978	50.23 115.32	Replaces original 1971 contract and interim short-term contract of 1974
Algeria-United States (El Paso)	1978-2003	10.0	1.24 1.90	 1978	63.56 97.38	Initial contract signed in 1969
Total trade		47.3				

[1]Cost, insurance, and freight.

Sources: Petroleum Economics Limited, Oil Industry Developments, September-October 1978 (London, 1978); "Prospects seem brighter for U.S. LNG and SNG," Oil and Gas Journal, 16 January 1978; "IGT: LNG trade could hit 28 billion cfd," Oil and Gas Journal, 14 August 1978; "The LNG market: Potential for the future," Petroleum Economist (November 1978).

TABLE 30-6. LNG projects firmly committed or under construction

Project	Term from - to	Volume ($10^9 m^3$/year)	Comments
Algeria-Belgium	1982-2002	5.0	Belgium to loan $325 million to Algeria toward LNG plant cost
Algeria-W. Germany/ Netherlands	1984-2004	8.0	Half of the volume is to be imported jointly with the Netherlands
Algeria-W. Germany	1984-2004	4.0	
Algeria-France	1980-2000	5.15	
Algeria-USA	1980-2000	4.5	
Indonesia-USA	1982-2002	5.5	
Malaysia-Japan	1983-2003	8.5	
Iran-Japan	1982-2002	4.0	
Total trade		44.65	

Sources: Oil Industry Developments, September-October 1978, Petroleum Economics Limited; "Prospects seem brighter for U.S. LNG and SNG," The Oil and Gas Journal, January 16, 1978; "IGT: LNG trade could hit 28 billion cfd," The Oil and Gas Journal, August 14, 1978; "The LNG Market: Potential for the Future," Petroleum Economist, November 1978.

TABLE 30-7. Probable LNG projects

Project	Term From - To	Volume ($10^9 m^3$/year)	Remarks
Alaska-USA	1980-2000	4.0	Initial volume will be 2 billion m^3 per year
Chile-USA	1982-2002	2.5	
Iran-USA	1982-2004	3.0	
Algeria-USA (via Canada)	1982-2002	10.6	U.S. Government denied approval for the project, as reported in the New York Times, 19 December 1978
Sarawak-Japan	1983-2003	10.0	
Algeria-USA (El Paso II)	1984-2004	10.3	U.S. Government denied approval for the project, as reported in the New York Times, 22 December 1978
Algeria-Sweden	1985-2005	2.0	
Nigeria-USA/ Western Europe	Mid-1980's for 20 years	16.0	
Trinidad-USA	Mid-1980's for 20 years	5.0-10.0	
Australia-Japan/ USA	Mid-1980's	9.0	
Canada (Melville Island)-Canada (East Coast)	Mid-1980's	2.5	
Qatar-Japan	Mid-1980's	8.0	
Colombia-USA	Mid-1980's	1.5	
USSR-USA	Mid-1980's for 25 years	10.0	
USSR-Japan	Mid-1980's	10.0	
USSR-USA/Europe	Mid-1980's for 25 years	20.0	Volume: USA-15 billion, France-5 billion

TABLE 30-7. Probable LNG projects--Continued

Source: Oil Industry Developments, September-October 1978, Petroleum Economics Limited; "IGT = LNG trade could hit 28 billion cfd," The Oil and Gas Journal, August 14, 1978; "The LNG Market--Potential for the Future," Petroleum Economist, November 1978; "U.S. Energy Future--Higher LNG imports will be needed," World Oil.

constantly growing. Japan imports about 80 percent of its gas in liquid form and Western Europe, 5 percent.

The countries with significant quantities of gas resources for export consist of such developing countries as Abu Dhabi (UAE), Algeria, Colombia, Indonesia, Iran, Nigeria, Saudi Arabia, Venezuela, Qatar, and others. Australia, Canada, and the USSR also possess significant amounts of exportable gas. Currently, the OPEC countries are supplying over 60 percent of internationally traded LNG and those countries have commitments which will increase their share to 85-90 percent by 1985.

<u>LNG Projects Around the World</u>

In 1964, the first regularly scheduled shipments of LNG to the United Kingdom and France began from an LNG plant built in Algeria. This project has been in continuous operation for nearly 15 years and has delivered 1 billion m^3 per year to the British Gas Council and 0.5 billion m^3 per year to Gas de France at Le Havre each year. Since 1964, there have been over 2,500 ocean shipments of LNG without serious accident.

From its beginning, LNG trade has grown dramatically and is still growing (Tables 30-5 through 30-7). But to many experts, LNG has not lived up to the expectations and many projects have been delayed and some cancelled.

There are adequate proven reserves of natural gas to support the projected LNG exports, which are summarized in Table 30-7, but conditions affecting the key exporting and some importing countries are highly influenced by political considerations and other factors.

The OPEC countries in general are reluctant to allow significant increases in LNG exports. Algeria, the pioneer of LNG, has committed virtually all its available gas. Saudi Arabia does not seem willing to appear on the LNG export market until its extensive domestic gas distribution and utilization plans are realized. Iran, which has the second largest gas reserves in the world after the USSR, has followed a recent policy to permit limited additional gas exports but to use gas primarily for local needs and for massive reinjections to maximize oil production. Iran's position was being reconsidered at the time (late 1979) of the writing of this report. Nigeria is the only OPEC member actively promoting gas exports to the United States and Western Europe.

The USSR is willing to increase gas exports to Western Europe, Japan, and the United States. Western Europe is using this opportunity by increasing its pipeline imports. Political considerations, such as granting by the USA of the "most favored nation" status to the USSR, and long-term credits apparently have to be resolved before a commercial LNG deal can be realized.

Mexico might appear in the future as a possible source of LNG supply, if its hopes to expand natural gas export by pipeline to the USA are not realized.[3]

The United States Congress has recently passed an energy bill and an important consideration for LNG trade is the American decision to remove new natural gas from price controls by 1985, about the time of the start-up dates of many possible or probable LNG import projects (Table 30-7).

Western Europe is actively on the market for LNG. Between 1985 and 1990, Netherlands' gas export contracts terminate and will not be renewed (totalling about 145 billion m^3 per year). Western European countries are therefore expected to increase their dependence on imported gas from the USSR and the Middle East countries. It will be seen from Table 30-8 that Western Europe might increase significantly the volume of LNG trade by 1985.

TABLE 30-8. Estimation of market economies' LNG demand potential to 1985
($10^9 m^3$/year)

Consumer country/area	By 1980	Percent of total	By 1985	Percent of total
USA	15.7	28.0	98.1	55.4
Japan	22.2	39.7	33.3	18.8
W. Europe	18.1	32.3	45.7	25.8
Total	56.0	100.0	177.1	100.0

Source: Based on data provided in Tables 30-5, 30-6, and 30-7.

[3] On September 23, 1979, an agreement was reached for importation from Mexico into the USA of 8.4 million m^3 per day. It is about 10 times less than was originally planned.

Japan continues to develop the importation of LNG although its share in world trade might decrease by 1985. The potential sources of supply for Japan are the USSR, Malaysia, Australia, and the Persian Gulf countries. The main end users of LNG are the electric power companies. Japan may attempt to redirect its efforts over the next decade by switching to some extent from existing LNG imports to the use of liquid petroleum gases (LPG). The large surplus of LPG production during 1980-1982, clearly points out the advantages for making such a policy switch for Japan.

Against this background, what are the prospects for LNG over the next decade? Table 30-8, based on data provided by an overview of firmly committed and probable LNG projects (Tables 30-6 and 30-7), suggests that total LNG imports are projected to more than triple. The potential markets should not change significantly. The prospects for LNG in relation to imported oil will probably improve in the future. Security of supply, diversification of supply sources, technical reliability, environmental problems, governmental regulations, sources of financing, and LNG prices will be the main considerations with regard to the LNG option. Some of the most important LNG considerations will be analyzed below.

Prospects for LNG Technology and Problem of Regulation

Although LNG facilities have been in extensive commercial operation for about 15 years, the evolution of liquefaction equipment, storage tanks, LNG tankers, and regasification equipment is likely to continue unabated.

It is not possible here even to summarize the main trends in the very dynamic and fast-changing LNG technology. [8] Apart from the natural desire dictated by economies to improve the means of transporting and storing this rather different product, the possible consequences of a spillage have increasingly affected various LNG equipment designs in recent years. More than 100 LNG storages and some 20 import/export terminals have been constructed since LNG commercial inception in 1964. Large volumes of LNG were handled worldwide during the past 14 years with a remarkable safety record. [9] But the problem of obtaining suitable sites in the USA and European countries has become one of the difficulties affecting the governmental approval of LNG projects.

On the basis of studies that have been made of the hazards of liquefied natural gas and the experience that has been gained in several countries in handling the fuel on a large scale, it was estimated that for one particular LNG site near New York, USA, the overall risk is about 10 times less than the risk of dying from a fire at home and about the same as the risk of being struck by lightning. [10]

Similar examples reveal that society tolerates certain highly improbable risks--risks that cannot feasibly be eliminated by design--in order to reap the benefits of energy, consumer products, transportation, central water-supply systems, and so on.

It seems reasonable to say that hazards involved in handling other fuels such as gasoline, propane (liquefied petroleum gas), and natural gas are similar to those in LNG. Moreover it can be expected that further experience in handling liquefied natural gas and further research by industry and regulatory agencies will raise the level of safety in its use even higher. It seems that technology and environmental regulations are not real obstacles for LNG expansion.

International Dimension of Financing LNG Projects

In spite of rapid expansion, the rate of the international LNG trade development has been significantly behind predictions made a few years ago. [11] One of the main reasons for that is the complexity of LNG financing. LNG ventures differ in several ways from the more traditional investments made in the oil industry. The required financing of LNG is massive, with total costs for one complete system amounting to billions of U.S. dollars. This invariably means that several investors may be needed for one project.

The potential market for LNG supply up to 1985 (Table 30-8) is mainly considered in the major industrialized countries and a substantial contribution to the financing of LNG will presumably be made by those countries. But a significant part of the network will be located and part of the finance will be sought within the boundaries of the producing countries.

Present analysis suggests that 2.5 billion m^3 yearly represent the minimum economic size for an LNG project and a capacity of 10 billion m^3 yearly is regarded as the most attractive scale. For a system having the latter capacity, as much as U.S. $2.5 billion may be required in capital outlays. The total may be expected to break down approximately as follows: $300 million for gas production and delivery to the liquefaction plant; $1 billion for the liquefaction plant and marine terminal; $900 million for marine transportation; and $300 million for the receiving terminal and regasification plant.

The length of the voyage is the key variable in an LNG investment. For a relatively short link, as from the Caribbean to North America, some six cryogenic tankers (costing perhaps $600 million) would be sufficient to assure the service for a project of an average size. But for a longer voyage, as from the Middle East to Japan, two or three times as many tankers will be required, calling for an investment of $1.25 billion or more in marine transportation alone. The cost of shipping liquefied gas is 6 to 8 times higher than that for equivalent volumes of crude oil. Of the total LNG investment, the portion representing the receiving and regasification plant, and usually the entire tanker fleet, would have to be financed in the importing country or in the country in which the tankers are built. It could be expected that part of the gas production, liquefaction, and transportation investment would have to be financed by the exporting country.

The sources of finance for these many and massive investments in the natural gas industry have to be analyzed in the context of individual national circumstances and the possible speed and scale of each producing country's advance. For the front-end investment it may be in the national interest to provide indigenous financing as far as possible by the producing countries themselves. But budgetary restraints may often make it necessary for them to seek the maximum degree of international participation.

In many cases, private investors are not willing or capable of taking the risk, and government guarantees for export credits or in other forms will be sought in developing and industrialized countries alike. Often, such international guarantees are very difficult to negotiate, since huge front-end investments will depend on contracts written 5 to 7 years before the project can come on stream.

Furthermore, the conditions of the contracts should then remain fair to all parties for 20 years or so. In contrast, crude oil contracts between buyer and seller today seldom cover more than 1 or 2 years, and usually do not require such large front-end investments. Moreover, LNG is a capital-intensive industry but provides little employment and to justify the benefits, long-term purchase commitments with an acceptable unit price are needed.

International cooperation seems to be essential for promotion of LNG trade through an exchange of experience, joint research and development, standardization, joint infrastructure projects, and investment sharing, based on internationally accepted strategy for development and guarantees for LNG long-term price mechanism to be competitive with imported oil or any other alternative fuel.

The Problem of LNG Prices

The landed prices of LNG currently moving in international trade on a regular basis (Table 30-5) vary significantly but stay below the equivalent cost of low-sulfur crude. The importing countries are objecting to LNG price linkage with imported oil.

Refusal by the U.S. authorities to admit the principle of equivalency between imported oil and LNG has so delayed plans to import Algerian gas that for some new contracts, Algeria has turned to European customers instead. Japan, given its specific situation, by and large accepts that, in terms of energy equivalent, LNG should be delivered at much the same price as competitive crude. Nevertheless, the average price for LNG delivered to Japan from all sources in 1977 and 1978 was still below the value of crude oil.

Estimates by the Iranian Gas Company (NIGC) indicate that LNG from Iran landed on the east coast of the United States would cost $3 to $3.25 per million Btu, and in Western Europe and Japan $2.65 to $2.90 at 1976 price levels, allowing Iran revenues of from 16 cents to 26 cents per million Btu. These figures cover the processing, at both ends, and transportation costs. The prospective annual revenue for LNG exported from Iran to Japan, assuming a start in 1980 at, say 20 cents per million Btu to the exporting country, would be about U.S. $30 million net a year for a project capacity of 4 billion m^3 yearly, while capital outlays for LNG projects of such capacity would exceed U.S. $1 billion.

This indicates how expensive it is to move natural gas internationally and why all natural gas exporting countries demand that importing countries facilitate the arrangements. The gap between significant oil revenues and relatively small LNG receipts to the exporters explain why only a limited number of oil- and gas-producing countries feel any urgency to the rapid expansion of LNG deliveries. In the absence of international agreements on

LNG prices and guarantees, which is a very complex issue, most of the oil- and gas-producing countries (it may be reasonably predicted on the analogy of oil in the first half of the 1970's and worldwide concern for conservation) might seek to use any serious gas shortages in the future to lever upward the producer's share of revenues.

SOME CONCLUSIONS AND POLICY IMPLICATIONS

As indicated above, a substantial part of the growth of conventional natural gas production during the next 10 to 20 years is likely to occur in the developing countries, where gas resources and their utilization have hitherto been inadequately investigated and developed. For natural gas producers with extensive resources but limited local usage owing to infrastructure limitations or other factors, export markets--by overland pipeline if possible, and otherwise in the form of LNG--could be considered to benefit from the scale effect. In some cases, especially when expected market is very far away, it may prove advantageous to pursue gas conversion into methanol versus the LNG approach, but expansion of local gas utilization in developing countries should be considered as a main policy objective.

The technology for LNG transportation on the seas and for storage is commercially available, technically advanced, and probably does not represent a significantly higher level of risk to the public and the environment than handling of conventional fuels, when internationally recognized safety measures are taken.

To facilitate natural gas expansion in developing countries, especially through modern technological developments such as the transportation of LNG, international cooperation assistance should be made available on a priority basis for the transfer of technology, training, advice on national policies, etc.

The very nature of the highly capital-intensive and technically complex LNG projects calls for an international approach to their financing and for providing joint responsibilities to the producing and consuming countries for successful and efficient operation. One of the most important aspects in the reliability of these projects to both sides is that they be economically sound.

REFERENCES

[1] Petroleum and Gas in Non-OPEC Developing Countries: 1976-1985, World Bank Staff Working Paper No. 289, Washington, D.C., April 1978.

[2] Rapid Development in Ocean Transport of LNG and LPG to Europe: International Petroleum Times, 15 June 1978.

[3] Gas Supply/Demand Gap Forecast for W. Europe: Oil and Gas Journal, 14 August 1978.

[4] Petroleum Industry Research Foundation, Outlook for World Oil into the 21st Century, prepared for Electric Power Research Institute, Palo Alto, California, May 1978.

[5] Brookhaven National Laboratory, Energy Needs, Uses and Resources in Developing Countries, prepared for the United States Agency for International Development, Upton, New York, March 1978.

[6] Mexico: Further Upgrading of Reserves: Petroleum Economist, October 1978.

[7] R. Sumpter, US gas supply/demand seen nearing balance: Oil and Gas Journal, September 25, 1978.

[8] There are extensive bibliographies on technical and economic aspects of LNG. For example, the U.S. Congress Office of Technology Assessment published in 1977 a comprehensive review, which analyzed current LNG technology and possible trends in LNG use and also identified and discussed some major policy issues surrounding LNG use in the USA. (U.S. Congress Office of Technology Assessment, Transportation of Liquefied Natural Gas, Washington, D.C., 1977).

[9] S. Wolf, Liquefied natural gas: The Bulletin 21, December 1978.

[10] E. Drake and R. Reid, The importation of LNG: Scientific American, 236, no. 4, April 1977.

[11] N. Proes, Is LNG still a viable option?: Hydrocarbon Processing, April 1977.

CHAPTER 31

FRONTIER TRANSPORTATION

H. Oliver[1]

Canada's northern areas are a vast cold and hostile environment. Once beyond the tree line we are faced with an area of some 1.3 million km^2 (one-half million mi^2) which consists of a polar desert with very little vegetation and limited wildlife. Included in this area are the Canadian Arctic Islands, which extend from the Canadian mainland to the polar ice cap. The wildlife that exists in the area has adapted to the winter climate and includes the polar bear, arctic fox, wolf, muskox, caribou, and, in the summer, migratory wild fowl.

The area presented a challenge to earlier day explorers searching for a northwest passage or striving to reach the "pole." Initially, John Cabot set out from Europe in 1497 to find the northern passage to India. He managed to reach Cape Breton. In 1576, nearly a century later, Martin Frobisher explored the coast of Baffin Island. Sir John Franklin's three-ship expedition vanished in 1845; this led to about 40 more expeditions seeking to discover the fate of Franklin. These voyages gradually broadened our knowledge of the northern areas. Others followed, pushing gradually across the top of Canada or toward the North Pole. Amundsen was first to sail the Northwest Passage in 1903, taking 3 years to complete the journey. In 1944, an 80-ton Canadian Mounted Police vessel, the St. Rock, made it in a single year.

Land transportation during this time was fairly restricted--eskimo dog train during the winter or canoe during the summer along the coastal areas or the rivers.

The early 1950's saw a greater emphasis on northern development. This derived from the building of the D.E.W. line (Distant Early Warning System), which gradually brought about the development of vehicles that could be used in the northern climate and from the petroleum geologists who saw the Arctic as a potential hydrocarbon-producing area. In 1962, the first well was drilled, at Winter Harbour. Transportation was a key factor in this

[1]Petro-Canada, Calgary, Alberta.

development. Initially the Douglas DC-3 from World War II was used as the work horse, with smaller planes as backup; vehicles such as the Bombadier and Skidoo were also developed as ground transportation. As activity increased, the DC-4, DC-6, and Convair Electra were used for ferrying men and materials to these northern sites. The Twin Otter, with wheels, skis, or floats, has proven to be a reliable aircraft for short-range transport. More recently, the use of the Boeing-727 for passengers and freight, and the Hercules for heavy freight, meant that a continuous, reliable air transportation system was in operation.

Following the early exploration at Winter Harbour on Melville Island in the early 1960's, work went on until 1969, when Panarctic Oils made the first major natural gas discovery, at Drake Point on Melville Island. Other gas discoveries followed, two of them being amongst the largest gas fields in Canada. Discoveries at Ellef Ringes and King Christian Island added to the potential of the area. Oil at Bent Horn, while not living up to expectations as yet, may prove to be another source of hydrocarbon energy. Discoveries in the Mackenzie Delta area and the Beaufort Sea have added to the energy resources of the Far North.

Panarctic Oils set up a main staging camp at Rae Point in the early 1970's in order to centralize its operations. This camp is supplied by a ship convoy in late summer each year. Fuel supplies, buildings, and ground transportation vehicles are taken in by this mode. Most ships utilized for this journey have some strengthening of their hulls in order to help them through icy water. The use of ice-breakers from the Canadian Coast Guard helps the convoys in and out of Rae Point, Resolute, and other settlements during the short summer weather window.

Figure 31-1 is a map of Northern Canada with the locations of the various areas where natural gas or oil have been found. With the increase in the price of energy during the mid-1970's, the need for energy from the High Arctic for Canada's economic well-being in the 1980's and beyond is clearly seen. Current estimates of the gas reserves amount to some 360 million m^3 (12.7 trillion ft^3) in the High Arctic and 158 million m^3 (5.6 trillion feet) in the Mackenzie Delta area. These figures do not include other sources, such as the Beaufort Sea or the White Fish development in the Sverdrop Basin. With Canada's total demand for energy, including exports, increasing, methods for bringing these resources to market have to be developed.

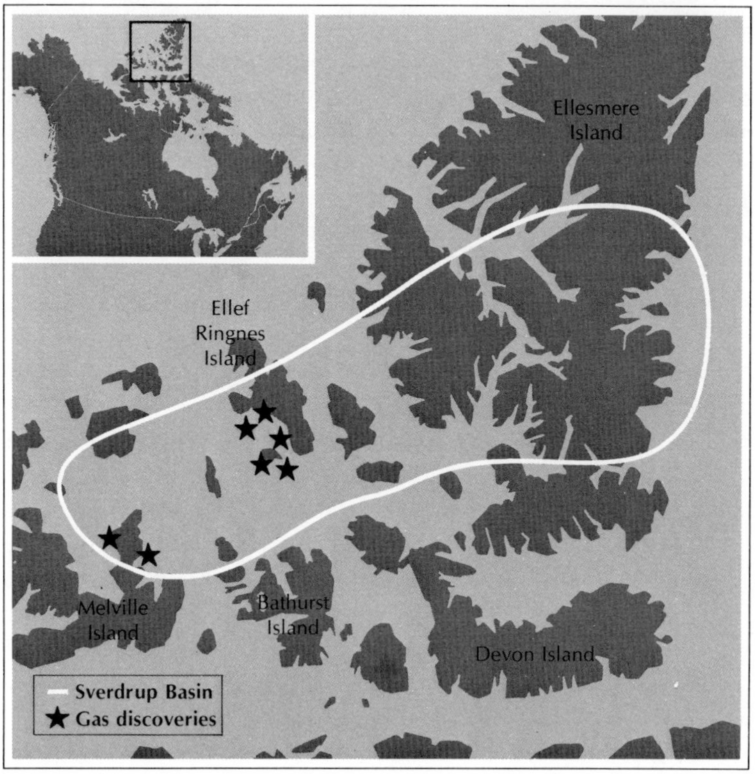

Figure 31-1. Known occurrences of natural gas and oil in northern Canada

RANGE OF OPTIONS

The physical location of the resource areas presents problems that would be huge by any stretch of the imagination. Coupled with the environmental conditions that must be faced, the task is immense. There are several forms and modes by which gas may be transported; these include:

1. Large diameter pipelines
2. Liquefied Natural Gas (LNG) tankers
3. Large, specially built aircraft carrying LNG
4. Converting the gas to methanol for transportation by either pipeline or tanker
5. Electrical generation and transmission.

A cursory glance at these options shows the following.

In order to bring electrical power some 3,700 km (2,300 mi) from Melville Island to the Winnepeg area of Canada for distribution, the generating and line losses would result in about a 70 percent loss of energy. The production of methanol from the gas, while converting it to a relatively easy product to handle, results in losses of 65 percent. The use of aircraft or dirigibles to carry methanol or LNG offers another unique transportation method which would require years of research and development. This leaves the most likely alternatives--pipelines and LNG tankers. Of these, a pipeline system would be the most efficient, delivering about 92 percent of the gas to market, while LNG tanker would deliver about 85 percent. Figure 31-2 shows the energy efficiency curves for all these methods of transportation.

The two alternatives currently proposed are the Polar Gas Pipeline System and the Arctic Pilot Project. These two systems are really complementary, the Arctic Pilot Project complementing the Polar Gas Pipeline.

Figure 31-3 shows another map of Canada and the routes proposed by Polar Gas, together with the shipping route proposed by the Arctic Pilot Project. The Polar Gas route enables the pipeline to pick up gas from the Mackenzie Delta area and transport it, as well as the High Arctic gas, to southern markets with only two water crossings. Recent deep-water pipeline technology makes feasible the crossing of McClure Strait, between Melville Island and Victoria Island. The water depth there is some 500 m.

Any undertaking in the Arctic region is large by any standard. Polar Gas, which was started in 1972, has been responsible for numerous technological

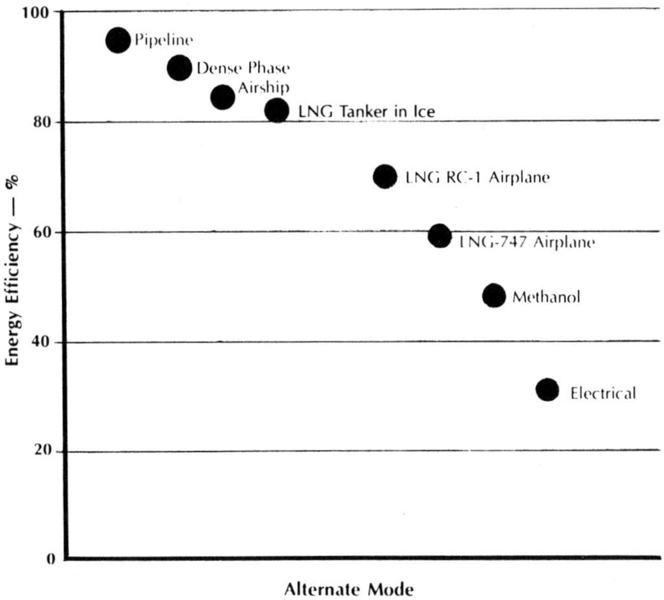

Figure 31-2. Energy efficiency curve for various methods of transportation

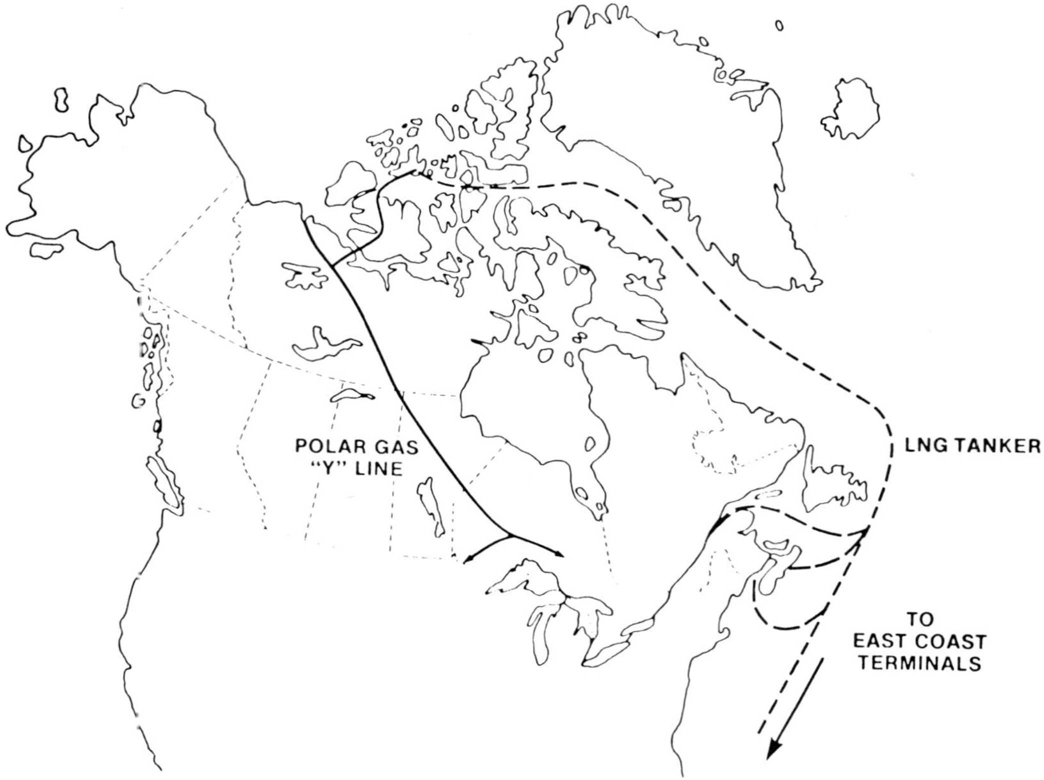

Figure 31-3. Transportation routes from the Arctic

developments as well as a deeper comprehension of the Arctic environment and especially the impact of a major construction project on the Arctic tundra and permafrost areas.

The Arctic Pilot Project, which is a joint venture of Petro-Canada, Alberta Gas Trunk Line, and Melville Shipping, is an alternative method to transport the Arctic resources to southern markets. The project is designed to produce and liquefy 7 million m^3 per day (250 million ft^3) of gas and move it to eastern Canadian markets in ice-breaking tankers. The ships for such a system would carry 140,000 m^3 of LNG and would be of Arctic class 7 design with 180,000 shaft horsepower. Thus, the vessel would have the capability to move through ice 7 ft thick. The project is unique in that it is intended to be a year-round operation, putting Canada in the forefront of Arctic shipping expertise and permitting development of other resources in the northern areas, such as iron, lead, and copper.

Shipping transportation in the Arctic will require present knowledge to be pushed further and further. Extensive weather navigation and ship design is currently being undertaken by the Arctic Pilot Project team. Additional research will continue in the areas of sensing, positioning, and communication, to enable these large vessels to navigate in Arctic ice on their 4,500 km (2,800 mi) journey.

In discussing frontier transportation, mention should be made of the possible use of large air-cushion vehicles. Small hovercraft have been part of the Arctic scene on and off for the past 10 years. The possible use of air-cushion vehicles capable of moving 1,500 to 3,000 tons of material at a time over the ice may also contribute to the transportation picture in the near future. The concern of all parties is for a safe, economical, and environmentally benign system. The dedication of the large teams of scientists, engineers, and environmentalists who have been and are working on these northern projects is highly commendable. It is hoped that they will see their hard work soon come to fruition.

REFERENCES

Department of Energy Mines and Resources. An Energy Strategy for Canada. Polices for self-reliance, 1977.

Petro-Canada Public Affairs Department. Arctic Pilot Project, 1979.

Polar Gas. Combined Arctic Pipeline System, 1979.

SECTION V. SHALE OIL, TAR SANDS, HEAVY OIL, ENHANCED OIL RECOVERY

CHAPTER 32

REASSESSMENT OF OIL SHALE PROSPECTS

V. A. Ovcharenko[1]

HISTORICAL TRENDS IN OIL SHALE DEVELOPMENT

The exploitation of oil shale as a fuel and a source of liquid hydrocarbons on a substantial scale extends over a period of nearly 150 years, although what was the first oil shale patent ("A way to extract and make great quantities of pitch, tar and oyle out of a sort of stone") was granted to a group of English entrepreneurs in 1694. By the beginning of this century, oil shale industries were established in such countries as Scotland, France, Sweden, Germany, Spain, USSR, and, later, South Africa, Australia, and China.

Many of the oil shale plants were closed when exposed to competition from crude petroleum. It is interesting to note that in almost all cases, shale oil plants have been built as a result of scarcity of liquid fuel. At the present time, the only large-scale oil shale operations are those in the USSR and China. The current production in the USSR is about 35 million metric tons of shale per annum, the main part of which is produced from the Estonian deposits. Two-thirds of the oil shale is used directly as fuel for electric power generation, the remainder for petrochemicals and gas. No gasoline or diesel fuel is produced, as this is not competitive with crude oil in the USSR, while direct utilization of oil shale is economic in comparison with coal, which has to be transported from distant coal mines (Kuznetsov, 1975).

In China, most oil shale is burned directly as fuel. Oil shale is mined in Turkey in connection with a lignite strip-mining operation, and both are burned as fuel for power generation.

The total production of the oil shale in the world is minuscule compared with the size of the petroleum industry. But it has been estimated that known oil shale resources, recoverable by present-day technology and converted into oil, contain a potential of more than 400 billion tonnes, about four times more than often-quoted world oil reserve figures. Production of oil from oil shale

---

[1]Energy Section, Technical Cooperation Department, United Nations.

could extend the oil era well into the next century. Unfortunately there are many practical and economic problems in oil shale utilization.

The term "oil shale" has been used for many years to describe certain fine-grained sedimentary rocks from which oil, resembling but not identical to crude petroleum, can be obtained by heating the rock to temperatures of 350°C to 500°C. In some cases oil shales contain small amounts of organic matter (bitumen) that can be extracted with solvents, but most of their organic content is in the form of insoluble complex organic compounds known as kerogens. These compounds appear to be the fossilized remains of simple plants such as algae, which grow at the surface of shallow bodies of water. Most shale oil was formed throughout recorded geological history during periods when large areas of the earth were covered by lakes and shallow seas.

Kerogens have a ratio of atomic hydrogen to carbon which is closer to that of petroleum than in the case of the higher plants, from which coals are largely derived. Consequently, organic compounds in oil shales are readily broken down by heat to produce oil, while those in coal more readily yield hydrocarbon gases.

The mineral content of oil shale can be composed of sand, clay, or lime, or a mixture of these, and in some cases, for example, the Green River shales of the western United States, the oil shales contain appreciable quantities of natural alkali minerals.

Major deposits of oil shale occur on every continent and have been worked at one time or another. Oil shale with 50 percent organic matter is known to exist in Asia, while deposits with 37 percent organic content are known in Africa. In South America oil shale deposits appear to exist in every single country, though systematic work has been carried out only in Brazil.

RESERVES AND GRADE ESTIMATES

Past estimates of world oil shale resources have been closely associated with estimates of organically rich shales. The organic content of known shales varies from near 0 to a maximum of about 50 percent by weight, although shales as rich as this are rare. However, for the purpose of evaluating the oil producible from a given shale, the organic content does not have a direct relationship to the amount of extractable oil. The percentage of conversion is directly related to the origin of organic material from which the oil shale was derived and the method of retorting. Conventional pyrolysis (retorting) can convert only 26 percent of the organic content of the oil shale found in Sweden into oil, while the same method can provide 71 percent conversion for the Mae Sat deposits in Thailand. A "rich" shale is generally regarded as one which will yield 100 litres or more of oil tonne of shale, when it is heated to approximately 540°C. This is the traditional way to evaluate the grade and productivity of oil shales, and it is commonly referred to as the Fisher Assay. This method of evaluation has been criticized recently, because it was demonstrated that for some lean shales, pyrolysis in a hydrogen atmosphere can increase oil yield by 2 to 2.5 times, compared to Fisher Assay predictions for the same oil shales. The organic carbon content in shale seems to be essential for oil yield.

Practically all present estimates of the quantity of oil extractable from oil shale are based on the Fisher Assays. Deposits of oil shale thinner than 3 m and averaging less than 40 litres of oil per tonne of shale were not generally included in these appraisals, although they may be exploited economically if associated with commercial coal beds or if they contain potentially valuable minerals and metals such as gold, vanadium, or uranium that can be extracted as a byproduct.

Table 32-1 shows estimates of the known resources of oil shales in the world in deposits yielding more than 40 litres of oil per tonne fron conventional retorting.* What is identified as known deposits have been sampled, mapped, and assayed, at least in part. The oil shales that are being mined or may be mined under present economic and technological conditions are considered recoverable under present conditions. Lower grade oil shales or higher grade oil shales that are more difficult to mine are included in the marginal and submarginal resources category.

The totals, which are given in Table 32-1, are minimum by their nature of approximations and to these must be added reserves in shales leaner than 40 liters per tonne and an allowance for unidentified or poorly investigated reserves. A compilation of global oil shale resources is hampered by limited exploratory activity and a lack of common terminology of reporting. Many of the countries having oil shale deposits have analyzed some of the oil shales but offer no information on the thickness, character, and oil content of the

*The information on oil shale resources on a country-by-country basis can be found in the United Nations Publication, "Utilization of oil shale" (1967), and in a number of more recent publications, see for example Donnell (1976) or Marland (1979).

TABLE 32-1. Estimations of the world oil shale resources (billion tons of oil)

Continents	Known resources Recoverable under present conditions	Marginal and submarginal
Africa	1.4	12.3
Asia	3.3	11.5
Australia and New Zealand		0.14
Europe	4.1	6.3
North America	11.0	290.4
South America	6.8	102.7
	26.6	423.34

Source: Modified from Duncan and Swanson (1965).

shale. Oil shales are found in varied depositional environments and in rocks of many ages; however, the oil shale endowment has not been reported in many countries, presumably because exploration efforts have not been sufficient to locate such deposits. India is one country in this category. It is inconceivable that such a large part of the earth's crust as India, containing rocks formed in many differing environments of deposition, would not have oil shale. Future exploratory efforts will undoubtedly delineate as yet unrecognized oil shale deposits in many areas of the world, and total exploitable oil resources in oil shale are probably many times the figures shown in Table 32-1. But even from the figures given, it can be seen that the physical amount of known shale oil reserves is significantly larger than that of known crude oil reserves, and that efforts in the improvement of shale oil extraction technology are justified.

STATE OF OIL SHALE TECHNOLOGY

Oil shale can be burned as a solid fuel, for example for electric power generation; it can be processed to yield shale oil or a combustible hydrocarbon fuel gas for residential or commercial use. Crude shale oil can be upgraded and refined into the following fuel products: gasoline, diesel fuel, distillate fuel oil, jet fuel, residual fuel oil, and liquefied petroleum gas. The following byproducts can be recovered from refining operations: coke, pitch, asphalt, ammonia, sulfur, paraffin wax, specific organic chemicals, and in some cases, a number of metals, including gold, vanadium, and uranium. Fine-ground ash and fly ash from direct burning of oil shale have been used for cement manufacturing and in the brick industry. But before this broad range of products can be produced, shale oil has to be extracted from shale.

The technology of the production of oil from shale falls into two distinct types: mining, followed by retorting in a surface plant, and "in situ" oil production. To date, mainly the first method has been applied in practice, with mining accounting for up to 50 to 60 percent of the total processing costs. This cost can be reduced by in situ retorting. This method is still in the experimental stage, although on a very substantial scale. For example, the $900 million modified in situ project of Occidental Petroleum Company in Colorado, USA, is supposed to start producing at a rate of 125,000 tonnes of oil per year in the mid-1980's and eventually reach 2.5 million tonnes per year (about 50,000 b/d) by 1990, if the economic factors remain reasonable.

The conventional method of first mining the shale and then retorting it in a separate nearby plant has been the one employed to date in all industrial-scale developments.

Mining

Underground mining of oil shale is similar to mining coal and varies rather widely, depending on the characteristics of the oil shale deposits. Where feasible, open-pit mining has been employed because of lower costs, but the bulk of the operations to date has involved underground mining. The cost of underground mining of shale from thick seams, often exceeding 3 m, with modern mechanized equipment may in some cases be lower than from open pits, because no overburden has to be removed and less land restoration is required.

Most efficient in many cases, room-and-pillar underground mining techniques can provide over 50 percent removal of shale in place. The more expensive longwall and other techniques, requiring artificial roof supports, may provide up to 80 to 85 percent removal of shale. In strip mining, with blasting of shale, an extraction of 80 percent of the shale in place is typical.

For small-scale mining with a minimum of mechanical equipment, as described by Savage (1978), the open-cast mining is essential to provide reasonable output per man per day. The mechanization of oil shale mining in the USSR has resulted in a threefold increase in average productivity for the period from 1960 to 1975. The average output in oil shale mining per man per month has reached over 160 tonnes to date and in strip mining about 550 tonnes.

Preparation

The oil shale preparation after mining is determined by the requirements of the consumer. In the USSR, for direct burning of oil shale, it should be crushed to a size not exceeding 25 mm and in some cases the shale is finely pulverized before being fed to the burner. For oil and tar extraction, lumps of 25 to 125 mm are selected from the mine output. The separation of oil shale from limestone and other nonorganic impurities is provided by using dry or wet processes in the mine or at the consumption site. The techniques are similar to coal beneficiation.

The pulverization of oil shale for direct burning or in some cases for retorting is a well practiced art. Several types of crushers such as jaw, gyratory, impact, and roll, are utilized. Although the crushing step is not considered to involve any critical problem, some shales with high organic content are resistant to crushing and this should be taken into account in designing and selecting the most suitable commercial crushing plants. Crushing is an important but relatively small part of the overall cost of oil shale utilization.

Retorting

Many above-ground processes for oil shale have been patented over the years, and new patents continue to be issued. But all retorting processes have one fundamental characteristic in common, namely, heating the oil shale to at least the pyrolysis temperature.

A wide variety of types of retorting kilns has been used, ranging from the most primitive sort of batch retort in the early years of the industry, through semi-continuous and continuous process kilns; vertical, inclined, rotating, stationary, and fluidized-bed kilns; and tunnel ovens. The more advanced experimental plants are all of the continuous-process type because of the desire to reduce labor costs as far as possible. So many variations on the basic retorting process have been devised that it would be pointless to describe them in detail. Some systems involve partial combustion of the shale with air, others heat the shale in a closed retort and use the gases evolved for heating purposes; still other systems employ indirect heating of the shale by contacting it with hot, spent shale, or ceramic balls heated to a high temperature. The choice of any particular retorting method is dictated by a combination of various technical factors, including the characteristics of the individual shale deposit and the nature of the end-products required, as well as economic factors. Retorting temperatures vary from a minimum of 350°C up to 1,000°C, depending on the process used and the range of end-products required. In general, the higher the retorting temperature, the lighter are the products of distillation. Above-ground retorting causes a waste-disposal problem, because the volume of spent shale after retorting is almost equal to the volume of the unprocessed shale. The in situ concept of obtaining oil from shale is being emphasized not only because it could reduce the cost of mining but also because it would greatly extend the amount of recoverable resources and probably offer important environmental advantages. However, in situ retorting is not adaptable to concurrent recovery of associated minerals and, as in the case of in situ gasification of coal, it is very difficult to control the underground process with sufficient accuracy from the surface. This contributes to the reduction of thermal efficiency and to waste of a significant part of the resource in terms of the percentage of the total reserve recovered.

Such processes nevertheless offer an economic advantage for extracting thick deposits of deeply buried shales or thin beds at various depths interspersed with layers of barren rock which otherwise would never be mined.

Field tests on in situ shale oil recovery have been conducted in a number of countries by oil companies and government research institutions. Some of the techniques are based on underground combustion, with hot combustion gases being swept horizontally ahead of the fire zone through a retorting zone. Others involve introduction of hot natural gas or steam into an artificially fractured zone. It has been found that there is insufficient naturally occurring permeability to allow passage of gases and liquids. Underground experimentation on fracturing of oil shale by using electrical, hydraulic, and explosive fracturing techniques has been conducted or studied.

In general, in situ technology is in the research stage to date and the scaling up of pilot field testing is not easy to predict, as demonstrated by the fault found in retort no. 6 of Occidental Petroleum Company last September.

To create the huge underground retort no. 6 (about 60 m square and 85 m deep) for demonstration of the Occidental "modified in situ" method of oil extraction, at least 20 percent of the retort's volume was mined out, leaving voids at regular intervals. Explosives turned the rest of the shale to rubble, which expanded into the voids. Combustion was initiated at the top of the affected shale column and progressed to the bottom with retorting taking place below the combustion zone. Air was injected at the top of the shale column, and oil was withdrawn from a sump at the bottom. During the experiment, part of the retort collapsed, damping the oil-producing fire. The retort no. 6 produced approximately two-thirds of what was expected despite the fault, and some experience has been gained. This experience will be helpful in Occidental's preparation of retorts nos. 7 and 8--the next experimental step toward the commercial operation of in situ shale oil production.

Special Features of Shale Oil Refining and Main End-Products

The oil produced by heating the shale is not identical with crude petroleum, and varies to some extent according to the nature of the deposit from which it is produced, and the type of distillation process employed. Raw shale oil is characterized by a relatively high wax content giving it a high pour point (the point at which the oil becomes semi-solid). In practice, the quality of the products can be varied by changing the retorting temperature, since higher temperatures "crack" or break up the heavier components into lighter products. Shale oil contains nitrogen compounds as well as sulphur, and the hydrocarbons have a different chemical structure from those of crude petroleum. This renders them both somewhat unstable in storage, and also more reactive in petrochemical processing. By a relatively simple refining process, which usually involves the addition of hydrogen, the raw shale oil can be upgraded (improving the viscosity and removing some nitrogen and sulfur) to make a fairly good quality synthetic crude oil which can be pipelined to a refinery and processed in the same manner as crude petroleum of comparative quality. However, additional research may be required to develop better catalysts specifically designed to remove nitrogen and sulfur from shale oil and to increase the yield of end-products.

It is also possible to produce an entirely gaseous product by retorting oil shale at higher temperatures (up to 1,000°C). The choice of a partial combustion process or an indirect heating method for retorting would determine whether the gas will have a lower or a higher calorific value. Attempts to obtain high output of gaseous products by hydroretorting and hydrogasifying of oil shale have been reported by the Institute of Gas Technology, USA, since 1966. The technical feasibility of this process has been established and extraordinarily high yields have been achieved. In some experiments, 97 percent of in situ crude shale oil was converted to gas with a methane content of 75 volume percent (Baker, 1976). The studies showed the effects of hydrogen content, pressure, and grade of shale, and indicated that substitute natural gas from oil shale could be a viable route for oil shale utilization.

Shale oil has not been utilized for commercial production of gasoline or diesel fuel because it was not competitive with crude oil until now. Direct utilization of oil shale for electrical power generation is economic in the specific local conditions of oil shale producing areas in the USSR and China and it might be economic to use oil shale to generate local power in many developing countries. Petrochemical feedstock derived from oil shale has proved to be another viable way of oil shale utilization, which improves significantly the economics of the oil shale industry in the USSR.

Shale-Derived Petrochemical Feedstock

In addition to conventional petroleum fuels, the chemical nature of shale oil is such that it easily forms a reactive petrochemical feedstock such as ethylene, propylene, benzene, naphtha, and others. Products not commonly obtained from crude oil, at least not in large quantities, also may be produced by properly adjusting oil shale retorting and further refining. Among these are tars for wood preservation and also tar acids and tar bases (certain families of organic compounds containing oxygen and nitrogen). They are valuable feedstock for manufacturing various resins. Owing to the nitrogen and sulphur content of shale kerogen, sulphurous and ammoniacal byproducts are also produced. It is also possible to use the gas from oil shale retorting for an atmospheric nitrogen fixation plant.

PHYSICAL AND ENVIRONMENTAL REQUIREMENTS FOR AN OIL SHALE INDUSTRY

Apart from the obvious necessity for an adequate reserve of oil shale, there are several other physical requirements which must be met before an oil shale industry can be established. Foremost among these is an adequate supply of water, which is required at almost all stages of mining and processing. Since the shale consists principally of inert mineral matter, it is intrinsically a low-value material and cannot be transported economically for more than a short distance. For this reason the processing plant must be located as close to the mine as possible, and in arid or semi-arid areas the availability of an adequate water supply near the mine may be the limiting factor in development. Based on the technology developed under conditions prevailing in Colorado, USA, mining and processing 1 tonne of shale in a conventional underground mine and surface retorting plant would require about 0.35 m^3 of water. In situ retorting of the shale would require about half this much.

Under the abundant water supply conditions of Estonia, USSR, the specific water consumption in underground mining is about 0.8 m^3 per tonne of shale. The main part of the water requirement is for technological processes (0.66 m^3) and is supplied by ground water.

A factor which can be of considerable importance is the disposal of the shale ash after retorting. Owing to the relatively low organic content of most shales, the weight of spent shale is about 85 percent of that of the raw shale mined, but because of changes caused by retorting, spent shale ash occupies about the same volume as the original raw shale.

The spent shale often contains appreciable quantities of caustic alkali compounds and other reactive constituents, including small quantities of carcinogens, which can be leached out of a surface tip and entered into the local surface drainage channels. This can have serious effects in semi-arid areas where downstream water users may be adversely affected. The shale ash also tends to blow in the wind, which can cause further problems. In the case of underground mines, the shale ash can be disposed of by using it to backfill worked-out areas underground, but this adds to the cost of the operation. Shale ash can also be used for backfilling strip mining excavations, but here the problems of leaching and blowing may still occur.

Fuel for an oil shale mining and extraction plant can usually be obtained as a byproduct of the operation, either in the form of gas from the kilns, or by burning residual carbon left in the shale after distillation. Since the total heat requirements of the system are considerable, failure to ensure the maximum thermal efficiency imposes a severe economic load on the project.

SHALE OIL COST ESTIMATIONS

The last 10 to 20 years have demonstrated that the cost of oil from oil shale is not easy to predict with certainty. It has been climbing steadily, remaining all the time above the rising crude oil prices. The 1968 United Nations Symposium on the Development and Utilization of Oil Shale Resources, which was held in Tallin, USSR, concluded that oil from the richer and more accessible oil shale deposits was about to cross the threshold to economic competition with crude oil. Allsman (1968) estimated, for example, that oil, retorted from oil shale in Colorado, USA, would cost $2.85 per barrel, which was very close to the price range of crude oil at that time. In the 1973 National Petroleum Council's study, the range of $3.90 to $6.35 per barrel of shale oil was presented. In 1979, the President of the Gas Research Institute, Henry Linden, estimated the price of shale oil at the point of production to be $23 to $34 per barrel, while the recent OPEC ceiling price for oil is $23.50.

Since the beginning of the 1970's, nearly every major oil company in the world has been researching profitable ways to turn coal and shale into oil and gas. Dozens of pilot and even commercial synthetic fuel projects were announced. Among the most promising commercial-sized projects were four oil shale plants (Table 32-2).

Under U.S. conditions, it is considered that the minimum economic size for an oil shale plant would be 50,000 barrels per day, equivalent to about 2,400,000 tonnes of oil per year, and requiring a mine production of some 70,000 tonnes of shale per day, or nearly 26 million tonnes per year. The capital cost of such a mine and plant would be in the order of $1 billion. It is about fourfold higher than estimations made in 1974. Construction time was estimated to be 4 years. The labor requirements for an underground mine and plant of this capacity are about 2,400 man-years during the construction period, with an operating labor force of 1,400 men.

Under conditions prevailing in most developing countries, it is likely that the economically viable, minimum size of plant would be lower, and that labor requirements would be higher. While each project would require individual study and detailed costing, it is probable that at current international selling prices for crude oil, many indigenous shale oil projects might be considered economically viable. However, there is no question that the cost of producing crude oil is only a small fraction of the cost of producing shale oil, and so any significant shale oil project would be a major undertaking,

TABLE 32-2. Oil shale commercial-size projects

Company	Location	Investment (millions of $)	Production (thousands of bbl per day)	Cost per bbl ($)	Status
Occidental Petroleum	Colorado	900	50	20	In situ production. First unit under construction for operation by 1985.
Union Oil	Colorado	780	54	24	First phase awaiting government subsidy.
Gulf/Standard (Ind)	Colorado	NA	200	25	One unit under way, another planned.
Atlantic Richfield and Tosco Corp.	Colorado	1,300	45	25	Awaiting tax incentives.

Source: Business Week, July 30, 1979, Special Report.

the cost of which might impose a serious burden on the whole economy of the country.

CONCLUSION: WILL IT BE A REALITY?

Oil shales are widely distributed geographically and the known reserves are very large indeed. They may be further increased if systematic exploration for further deposits is undertaken. Oil shales have been exploited as a source of solid fuel and synthetic liquid hydrocarbon for nearly 150 years, and the technology for extracting oil from shales is well understood, although what may work in a small pilot plant may be difficult to apply commercially. In addition to liquid fuels, oil shales yield a number of valuable byproducts, among which are metals, waxes, resins, tars, and ammoniacal and sulphurous compounds. The shale ash can sometimes be used as a building material or as cement. Oil shale can also be used to produce pipeline gas, or may be burned directly as boiler fuel.

Oil shale is a low-grade resource which requires the handling of large quantities of inert mineral matter, usually of little or no economic value.

The economics of oil shale development are dominated by this factor, which inhibits transport to any distance from the mine and that requires processing as close to the source as possible. Disposal of spent shale ash can sometimes present technical and economic problems. Oil shale development and processing often requires large quantities of water, which in arid areas might be the limiting factor for development on a large scale.

Where there is a free choice between the two, oil shale as a source of liquid fuels is far less attractive economically than is crude oil on a cost basis.

Shale oil will probably never dominate the energy market or displace conventional petroleum, but will simply take its proper position along with synthetic fuels from coal and other sources in the total energy mix.

Present-day prices and availability of crude oil on the international market can probably justify the development of indigenous oil shale deposits in a number of countries, but so far very limited development seems to have been commenced, having probably been inhibited by the high capital cost of shale oil development combined with a measure of uncertainty as to the future economic climate.

Many times in the past, it was predicted that oil from shale would be an economic reality within a few years. And today it could be said that the shale oil industry may be a reality in 5 years, but because of the lead time requirements, steps should be taken now at the national and probably the international level in establishing a framework within which oil shale can contribute to an alternative fuel transition in a timely fashion.

REFERENCES

Allsman, Paul T., 1968, Proceedings of the Fifth Symposium on Oil Shale: Colorado School of Mines, USA,

Baker, L. K., 1976, Producing SNG by hydrogasifying in situ crude shale oil: Bulletin of Mines, RI 8011.

Business Week, July 30, 1979, Special Report.

Donnell, John R., 1976, Global oil shale resources and costs: UNITAR-IIASA Conference on Unconventional Oil Resources, Vienna.

Feldman, H. F., et al., 1966, Production of Pipeline Gas by hydrogasification of oil shale: Institute of Gas Technology, Res. Bull. no. 36.

Gallois, Ramnes, 1978, What price oil shales? New Scientist.

Kuznetzov, D. T., 1975, Oil Shales of the World: Moscow, Nedra.

Linden, Henry R., 1979, Energy Policy and the Gas Option, Pacific Coast Gas Association, Annual Business Meeting, Houston, Texas.

Marland, G., 1979, Shale Oil: US and World Resources and Prospects for Near-Term Commercialization in the USA: Oak Ridge Associated University, Oak Ridge.

National Petroleum Council, 1973, U.S. Energy Outlook; Oil Shale Availability: Washington, D.C.

Savage, John W., and Savage, Margaret W., 1978, The V-Bank process, a report on the feasibility of establishing small oil shale industries: Proceedings of the First Conference on Small Scale Mining of the World, Mexico.

CHAPTER 33

MODIFIED IN SITU OIL SHALE RETORTING

R. E. Lumpkin[1]

INTRODUCTION

The history of oil shale retorting, particularly in the United States, has been controlled by the discovery and development of conventional petroleum reserves. The very first oil well, drilled by Drake in 1859 at Titusville, Pennsylvania, showed that petroleum could be produced cheaper than shale oil, and doomed the fledgling retorting industry that had sprung up along the Ohio River in the 1850's. The possibility of shortages of oil also led to intensive U.S. oil shale development efforts following World Wars I and II, but both periods ended with the discovery of major new petroleum reserves, first in east Texas, then in the mid-East. The discovery of the huge mid-East reserves also led to the demise of the Australian and Scottish oil shale industries.

Today we face a crisis of tight world petroleum supply, and again oil shale appears to be the most attractive alternative source for liquid fuels. Shale oil still costs more to produce than conventional crude, but world oil price is no longer related to the cost of production. Occidental Petroleum Corporation has developed an in situ retorting process that we believe can produce oil at prices competitive with petroleum on the world market.

This chapter describes the OXY process, reviews the field experiments that have been run to date, and summarizes a commercial plant design. It was designed as part of Phase I of a cooperative agreement between Occidental and the U.S. Department of Energy. Capital and operating costs for this design are presented.

THE OXY PROCESS

To recover oil from shale, it is necessary only to heat the shale to 800°F. The organic component of the rock (kerogen) decomposes to produce a liquid fuel similar to petroleum and a hydrocarbon-containing gas. Residual carbon is left

---

[1]Occidental Research Corporation, Irvine, California.

on the mineral matrix. The key to the OXY process is to process large volumes of rock with a minimum of materials handling.

Typical oil shale is hard, nonporous, and generally unsuitable for in situ retorting without prior fracturing. The OXY concept is to create voids in the formation using conventional mining. These voids are then distributed through a large volume of rock by using explosives to fragment, or rubblize, the formation around the mined voids. This procedure creates a dimensionally controlled underground chamber with a high enough permeability to allow the flow of gases through the mass of broken shale. Generally, the mined void volume within a retort is in the range of 15 to 25 percent. Because some mining is required to create the necessary permeability, the process is usually called the modified in situ process (MIS).

The retorting operation is started by heating the rubble pile at the top, using an outside energy source. Once the top is hot, the use of fuel is discontinued and retorting is continued by passing air and a diluent gas down through the rubble. An operating retort contains four major zones (Fig. 33-1). Within the first, or preheat, zone, the inlet gas is heated by contact with hot spent shale. The hot gas then reaches a combustion zone in which the oxygen is consumed by burning residual carbon. Some oil and gas may also be burned. Below the combustion zone is the retorting zone in which the hot combustion gases thermally decompose the organic component of the shale to produce oil and hydrocarbon gas. In the final zone, the combustion and retorting gases are cooled as they flow downward, condensing some water and vaporized oil. The oil and water that flow from the bottom of the retort collect in a sump located behind a gas-tight bulkhead at the base of the retort. After initial oil and water separation takes place, the products are pumped to the surface through separate systems. The operation is continued until all of the shale has been heated and the oil recovered. This process has been labeled the vertical modified in situ retorting process.

Like all such ideas, the transition from concept to demonstration is a critical step. In this case, the transition was particularly challenging because both the mining and chemical processing operations were far different from the existing state-of-the-art at the time the process was conceived. From a mining viewpoint, the process requires a control of particle size and void volume distribution, a task never considered in normal mining operations. From a chemical processing viewpoint, the sheer size of the retort, the range

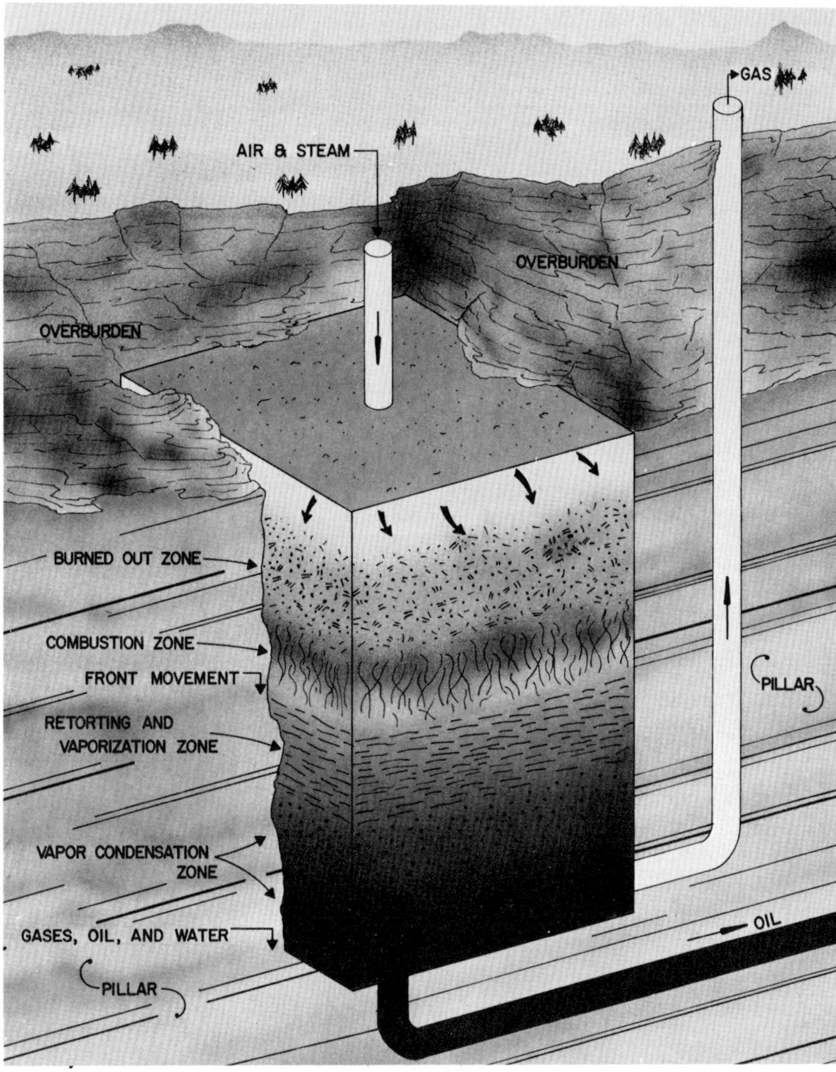

Figure 33-1. Occidental MIS retorting process

of particle sizes, and the low void volume all presented new dimensions in processing operations. It is one thing to measure and control conditions within a steel reactor above ground, but it is quite another to apply the same operating principles to a reactor that has an acre or more of cross-sectional area and is as tall as a 30-story building.

OCCIDENTAL FIELD EXPERIMENTS

We were convinced from the beginning that the most critical step in the development of a process would be to demonstrate in the field both rubblization and in situ retorting. No amount of modeling or pilot plant work could answer the basic question of field operability. The Logan Wash test site (Fig. 31-2) was acquired in the summer of 1972 and the field program was initiated. This property is at the southern edge of the Piceance basin in western Colorado. The average grade here is low, averaging 17 gallons per ton (gpt), but access to the formation could readily be achieved by using drifts into the outcropped shale. Site access work began in July and mining commenced in early September 1972.

Experimental Retorts

The first phase of the mining and retorting program was to prepare three in situ retorts, each approximately 1,000 ft^2 in cross-sectional area and varying in height from 72 to 114 ft. These were small in comparison to the envisioned commercial retorts, yet were large enough to test basic concepts and to permit scale up to the full size at a later date. They were also large by any normal process development standards. For example, the smallest retort contained approximately 4,000 tons of oil shale, 25 times the size of the largest experimental above-ground retort ever used to simulate in situ retorting.

Retort 1E was prepared in the winter of 1972 and startup began in May 1973. The void volume was mined in the form of a small room at the base of the retort with a vertical center raise having a circular cross section. Blast holes were vertical and parallel to this center raise, as shown in Fig. 33-3. Some difficulties were encountered during startup with respect to ignition, measurement, and control of temperature and pressure, and sealing of the retort. Actual retorting began in July, and the retort operated satisfactorily

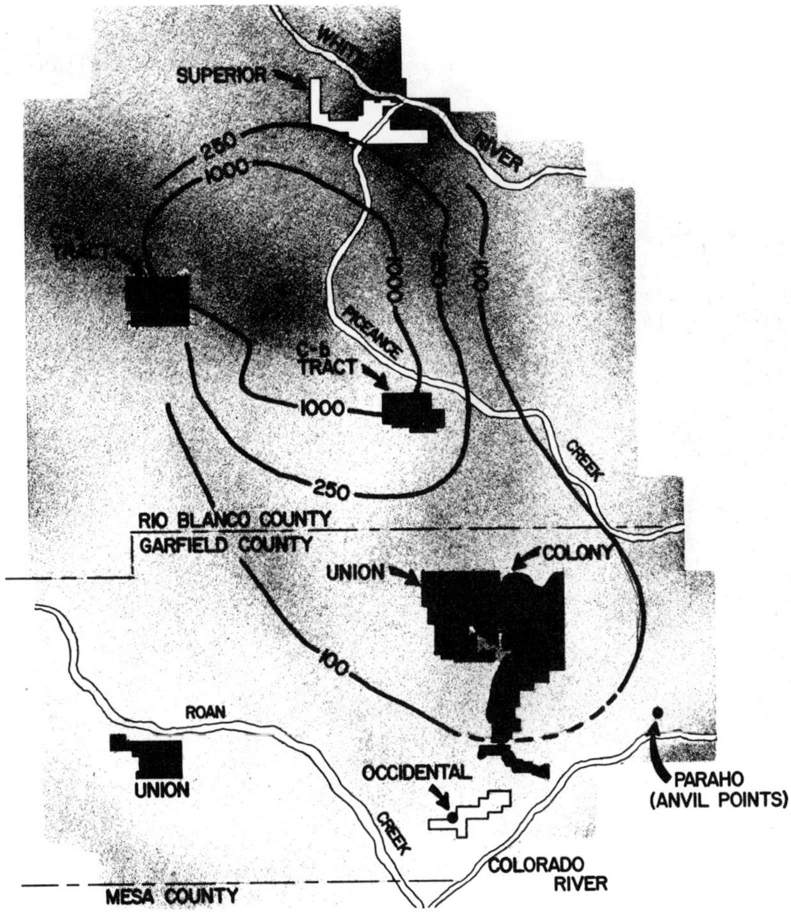

Figure 33-2. Oil shale tracts in the Piceance basin

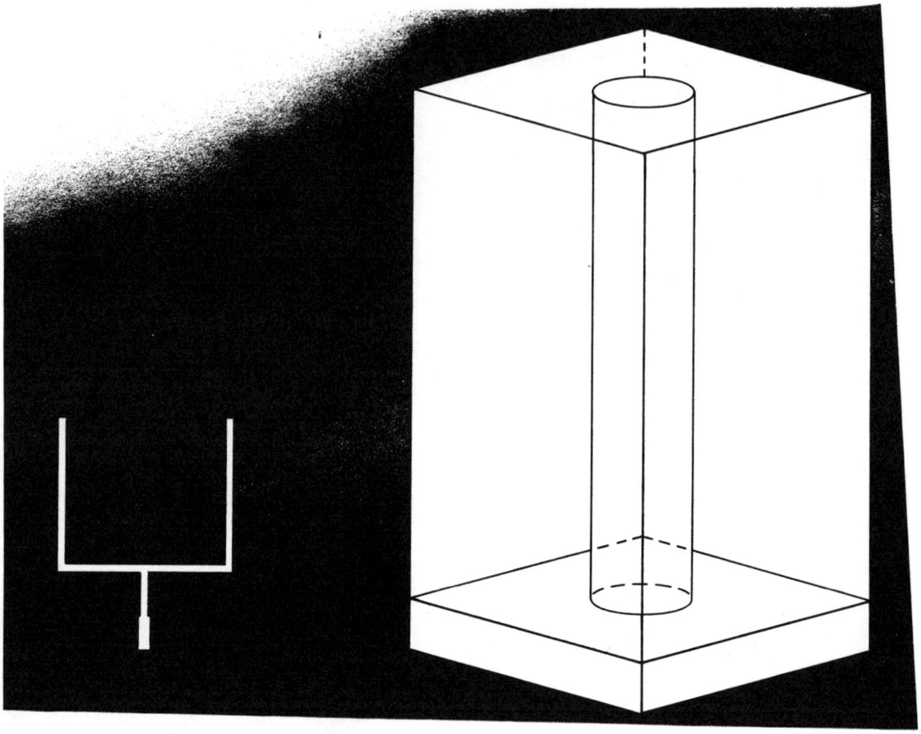

Figure 33-3. Retort 1E mining plan

from that point on. Over 1,200 barrels of oil were produced, a high recovery in relation to theoretical for this first experiment.

Retort 2E was prepared in 1973 and retorted in 1974, utilizing the same basic rubbling design. The void volume was lower and the retort height was increased by about 20 ft. The additional retort height consisted of oil shale that was considerably lower in grade than that normally processed in surface retorts. Retort 2E still performed satisfactorily and produced 1,400 barrels of oil.

While Retort 2E was in preparation, construction started for a new entry to the oil shale about 3,500 ft away. This area was designed to test at least one commercial-sized MIS retort and to provide access for commercial operations after testing was complete, if such a course of action were warranted.

Retort 3E, the final retort in the experimental mine, utilized an entirely different rubbling technique. As shown in Fig. 33-4, there were two sublevels mined above the retort floor using a short raise along side the retort. Blast holes were perpendicular to the free faces. Retort 3E was about 40 ft greater in height than Retort 1E and, again, the added height was in low-grade oil shale. The new fragmentation system led to different operating characteristics from the earlier experiments; although some problems were encountered, the retorting proceeded smoothly, with a minimum of operational difficulty. Retort 3E produced 1,600 barrels of shale oil. In terms of oil recovery from the rubblized shale, this represented the best result of the three tests. In fact, the actual oil recovery was quite close to the maximum predicted by our computer simulation of the retorting process for this particular grade of shale.

Retort 4

Retort 4, the first large-scale retort in the "commercial" mine, was originally designed as a scale-up of Retort 1E. However, this design was altered to test yet a third fragmentation system (Fig. 33-5). Two vertical slots replaced the center raise concept. These slots extended essentially from the floor of the retort to a mined-out room at the top, and blast holes were drilled parallel to the slots from the upper level.

Retort 4 was mined in the first half of 1975 and rubbled on August 10 of that year. It was ignited after extensive flow and tracer testing in December of that year. The cross section was 120 by 120 ft and the height was 270 ft. For the first time, the B-groove, a thick zone of barren rock within the oil

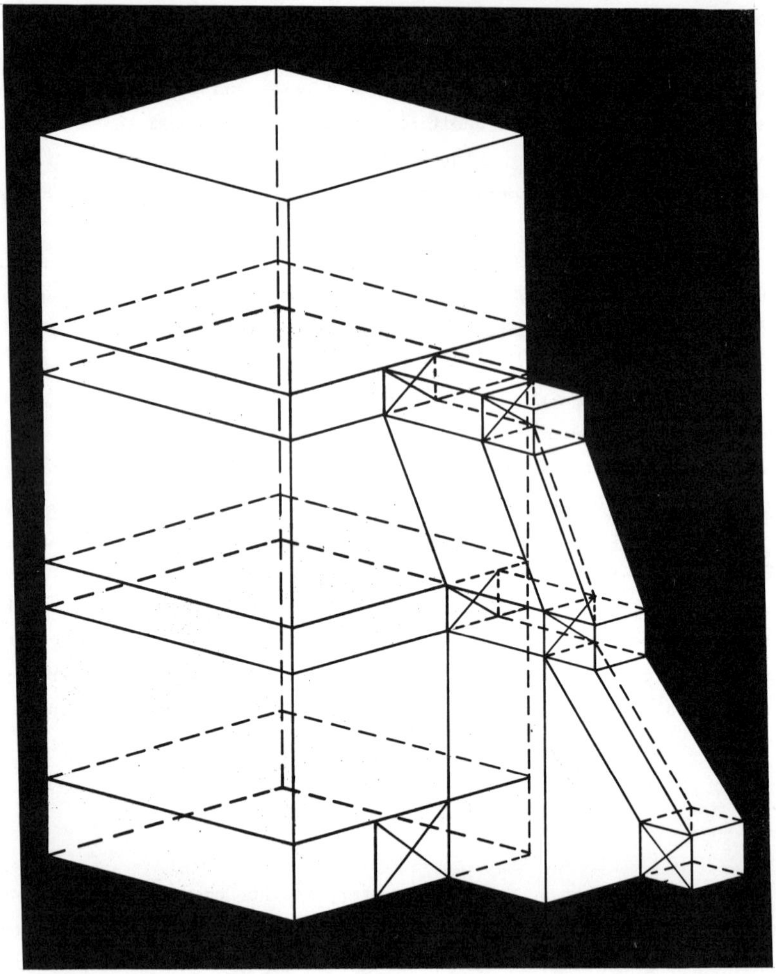

Figure 33-4. Retort 3E mining plan

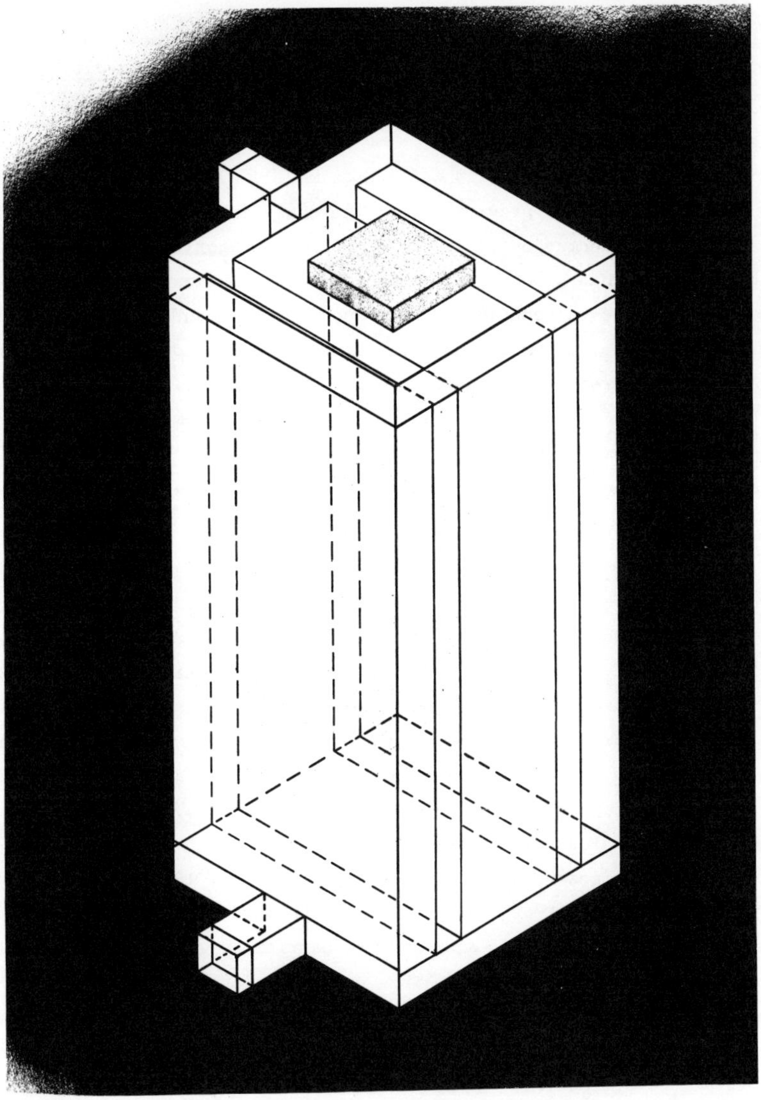

Figure 33-5. Retort 4 mining plan

shale formation, was included within the rubble. The Mahogany zone in this area contains a 32-ft section of 35 gpt rock. Thus, the retort contained a wide variety of shale grades, including those we expect to encounter in a commercial operation.

Oil production from Retort 4 was approximately 27,500 barrels, significantly less than would be predicted from the shale in place. The major reason was the geologic orientation of the retort that was aggravated by the mining techniques used. This problem prevented adequate rubblization of the ore body. On balance, Retort 4 was considered very successful because of the information gained. For example, a new approach to retorting was used for part of the run. All previous retorts had used an air/recycle system, whereas Retort 4 used this system at the start, but later switched to steam/air. This is now the operating system of choice for commercialization. In addition, valuable experience was gained in ignition and start up of large retort cross sections.

Retort 5

To accelerate development of advanced mining and processing technology for the OXY in situ process, Occidental entered into a cooperative agreement with the U.S. Department of Energy in September 1977, retroactive to November 1976. Under the first phase of this agreement, two additional large-scale retorts, 5 and 6, were to be prepared and operated to evaluate advanced alternative rubblization techniques. In addition, a definitive design for a 2,500 bbl/day facility was to be prepared. From this definitive estimate, a 50,000 bbl/day typical commercial plant was to be designed.

Retort 5 was mined between November 1975 and April 1976 using a design similar to that of Retort 4. The design incorporated a single slot in place of the dual slot of Retort 4, and also a 40-ft barrier or sill pillar directly above the section to be rubblized. This configuration is shown in Fig. 33-6. In practice, the vertical slot was pulled only to within 40 ft of the air inlet level. Blast holes drilled through the sill pillar were not loaded in the top 40 ft so the sill pillar was not fragmented during rubblization. The blast holes then served as a distributor plate for the retorting gas. Retort 5 also incorporated a significantly larger blasting pattern, an altered blasting system, and aimed at a lower void fraction than obtained in Retort 4. The low void fraction was tested to see if commercialization of the low-grade Logan

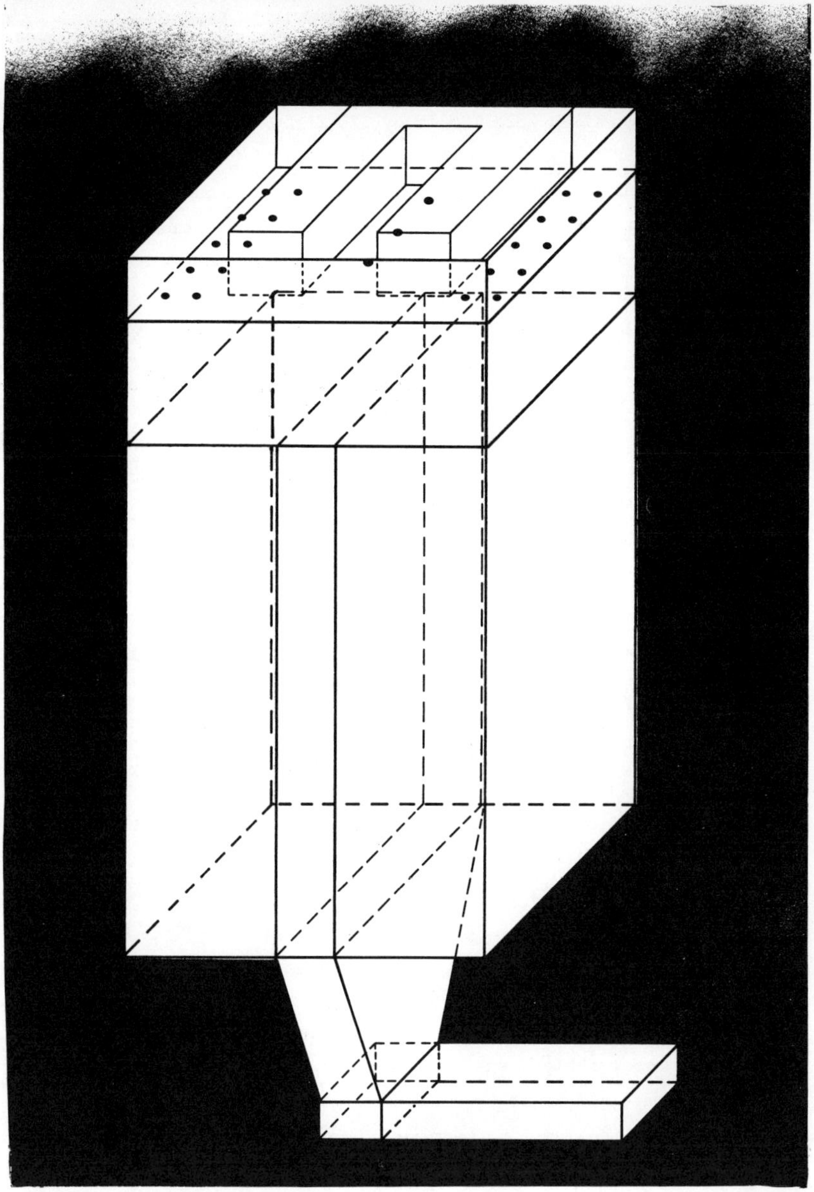

Figure 33-6. Retort 5 mining plan

Wash property were possible. We recognized the risk but felt it was justified. The overall dimensions were 120 ft by 120 ft by 200 ft, including the sill pillar.

It became evident from the tracer tests that a good distribution of the void volume was not obtained overall, and this was further confirmed when channelling occurred during retorting. The retort parameters held fairly close to the desired conditions for about the first month after the April 1977 start up, then the performance began to deviate. Corrective procedures were applied to the retort by varying the operating conditions, but none of the procedures was totally effective. Retort 5 was shut down in November, at which point the total oil production was approximately 11,000 barrels. This was less than had been anticipated and was attributable to the nonuniform distribution of voids across the retort cross-section.

Retort 6

Retort 6 was a scale up of the Retort 3E design using two intermediate mining levels between the top and floor levels. It utilized the sill pillar concept of Retort 5 and had a 23 percent design void volume. Figure 33-7 shows the mined-out areas prior to rubblization. The retort is 160 ft by 160 ft in cross section and 269 ft high. Although the rubble void volume was the same as the design, Retort 6 did not fill with rubble. Figure 33-8 shows the configuration of the retort after the rubblization event. A continuous void between the bottom of the sill pillar and the rubble pile resulted because the lower level was slightly overmined and some of the broken rock flowed out into the mine drifts at the intermediate and product levels during rubblization.

After an exhaustive series of flow and tracer tests, the retort was ignited on August 28, 1978. The start-up phase ended 16 days later when a portion of the sill pillar fell into the void between the pillar and the rubble pile. This event fully tested the built-in safety features of the process. It took approximately 2 weeks to erect a permanent bulkhead to seal up the top of the retort from the rest of the mine. During that time the retort continued to operate, although under unusual conditions. The OXY process uses blowers to draw air through the rubble pile from the bottom, and the retort is therefore always maintained under negative pressure. Thus it was possible to maintain operation during this upset. After the new permanent bulkheads were installed, an additional 2 weeks were required to restore "normal" operation. This was

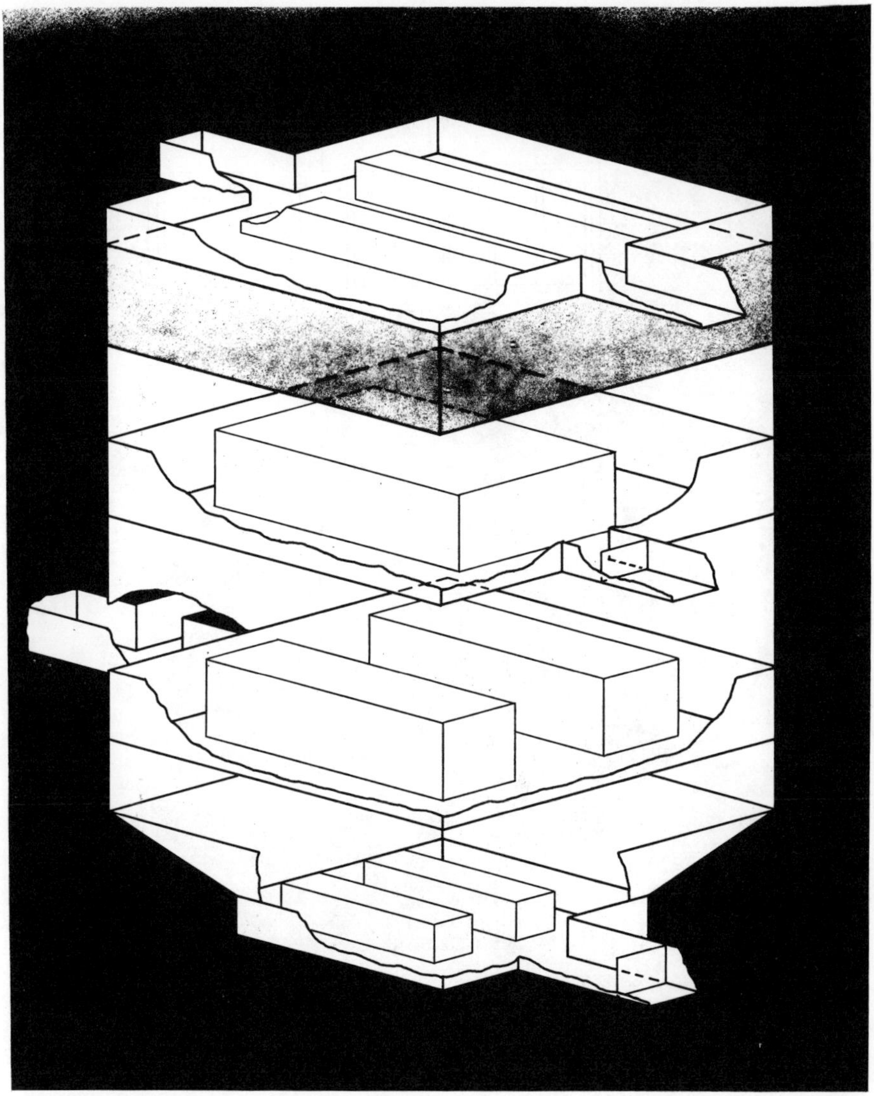

Figure 33-7. Retort 6 before rubblization of shale

Figure 33-8. Retort 6 after rubblization of shale

defined as supplying air and steam through the original gas distribution system to the top of the retort.

I don't want to minimize the problems caused by the partial failure of the sill pillar. Commercial mine designs will certainly use this experience to avoid any reoccurrence. However, no one was injured and control over the safety hazard was established very quickly. Having gone through such an experience improves our chances of avoiding similar mishaps in the future, when the scale of operations will be larger and the hazards therefore more severe. The only permanent consequence of the accident was a loss of control of the gas distribution at the top of the retort. At the time the bulkhead was closed, this loss did not seem to be serious.

Our production rates in February 1979 were the first indications that problems were developing in the retort. Final oil production was 54,000 barrels, about three-fourths of the expected yield. We now believe that significant amounts of oil were lost due to bypassing or coking, and that this behavior is attributable to the sill pillar failure and subsequent loss of control. The regions of the retort that operated normally apparently produced the expected amounts of oil.

An idea of the scale of the operation described may be gained from Fig. 33-9, in which the six retorts are compared with each other and with the Statue of Liberty with respect to volume.

COMMERCIAL PLANT DESIGN

Part of the Phase I agreement with the U.S. Department of Energy was the design of a conceptual 50,000 BPD commercial facility. The Ralph M. Parsons Company prepared the design under a subcontract to this agreement. The following information is from that design.

The basic component of the commercial facility is the retort. Retorts are 155 ft by 310 ft in plan and 395 ft high. (This retort size was used for the Parsons cost estimate. Like all dimensions given below, the size may be changed before a commercial facility is built.) For operational convenience, retorts are arranged in clusters, with one cluster regarded as a common operational entity. Clusters are defined as groups of 8 or 16 retorts, arranged in 2 rows of 4 or 2 rows of 8, depending on location within the mining/operating plan for the tract.

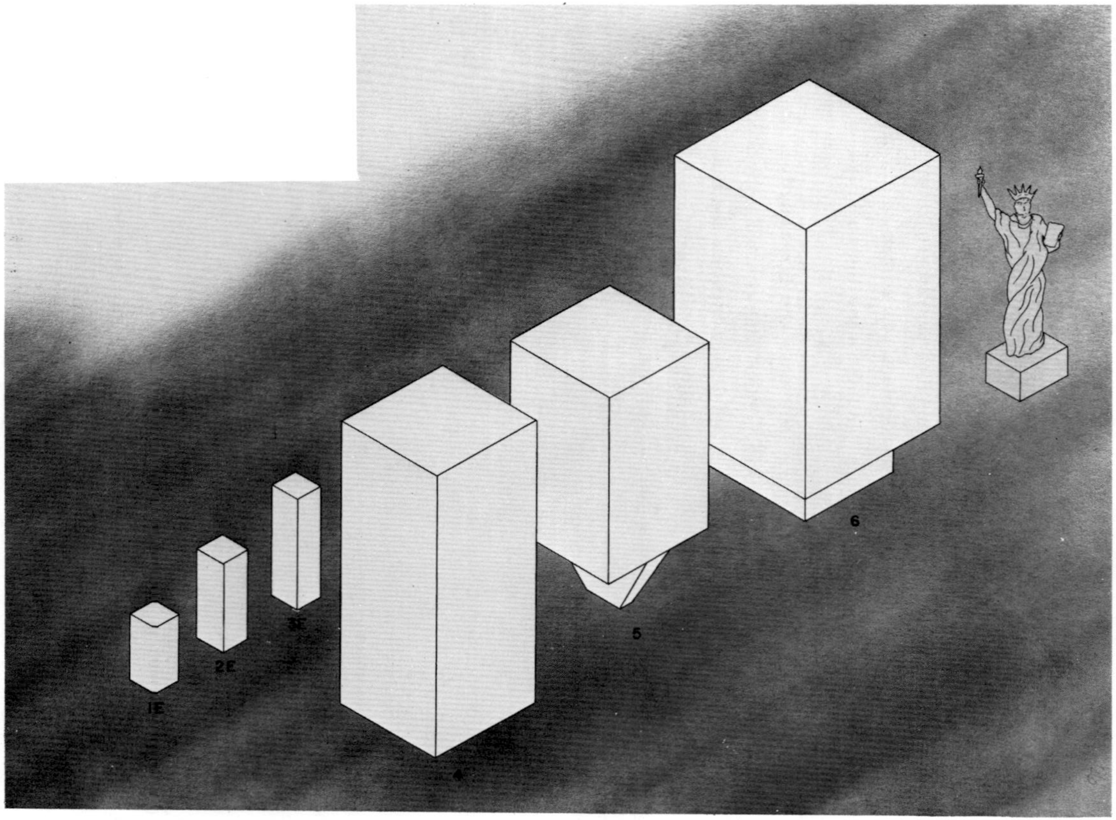

Figure 33-9. Size comparison, Retorts 1-6

A 50-ft barrier pillar is left between each retort to obstruct gas and/or heat flow so that each retort may be individually controlled. While the clusters are in operation, at least one more adjacent cluster already will be prepared and ready for ignition, and others will be under preparation. The prepared cluster provides a buffer for mine-development activity. Prior to rubblizing, a production level below the retort is established to handle liquid products of the retorting process. Also, a gas drift is developed below the production level; an independent system draws gases down through raises that connect the gas level to the production level. Once retorting begins, the production section under the retorts and the entire gas drift system are inaccessible for the remainder of the life of the mine.

At the surface, process gases are first cooled by contact coolers; this is primarily to condense water and to recover entrained hydrocarbons and ammonia ahead of the subsequent sulfur-removal process. Hydrogen sulfide is then removed from the cool gases by a Stretford unit. Products are elemental sulfur and a treated gas containing about 10 ppm total sulfur. Flaring capacity at 25 percent of the total gas flow is provided in case of process interruption. Gas flow can be reduced to this level by shutting off air to the retorts.

About 50 percent of the treated gases are burned in high-pressure boilers to produce steam for utility requirements and for retorting. High-pressure steam is used to drive some of the process gas blowers, with exhaust steam then flowing to retorts and general utilities. Retort steam is provided from gas blower turbine-drive exhaust and from process-water steam generation. The utility requirement is also provided by the process gas blower turbine-drive exhaust steam.

The balance of the treated process gases not burned in the plant boilers is ducted to an adjacent utility plant, which would be built and operated by others.

Figure 33-10 shows the plan view of the mine. Retort clusters are separated by 150-ft pillars and arranged in panels. These panels consist of parallel clusters extending outward from the gas shaft to the tract boundary. The 190-ft panel pillars separate the various panels and contain all development drifts.

Six shafts are provided on the property: a 29-ft diameter production shaft capable of handling 60,000 tons of rock per day; a 34-ft diameter service

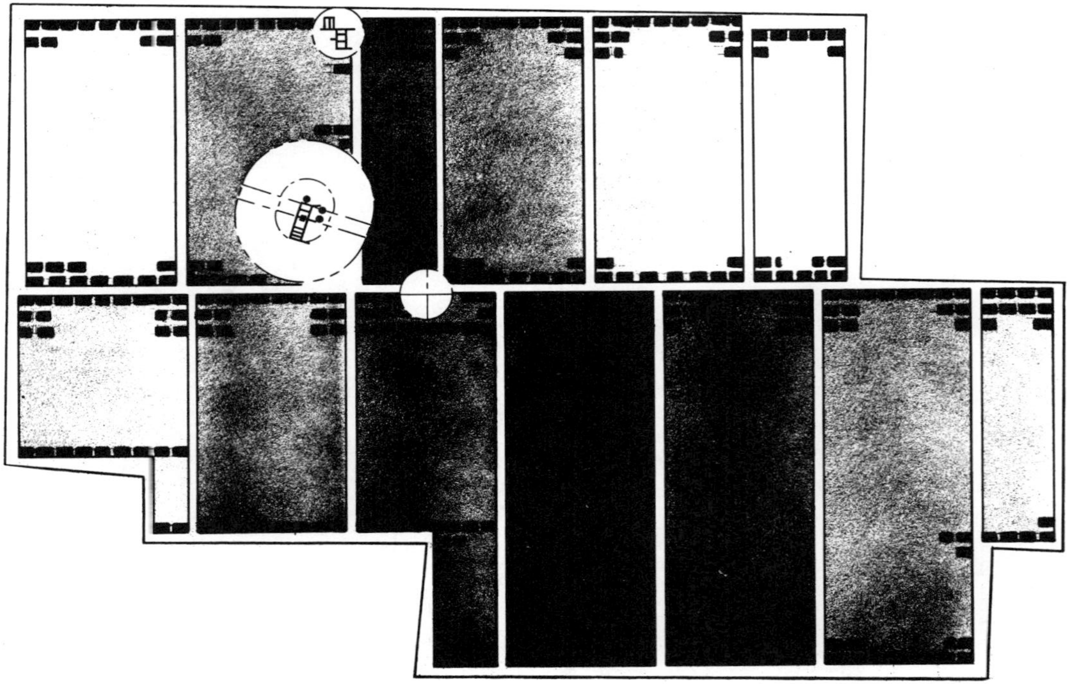

Figure 33-10. Proposed commercial mine plan, modified in situ retorting

shaft; a 34-ft diameter off-gas shaft; and three ventilation shafts of 28-ft, 27-ft, and 15-ft diameter. The latter also doubles as an escape shaft.

A time span of 3½ years is estimated for preproduction mine development after completion of the ventilation/escape shaft. This assumes that all systems and designs are proven and that evaluation and selection of project alternatives will not interfere with the schedule. It also assumes mining productivity begins at a low level and then increases during the preproduction development. Preproduction mine development is continuous. It includes all mine and retort preparation costs up to ignition of the first retort. Preproduction then continues through an intermediate or startup period until full plant capacity of 50,000 barrels per day is reached. During the time of "start up," oil is produced and income is generated to offset operating costs.

The estimated capital costs are shown in Table 33-1, including the costs to get from ignition of the first retort to full production. All costs are based on Feburary 1979 constant dollars. The estimated annual operating costs are given in Table 33-2, which shows a typical year's operating costs for the mine, process, and general facilities.

Definitive cost estimates for a commercial oil shale project can be made only after further experience has been gained and more information is gathered on site characteristics. Estimates have been made to provide a general projection of capital and operating costs which are expected to occur over the life of the project. These estimates assume favorable economic and political climates which allow proceeding on the plan schedules. Individual components of cost may vary significantly or may be eliminated in actual practice.

Several options may present economically attractive alternatives as a supplement to the basic program described here. For example, surface retorting of mined-out shale may prove to be attractive at some future date.

CURRENT ACTIVITIES

Within the last 2 years, Occidental has passed a number of significant milestones. We received substantial funding from the United States Government to demonstrate our technology on a large scale. Retorts 5 and 6 were completed with mixed success. Construction has started on the next demonstration phase, also with DOE support. This phase plans to include operation of two retorts simultaneously.

TABLE 33-1. Estimated capital cost

Initial gross investment	$628,255,000
Contractor's home office and managing operator's cost	88,638,000
Total initial capital	$716,893,000
Additional cost to reach full production:	
Interim capital and operating cost	$318,366,000
Oil sales: 22,287,000 bbl at $14.25/bbl	(317,590,000)
Estimated ammonia sales	(6,000,000)
Net cash outlay	$711,669,000

TABLE 33-2. Estimated annual operating cost

Labor: salaried, hourly, and contract maintenance	$ 54,172,000
Materials and supplies:	
Operating	33,282,000
Maintenance	10,796,000
Utilities:	
Electric power	25,128,000
Natural gas	4,100,000
Total estimated annual operating cost	$127,478,000

We have a new partner in our commercial prototype plant being constructed on the C-b Federal Prototype Lease Tract in western Colorado. Our original partner, Ashland Oil, withdrew from the project, leaving control of the oil shale property to Occidental. Tenneco has purchased a half interest in this project. We have licensed Rio Blanco Oil Shale Company, a partnership of Gulf Oil and Standard Oil (Indiana), to use our technology in their development program. And finally, we have signed an agreement with Morocco to investigate the possibility of using OXY oil shale technology to develop an oil shale facility in that country.

There is increasing recognition that vertical modified in situ retorting is the best available method for recovering oil from shale. Occidental Oil Shale, Inc., intends to commercialize this process on the C-b Prototype Oil Shale Tract in a prudent, businesslike manner. Shaft sinking has already started and is proceeding on schedule. Under the current schedule, shaft sinking would be complete in mid-1982, the first two-retort module ignited in the fourth quarter of 1984, and full production reached in 1986.

In addition to the mass of mining and construction activities that we are undertaking, the project will be under extremely close scrutiny by government agencies as well as a host of environmental groups to ensure that the operation can be conducted in an environmentally controlled manner. To be responsive to this, OXY will perform its work consistent with the various permits that must be issued.

In summary, technology to economically recover oil from oil shale has now been developed by Occidental Petroleum Corporation and we are moving forward aggressively to build the first commercial oil shale recovery plant in the United States.

CHAPTER 34

CHARACTERISTICS OF OIL SHALES AND BITUMINOUS ROCKS OF THE USSR

N. S. Beskrovnyy, V. F. Krymov, and M. V. Tolkachev[1]

1. According to academician N. V. Melnikov, 57 percent of world coal reserves, more than 25 percent of potential natural gas resources, more than 60 percent of peat resources, and 50 percent of oil shales reserves, and a significant portion of world oil reserves and nuclear fuel are concentrated in the Soviet Union. However, the rates and economic indices of developing energy resources in the USSR are greatly influenced by the fact that 90 percent of potential reserves (about 75 percent of proved reserves) of power resources is in the Asian part of the country, while 80 percent of total annual production of power resources is consumed in the industrially developed regions of the European part of the country and in the Urals (Melnikov).

Further progress in the USSR fuel-power complex depends not only on the development of the Eastern areas, but also on discovery and production of new types of natural power resources in the European part of the country--oil shales and natural bitumens, which play as yet insignificant roles in the fuel-power balance of the USSR.

2. The term "oil shales" is used with different meanings. In a broad sense these are high-ash, combustible, solid deposits, containing organic matter of various origins: coal-type bituminous, pyrobituminous, and liptobiolithous shales; high-ash sapropelites; and other lithified rocks with increased content of organic matter (kerogen), which yield large amounts of bitumen during heat treatment and a restricted amount of bitumen under normal temperatures.

Oil shale is a shaly carbonate rock, which differs from the enclosing barren rock matrix by higher kerogen content (10 to 80 percent). Depending on the character of organic matter and its content oil shales grade into rocks containing oil-range bitumens, into sapropelic and humus-sapropelic coals, and into high-ash coal-bearing and liptobiolithous shales (Kalmykov).

---

[1]USSR.

The oil shales are distinguished from coals by three features: (1) the nature of the organic matter (sapropelic but not humic organic matter prevails); (2) the quantity of the mineral substance is greater than that of the organic; and (3) the close microdispersional ties between the organic matter (kerogen) and the rock matrix.

In the USSR, when estimating geological reserves, the boundary between the types of resources is drawn conventionally (Pogrebnov):

- Between oil shales and coal-bearing bituminous rocks according to the maximum calorific value (Q_o^c) 1,200 kcal/kg;
- Between the oil shales and coals according to the ash content, 50 percent, including the content of carbonic acids in carbonates (A^c+CO_2);
- Between oil shales and coal-bearing shales, by the outcome of resins (T^r), 10 percent.

In the USSR, oil shale deposits are known in the formations of the following ages:

- Lower and Middle Cabrian (the Olenek basin);
- Lower and Middle Ordovician (the Baltic basin);
- Upper Devonian (Domanic formation of Timan-Pechora province and Famenian of Pripyatsk basin);
- Upper Carboniferous-Lower Permian (East Kazakhstan);
- Upper Jurassic (the Volga and Timan-Pechora basin);
- Middle Jurassic (the Caucasus deposits);
- Middle-Upper Cretaceous (deposits of the Irhkutsk basin);
- Paleogene (South Tajik basin);
- Neogene (the Ukraine and the Caucasus).

Ultimate in-place reserves of the oil shales in the USSR exceed 190 billion tons and proved reserves, 6.8 billion tons, at the end of 1978 (Pogrebnov). The greatest deposits are in the Estonian Republic, Leningrad region, and Volga basin.

At present 18 deposits of oil shales are identified in the USSR, with in-place reserves of 820 million tons in 1978. In 1978 oil shale production was 35.5 million tons.

3. Accumulations of natural bitumens are reported in most geological regions of the USSR (Beskrovnyy, 1975 and 1979; Valeyev). Among them, heavy, highly viscous oils, malthites, asphalts, and asphaltites are prevalent. They are natural mixtures and concentrates of asphaltic-resinous components of oil,

with a different balance of lighter hydrocarbons. By the VNIGRI classification they are classed together with oil and solid paraffin hydrocarbons as naphthedes (Uspenskiy).

The largest and the most prospective bitumen-petroliferous basins of the USSR are (Valeyev):

- A group of platform basins of the pericratonic syneclises, for instance, the North Caspian basin, with 40 billion tons of bitumen resources, and the Timan-Pechora basin, with 6.5 billion tons of bituminous rocks;
- Orogen basins of the foredeep type (Lena-Angarsk, with ultimate resources of 10 billion tons of bitumens);
- The basins of the intermontane areas and graben syneclises (Tungussk, with resources no less than 2.5 billion tons of bitumens).

According to the structural and geological setting, the following types of bitumen accumulations are defined:

a. Surface effusions, asphalt lakes, and impregnations of outcropping rocks;

b. Rock beds selectively or completely bitumen-impregnated;

c. Vein, pocket, and lens-like accumulations of asphaltites and ozokerites;

d. Accumulations of insoluble carbonized bitumens (kerites, antroxolites) in the form of veins and scattered pockets in the rocks.

The first type of bitumen accumulation is known in regions of present-day tectonic activity and is usually associated with surface oil seepages in fault zones. The known deposits of this type in the USSR are usually small in size and reserves; as a rule, they are surface mineable. Most of them are located in the southwestern part of the USSR and Middle Asia, Azerbaijan, Georgia, Turkmenistan, and Uzbekistan. Outcrops of brea-impregnated rocks, confined to relatively uplifted flanks of salt domes, are as a rule associated with major graben faults and are linearly extended in the direction of these faults in the form of erosional remnants. This fact speaks for genetic association of these phenomena with the breaking up of oil accumulations.

The second type of bitumen concentration, selectively or completely saturating outcropping or shallow rocks, is one of the most widespread. Accumulations of this type often contain significant bitumen reserves in sandstones and carbonates (dozens or hundreds of millions, even billions of tons). Most often they are associated with the flanks of large uplifts (Anabarsk and other shields), within the confines of which oil-bearing deposits either crop out or are brought close to the surface. This type of bituminous rock has a wide stratigraphic range, from Riphean to Quaternary. The main feature of this type of accumulation is its confinement to beds having fair reservoir properties. The Siliger-Markhinsk bitumen accumulation in Western Jakutiya serves as an example of this type deposit. In this locality, in the Upper and Middle Cambrian carbonate sequence, heavily saturated bitumen strata, having a thickness of 0.3 to 5 m, and a bitumen content of more than 5 percent, alternate with selectively saturated horizons (bitumen content 1 to 3 percent) of 0.1 to 1.5 m in thickness and with rocks having a bitumen content less than 1 percent and thickness of 0.4 to 15 m. Similar bitumen distribution is observed in the Melekess (Tatar ASSR and Uljanovsk region) and in the Oleneksk (Jakut ASSR) deposits, with approximate bitumen resources of about 19 and 1.3 billion tons, respectively (Beskrovnyy, 1979).

Stratified bituminous rock accumulations having much smaller reserves are known in Timan-Pechora and Volga-Urals provinces, and also in Mangyshlak, Fergana, Turkmenistan, and Pre-Caucasus. The great advantage of these deposits is thin overburden (from a few tens to a few hundreds of metres).

Blanket, lens-like, and irregular-shaped deposits are typical of many deposits of ozokerite. The deposits of Chelrken, the West Ukraine, Fergana, and the others are included here.

The third type of accumulation (vein, pocket, or lens-shaped solid bitumen deposits) is widespread, but has small reserves. Similar accumulations occur under various geological conditions, mainly in zones of rock crushing and faulting. Their specific feature is the purity of the organic substance and low mineral impurity. For instance, the Sadkinskoe deposit (Orenburg region) is represented by an asphaltite vein and is confined to almost vertical joint crossings in sandstones, limestones, and marls of the Tatar stage. The vein length is 680 m and the width of outcrop is not more than 20 m. Downdip, it is traced to a depth of 150 to 200 m, where its thickness is 2 to 5 m. The depth of the producing mines reaches 120 m and the vein wedges out in the depth interval from 230 to 300 m.

The Boreslav ozokerite deposit in the West Ukraine is composed of a series of bed-like bodies of ozokerite in the Neogene sandstones. It forms veins and pockets in Miocene and Upper Oligocene clays and shales. The vein thickness reaches several metres, can be traced to a depth of more than 330 m, and the ozokerite content in the rocks is 2 to 3 percent. The depth of surface mining at present reaches 140 m.

Vein form is also typical for many other ozokerite deposits.

Accumulations of hard, insoluble bitumens-kerites, anthraxolites, and shungites, which occur as veins and layered impregnations, or in a form of beds among metamorphosed rocks and in geothermal formations--are grouped in a special type of deposit. As an example, the kerite accumulations in Paleozoic rocks of Turkmenistan and Alay Ridges, and Upper Cretaceous limestones of the Gaurdak sulphur deposit, can be mentioned.

Shungite accumulations, usually in the form of beds, are reported from Proterozoic sediments of North Onega region. The carbon content reaches 50 to 60 percent, and small veins in these strata are almost entirely made up of practically pure shungite.

Despite the wide occurrence of bitumens in many areas of the USSR, most of them are still poorly known. Commercial reserves are evaluated only for three deposits (Valeyev):

1. Pervomaysk (Kuybyshev area)--1.876 million tons
2. Notanebsk (Georgia)--0.438 million tons of bituminous rocks
3. Sadkinsk (Orenburg area)--0.194 million tons of asphaltites. This deposit is being produced at present.

REFERENCES

Beskrovnyy, N. S., Goldberg, I. S., 1979, Geology and Resources of Natural Bitumens; First All-Union Seminar: Geology of Oil and Gas, 1979, no. 6, p. 61-62.

Beskrovnyy, N. S., and others, 1975, The Natural Solid Bitumens in the USSR--An Important Raw Material Reserve of the Natural Economy: Geology of Oil and Gas, no. 4, p. 14-20.

Kalmykov, G. S., 1979, Oil-shales as the resource of liquid raw material, in Ditmar, V. I., ed., Geology of bitumens and their host rocks: Moscow, Izdatel'stvo Nauka, p. 88-92.

Melnikov, N. V., and Shelesy, V. A., 1975, A fuel-power complex of the USSR: Planned Economy, no. 2, p. 8-19.

Pogrebnov, N. I., ed., 1978, Geology of coal and oil-shale deposits in the USSR, v. 12: Moscow, Izdatel'stvo Nedra.

Uspenskiy, V. A., and others, 1964, The bases of the genetic bitumen classification: Vsesoyuznyy Neftyanoy Nauchno-Issledovatel'skiy Geologorazvedochnyy Institut (VNIGRA) Trudy, v. 230, 267 p.

Valeyev, R. N., and others, 1979, The bitumen-petroliferous basins and estimation of natural bitumen resources of the USSR: Sovetskaya Geologiya, no. 5, p. 34-45.

CHAPTER 35

OIL SHALE AND OIL SAND IN ZAIRE

Manzukula Yimbu[1]

OIL SHALES

Oil shales were discovered in territory that is now part of Zaire in 1909--in Haut-Zaire, south of the village of Kisangani (Fig. 35-1). But it was not until the 1950's that the idea of exploiting these resources occurred and experimental projects were conducted by the Grands Lacs Railroad in conjunction with a Swedish firm. However, research into oil shales practically ceased almost 20 years ago. How can we explain the little interest displayed up to the present time in the exploitation of these energy resources, lying as they do in an isolated region confronted with so many energy problems? This report will show that the reasons are both technical and economic.

The Reserves and Physical and Chemical Nature of the Oil Shales in Zaire

The Kisangani basin has been divided into three blocks called Lualaba, Oviatuku, and Haute-Lilu. The deposits are situated along the Zaire (Congo) River and its tributaries, the Oviatuku and the Lilu. There are probably about 117 million tonnes of reserves in this basin, distributed as follows:

Deposit	Reserves (1,000 tonnes)	Max & min oil content (ℓ/tonne)	Sterile/ mineral ratio
1. Lualuba			
Upper	900	100/75	3/1
Lower	3,700	-	-
Left bank	10,600	100/75	5/1
Kewe River	2,900	100/75	3/1
Iku and Songa River	1,550	100/75	3/1
2. Oviatuku	3,000	64/60	3/1
3. Haute-Lilu	94,400	100/60	1/1
Total	117,050	-	-

[1]Service Presidentiel d'Etudes, Kinshasa, Zaire.

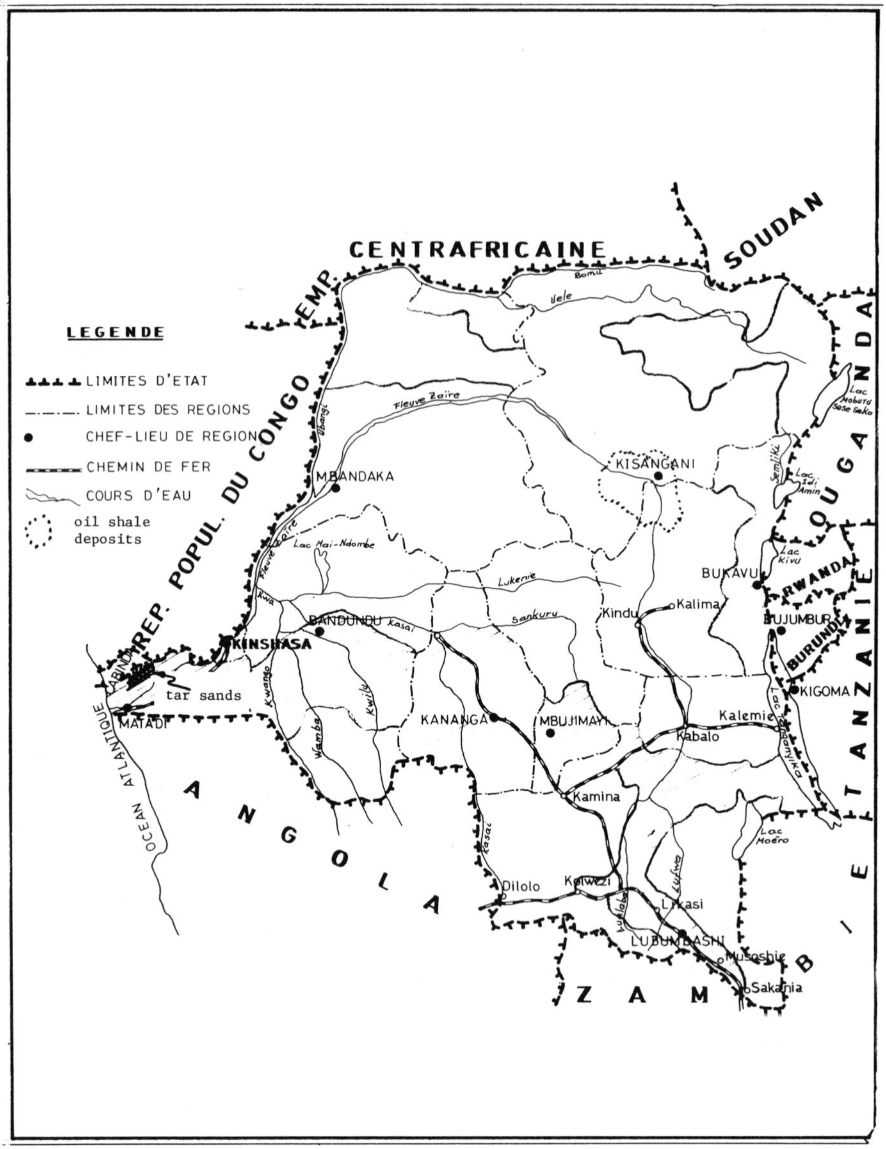

Figure 35-1. Oil shale and tar sand deposits in Zaire

The main physical and chemical characteristics of the oil shales in Zaire are:

oil content (ℓ/tonne):	100/60
water content (% by weight):	6.0
residues (% by weight):	80.2
liberated gas (cu Nm/tonne):	27.0
calorific value (kcal/kg):	1,321.0

The rate of kerogen-to-oil conversion, the specific gravity, and the hydrogen, nitrogen, and sulfur contents have not yet been determined.

Although the total amount of the reserves is significant, it should be noted that the individual deposits are small. The mineral-to-sterile ratio may also lead to problems in making the deposits pay off.

In comparison with other regions of the world it seems that the oil content of the Zaire shales is very near that of shales from Colorado and Autun, the water content close to that of shales from Irati in Brazil, the amount of solid distillation residues about the same as that observed in the oil shales of Nova Scotia, and the amount of gas close to that observed in Brazil and France.

Utilizing the Oil Shales of Zaire

The oil shales of Zaire will be capable of a relatively important part in absorbing the energy crisis in eastern Zaire. Whatever method is employed should enable large amounts of raw oil to be obtained, because there are not many uses for gas in this region.

A project for utilizing these resources is already in existence. In its final stage, the outputs of the three products in greatest demand in this basically rural and agricultural environment will be:

gas oil:	60,480 m^3
gasoline:	60,480
kerosine:	45,360

If consumption for 1975 is taken as a base, this output would make the Northeast completely self-sufficient in these products.

In 1974, the investment required to construct a plant producing these amounts of the petroleum products in question was estimated at 7 million zaires, equivalent to U.S. $14 million. Such a plant would utilize the Lurgi process.

The resulting gas could be used to help run the plant, and the ash utilized in the production of pozzolan cement.

For reasons that are mainly financial, this project has not yet been realized, and should not be, in fact, until detailed studies of the exploitable deposits have been conducted. Another factor to be taken into consideration is the danger of flooding in the Kisangani basin if the hydroelectric plans for the Zaire River between the Kisangani and Ubundu are approved and carried out.

OIL SANDS

There are rocks impregnated with bitumen in Bas-Zaire. They consist mainly of sandstone and oil sands and of asphaltic limestone of Lower Cretaceous age.

The exploitation of these deposits was not considered before the colonial Ten-Year Plan (1950-1960) was put into effect, because of the lack of demand in the Belgian Congo, as Zaire was then known.

In 1950, the firm of Forminière was requested by the colonial government to make plans for the exploitation of these rocks. Together with the Belgian Asphalt Corporation (ASPHALTCO), Tele, and the Congo Corporation for Commerce and Industry (CCCI), Forminière then constituted the Congo Syndicate for Bitumens and Asphalts (SYBICO), which later became the Congo Corporation for Bitumens and Asphalts (SOBIASCO), which exploited the rocks until 1958.

The Reserves and Physical and Chemical Nature of the Oil Sands of Zaire

Based on experimental borings carried out by the Zaire Petroleum Research Corporation (SOREPZA), the President's Research Service (SPE) has made calculations of the potential reserves. They amount to more than 300 million tonnes of oil sand, assaying at 11 to 20 percent bitumen.

Nevertheless, it seems that the reserve potential of all sites south of the Lubuzi River along the Zaire-Cabinda border would also be very significant, amounting to 2.5 billion tonnes. The amount of extractable oil has been estimated at 3 million tonnes.

The mean volumetric content of the oil sands in Zaire is sand, 65.4 percent, bitumen, 23.4 percent, and void space, 11.2 percent.

The sand is quartz, of a light reddish-brown color, with a density of approximately 2.65.

Utilizing the Oil Sands of Zaire

Up to 1958, when SOBIASCO ceased operations, 67,313 tonnes of oil sands were taken out. These sands were used to asphalt the airport at the coastal town of Muanda and the Boma-to-Tshela and Muanda-to-Banana roads. Thus, all of these projects were carried out in Bas-Zaire, within 80 km of the deposits.

SOBIASCO has attempted to extract the bitumen content of these sands, but operations were abandoned because the bitumen extracted cost 5 or 6 times as much as that obtained from the refinery.

In the hope of bringing about a substantial improvement in the foreign exchange situation by utilizing local bitumen and of completing its plans for asphalting the road system as inexpensively as possible, the Zaire Bureau of Roadways has asked the President's Research Service (SPE) to carry out economics research to determine the feasibility of exploiting the country's oil sands.

Major obstacles to following up the exploitation of the Bas-Zaire oil sands may turn out to be the disadvantageous sterile-to-mineral ratio and clogging of the material, which makes for serious difficulties in working the minerals.

Within a perspective where appropriate solutions to the problem of clogging in oil sands will be found and as a consequence render the exploitation of these resources feasible, the utilization of oil sands like those used to asphalt the roads of Bas-Zaire is conceivable. The extraction of bitumen from the oil sands of Zaire by the Clark hot-water process is also possible and would enable this material to be used in asphalting the roadways of other regions of the country as well.

REFERENCES

G. Passau, La géologie du bassin de schistes bitumineux de Stanleyville - A.S.G.B. + 45 (1921-1922).

Compagnie Belge D'Entreprises Minieres, p/c, SICAI, Schistes bitumineux de Kisangani, dossiers 1 et 2 (1971).

Italpianti, Studio di pnefattibilita per Impianti di produzione cemento in Zaïre B M C /2 473, (June 1974).

R. Brosius, Exploitation des roches bitumineuses et asphaltiques du Con go Belge, Tomme XXIV fasc. 4, Bruxelles (1953).

R. Van Ganse, Notes sur les sables asphaltiques du Mayumbe (1951).

G. Fortems, Etude des couches crétacées de l'entre Lukula - Lubuzi et Lundu (1958).

CHAPTER 36

CURRENT IN SITU OIL SANDS PILOT PROJECTS IN ALBERTA, CANADA

D. E. Towson[1]

Alberta's Oil Sands contain one of the world's largest accumulations of crude oil. To date, however, only a very small amount of this oil has been produced, primarily because of lack of the required recovery technology. In this chapter, I would like briefly to review some of the field pilot work currently being carried out in an effort to develop this technology.

I will begin by reviewing some of the facts relating to the Alberta Oil Sands. As shown in Fig. 36-1, the oil sands are divided geographically into four main deposits--Athabasca, Wabasca, Peace River, and Cold Lake. The bitumen in these deposits is contained in complex sand reservoirs of Lower Cretaceous age. Some important points to note about the oil sands are:

1. The deposits are estimated to contain close to 160 billion m^3 of bitumen in place. This compares to the estimated 8 billion m^3 of light and heavy conventional oil remaining in known reservoirs in Western Canada.
2. The depth of the deposits varies from 0 to 900 m.
3. Less than 10 percent of the oil sands is considered amenable to surface mining methods. It is generally accepted that some form of in situ recovery method will be required for the remaining 90 percent.
4. The bitumen contained in the oil sands is essentially immobile at initial conditions, even in the deeply buried deposits. It is really an asphaltic substance which is used by the Indians to patch canoes and which the City of Edmonton used as a road-paving material as early as 1915.

Returning now to the question of recovery technology, various experimental projects have been carried out in the Alberta Oil Sands during the past 75 years. The initiation and continuation of these projects has tended to be cyclic in nature and it wasn't until 1967 that the first commercial production was realized, when the Great Canadian Oil Sands surface mining operation went

---

[1]Manager, Technical Services, Petro-Canada, Calgary, Alberta.

Figure 36-1. Oil sands deposits of Alberta

on stream. This was followed 11 years later by the Syncrude project, using similar technology. As was pointed out earlier, however, less than 10 percent of the oil sands is considered amenable to surface mining. Most requires the application of in situ recovery techniques.

The mining schemes have tended to dominate the press coverage of the oil sands because they are commercial projects. What is not so commonly known outside the industry is the extensive program of field pilot projects being carried out in an effort to develop in situ processes. This changed in 1978 with the announcement that Esso Resources received conditional approval to build a commercial, in situ project to produce 22,000 m^3 per day of synthetic crude at Cold Lake. The scheme involves production of bitumen by steam stimulation. Esso has been experimenting in Cold Lake since 1964. Most of the work has been directed to development of the steam-stimulation process. This involves injection of steam into a well completed in the oil sand reservoir in order to reduce the bitumen viscosity and provide some drive energy, followed by production from the same well. This cycle is repeated until insufficient additional bitumen is produced to offset the steam costs.

In addition to Esso, a number of other companies are evaluating steam stimulation at Cold Lake. It should be noted that many of these projects are funded by more than one company and in some cases part of the funding comes from government agencies such as the Alberta Oil Sands Technology and Research Authority. For simplicity, in this paper I will usually refer only to the operator in discussing the pilots. Chevron and Norcen at Cold Lake and Gulf Oil at Wabasca are evaluating steam stimulation or a modification of steam stimulation.

Gulf also recently completed an evaluation of steam stimulation at Cold Lake and British Petroleum is evaluating a combination steam-stimulation and combustion process at its Cold Lake pilot. Other steam-stimulation pilots include Union Texas and World Wide Energy at Cold Lake and Union Oil at Athabasca.

In addition to steam stimulation or huff and puff, steam can also be used in a drive process. Here steam is injected into a well to heat and drive the bitumen to a second well. The main problem with this process is injectivity, that is, the semi-solid bitumen in the oil sands prevents the injection of sufficient steam to accomplish a drive. Shell has been able to overcome this problem at Peace River by utilizing a thin water layer at the base of the oil sands. Steam is injected into this water zone and the pressure cycled, moving the steam up into the reservoir and the bitumen down into the flow channel where it is carried to a production well. The application of this process is of course limited to areas having a thin underlying water zone. Two modified steam-drive processes are being tested by Texaco in shallow oil sands in the Athabasca deposit.

In addition to injecting steam, lowered bitumen viscosity can also be produced by injecting air and burning part of the bitumen to heat the remainder. Amoco is currently testing this in situ combustion process in Athabasca.

The in situ processes described so far have two things in common; they all rely on the addition of heat to reduce bitumen viscosity and they all involve injection of a fluid to achieve the heating. As pointed out previously, fluid injection is seriously hampered by the bitumen itself blocking the flow channels. In an effort to overcome this problem, Petro-Canada, as operator for a group consisting of Petro-Canada, Canada-Cities Service, Esso Resources, and Japan Canada Oil Sands, is constructing a plant to evaluate electric preheating in the Athabasca Oil Sands. In this case electric resistance heating is used to preheat the bitumen and make it mobile. Steam or some other fluid can then be used to drive it to a production well.

Another constraint on oil sands development is the small area influenced by a production or injection well. This results in the need for a large number of wells and consequently high capital investment. One means of reducing well costs is the Mine Assisted In Situ Recovery Process or MAISRP. Here the wells originate not from the surface but from mine tunnels in or below the oil sands. Petro-Canada is also operating a pilot test to evaluate the feasibility of drilling and completing long horizontal wells into the oil sands. Canada-Cities Service, Esso Resources, Gulf Canada Resources, and Husky are coventurers in the project. The initial work is being carried out at a site where the oil sand is exposed at the surface. A successful test here will lead to an underground test.

The in situ oil sands pilot projects are summarized in Table 36-1. The length of the list indicates the high level of activity in the area of in situ recovery pilot projects. This demonstrates the dedication of the industry to developing the technology to unlock this vast resource. Esso Resources' scheme is, I am sure, only the first of many commercial ventures. It should also be noted, however, from the cost and life data that the road to commercialization is both long and costly.

TABLE 36-1. In situ oil sands pilot projects

Operator	Location	Estimated cost ($ million)	Start date	Projected life years
Amoco	Athabasca	71	1976	8
B.P.	Cold Lake	18	1978	7
Chevron	Cold Lake	?	1977	5
Esso	Cold Lake	50	1972	8+
Gulf	Cold Lake	5	1977	2
Gulf	Wabasca	8	1975	5+
Murphy	Cold Lake	2+	1974	6+
Norcen	Cold Lake	10	1976	4+
Petro-Canada	Athabasca	5	1979	1
Petro-Canada	Athabasca	35	1980	5
Shell	Peace River	58	1977	10
Texaco	Athabasca	?	1973	7+
Union Texas	Cold Lake	2.5	1975	5+
Weco	Cold Lake	10	1974	6

CHAPTER 37

THE MINE-ASSISTED IN SITU PROJECT--AN UPDATE

P. C. Quinn

In an earlier paper I described in detail a feasibility study that was conducted to determine if the mine-assisted in situ oil recovery technique could be applied to the extraction of bitumen from the Athabasca deposit. The results of that study, summarized in Table 37-1, indicated that bitumen could be produced at a cost of less than $7 per barrel (1976 dollars).

TABLE 37-1. Summary of design criteria

	Case A	Case B
Net bitumen production--BPD	100000	100000
Gross bitumen production--BPD	122280	152800
Oil in place--MB/acre	300	300
Steam/oil ratio	2/1	4/1
Recovery--percent	45	30
Cell life--years	5	10
Production/cell--BPD	10190	3400
Cells on production	12	45
Replacement cells/year	2.4	4.5
Drilling/cell--M feet	610	75
Tunneling/cell--M feet	16	16
Cost of bitumen--1976 $/BBL	5.03	6.06

The most significant variable in the mine-assisted recovery technique is the drilling and completion of horizontal injection/production wells in the oil sand formations.

Petro-Canada, Canada Cities Service Ltd., Gulf Oil Canada Ltd., Husky Oil Operations, Ltd., and Esso Resources Canada Ltd., entered into a joint venture agreement to test the feasibility of horizontal drilling and completion, through a surface demonstration test in the Athabasca River valley.

The Mildred Lake site was chosen as the test site because of the accessibility to a limestone/oil sand interface. This site had the added advantage of being serviced by an all-weather road. Power and communication were also readily available.

On September 5, 1978, a contract was awarded to engineer and construct the surface demonstration test facility.

The Energy Resources Conservation Board of Alberta approved our application to construct and operate a pilot facility to test horizontal drilling and completion on November 16, 1978.

Field construction started with site preparation at Mildred Lake on November 20, 1978, and was completed on March 30, 1979. A total of 172,000 m^3 of overburden was removed to expose the limestone oil sand interface.

The drilling operation began in May 1979. Three horizontal wells were drilled and completed.

Figure 37-1 shows the well locations. Six vertical observation wells were also drilled as part of our program. The wells were collared in the Devonian limestone, which underlies the McMurray formation, and the wells were drilled at an angle of +5 degrees from horizontal.

The wells were drilled into the limestone with a 219 mm bit for a total distance of 210 m and 168 mm casing was cemented. The oil sand was then drilled with a 149 mm bit to a total length of 310 m. The oil sand portion of the well was completed with a 115 mm slotted liner. One well was drilled a total distance of 543 m. The wells were spaced approximately 8 m apart. The well drilling portion of the test program was completed in June 1979.

In drilling the horizontal wells, we experimented with sonic drilling without mud circulation and sonic drilling with mud circulation and a rotary drilling system known as "The Journal Bearing Technique."

Figure 37-2 is a simplified flowsheet of our steam injection/bitumen handling facilities. Construction of these facilities was completed in September 1979 and the first steam was injected into the formation at the beginning of November.

The steam injection phase of the project is designed to evaluate the following:

1. Effects of thermal expansion in the casing
2. Movement of the well bore during production
3. Success of the high temperature cement used

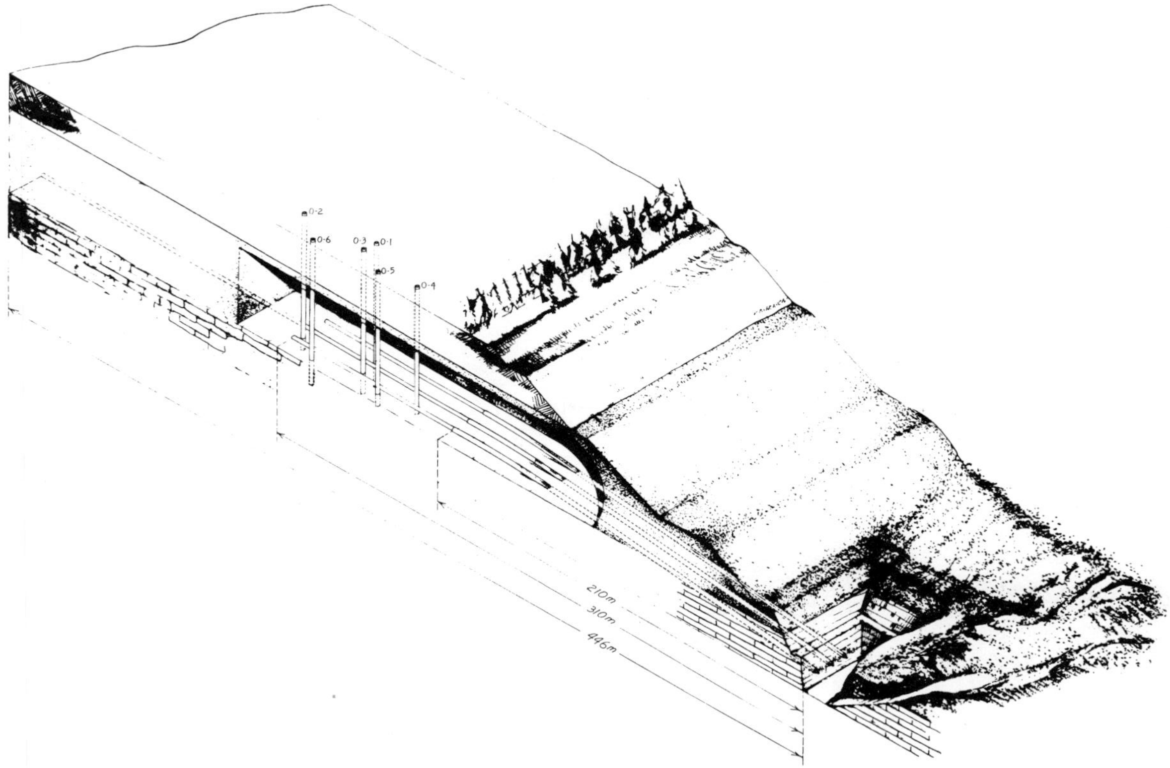

Figure 37-1. Mine-assisted in situ project well locations

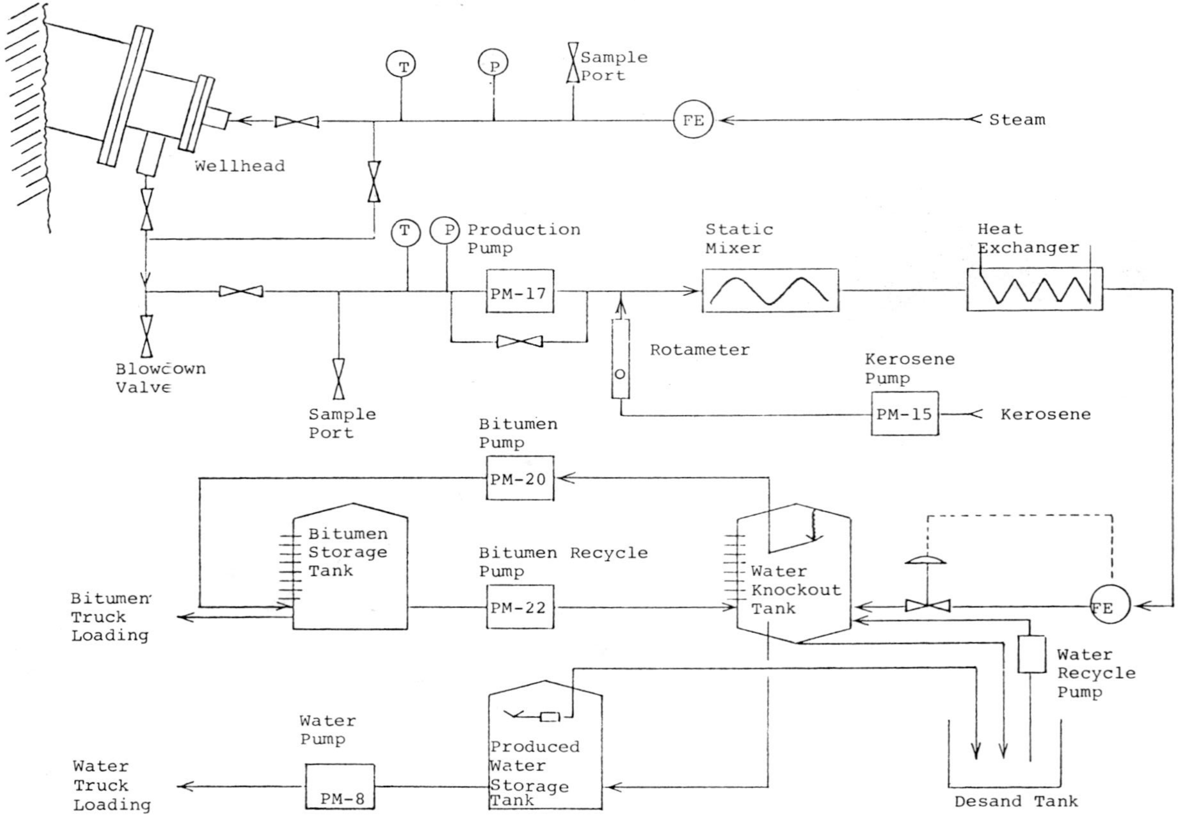

Figure 37-2. Simplified flowsheet of skam injection and bitumen handling facilities

4. Amount of sanding in the well
5. Extent of heat dispersion and the thermal characteristics of the oil sand
6. Time requirements for each cycle of the initial production--injection scheme
7. Initial production--injection rates and volumes

The present schedule calls for a 1-year steam injection/production test, at which time the first phase of the project will be completed.

The test results to date have indicated that we can successfully drill and complete horizontal wells in the oil sands. Results from the steam injection portion of the tests will enable us to design thermal completions for an underground test.

REFERENCE

Quinn, P. C., and Duncan, G., 1980, The mine-assisted in-situ project, in Meyer, R. F., and Steele, C. T., eds., The future of heavy crude oils and tar sands: New York, McGraw-Hill, in press.

CHAPTER 38

THE AVAILABILITY OF HEAVY CRUDES IN LATIN AMERICA

Francisco Gutierrez

INTRODUCTION

In 1976, UNITAR sponsored an international conference in Laxenburg, Austria, to analyze the availability of hydrocarbons worldwide. In 1977, Venezuela and Canada presented a symposium in Edmonton, Alberta, on heavy crudes and bituminous sands. In June 1979, UNITAR held the First International Congress on Heavy Oil and Tar Sands. All these subjects were also discussed at the World Congress on Energy, Istanbul, Turkey, in 1978 and at the World Conference on Petroleum, Bucharest, Romania, in September 1979. At all these events the present author presented papers on the study and investigation of the Orinoco Petroleum Zone and on the presence of heavy crudes in Latin America. The main purpose was to call the world's attention to the importance of heavy crudes in the international energy balance.

Mainly for economic reasons, heavy crudes have not been adequately enough developed to enter into the commercial hydrocarbon picture on a large scale. Nevertheless, the increasing difficulties encountered in finding new reserves of conventional crudes, and their increasing price, should lead, as it already seems to be doing, to more attention being paid to the study and exploration of the enormous deposits of heavy crudes that exist throughout the world.

Locations in Latin America

In Alberta last June, the present author spoke about the geological and geographical distribution of heavy crudes and petroleum lutites in Latin America.

From Mexico to Argentina, with the exception of the Central American countries, the Latin American nations have reported the presence of major and minor deposits of heavy crudes and bituminous lutites.

Venezuela has the largest accumulation of known liquid heavy crudes, and Brazil the largest volume of petroleum lutites. Nevertheless, Peru and Bolivia have, in the Amazon region, and Argentina in the North, used borings to confirm the presence of heavy crudes in significant sedimentary areas. Argentina and

Chile have bituminous sands, and Brazil has a large number of deposits of lutites and aligned bituminous schists in the North, East, and South.

Figure 38-1 shows the areas where Mexico has reported the presence of appreciable amounts of heavy crudes. The Mexican oil industry has promoted studies on the processing of heavy crudes at centers in Tampico and Salamanca, and the Mexican Petroleum Institute has been conducting research on catalysts for several years.

Figure 38-1. Locations of heavy crude deposits in Mexico

Figure 38-2 shows the location of deposits of heavy crudes in Venezuela, the most familiar of which is the Orinoco Petroleum Zone. The figure also shows the Guanoco Region in the Northwest, the Barinas Region in the Southwest, the Urdaneta and Boscan Region west of Lake Maracaibo, and the Costa Bolivar Region east of the lake. The total estimated volume of all these zones together is between 2.3 and 2.5 X 10^{12} barrels of petroleum in situ. Note that I do not refer to these volumes as either resources or reserves. They are accumulations, estimated in accordance with available methods and knowledge, that the Venezuelan oil industry is developing medium- and long-term plans to explore and quantify.

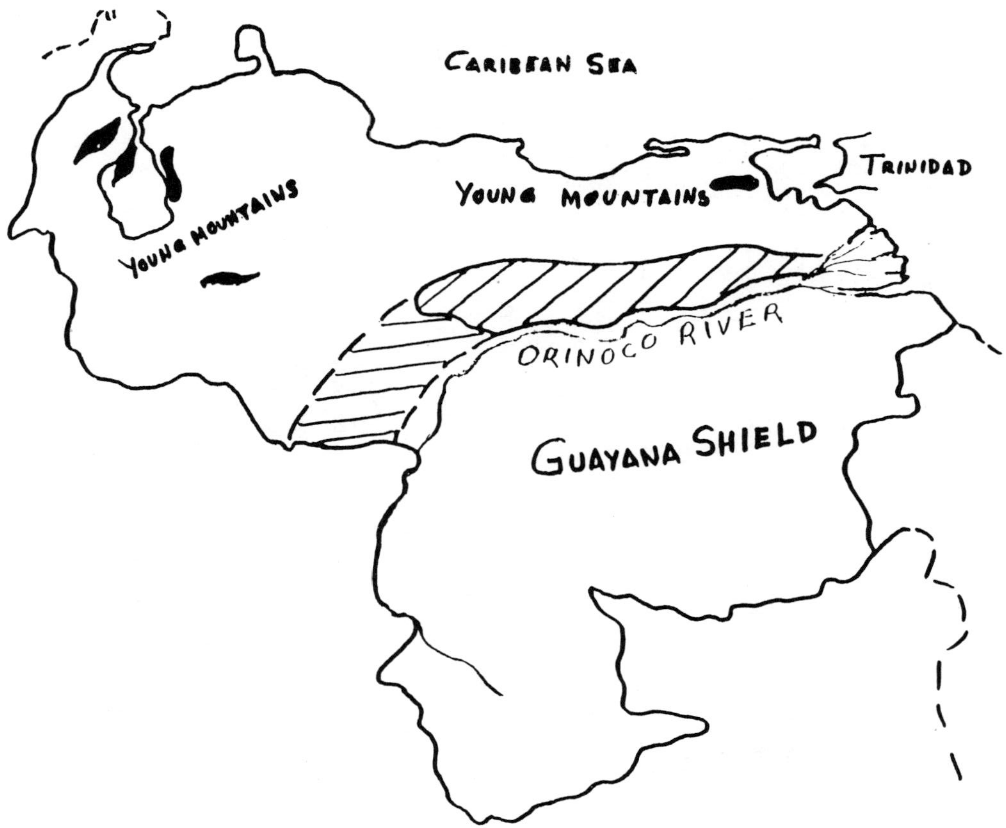

Figure 38-2. Locations of heavy crude deposits in Venezuela

Figure 38-3 shows the different regions in which the presence of heavy crude has been detected, either in sands or calcites or in the form of lutites or bituminous schists. There are large accumulations of these lutites and schists along the piedmont extending from the cordillera to the shield area in Colombia, Ecuador, Peru, and Argentina, bordering the Brazilian Shield on the Northeast and South.

As for the regions outside of Venezuela, I estimate volumes of 8×10^{11} to 1×10^{12} barrels of petroleum in situ.

CONCLUSIONS

1. The author believes that the so-called energy crisis has several components, only one of which may be the decreased deliveries of and not the lack of hydrocarbons.
2. Energy, especially energy from the hydrocarbons, must increase in cost in order to be produced, and therefore in price in order to be sold. The economies of both the developed and developing countries will depend on the incremental costs of development demanded by the incorporation of new technologies.
3. The factor of accelerated incremental energy costs incidental to the incorporation of new technologies will necessarily be influenced by the effective rates of conservation as well as by the means employed to preserve the environment.
4. For heavy crudes to enter definitely and effectively into the commercial energy picture it will be necessary
   a. for production methods to develop substantially, which will lead to improved recovery factors,
   b. for research to proceed on upgrading heavy crudes, which will lead to reviving transportation and to the subsequent utilization of such crudes,
   c. for economic incentives to be established to induce such development if the latter is to be possible.

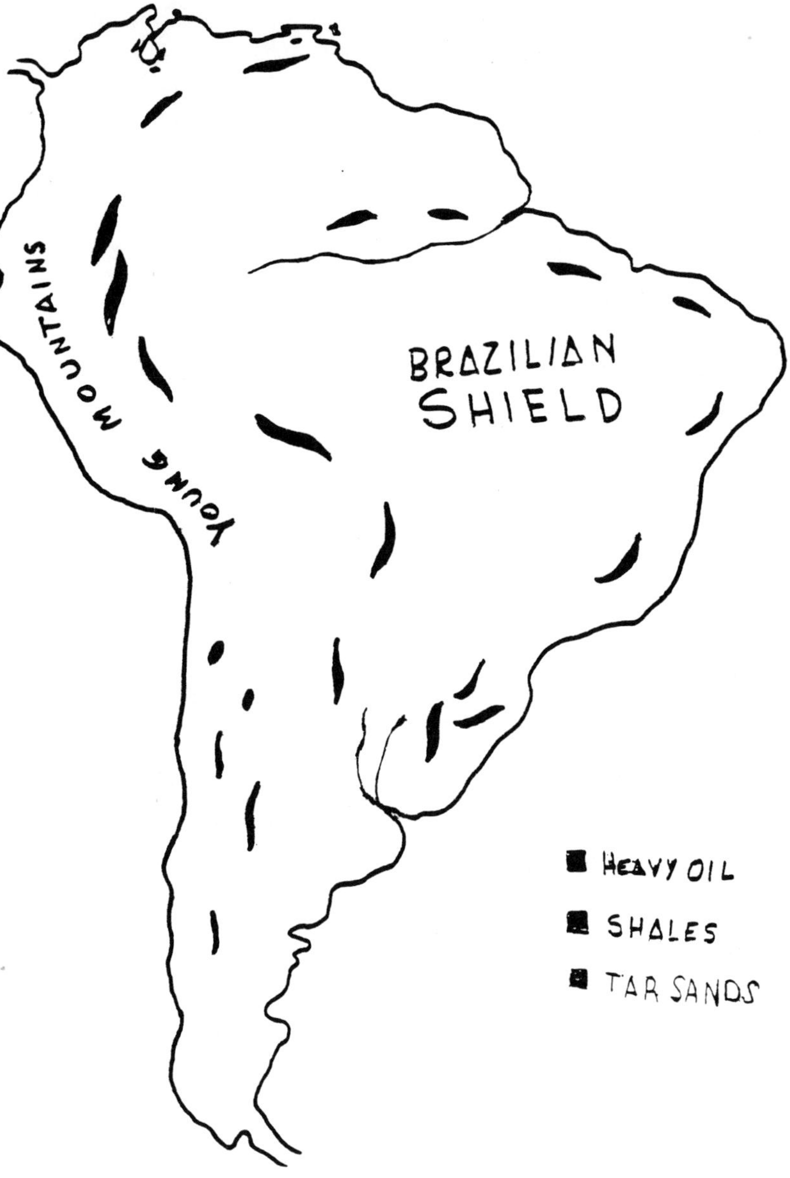

Figure 38-3. Heavy crude occurrences in South America

CHAPTER 39

EVALUATION AND DEVELOPMENT OF THE ORINOCO OIL BELT, VENEZUELA

Carlos J. Borregales L.[1]

SUMMARY

The Orinoco Belt is considered one of the largest untapped oil accumulations in the world. Production tests performed in about 15 wells have shown that the oil could be produced at high rates by primary means and therefore, from the exploitation standpoint, the reservoirs can be considered conventional. The Orinoco Belt will be a very important oil supply source in the future.

Petroleos de Venezuela in 1979 initiated an intensive exploration program, aiming primarily to delineate and quantify hydrocarbon resources by 1983 and to select on a shorter term basis the most prospective areas for exploitation. Present plans call for the production of about 1 million B/D by the year 2000. Two definitive projects are already planned. The first has the objective of producing, by 1988, 125,000 B/D of upgraded crude from an area, located in the Monagas State, which is under responsibility of Lagoven, an affiliate of Petroleos de Venezuela. For this particular project, the firm Bechtel has been contracted as planning advisor and other experienced and reliable companies are being used for specific studies and designs. The second project comprehends the development of 75,000 B/D of conventional heavy oil using the existing production infrastructure in the south of the State of Anzoátegui. In both projects, existing proven technology and experienced contractors will be used for engineering, construction, and project management.

Since most of the produced oil from the Orinoco Oil Belt may require upgrading, a significant effort will be required from the consumer countries to build enough refining capacity for this type of crude.

ORINOCO OIL BELT

The Orinoco Oil Belt is located in the southern part of the Eastern Venezuela basin and contains a large complex of hydrocarbon accumulations covering a total area of about 42,000 km^2 (Fig. 39-1). It includes a wide range

[1]Petroleos de Venezuela.

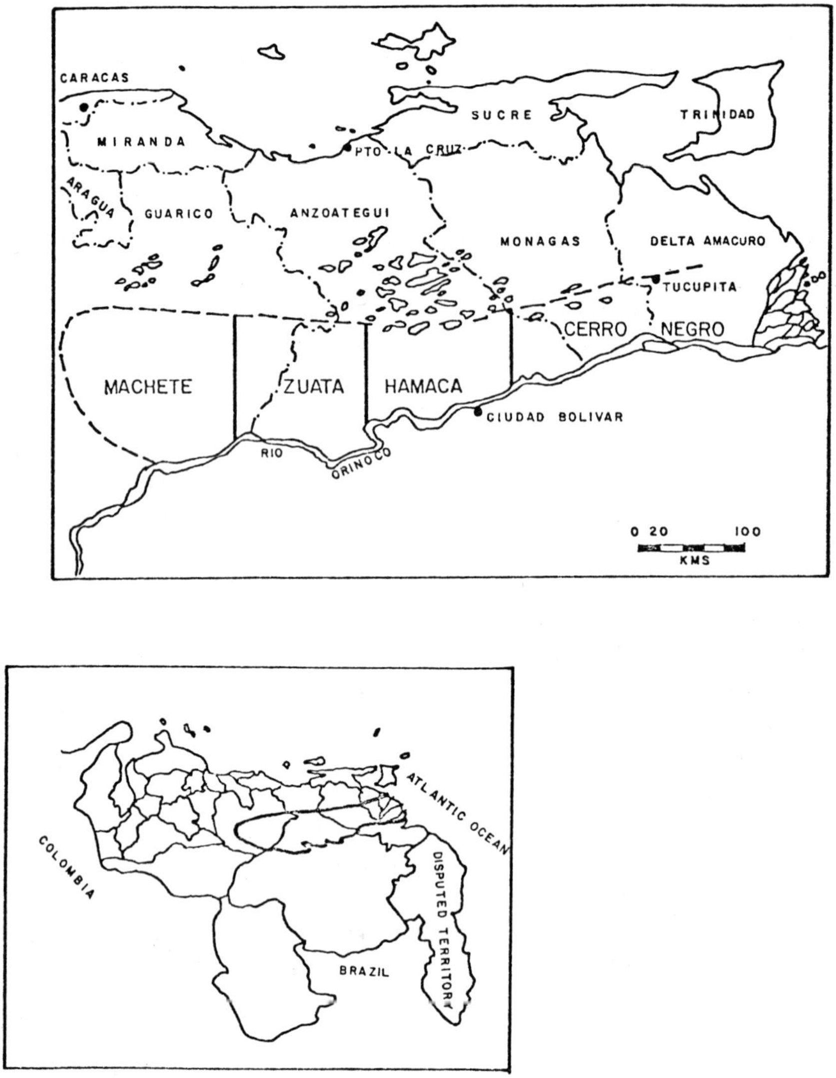

Figure 39-1. Orinoco Oil Belt location map

of reservoir characteristics and crude properties. Some crude has a gravity higher than 15° API, a fair amount ranges between 10° and 15° API, and most of it is less than 10° API.

The estimation of the oil in place is uncertain due to a paucity of information, lack of knowledge of the extent of the reservoirs, and the geological complexity of the area. However, it is estimated that the resources could exceed 1 trillion barrels of oil.

The oil is mobile at reservoir conditions due to the high reservoir temperature and therefore is producible by primary conventional means. Production tests performed in some wells have resulted in productivities between 50 and 600 B/D/well per oil-bearing sand and each well may have up to 10 different productive intervals. It has also been determined that the heaviest oil found in the area (8° API) is producible at relatively high rates, between 50 and 500 B/D, depending on pay zone thickness.

The oil recovery will basically depend on production mechanisms. It is considered that solution gas drive and compaction will be the main driving forces.

If solution gas drive is the only active production mechanism, the average primary recovery plus steam soak recovery will be about 8 percent of the STOIIP but it may be as high as 24 percent if compaction happens to be an important drive mechanism.

EVALUATION PROGRAM

The estimated volume of oil in place depends on the assumptions made by the geologists regarding the structural pattern, trapping characteristics, and source and migration of oil. The first estimate of oil in place was made by Galavis and Velarde [1] and was presented at the VII World Petroleum Congress held in Mexico during April 1967. With the information available at that time and assuming a discount factor of 50 percent, they estimated that the oil in place would be about 700,000 million barrels of heavy oil. More recently, in 1977, Francisco Gutierrez [2] in a paper presented to the Tenth World Energy Conference at Istanbul, Turkey, indicated that the volume of oil in situ exceeds 2 trillion barrels. In any case, it is obvious that these oil accumulations can be considered as supergiant.

In order to make firm development plans for the exploitation of the Orinoco Oil Belt and at the same time to have a thorough evaluation of these resources, Petroleos de Venezuela started, during mid-1979, an intensive exploration campaign to complete the regional evaluation of the Orinoco Belt by 1983. During this program, about 16,000 km of seismic lines will be shot and about 520 wells will be drilled. Taking into consideration the size of the area of interest and the development program envisaged for the medium term, it was necessary, for the execution of the exploratory program, to establish priorities based on expected volume of oil per acre and oil quality, as shown in Fig. 39-2.

Simultaneously, another drilling campaign will be initiated during January 1980 with the objective of evaluating the most prospective areas already delineated and to define more precisely the areas where exploitation will be initiated. This program calls for the drilling of 380 additional wells in the period 1980 to 1984, which is the minimum number of wells required to evaluate the two areas already selected to start the exploitation.

The total investment associated with this program is estimated to be $430 million (1979 constant value dollars).

PRODUCTION TESTS

Since the oil-bearing formations are unconsolidated sandstones and the oil viscosity is high, the tools which are normally used to obtain representative fluid samples do not work properly. Therefore, production tests using portable pumping units and rod pumps are contemplated. Production tests made in about 20 productive intervals in 15 different wells have shown very high productivity, with production rates between 50 and 600 B/D/well without using any heating or diluent. This confirms that even the heaviest crude in the area (8° API) is producible by primary, conventional means and therefore, from the exploitation standpoint, this crude can be considered conventional. Furthermore, the oil productivity is very similar to Jobo, Morichal, Miga, Melones, Oveja, and Pilon fields located on the northern fringe of the Oil Belt, which are currently producing about 12,000 B/D of oil between 8° and 15° API.

ENHANCED RECOVERY PILOT PROJECTS

Based on the production tests performed in the belt, exploitation can be initiated by primary means and steam stimulation. The oil recovery to be obtained using only these production techniques will depend on the indigenous production mechanisms.

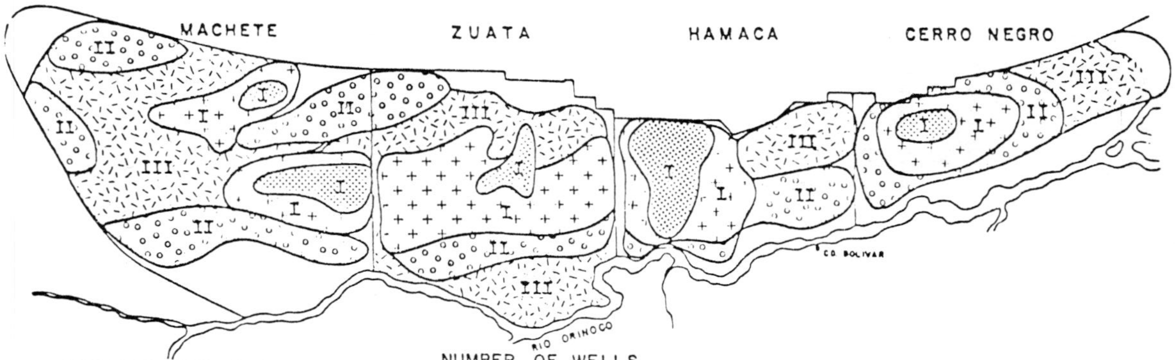

NUMBER OF WELLS

PHASE	I		II	III		TOTAL
YEAR	1979	1980	1981	1982	1983	
CERRO NEGRO	16	40	30	13	10	109
HAMACA	40	40	30	30	-	140
ZUATA	6	40	40	40	15	141
MACHETE	8	20	40	57	3	128
TOTAL	70	140	140	140	28	518
RIGS YEARS	1,9	5,0	5,0	5,0	1,0	17,9

Figure 39-2. Orinoco Oil Belt regional exploration program; 1979-1983 drilling program

C. Borregales [3], in a paper presented in the UNITAR Conference "The Future of Heavy Oil and Tar Sands," held at Edmonton, Canada, during June 1979, estimated that if solution gas drive were the only production mechanism, the average oil recovery by primary and steam stimulation could be 8 percent STOIIP, but this could increase to 24 percent of the STOIIP if compaction happens to be an important driving mechanism. In any case, this means that the initial development of the belt can be initiated using these well-known production techniques. This is confirmed by the excellent results obtained both with the primary production tests of the wells in the Orinoco Belt and with the steam-soak pilot projects performed in the fields located in the northern fringe of the Oil Belt.

In the prospective areas which have been already selected to start the exploitation phase, the following recovery processes will be tested:

- Steam stimulation
- Continuous steam injection
- In situ combustion

In addition, other nonthermal processes will also be evaluated, depending on laboratory test results.

A steam-soak pilot test and a steam-drive pilot test will be initiated during 1980.

DEVELOPMENT OF PRODUCTION

As it was concluded during the First UNITAR Conference [4], the exploitation of these resources is not a technological problem but one of economic and political decisions; it is believed this oil can make a significant contribution to energy requirements in the medium term.

As far as the Orinoco Oil Belt is concerned, the program established by Petroleos de Venezuela calls for the development of about 1 million B/D of oil by the end of this century. This program assumes the following:

1. The conservation policy defined by the government of Venezuela aiming at the development of a production capacity of only 2.8 million B/D.
2. Very limited success in the search for medium and light oil in other prospective basins in Venezuela.

Therefore, future development will be determined by both political and economical considerations defined by the Venezuelan government and, of course, by the length of time required for development.

In the short term the following objectives have been defined:

1. Develop 125,000 B/D of upgraded crude by 1988 in the south portion of the state of Monagas. This project has been assigned to Lagoven.
2. Establish a production capacity of 50,000 B/D of commercial non-upgraded crude by 1983 and 75,000 B/D by 1988 in the south portion of the state of Anzoátegui. This volume of crude will be handled utilizing currently idle production capacity in the Menven system (another affiliate of Petroleos de Venezuela).

The planning studies for the development of the 125,000 B/D of upgraded crude in the south portion of the state of Monagas were initiated by Lagoven in 1978 and the project is expected to come on stream by 1988. Construction of the infrastructure and the production/upgrading facilities will be started in 1982 and 1983, respectively.

For this project, Bechtel was contracted as planning advisor and all the necessary engineering, construction, and project management will be contracted with available experienced and reliable contractors. So far, companies such as Exxon, Foster Wheeler, C. F. Braun, Chem System, A. D. Little, and Pyra have been used for different specific tasks.

The development of 50,000 B/D of nonupgraded oil in the south portion of the state of Anzoátegui deals mainly with less viscous oil (12° to 15° API) and will use the existing infrastructure of the area. Meneven has already initiated the design studies for such a project and again, available technology and experienced companies will be utilized.

UPGRADING OF THE CRUDES

Since most of the Orinoco crude is viscous and contains a high concentration of metals (200 to 500 PPM of nickel plus vanadium), it requires upgrading to make the product acceptable to conventional refineries. The first project being designed by Venezuela is a production-upgrading module of 125,000 B/D. However, because of the trend of world production to heavier crudes and the importance of giant heavy oil accumulations, such as the Orinoco Oil Belt, the future supply of oil calls for a strong effort from the consumer countries to construct required upgrading facilities.

CONCLUSIONS

1. The expected reserves in the Orinoco Oil Belt are very large and they could contribute significantly to oil demand in the medium term.
2. The heavy oil contained in the Orinoco Belt is producible by primary conventional means and from the exploitation point of view it is considered conventional.
3. Venezuela initiated in 1979 an intensive exploratory campaign to evaluate the Oil Belt and define more precisely these vast resources.
4. The development program for the Oil Belt calls for the generation of 1 million B/D of oil by the year 2000. This program will depend on the results of the current search for light and medium crudes in other areas and on Venezuelan energy policy.
5. Two different projects are already defined. The first calls for the production of 125,000 B/D of upgraded crude by 1988 and the second, 75,000 B/D of conventional heavy oil by the same year. Both projects are currently in the design phase.

REFERENCES

[1] Velarde H., and Galavis, J. A. Estudio Geologico y de evaluation preliminar de reservas potenciales de Petróleo pesado en la Faja Bituminosa del Orinoco; Cuenca oriental de Venezuela: VII World Petroleum Congress, Mexico, April 1967.

[2] Gutierrez F., et al., Technological needs for the Production-Upgrading of the Orinoco Extra Heavy Crude Oils: Tenth World Energy Conference, Istanbul, Turkey, September 1977.

[3] Borregales, C. J., Production Characteristics and Oil Recovery in the Orinoco Oil Belt: UNITAR Conference, The Future of Heavy Crude and Tar Sands, Edmonton, June 1979.

[4] UNITAR First International Conference on the Future of Heavy Crude and Tar Sands, Conference accomplishments, conclusions and future plans, Edmonton, June 1979.

CHAPTER 40

VENEZUELA'S HEAVY OIL DEVELOPMENT PROSPECTS AND PLANS

A. Volkenborn[1]

HEAVY OIL DEVELOPMENT TO DATE

Venezuela's long oil production history has included heavy oil development from an early date. Some of the world's largest heavy oil fields started production more than 40 years ago along the shores of Lake Maracaibo. Today about a third of the total country production of 2.3 million B/D is of heavy oil and this proportion is increasing in line with a greater availability of reserves of this type and a production policy oriented toward conservation of more limited light oil reserves.

The development of heavy oil production initially was concentrated mainly in the very large heavy oil fields along the Eastern shore of Lake Maracaibo. It is worthwhile to recapitulate some aspects of this development because it is very relevant to the future of heavy oil production in Venezuela and in particular to that of the Orinoco Oil Belt. The fields involved are Tía Juana, Lagunillas, and Bachaquero, which contain a total of some 34 billion barrels of oil in place. It was appreciated at an early stage that conventional primary techniques would only recover about 11 percent of the oil in place in these very heavy (less than 15° gravity API) oil fields. From about 1955 onward, much effort was devoted to increasing this figure and this led to the eventual development of some very successful thermal recovery techniques. Today these techniques have evolved into a two-stage--steam-soak, followed by a steam-drive--supplementary recovery program which has increased the expected recovery from these fields to about 32 percent of the oil in place.

In this program, primary cold production is followed by steam soaking, in which the wells are injected with a slug of steam, generally about 5,000 tons, during, say, a 2-week period, and then produced at a much increased rate, mainly because of oil viscosity reduction, for a period of 1 or 2 years. Steam soaking has been applied on a large scale in these fields and it is presently estimated that it will increase recovery from an original primary figure of

---

[1]Maraven S.A., Filial de Petroleos de Venezuela.

11 percent to about 18 percent of the oil in place. To date some 1,650 wells have been injected with a total of 20 million tons of steam. So far, 500 million barrels of additional oil have already been produced and this represents an additional recovery of 4.4 percent of the oil in place in the areas under development. In short, steam soak is a well proven technique in Venezuelan oil fields.

The steam-soak process is extremely efficient, as it yields about 25 barrels of oil per ton of steam injected. To raise 1 ton of steam requires about 0.6 barrels of oil, resulting in a thermal efficiency of about 40 barrels of oil produced per barrel of oil burned. The reason for this high efficiency is to be found in the production mechanism of these reservoirs which are basically of a very unconsolidated nature and where reservoir compaction results in pressure maintenance and oil displacement. This somewhat unusual compaction feature is extremely important and made really effective in combination with steam injection. The result is low-cost oil often cheaper than by primary means. The existence or not of such a production mechanism in the Orinoco Oil Belt area will obviously have very important implications on its development program, reserves, and the cost of oil produced.

When reservoir compaction is almost complete and no additional oil can be recovered by steam soak, a follow-up stage of steam drive is applied. Steam is injected continuously through injection wells to drive the oil to the production wells. This technique has been demonstrated in a very large pilot test, where a total recovery of 45 percent of the oil in place was obtained. At present the first commercial project of this type is underway and performing entirely according to expectations. It is one of the largest projects of its kind in the world, with an injection of some 8,000 tons of steam/day. Production has increased from 7,000 B/D initially to about 19,000 B/D at present. It is estimated that with steam drive, the recovery from these fields can be increased on average from the 18 percent attainable by steam soak to a total of 32 percent of the oil in place. The process however is far less efficient than the steam-soak/compaction mechanism and results in only about 2 barrels of oil per ton of steam injected, or a thermal efficiency of 3 barrels of oil produced per barrel of oil burned. This background experience in existing fields in the Lake Maracaibo basin points to the likelihood of higher recoveries as well in other heavy oil producing areas in Venezuela. Indeed, although on a lesser scale, very favorable results have been obtained with steam soaking in other fields in both Eastern and Western Venezuela, generally resulting in the same high efficiency of about 25 barrels of additional oil per ton of steam injected. Other thermal recovery techniques, such as in situ combustion, have been tried, in some cases with success, but are not now considered for large-scale application.

THE ORINOCO OIL BELT

The Orinoco Oil Belt is probably the largest essentially untapped oil accumulation in the world. It covers an area north of the Orinoco River some 700 km long and 50 to 80 km wide, with a total surface area in excess of 40,000 km^2 (about the size of Denmark). Oil-bearing sands range in depth from 600 ft to 7,000 ft and consist of unconsolidated, highly porous, highly permeable sands of Oligocene and Miocene age, with characteristics very similar to those found in the heavy oil fields in the Lake Maracaibo basin. To date some 120 wells have been drilled in the area and the gravity of the oil found ranges from 7° to 18° API. Subsurface temperatures are such that the oil flows under reservoir conditions; this has very significant implications on recovery. In other words, the accumulations are really oil reservoirs and not tar sands, and this allows for the application of known recovery techniques.

One little-realized fact is that oil fields belonging to the northern fringes of the Oil Belt have been in production for many years and today a total of some 80,000 B/D are produced from the Oil Belt. In some of these fields, such as Jobo and Morichal, thermal recovery techniques, and specifically steam soak, have been applied with the same degree of success as observed in the Lake Maracaibo area. Production increases of two to five times primary production rates have been achieved and typical hot-well production rates range from 200 to 1,000 B/D per well. Again efficiencies of the order of 25 barrels of oil per ton of steam have been achieved.

The question now arises as to how much oil is contained in this Oil Belt and how much may eventually be recovered. About the oil in place, various rather different estimates have been made. A heavily discounted estimate, based on information from a rather small number of wells drilled in relation to the size of the Belt, gives an oil in place figure of around 700 billion barrels. A more realistic estimate is probably around 2,000 and possibly up to 3,000 billion barrels.

Primary recovery would obviously be very low (no more than a few percent) but supplementary thermal methods are very likely to substantially increase this figure. One critical factor is whether compaction is, or is not, an important recovery mechanism. From an apparent lithological similarity with other Venezuelan reservoirs which do compact and from the excellent results obtained by steam soaking fields on the fringes of the Oil Belt, one would be inclined to judge that compaction is likely. However, it has not yet been confirmed, possibly because of lack of precise surface level measurements or because threshold pressures to trigger the compaction/subsidence process have not yet been reached.

At any rate, if compaction does occur, and assuming the compressibility of the rock is fairly high (say 40 10^{-6} psi^{-1}), then a high recovery by steam soak may be expected, perhaps of the order of 15 to 20 percent, as in the case of the Lake Maracaibo heavy oil fields. In addition to this, steam drive with or without compaction could be expected to lead to an average total recovery of at least 30 percent. Hence, combining oil in place with recovery figures, a range in recoverable reserves between 70 (10 percent of 700 billion) and 900 (30 percent of 3,000) billion barrels recoverable oil may be surmised, with perhaps something of the order of 500 billion barrels a realistic planning figure for the time being.

Reserves of this magnitude could effectively remove resource considerations from production decisions. From a technical standpoint they could sustain a very high production level for an extremely long time.

DEVELOPMENT PLANS

To maintain its production potential, Venezuela will need a significant contribution from areas not presently under development. This additional oil can come from either new areas now being explored, such as the offshore continental shelf, or from the Orinoco Oil Belt, or both.

To meet requirements as presently envisaged, the Orinoco Oil Belt is planned to contribute on the order of 200,000 B/D by 1988, 600,000 B/D by 1995, and 1 million B/D by the year 2000.

Alternative development schemes for the country coupled with world energy demand could also lead to production rates of say 2 or 3 million B/D by the turn of the century.

As a first stage, a comprehensive evaluation program is required. This involves better definition of the Oil Belt, and identifying the most attractive development prospects, reservoir production mechanisms, and appropriate production schemes. For this purpose the oil belt has been divided into four areas, where each of the four Petroleos de Venezuela operating companies and the research institute INTEVEP, are to carry out a comprehensive evaluation program. Until 1984 this comprises some 1,600 km of seismic lines, the drilling of about 500 exploration and 1,100 evaluation wells, and a nonconventional exploration program.

At the same time plans are being laid for pilot production projects where various thermal recovery techniques will be tested. A first production module of 125,000 B/D of upgraded crude oil is planned to come on stream by 1988, when production will most likely be by steam soaking and upgrading most probably through a coking unit with coke to be used for both steam and electricity generation. Other pilot tests and development modules are planned for prospective areas to be defined by the exploration program.

As regards upgrading of Venezuelan heavy crude, commercial projects for a flexicoker and an experimental hydrodemetalizing plant are already underway in existing refineries. Other processes are in various stages of research and evaluation. One interesting aspect of deep conversion directly in the production areas is the fact that heavy residues or gases generated by these means would be used for steam generation in turn required for the production process. Also transport of an upgraded or partially upgraded crude through long pipelines to coastal ports would be much easier than that of a virgin, extremely heavy and viscous oil. Finally, from the commercial point of view, some upgrading in the production areas may be necessary.

In summary, it may be said that plans for the Orinoco Oil Belt are well underway. The production technology required should pose no major obstacle, for the most likely applicable technique--steam soak--has been extensively used in very similar areas and much experience exists with the alternative or follow-up process of steam drive. Upgrading may take various forms but commerical projects are also already underway.

THE COMMERCIAL FUTURE

Investment estimates for development of the Orinoco Oil Belt range from $3,000 to $10,000 per B/D potential, dependent on whether a highly effective compaction mechanism exists or whether steam drive will be the main production method. Dependent on the same alternatives, overall production costs could range from \$2 to \$8 per barrel and upgrading, \$3 to \$5 per barrel of product; hence, in the end a high quality synthetic crude costing in the range of \$5 to \$13 per barrel would be obtained.

Quite apart from deep upgrading, large volumes of Venezuelan heavy crude are already being used for such specialty purposes as naphthenic lubricant and asphalt manufacture. Indications are that the oil belt crude is also suitable for these specialty uses and in view of decreasing availabilities of naphthenic crude oils elsewhere in the world, supplies from the Orinoco Oil Belt will be extremely important in this market.

Thus, all cost figures indicate levels well below those of other nonconventional sources of energy such as tar sands and shale oil. As such, the Orinoco Oil Belt with its vast potential, of the same order of magnitude as present-day conventional world oil reserves, could play a major role in the world energy scene. It fits neatly into a transition from dwindling conventional oil reserves to much higher cost alternatives, while enjoying the same unique applications as conventional oil.

The degree to which the Orinoco Oil Belt will play this role is a matter of great importance. It clearly represents an option for significant additional oil production for Venezuela.

CHAPTER 41

PROSPECTS FOR THE DEVELOPMENT AND PRODUCTION OF HEAVY CRUDE OIL RESERVES IN NIGERIA

J. B. Owokalu[1]

INTRODUCTION

The search for petroleum in Nigeria commenced in 1937. Except for the interruption during the Second World War, petroleum exploration activities progressed steadily, culminating in the commercial discovery of the Oloibiri field near Port-Harcourt, in 1956. Commercial production started shortly thereafter at about 5,000 barrels of oil per day (bopd), rising steadily to the current rate of around 2,200,000 bopd. Production of hydrocarbon gas is now 810 billion standard cubic feet (SCF) per day, all or nearly all of it associated gas. Table 41-1 shows Nigeria's production of oil and associated gas.

PETROLEUM PROVINCE OF NIGERIA

The Niger Delta basin is the main petroleum province of Nigeria (Fig. 41-1), where intensive exploration efforts have hitherto been concentrated. Recently, however, interest has been generated in other sedimentary basins, notably Anambra basin and Chad basin, both of them located north of the Niger Delta basin. Concurrently, attention has also been extended to the definition of the gas prospects of the Niger Delta and other basins.

HEAVY OILS PROSPECTS

Heavy oils, which are known to exist in Nigeria, generated even less interest in the past for a variety of reasons not unconnected with the economic and technical viability of exploitation. But considerable reserves of such oils are located mainly in the western fringe of the Niger Delta and at shallow depths within the present main producing province.

[1]Nigerian National Petroleum Corporation, Lagos, Nigeria.

TABLE 41-1. Nigeria's total oil and gas production from inception

Year	Oil (10^6 barrels)	Gas (10^9 SCF)
1958	1.9	1.6
1959	4.1	4.9
1960	6.4	5.1
1961	16.8	10.9
1962	24.6	17.2
1963	27.9	22.1
1964	44.0	36.3
1965	99.4	79.5
1966	152.5	102.7
1967	116.6	93.1
1968	51.9	51.7
1969	197.5	145.7
1970	395.0	283.9
1971	560.0	458.2
1972	667.0	604.6
1973	753.0	773.3
1974	825.0	960.4
1975	652.0	660.3
1976	757.0	752.1
1977	766.0	774.4

LOW GRAVITY OILS OF THE MAIN PRODUCING PROVINCE

In this paper, only the heavy bituminous oil reserves shall be discussed since the low gravity crudes of the main producing region do not come within the purview of unexplored deposits. Nor do they strictly qualify to be described within the same context as the bituminous "tar sands" oil, the main subject of discussion. Usually, such low-gravity oils are encountered at shallow depths in the Niger Delta basin; they are free-flowing, viscous (450 to 700 centipoises at 60°F), low-gravity (8° to 16° API) oils occurring in geological and structural conditions of the typical Niger Delta mode of loosely

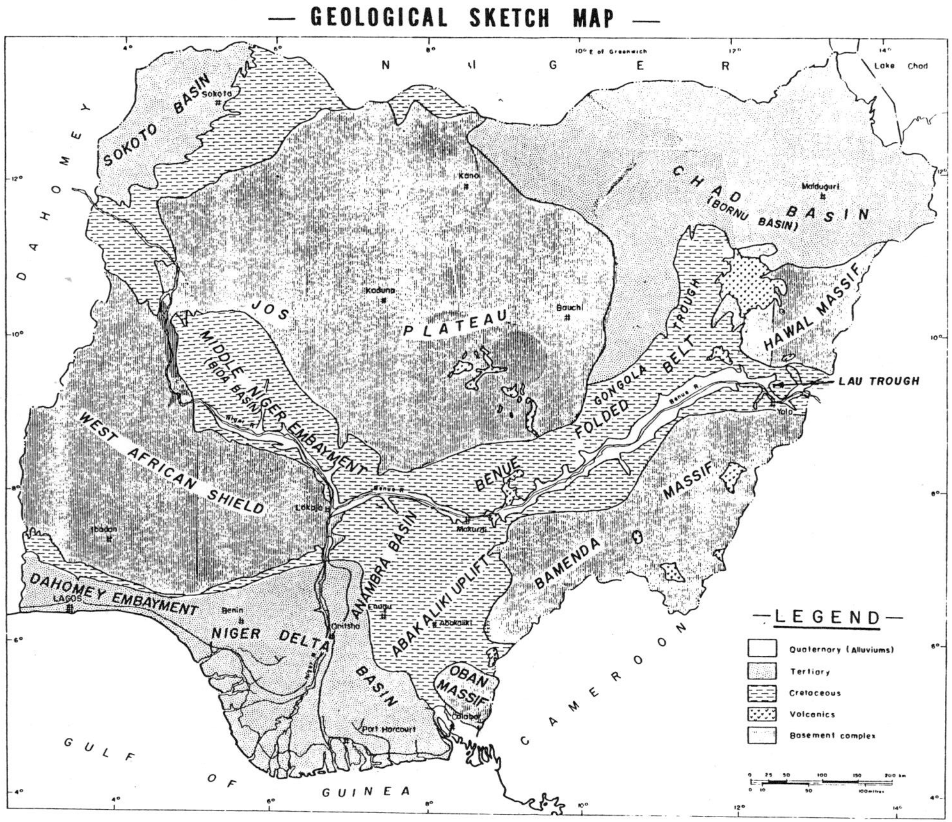

Figure 41-1. Geological sketch map of Nigeria

consolidated sands and sandstone of Tertiary age within roll-over anticlinal structures associated with growth faults. Exploitation of such oils through conventional means is plagued by low well productivity and inflow, adverse vertical lift characteristics, as well as sand and water problems. Generally too, when it has been successfully produced, there is reluctance to inject it into the main crude production streams whose qualities, 28° API, 35° API, and 42° API, it would depress. On the other hand not enough of it exists in geographically contiguous or convenient locations to justify the expenditure for separate facilities. Therefore such deposits have tended to be deferred for production in the future.

BITUMINOUS TAR SANDS DEPOSITS

Early evidence of the occurrence of the tar sands comes from reported bitumen seepages in Ijebu and Okitipupa provinces of Nigeria. Explorationists determined that the tar seeps occur in a slim belt parallel to the contact with the crystalline basement complex while the drill confirmed the existence of free-flowing, tarry oil in sands juxtaposed with bituminous tar sands.

GEOLOGY OF OCCURRENCE

The tar sands and heavy oils occur within the eastern extension of the sedimentary basin called the Dahomey Embayment. The basin is late Cretaceous in age; it is bounded and underlain by block-faulted crystalline rocks on which sediments (shales, sands, limestones) of late Cretaceous (Maestrichtian) to Tertiary (Paleocene and Miocene) age were deposited. The basinal stratigraphic column is illustrated in Table 41-2 and Fig. 41-2. The crystalline basement rocks slope southwards, as do the gently dipping sediments, whose thickness reaches 7,000 feet in Afowo well drilled in the west of the basin, and 6,200 feet in Ise-2 well (Eastern basin). Immediately overlying the basement rocks is the Abeokuta Formation, which actually contains the bituminous oil sands member. Channel deposits of variable thicknesses are abundant and appear to form the main reservoir rock. Well and borehole evidence indicates that the formation thickness reaches 900 feet inland.

Structure does not play a dominant role in the trapping of the oil as the homoclinal nature of the sediments demonstrates. The basement-controlled faults and stratigraphic conditions, however, account for the retention of the

TABLE 41-2. Basinal stratigraphic column

Name of formation	Description	Age	
Benin Ijebu	Topmost--alternation of sand, shale, and lignite films	Recent through Upper Miocene	TERTIARY
Afowo	Submarine channel fill	Middle to Lower Miocene	TERTIARY
Oshosun	Thin beds of limestone, presence of gypsum and phosphate	Middle to Lower Miocene	TERTIARY
Imo Shale	Associated with soft, very thin laminated green clay	Middle to Lower Miocene	TERTIARY
Ewekoro	Fossiliferous alternation of sand and shale with limestone	Middle to Lower Miocene	TERTIARY
Upper Nkporo Shale	Itori sand locally present	Upper Cretaceous	CRETACEOUS
Abeokuta	Thick, massive continental sand; main target sand	Lower Cretaceous	CRETACEOUS
Basement	Faulted step-like faults	Pre-Cambrian	

oil within the sands; but where these controls are lacking, oil movement and, consequently, surface seepages result. Trapping has been largely due to lithologic and porosity changes as well as the stoppage of migration paths by thick mats of "solid" bitumen.

PROSPECTIVITY

Previous Work

A total of 15 wells have been drilled in the basin. Most of them, located in the western part, reached depths between 5,000 feet and 8,000 feet, although a few exceeded 9,000 feet. Four of the wells flowed appreciable oil and gas, which were encountered at varying depths between 2,000 feet and 7,000 feet. While this would suggest the western basin to be somewhat unprospective, the recent announcement of production plans by Benin Republic in its border region belies that conclusion. On the Nigerian side, that is, in the eastern part of the basin, 15 shallow wells drilled between 1908 and 1914 bottomed

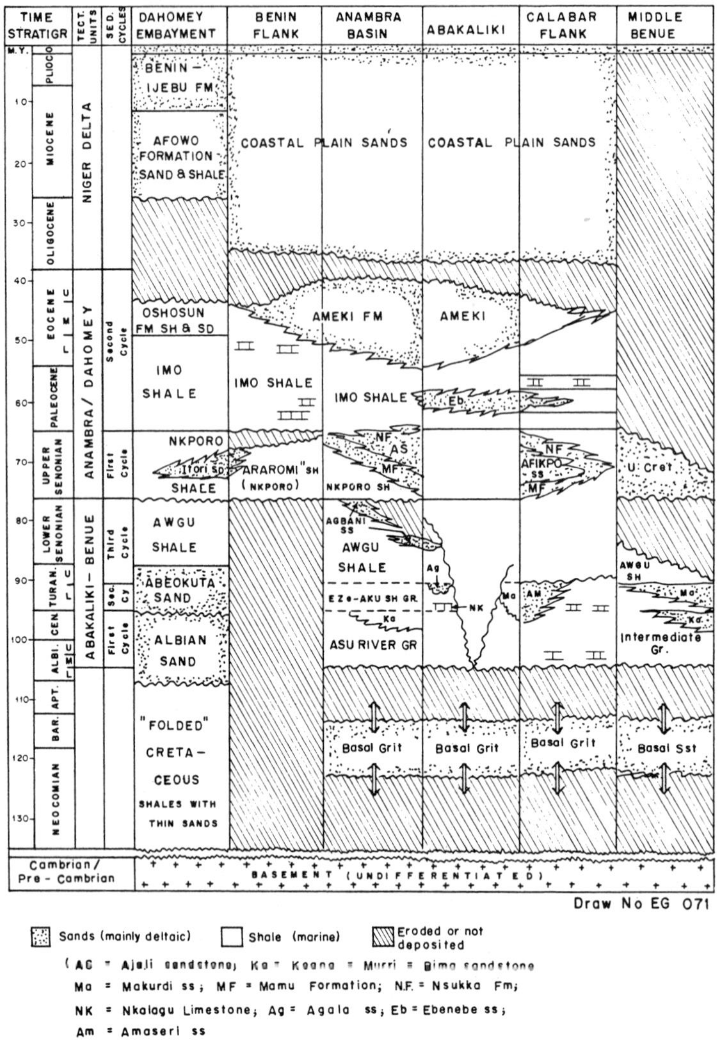

Figure 41-2. Stratigraphic synopsis

in the basement at depths between 90 feet and 2,300 feet. They all encountered the oil sand and some flowed various rates of oil and gas. Further and more recent work (1976), in which six shallow wells were again drilled and studied, has permitted more definitive evaluation of the hydrocarbon prospects of the basin and led to favorable conclusions about the presence of exploitable reserves of producible oil and bitumen.

RESERVES

Recoverable Free Oil

Analysis of all available information indicates that about 1 billion barrels of heavy, highly viscous, free-flowing oil could be recovered. But low flow rates, below 150 barrels per day, are common and will have to be contended with.

Producible Bitumen

The bituminous (tar) sand stretches through a zone approximately 5 miles wide and 80 miles long. The primary bituminous interval has a thickness varying from 20 feet near outcrop points to 170 feet to the south. Regional analysis suggests only localized discontinuities. The tar sand reserves are zoned into two parts--namely, reserves in outcrop areas, where average thickness is 20 feet, and reserves downdip of the outcrop, in places where the overburden does not exceed 160 feet--and are estimated at between 3,000 and 5,000 million barrels.

PRODUCTION TECHNIQUES

The free-flowing, viscous reserves producible from conventional wells can be recovered by any of the techniques that ensure in situ viscosity reduction, which in effect improves the inflow performance as well as the vertical lift characteristics. It is estimated that for such production to sustain viable operations a minimum well flow of 150 to 200 bopd is needed. Substantial in-place oil in excess of 1,000 million barrels exists. With a recovery factor reckoned at 40 to 60 percent, the oil yields will be around 400 to 600 million barrels. The measured oil gravity ranges from 5.3° API to 14.6° API; sulphur content is as much as 6.5 percent; while the water saturation of the oil sands ranges from a low of 14 percent to a high of 75 percent. By far the most significant recovery is obtainable by either solvent or hot-water extraction of hydrocarbons from the downdip, heavily impregnated tar sands. The average bitumen content of the tar sands is 20 percent by weight, although sample measurements indicate a range of 10 percent to 80 percent. Water content is minimal, while the fine-sand content ranges from 1.7 percent to 5 percent by weight. Mineral impurities identified as pyrites are present in discrete but insignificant quantities. The estimate of the recovery of bituminous oil is in excess of 4 billion barrels, while the operation is believed to be capable of profitable execution with a minimum of two plants of 50,000 to 60,000 barrels-per-day capacity.

CONSTRAINTS TO DEVELOPMENT

The problems attendant to production through conventional well operations have been mentioned elsewhere. They relate to low productivity and the inadequacy of known methods for in situ viscosity reduction. In fact, the conclusions from early exploration of the area appeared to discount the technical feasibility of thermal recovery processes on account of inadequate thickness of the oil sands (average thickness is 35 feet); insufficient depth to the target sands (overburden of zero to a few thousand feet); low oil saturation, generally over 50 percent; and unavailability of fuel gas.

Recent work based on new and more widespread core data invalidates some of these conclusions and shows that oil saturation is actually a high average, while the gas availability problem is no longer a deterrent because nearby projects are planned that will require gas from the main producing province. Additionally, current day economics of production show better profit indices for low productivity wells generally.

In respect of bituminous oil extraction, proven technology now exists, and the reserves of the order of 4,000 million barrels are believed to be capable of supporting two or more extraction plants each of 50,000 to 60,000 bopd capacity.

CONCLUSION

In comparison with other known deposits of tar sands, the Nigerian occurrence appears meager. It is nevertheless a possible source of new oil, whose exploitation could be enhanced if the identified technical and operational constraints could be removed.

ACKNOWLEDGMENT

Acknowledgment is due to the Petroleum Inspectorate, Nigerian National Petroleum Corporation for permission to utilize certain in-house reports in the preparation of this paper.

REFERENCES

Annual Reports, 1962-74, Petroleum Division, Federal Ministry of Mines and Power, Lagos.

Annual Report, 1975-77, Department of Petroleum Resources, Federal Ministry of Petroleum, Lagos.

Various unpublished confidential reports of the Petroleum Inspectorate, Nigerian National Petroleum Corporation.

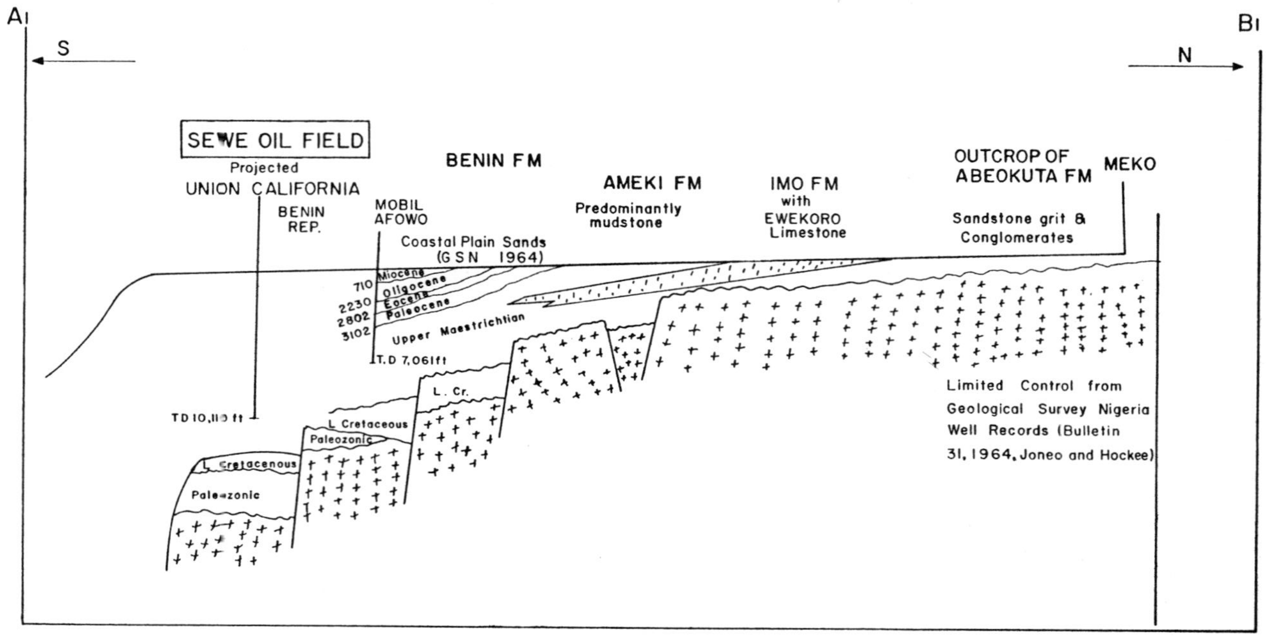

Figure 41-3. North-south cross section through the Dahomey Embayment

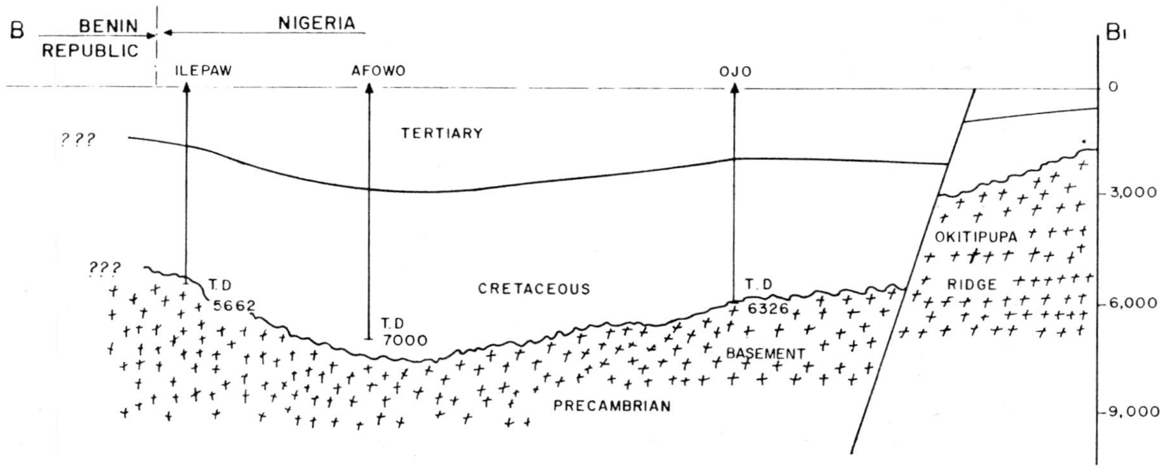

Figure 41-4. East-west cross section through the Dahomey Embayment

TABLE 41-3. "Deep" well drilling in the western Dahomey Embayment

Well name	Total depth	Status	Hydrocarbon interval	Year drilled	Location	Result
A1		Exploratory	Data not available	Not available	Offshore	Oil
A2	9,076'	Exploratory	Data not available	Not available	Offshore	Dry
A3	9,225'	Exploratory	Data not available	Not available	Offshore	Trace
A4	6,948'	Exploratory	Data not available	Not available	Offshore	Dry
B1	7,400'	Exploratory on structural closure	Minor oil shows at 5,410'-5,450' 5,570'-5,690' 6,240'-6,370'	1970	Offshore 469650E 697510N	Abandoned dry 26/10/70
D1	Proposed T.D. 11,000' Reached T.D. 7,058'	Exploratory	Data not available	1969	Offshore water depth 159'	Dry junked
D2	Reached T.D. 8,923' Proposed T.D. 10,000'	Exploratory structural closure	Minor oil shows at 7,926'-7,945' 8,335'-8,360'	1970	251,565 mN 41,771 mE coordinates	Abandoned dry 28/9/70
D_o - D-2	10,721'	Exploratory	Data not available	Not available	Offshore	Trace
D_o - D-3	6,683'	Exploratory	Data not available	Not available	Offshore	Dry
D_o - F-1	7,683'	Exploratory	Data not available	Not available	Offshore	Dry
Bodeshe	4,954'	Exploratory	None	1961	Land	Dry
C_1	7,365'	Exploratory	No data	Not available	Offshore	Dry
Illepaw	5,667'	Exploratory	None	1960	Land	Dry
Afowo	705'	Exploratory	None	1959	Land	Dry
Ojo-1	6,326'	Exploratory	Oil gas show at: 2,420'-2,520' 2,790'-2,860'	1960	Land	Trace

TABLE 41-4. Analysis of some available corehole data

Lekki Coreholes. Results of analysis of data from some of the coreholes are shown below. The results typify the general situation.

Corehole A

Coordinates: E. 197,075 N. 295,570
Total depth: 266'
Lithology: 200' of unconsolidated sands and clays overlying crystalline basement.
Oil occurrence: Top of tar at 27'. Strong water flow from interval carried much of the tar to surface.
Summary: Heavy tar saturation 27'-140', spotted tar saturation 140'-200'. Gross interval = 113'.

Corehole B

Coordinates: E. 193,805 N. 294,485
Total depth: 375'
Lithology: Sands, shales, and clays of Cretaceous-Tertiary age. (Ijebu beds and red sands and clay.)
Oil occurrence: Heavy tar saturation between 225'-280' with tar traces from 260'-300'.
Total net pay sand: 24.5' in seven intervals. Gross intervals: 227'-324' = 97'.

Core Sample Analysis

Depth	Net sand	Porosity %	Residual oil %	Water saturation %	Remarks
227'-229'	2'	30.8	64.0	29.5	-
239'-242'	3'	26.4	21.6	71.9	High clay content
		31.3	27.1	67.2	
268'-271'	3'	32.8	69.8	22.2	10.5° API gravity

Corehole C

Coordinates: E. 197,640 N. 291,545
Total depth: 541'
Lithology: Sands, shales, and clays of Cretaceous-Tertiary age.
Oil occurrence: Spotted light tar occurred between 375'-400'.
Total net pay sand: 22.5' in six intervals. Gross interval: 91'.

TABLE 41-4 (continued)

Core Sample Analysis

Depth	Net sand	Porosity %	Residual oil %	Water saturation %	Remarks
368'-371'	2.5'	29.8	47.9	46.2	6.4° API gravity
372'-376'	4'	30.3 36.7	18.8 73.0	74.3 14.4	High clay content
382'-384'	2'	36.9	30.1	63.2	-
390'-395'	4'	34.9	46.7	45.6	-
431'-439'	6'	39.1	50.9	35.0	5.3° API gravity

Corehole D

Coordinates: E. 201,202 N. 288,992
Total depth: 673'
Lithology: Sands, shale, and clays of Cretaceous-Tertiary age.
Oil occurrence: Spotted heavy tar sands between 540'-550'.
Total new pay sand: 17.5' in six intervals. Gross interval: 464'-541' = 83'.

Core Sample Analysis

Depth	Net sand	Porosity %	Residual oil %	Water saturation %	Remarks
465'-467'	2'	40.3	56.6	17.6	-
469'-470'	1'	43.3	21.0	60.3	High clay content
		38.7	9.6	78.3	High clay content
533'-540'	6'	37.4 43.3	41.5 42.3	26.4 40.2	- -

TABLE 41-5. Analysis of information from well logs, bitumen seepages, and shows in some boreholes

Bore-hole	Location	Height above sea level	Depth of oil sand	Depth of crystal-line rock	Remarks
1	Erimo	40'	-	95'	Bitumen seepage at Erimo
2	Sumoge	30'	548'	763'	-
3	Banjo	30'	Between 400'-500'	590'	-
4	Agbabu	70'	About 860'	896'	-
5	1 mi south of Idakun	20'	Slight bitumen, grey sand at 712'	725'	Water with oil between 715'-725'. Well flowed over casing at up to 86 bbls of oil a day and 68 bbl/day of water.
6	Makun	10'	Several bitumin-ous hori-zon to 1,741'	1,996'	0-887' sands, clays, and mud-stone. Small amount of oil and gas rising from well. Ground around borehole covered with hardened bitumen from well.
7	Alo Village (8 km east of Oni)	50'	657'	739'	Clayey grit and white quartz, browning grey and with sulphur. Thick black bitumen.
8	3.2 km south of Idekan	50'	-	1,110'	Oil sand under-lain by dark grey clay shale smelling of bitumen and sulphur.
9	South side of Ebba Island	10.9'	-	2,252'	Oil show at 1,440'-1,460' and bitumen between 2,188' and 2,218'.

CHAPTER 42

ENHANCED OIL RECOVERY--APPLICATION AND FUTURE PROSPECT

Robert L. Whiting[1]

INTRODUCTION

The purpose of this chapter is to present a review of the past, present, and the future possibilities for increasing oil recovery through the application of enhanced oil recovery processes. Due to the nature and scope of the subject, the author has drawn freely on the published literature and private communications, particularly with respect to the present state of the art and the future potential for these processes. Much of the material is the result of studies made by Committees of the American Petroleum Institute, the Society of Petroleum Engineers of AIME, the National Petroleum Council, the Gulf Universities Research Consortium, and independent studies by the author. The committee results are all in the public domain but most of the data from which the results were derived are proprietary.

CHARACTERIZATION OF RESERVOIR ENERGY AND PROCESSES

During the past two decades, much confusion has existed due to the terminology associated with natural petroleum reservoir drive (energy) mechanisms and supplemental reservoir drive mechanisms. Recently, within the USA, a reasonable standardization of terms associated with reservoir energies has been achieved--due in considerable part to the insistence of the USA Federal Government. Now the generally accepted classifications in the USA are those shown in Table 42-1 and will be those used in this chapter.

Historically, oil has been produced initially by utilizing natural reservoir energy. Such production is called primary oil recovery. Secondary oil recovery is obtained by pressure maintenance, either partial or complete, through injection of water, gas, or a combination of gas and water after some degree of primary recovery. Thereafter, one or more of the many new and/or exotic oil recovery processes have been used to obtain what has been called in

[1]Department of Petroleum Engineering, Texas A&M University, College Station, Texas.

TABLE 42-1. Classification of oil production methods

PRIMARY RECOVERY
Types of drive
Solution gas
Solution gas-gas cap
Water
Gravity
Combination
SECONDARY RECOVERY
Immiscible drives
Waterflooding
Gas injection
Foam drive
Attic, alternate injection and production, etc.
ENHANCED RECOVERY
Polymer flooding
Miscible drives
High-pressure dry gas miscible
Enriched gas miscible
Miscible slug
Water driven
Gas driven
Alcohol
Inert gas
Carbon dioxide, nitrogen, flue, air, inert
Water driven inert
High-pressure inert
Cyclic
Partially miscible-gas-solvent
Carbonated water, etc.
Thermal drives
In situ combustion
Forward and reverse
Heat injection
Downhole heating (heaters, electrical, etc.)
Hot fluid injection (hot water, steam, etc.)
Cyclic steam (huff and puff, etc.)

the past, tertiary oil recovery. However, the term enhanced oil recovery is now being used to replace the term tertiary oil recovery because supplemental processes increasingly are being applied to reservoirs early in their life, sometimes before either primary or secondary methods have reached economic limit. Hence, the term enhanced oil recovery used in this chapter has a broader meaning than tertiary oil recovery.

Due to the fact that the aforementioned classifications have not been adopted internationally, short descriptions of the processes are included here. For detailed process descriptions the reader is referred to the bibliography.

The potential for secondary and enhanced oil recovery from a particular reservoir is dependent upon the rock and fluid properties, the prior fluid production and pressure history, and the overall reservoir engineering management of the reservoir in the past. Of primary concern is the identification of the type of natural reservoir energies present and the extent to which these energies have been utilized in the past.

PRIMARY RECOVERY

Results to date have permitted the identification of the types of natural reservoir drives shown in Table 42-1. The sources of energy for such drives and the range of oil recoveries from each are as follows.

Solution Gas (Depletion) Drive. These reservoirs contain either gas-saturated or gas-undersaturated crude oil. Energy for production of the oil is derived from liquid expansion and/or dissolution of the gas from the crude oil upon reduction of reservoir pressure. Oil recovery by this mechanism is relatively low, ranging from 20 to 40 percent of the oil originally present.

Solution Gas-Gas Cap Drive. This type reservoir contains a gas-saturated crude-oil zone overlain by a free-gas zone, called a gas cap. Expulsion of oil from the reservoir results from the combined energy of dissolution of gas from the crude oil and the expansion of the gas cap. The gas cap contributes to the maintenance of reservoir pressure, thereby reducing the dissolution of gas from the crude oil. Hence, higher oil recoveries than by depletion drive are obtained, ranging from 30 to 60 percent of the original oil in place.

Water Drive. This type reservoir contains either a gas-saturated or gas-undersaturated crude oil zone which has an associated water aquifer. Energy for oil production is derived from natural water encroachment from the aquifer into the oil zone upon pressure reduction in the latter. Depending upon the degree of pressure maintenance in the oil zone, the oil recovery ranges from 40 to 80 percent of the oil originally present.

Gravity Drive. Primarily because of the difference in density of crude oil and natural gas, gravity may be an effective force in increasing oil recovery from reservoirs having the natural drives discussed previously. The range of increased recovery is from 0 to 20 percent of the original oil in place.

Combination Drive. All possible combinations of the aforementioned drives have been identified in producing reservoirs. The oil recovery from such reservoirs is dependent upon the relative effectiveness of the various drive mechanisms.

SECONDARY OIL RECOVERY

Historically, at some time during the dissipation of the natural reservoir energy, supplemental energy in the form of water and/or gas injection may be used to increase oil recovery from the reservoir. Economic application of such techniques necessitates identification and evaluation of the effectiveness of the primary (natural energy) oil recovery mechanism. Secondary recovery methods necessitate the use of both injection and producing wells. Modern technology utilizing reservoir stimulation permits the optimization of the number and location of wells, fluid injectivity and productivity, etc. Both regular and irregular spacing of wells (called flood patterns) are employed.

Water Injection

The factors controlling water injection oil recovery may be conveniently divided into primary variables, namely: (1) primary recovery efficiency, (2) connate water saturation, (3) sweep efficiency, (4) residual oil saturation, and (5) crude oil saturation; and secondary variables--with an indication in parentheses of the primary variables which they affect--oil viscosity (1,3,4); permeability (1,3,4); structural considerations (1,3); uniformity of reservoir rock (3); type of flood (3); time of flood initiation (5); and economic factors (1,3,4).

The economic factors include such items as crude oil price, depth of reservoir, operating costs, well spacing, availability of water, and other considerations which may affect the economic limit to which flooding operations may be applied.

Due to the multiplicity of variables, waterflooding recoveries are seldom reported as such. The usual practice is to report combined primary plus waterflooding recovery. The usual range of such recoveries is from 20 to 50 percent of the oil originally in place.

Experience has shown that the optimum time for initiating a waterflood is prior to an appreciable reduction in the shrinkage of the crude oil. Delays in initiation for good prospects result in substantial recovery loss and are much more severe for poorer quality prospects. Further, the risk is much higher for reservoirs having high connate-water saturation and rarely are waterfloods initiated in reservoirs having water saturations exceeding 50 percent. Operators are also extremely cautious in applying waterflooding to reservoirs having high primary recovery. And finally, the sweep efficiency is critical in achieving successful waterfloods. Although sweep efficiency may be predicted by analogy and/or experience with representative data, inadequate rock- and fluid-heterogeneity information is the principal cause for waterflooding failures.

Frequently after waterflood, water containing chemicals is injected. This process is considered as an enhanced recovery technique and is discussed in another chapter.

Gas Injection

As discussed in waterflooding, the injected fluid increases the reservoir energy causing the crude oil to flow to the production wells. The success of such a process is dependent primarily upon the displacement efficiency and the sweep efficiency. Unfortunately, in such processes, gas is not nearly as effective a secondary recovery process as waterflooding because of the relatively low viscosity and density of gas. However, gas may be used in practice as the displacing fluid because of its availability and high injectivity potential.

Two types of gas injection processes are identified: (1) dispersed gas injection and (2) segregation. The former process is ordinarily applied to reservoirs which have been producing by a solution gas drive. The primary and secondary variables cited under waterflooding are applicable to both these processes. Experience has shown that this frontal drive process is most effective in relatively thin, low vertical permeability reservoirs having little dip. In reservoirs having a high vertical permeability and large vertical thickness, gas injection may result in high oil recovery because of gravity segregation. Ordinarily, gas is injected near the crest of the structure under favorable conditions.

As mentioned earlier, the low viscosity of gas injected is a major contributor to low oil recovery by this process. In order to partially alleviate this condition, a mixture of 90 percent gas and 10 percent water and a foaming agent has been injected. The resulting foam mixture has a much higher viscosity than gas alone. However, the stability of foam injected into oil reservoirs is difficult to control and it is difficult to achieve the desired oil displacement. Experience has shown that foams readily invade the large pores of heterogeneous rocks, and are frequently adsorbed on clays. Hence, foams offer some promise as selective plugging agents and may retard early breakthrough of oil displacing agents.

ENHANCED OIL RECOVERY

During the past 50 years of academic and industrial research and industrial operations many prospective enhanced oil recovery methods have evolved. For the processes proposed there have been many claims and counter-claims relative to their physical and economic feasibility. The situation was so confused that about 5 years ago it was decided that an in depth study and evaluation of these enhanced recovery processes was necessary. This study included USA participants from academic institutions, state and federal agencies, and the petroleum industry. The enhanced recovery processes listed in Table 42-1 were identified and classified. The consensus of the 22 major petroleum companies participating in the study was that four enhanced recovery methods would dominate the USA scene with respect to the additional amount of crude oil produced by enhanced oil recovery techniques. Those listed, together with the estimated percentage of recoverable oil attributable to the process, were:

1.	Micellar (water-driven miscible slugs, surfactant, or microemulsion flooding)	58%
2.	Thermal	29%
3.	Carbon dioxide	89%
4.	Hydrocarbon miscible	5%
		100%

It is recognized that these estimates are dependent upon the number and types of reservoirs, drive mechanism, oil recovery by primary and secondary processes, etc. A brief description of each process follows.

Surfactant or Microemulsion

The surfactant or microemulsion process utilizes multiple-slug injection, surfactant followed by a water containing a polymer. The purpose is to increase oil displacement efficiency through reduction of capillary forces and increase the sweep efficiency by mobility control. Since reservoirs differ in their rock and fluid properties, each requires a unique chemical formulation for the injected slug.

At this time, there is rapidly expanding interest in this process, resulting in accelerated laboratory, pilot, and field testing. As is evidenced by the industry survey, this process is expected to possess the greatest potential for achieving increased oil recovery by enhanced techniques in the long term.

Thermal Processes

Of the four principal enhanced recovery processes, thermal methods have thus far received the greatest application in the field. Such experience during the past decade, particularly in California, USA, has been effective in developing an annual production of approximately 250,000 barrels of oil per day.

In thermal processes, heat is added to the reservoir, decreasing the viscosity of the oil, partially vaporizing the oil, increasing the reservoir pressure, and thereby increasing oil mobility. The two principal thermal methods are (1) steam injection and (2) in situ combustion.

In the steam injection process a mixture of steam and hot water (the mix varying over a wide range of quality) is injected into the reservoir. The producing well may be stimulated with high-quality steam followed thereafter by steam drive of the reservoir. The steam soak (huff and puff) process, in which the same well is used successfully for injection and production, is also widely used.

In situ combustion has also received extensive field testing in California. In this process, air is injected into the reservoir, igniting the oil and producing the heat to reduce the oil viscosity and partially vaporizing the oil.

The crude oil is driven into producing wells by a combination steam-hot water-gas drive.

Experience has shown that the efficiency of the process may be improved by alternately injecting air and water (COFCAW process, combination of forward combustion and water). The injected water appears to increase the utilization of heat by transferring it from the hot rock behind the combustion zone to the rock ahead of the combustion zone containing the steam.

The availability of water and the energy necessary to generate hot water and/or steam for use in these processes and their ecological and environmental impact are major considerations in determining economic feasibility.

Hydrocarbon Miscible

This method utilizes high-pressure hydrocarbon gas and/or liquid injection to produce miscibility with the reservoir oil. Under reservoir conditions, the injected fluid serves as a solvent for the oil. Hence, interfacial and capillary forces are virtually eliminated and displacement of oil from the reservoir space contacted is essentially complete. Through appropriate means, the residual oil left after secondary recovery operations can be recovered by this process.

Laboratory and field tests have proved that the process will recover essentially all the oil in the pore space contacted by the solvent. However, in the USA, it appears that this process will not be widely used as an enhanced recovery technique because of the costs-to-benefits ratio. Further, these solvents are in short supply, with projected future increased costs. However, in areas where these conditions do not prevail, this enhanced oil recovery technique should be given consideration.

Carbon Dioxide

The success of the carbon dioxide fluid injection method is dependent upon the creation and maintenance of the miscibility of the mixture of carbon dioxide and the vaporized hydrocarbons with the reservoir oil. For each reservoir, this is accomplished by optimization of the volume and rate of fluid injection and the reservoir operating pressure.

This process has thus far proved most successful in carbonate rocks. Such operations are characterized by high reservoir operating pressures, and premature production of carbon dioxide. Consideration of the availability of carbon dioxide and the need for remedial workovers and separation and reinjection of produced carbon dioxide are necessary in order to evaluate the economic feasibility of this type of enhanced recovery process.

Chemical Waterflooding

Polymers, such as polyacrylamide and polysaccharide, have been added to injected water in waterflooding to improve the mobility ratio between the injected fluid and the crude oil. Other chemicals, such as sodium hydroxide, sodium silicate, and sodium carbonate, have been added to floodwaters to improve recovery through interfacial tension and wettability changes and emulsification.

Although polymer and alkaline flooding have seen some pilot testing and commercial application, at this time it is not believed that these processes will contribute significantly to increased oil recovery in the near future.

WORLD ENHANCED OIL RECOVERY PROJECTS

Enhanced oil recovery is receiving increasing attention in the USA and in 1978 there were 196 such projects in some phase of development, either pilot, field test, or active. Through December 31, 1978, the following distribution of active enhanced oil recovery operations in the USA was reported:

Micellar/surfactant	22
Steam drive	43
Steam soak	56
Combustion	16
Polymer	21
Caustic	3
Gas miscible	6
Carbon dioxide	14
Hydrocarbon miscible	15
Total	196

The approximate enhanced oil recovery production during 1978 in the USA was estimated to be 373,000 barrels per day, with a distribution as follows:

Steam	250,000
Combustion	10,000
Chemical	3,000
Gas	110,000
Total	373,000

Although continuing efforts are being made by a number of organizations and individuals to obtain worldwide data on enhanced recovery operations, such efforts have received limited response. However, the data available yield the following (stock tank barrels per day):

Country	Type of process	Daily production
Canada	Total	20,900
	Steam soak	6,000
	Steam drive	500
	Forward combustion	6,000
	COFCAW	400
	Hydrocarbon miscible	8,000
Germany	Steam drive	1,000
Hungary	Carbon dioxide	300
Indonesia	Steam soak	22,000
Libya	Hydrocarbon miscible	100,000
Trinidad	Steam and hot water	7,000
USA	Various	373,000
USSR	Steam drive	2,600
Venezuela	Steam soak	1,601,000
Total		1,686,800

FUTURE OIL RECOVERY AND PRODUCTIVITY FROM PRIMARY, SECONDARY, AND ENHANCED RECOVERY OPERATIONS

Since 1973, there have been many petroleum reserve, potential reserve, and productivity estimates by all segments of government, petroleum industry, consultants, etc. Enhanced oil recovery has received great attention for many years since there have been those who have claimed that substantial reserves were imminent from these processes and that the oil productivity therefrom would compensate for world declining oil production. Thus far, such has not been the case.

In any estimate of enhanced oil recovery potential for a country, region, field, or reservoir, an extensive detailed study and evaluation is necessary. Criteria, obtained from laboratory and/or field experience, are a necessity if one is to have confidence in the estimates. Since a number of different enhanced oil recovery processes may be applicable to a reservoir, it is necessary to ascertain which is most favorable for the particular reservoir.

For many years, on a continuing basis, the author has actively participated in oil and gas and other energy resource and reserve estimates. His procedure for making such estimates for crude oil reservoirs ordinarily necessitates the estimate of the original oil in place; the cumulative oil production; the oil recovery by the natural reservoir energy, primary recovery; the oil recovery by waterflooding and/or gas injection, secondary recovery; the ultimate oil recovery by combined primary and secondary recovery; the residual oil in place after primary and secondary recovery; the oil recovery by enhanced oil recovery techniques; and finally the total oil recovery by all processes.

At this time, the criteria shown in Tables 42-2 and 42-3, together with economic analyses, are used to select the most favorable enhanced oil recovery process for a particular reservoir.

The results of such an evaluation made by the author for the world crude oil reservoirs, and updated to December 31, 1977, for this Conference, are shown in Table 42-4. In order to quantify and characterize the crude oil resource and reserves, tables similar to this are prepared for crude oil reservoirs having various characteristics including API gravity, viscosity, gas solubility, etc. Also statistical studies are made of reservoirs to estimate the original oil in place and reserves related to time, cumulative crude oil production, annual crude oil production, etc. All of this information is combined to form the data base for predicting the original oil in place and reserves for each individual formation, type of rock (sandstone, limestone), field, county, district, state, and country. The accuracy of the data base is checked periodically by comparing results for the future obtained through its use with actual performance data. Modification of the data base has been made historically to improve accuracy of predictions.

Results of the latest (December 31, 1977) study for crude oil reservoirs in the world are shown in Table 42-4. To the author's knowledge, this is the first published summary of world resource data in such detail. Hence, there is little opportunity for comparison of these results with the efforts of other investigators other than in the totals shown for each column.

The original oil in place for the world falls within the range of estimates of other investigators. However, this provides little confidence, because the reported range extends from 2,800 to 3,600 billion stock tank barrels. The cumulative oil production for each country and the total cumulative

TABLE 42-2. Screening criteria for selection of EOR candidates

Screening parameters	Units	Chemical flooding processes: Surfactant	Chemical flooding processes: Polymer	Chemical flooding processes: Alkaline	Miscible processes: Carbon dioxide	Steam	Thermal processes: In-situ Combustion
Oil gravity	°API	≥25	...	≤35	≥27	≤25	≤25
Oil viscosity (μ)	cp	≤30*	≤200	≤200	≤10	≥20	≥20
Depth (D)	feet	...*	...	...	>2,300	>200<5,000	>5,000
Zone thickness (h)	feet	...	...	...	...	>20	>10
Temperature	°F	≤250	≤200	≤200	<250	...	...
Permeability, average ($\bar{K}$)	md	≥20*	≥20	≥20	...	...	...
Transmissibility (Kh/μ)	md/ft/cp	...	...	...	...	100	20
Salinity of formation brine (TDS)	p.p.m.	≤200 000	...	...	...	...	...
Minimum oil saturation at start of process							
In water-swept zones (S_{orw})	fraction	≥0.20	...	...	...	...	...
Mobile (S_{or}-S_{orw})	fraction	...	≥0.10				
Minimum oil content at start of process (S_{or})	B/AF	...	...	...	...	>500	>500
Rock type	...	sandstone	...	sandstone	sandstone or carbonate	sandstone or carbonate	sandstone
Other	...	**	**	**	**	**	**

*Used to estimate required well spacing for given project life.

**Other criteria including geological data were also considered. The specific criteria used depended upon the process. In general, reservoirs with continuous and uniform strata were ranked higher than were heterogeneous reservoirs with extensive faulting, discontinuous lens-type deposits, fractures, etc. Similarly, reservoirs with extensive underlying aquifers or overlying gas caps were ranked lower or excluded as EOR candidates in view of the decreased likelihood of a successful EOR project. From National Petroleum Council.

TABLE 42-3. Criteria for application of enhanced recovery methods

Process	Oil grav. API	Reser-voir temp., °F	Operating press., psia	Oil visc. CP	Perm. md	Thick-ness ft	Favorable factors	Factors which increase risk
High-pressure gas drive*	>40	--	>3,500	<1	--	--	1 Deep formation (see pressure) 2 High dip 3 Undersaturated crude with high C_2-C_6 concentration	1 Extensive fractures 2 Strong water drive 3 Gas cap 4 High vertical permeability in a horizontal reservoir 5 High permeability contrast 6 Natural gas supply limitations
Enriched gas drive* (condensing gas drive)	>30	--	>1,300 (900-1,000)	<3	--	--	1 At low pressure, gas required with high C_2 and low N_2 2 High dip 3 Thin pay, low k_v	1-6 same as above 7 LP-gas supply limitations
LPS slug	>30	--	<1,300 (800-900 with modif.)	<5	--	--	1 Distress (cheap) LP-gas available 2 High dip 3 Thin pay, low k_v 4 C_2 rich fluid for low pressure reservoir	Same as above
CO_2 (miscible)* (water or gas driven)	>30	--	>1,100	<3	--	--	1 CO_2 available 2 High dip 3 Thin pay, low k_v	1-5 same as above 6 CO_2 supply and transportation require high initial investment
Miscellar flooding*		≦200	--	<10	20-50 in contacted portion of reservoir	--	1 It is ESSENTIAL that good water be available (<5,000 pps total dissolved solids; <500 pps Ca^{++} and Mg^{++}) 2 Waterflood sweep <50%	1 Extensive fractures 2 Strong water drive 3 Gas cap 4 High permeability contrast 5 Highly saline (>30,000 pps TDS) connate or flood match
Steam drive	>10	--	<2,500	--	--	>20	1 Existing wells adaptable to steam injection 2 Available gas supply for steam generation 3 Available water which is cheap, slightly alkaline, free of H_2S, oil, dissolved iron, and turbidity	1 Strong water drive 2 Gas cap 3 Low net to gross pay fraction 4 Extensive fractures (not as serious as in other injection methods)
COFCAW	<45	--	250	--	--	>10	1 Formation temperature 150°F 2 Low vertical permeability 3 Available water which is cheap, and won't precipiate solids in presence of air 4 Existing wells in condition to withstand high pressure 5 Cheap gas supply for compressors	1 Extensive fractures 2 Gas cap 3 Strong water drive 4 Low net to gross pay fraction 5 Serious preexisting emulsion problems

*Requires laboratory tests to confirm suitability.

Source: Adapted from Guffen, Ted M. "Oil Production to Expect from Known Technology," Oil & Gas Journal 71, no. 19, p. 66 and subsequent data from Society of Petroleum Engineers Enhanced Recovery Data Publications and Gulf Universities Research Consortium Publications.

TABLE 42-4. World oil resource recoveries by countries, conventional methods and enhanced recovery, present technology and economics as of 12/31/77
(Billions of stock tank barrels)

Country	Original oil in place	Cumulative production	Ultimate recovery[1]	Reserves	Residual oil in place	Enhanced recovery	Total recovery
Saudi Arabia	522.4	30.8	194.6	163.8	327.8	40.5(7.75)	235.1
USA	445.0	113.8	143.3	29.5	301.7	25.8(5.80)	169.1
Kuwait	349.8	18.3	120.7	102.4	229.1	29.7(8.49)	150.4
USSR	316.2	53.3	128.4	75.1	187.8	13.9(4.42)	142.3
Iran	270.8	26.4	93.3	66.9	177.5	15.0(5.50)	108.3
Venezuela	238.7	33.6	51.6	18.0	187.1	20.0(8.38)	71.6
Iraq	181.5	12.3	62.6	50.3	118.9	10.0(5.50)	72.6
Abu Dhabi	111.1	4.0	40.1	36.1	71.0	10.0(9.00)	50.1
Libya	104.6	11.2	48.3	37.1	56.3	4.0(3.82)	52.3
China	68.5	3.8	24.7	20.9	43.8	2.7(3.94)	27.4
Nigeria	65.5	6.1	24.8	18.7	40.7	1.4(2.14)	26.2
Indonesia	57.3	7.3	20.9	13.6	36.4	2.0(3.49)	22.9
Algeria	52.3	4.6	17.8	13.2	34.5	3.1(5.92)	20.9
Mexico	46.0	6.2	19.6	13.4	26.4	1.1(2.39)	20.7
Canada	43.3	8.3	14.3	6.0	29.0	3.9(9.01)	18.2
United Kingdom	40.5	0.4	13.9	13.5	26.6	1.1(2.72)	15.0
Neutral Zone	38.0	2.8	12.2	9.4	25.8	3.0(7.89)	15.2
Norway	23.7	0.3	6.3	6.0	17.4	2.0(8.43)	8.3
Brunei-Malaysia	12.8	1.7	4.7	3.0	8.1	0.4(3.13)	5.1
West Germany	5.7	1.2	1.6	0.4	4.1	0.4(7.02)	2.0
Netherlands	3.7	0.3	1.1	0.8	2.6	0.2(5.41)	1.3
Pakistan	1.3	0.1	0.4	0.3	0.9	0.1(7.69)	0.5
Total	2,998.7	346.8	1,045.2	698.4	1,953.5	190.3(6.35)	1,235.5
Other	231.1	28.5	69.1	40.6	162.0	11.8(5.11)	80.9
Total world	3,229.8	375.3	1,114.3	739.0	2,115.5	202.1(6.26)	1,316.4
% original oil in place		11.6	34.5	22.9	65.5	6.3	40.8

[1] Sum of primary and secondary recovery processes.

oil production represent the best available data and are those used in most evaluations.

The reserves shown are for the most part considerably higher than those reported by others. Reserves are considered to be indicative of the crude oil which geological and engineering data demonstrate with reasonable certainty to be recoverable in future years from known reservoirs under existing economic and operating conditions. The author's higher reserve estimates are due to his acceptance of the applicability of secondary and/or enhanced recovery techniques to specific reservoirs prior to completion and/or operation of the method. However, pilot testing has indicated in such reservoirs that the techniques are economical. Similar comments are applicable to the ultimate recovery estimates. The ultimate recovery values shown represent combined primary and secondary recovery processes.

The residual oil in place values indicate the volume of oil remaining in the reservoirs after application of primary and secondary recovery operations. Hence, the residual oil volume is the target for application of enhanced oil recovery techniques.

The enhanced recovery estimates were made after detailed reservoir evaluation considering the variables discussed in another section of this chapter and shown in Tables 42-2 and 42-3. As might be expected, an extremely wide range of percentage recoveries was obtained due to the diversity of rock and fluid properties applied, the engineering management of the reservoirs, and economics. Enhanced oil is considered to be new oil in the United States and the estimates for recovery are based upon a world price of $25 per barrel in the future (constant dollars).

It is significant that of the 3,229.8 billion barrels of oil in place, only 1,114.3 billion are estimated to be recoverable by combined primary and secondary recovery methods. Hence, 2,115.5 billion barrels of oil will remain in the reservoir as residual oil. At the date of the report, the combined primary and secondary crude oil production was estimated to yield a 34.5 percent recovery based upon original oil in place. It is estimated that an additional 202.1 billion barrels of oil will be recovered by enhanced recovery methods, which would yield a 40.8 percent ultimate oil recovery total.

It should be recognized that the application of enhanced recovery techniques to petroleum reservoirs is dependent upon technology, economics, and government policy. Improved economics and government policy stability are essential if the residual oil resource is to be expeditiously exploited through application of enhanced recovery techniques.

SUMMARY AND CONCLUSIONS

A detailed study (dated December 31, 1977) of the petroleum reservoirs in the world has been made in an effort to estimate the potential for enhanced oil recovery. The results are shown in Table 42-4 and may be summarized as follows:

Original oil in place, billion stock tank barrels	3,229.8
Cumulative oil production, billion stock tank barrels	375.3
Ultimate oil recovery, billion stock tank barrels	1,114.3
Oil reserves, billion stock tank barrels	739.0
Enhanced oil recovery, billion stock tank barrels	202.1
Oil recovery by primary and secondary processes, percent of original oil in place	34.5
Oil recovery by enhanced recovery processes, percent of original oil in place	6.3
Total oil recovery by primary, secondary, and enhanced processes, percent of original oil in place	40.8

BIBLIOGRAPHY

Planning Criteria Relative to a National RDT&D Program to the Enhanced Recovery of Crude Oil and Natural Gas: Gulf Universities Research Consortium Report no. 130, November 1973.

Assessment of Enhanced Recovery Technology as a Means for Increasing Total Crude Oil Recovery in Texas: Gulf Universities Research Consortium Report no. 141, December 1974.

Assessment of Enhanced Recovery Technology as a Means for Increasing Natural Gas Recovery in Texas: Gulf Universities Research Consortium Report no. 142.

Preliminary Field Test Recommendations and Prospective Crude Oil Fields or Reservoirs for High Priority Field Testing: Gulf Universities Research Consortium Report no. 148, Feburary 1975.

U.S. Energy Outlook--Oil and Gas Availability: National Petroleum Council, Washington, D.C., 1973.

Enhanced Oil Recovery: National Petroleum Council, Washington, D.C., December 1976.

Project Independence Report: Federal Energy Administration, November 1974.

1976 National Energy Outlook: Federal Energy Administration.

The Potential and Economics of Enhanced Oil Recovery: Lewin & Associates, Inc., for the Federal Energy Administration, April 1976.

Improved Oil-Recovery Field Reports: SPE-AIME, vol. 1, no. 1-4, June 1975-March 1976 to present.

Proceedings of SPE-AIME Improved Oil Recovery Symposium, Tulsa, Oklahoma, April 16-19, 1972; April 22-24, 1974; March 22-24, 1976.

A Survey of Secondary Recovery and Pressure Maintenance Operations in Texas to 1968: The Railroad Commission of Texas, Austin, Texas, Bulletin 68; also Bulletins 70, 72, and 74.

Secondary and Tertiary Oil Recovery Processes: Interstate Oil Compact Commission, Oklahoma City, Oklahoma, September 1974.

Reserves of Crude Oil, Natural Gas Liquids and Natural Gas in the United States and Canada as of Decmeber 31, 1976; vol. 31: American Petroleum Institute, Washington, D.C., May 1977 (also previous volumes).

International Petroleum Encyclopedia, 1977: Petroleum Publishing Co., Tulsa, Oklahoma, 1977 (also previous editions).

The Texas Energy Perspective, Proceedings of Energy and Feedstocks: Challenges for Texas: College of Engineering, Texas A&M University, p. 31-40, 1976.

Increased Oil Recovery, Secondary and Tertiary; Application and Future Prospect: The Arabian Journal for Science and Engineering, Special Issue, p. 73, 1978.

CHAPTER 43

THE ROLE OF MICROBIOLOGY IN ENHANCED OIL RECOVERY

W. R. Finnerty[1]

Of the known "oil in place" only about one-third is considered economically recoverable with present technology. The location of the remaining 200 billion (10^9) barrels of oil in place in the continental United States is in known fields but is unrecoverable by existing technology. The dollar value of this unavailable resource is estimated to be in excess of $2,000 billion. The dollar value of organic matter in oil shales in the United States alone is considered to be in excess of $2,000 billion. Given these striking incentives, it is enigmatic that so little progress has been made in secondary and tertiary oil recovery technology.

A viable, broad-based enhanced oil recovery (EOR) technology program does not appear to exist at this time. The EOR potential of some billions of barrels of oil can be achieved with money, time, and manpower. The absence of a technical plan that coordinates the overall effort to find early basic answers through research relates directly to our ultimate national welfare. It is emphasized that EOR problems will not be solved with single projects and progress will be evolutionary, not revolutionary. Ironically, modern-day science can tell us more about the moon's behind than the ocean's bottom. We can make a better qualified description of the moon's surface than a quantified description of any oil reservoir. We are better able to receive, display, and process imagery of the Martian surface than to obtain, display, and analyze essential rock-fluid data from the oldest or largest oil fields.

As domestic oil production declines, the necessity and urgency to develop alternative energy resources is obvious. However, the economy of industrialized nations is based on petroleum and will continue to be, due to large capital requirements, environmental barriers, and time-lag factors that energy alternatives and supporting technology dictate. Oil and gas will continue to supply the bulk of near-term and mid-term energy resources.

Current research programs in EOR focus on reservoir rock properties and characteristics, fluid and rock mechanics, and high-resolution geophysics and

---

[1]Department of Microbiology, University of Georgia, Athens, Georgia.

logging. The purpose of this chapter is neither to review nor criticize the ongoing developmental research effort related to EOR. Rather, it is intended to present a review of supplementary technology that will aid in obtaining greater oil production and processing capabilities. This supplementary technology concerns biological processes and the application of these processes to specific and well-defined problems. Microbiological processes have comprised the subject matter of four separate workshops since 1975. The first, entitled "The Role of Microorganisms in the Recovery of Oil," was sponsored by the National Science Foundation in 1975. The second concerned "Microbial Energy Conversion," under sponsorship of UNITAR in 1976. The third, under the auspices of the Gulf Universities Research Consortium, took place in 1977. Recently, a Department of Energy-sponsored workshop addressed the topic of "Microbiological Processes Useful in Enhanced Oil Recovery." Almost 5 years have passed with little development of basic and/or applied research to interface the technology afforded by biological processes with problems associated with improved oil production capacities. The following discussion will outline scenarios for various biological processes that have been identified as promising areas for research development. These scenarios will be presented in context of identified problems recognized by and being addressed within the petroleum industry and by the Federal Government. It is apparent that for biological processes to become integrated into oil production research efforts, biologists must recognize and understand the limitations of current physical-chemical data bases about oil reservoirs as well as the constraints characterizing developments within the industry.

STATUS OF EOR TECHNOLOGY

Enhanced oil recovery has advanced to a stage where recovery processes are technically feasible in limited application to specific, highly favorable reservoir conditions. EOR technology remains, however, very much in a research mode from several critical vantage points. Areas needing immediate development include:

1. Advances in known oil-displacement technology;
2. Extension of oil-displacement technology to more complex reservoir conditions;

3. Development of the science and technology relating to subsurface geologic description and quantification of the geologic parameters that control the oil displacement process;
4. Generation of a better understanding of oil-displacement processes to enable technical design for and application to specific reservoirs.

Research programs designed for improved definition and quantitative description of reservoirs and reservoir rocks have been addressed by the petroleum industry. Some of the details of these programs are listed in Table 43-1.

TABLE 43-1. Programs for improved definition and quantitative description of reservoirs and reservoir rocks

I. RESERVOIR ROCK PROPERTIES

A. Petrography of 50 Selected Reservoirs with Residual Oil.
B. Diagenesis of Reservoir Rock.
C. Porosity and Permeability in Existing Reservoirs.
D. Genesis of Reservoir Rock--Inhomogeneity and Anisotropy.
E. Relationships Between Physical and Chemical Characteristics of Reservoir Rock.
   1. Residual Oil Saturation.
   2. Correlation of Mineral and Fluid Data.
   3. Correlation of Borehole Logs with Rock and Fluid Content.
   4. Porosity and Permeability.

II. FLUID AND ROCK MECHANICS

A. Effect of Rock-Fluid Interfacial Chemistry In Situ.
B. Prevention and Mitigation of Formation Damage.
C. Rock and Fluid Properties In Situ.
D. Directional and Point Properties and Spatial Distribution.
E. Interpretation of Tracer Flow Tests.
F. Improved Procedures for Numerical Modeling.

III. HIGH RESOLUTION GEOPHYSICS AND LOGGING

A. Applicability of Directional Drilling to EOR.
B. Downhole Microwave Scanning.
C. Application of High-Resolution Engineering Seismic Profiling EOR.
D. Application of Seismic Refraction Techniques.
E. Well-to-Well Seismic Methods.
F. Areal Seismic Survey Over EOR Field.
G. Evaluation of Borehole Gravimetry for EOR.
H. Examination of Geologic Models.

These programs indicate that greater sophistication is necessary in providing the technology for generally effective oil recovery. As a final commentary before discussing the application of biological processes to oil production, I would like to emphasize the interdisciplinary nature of research programs addressing both EOR and oil production. Such programs necessitate the involvement of chemists, engineers, physicists, geologists, economists, and last, but not least, biologists, if the full potential of available scientific manpower is to be realized in accomplishing national goals of energy independence and self-sufficiency.

Biological Processes and Oil Production

A detailed discussion on the application of microbial processes to oil production will necessarily fall short of precise or general solutions to the myriad of questions posed when addressing their direct utility and value. Even as the petroleum engineer lacks definitive knowledge supported by comprehensive data bases on the chemical and physical structure of the oil reservoir, the microbiologist also lacks information necessary to critically address possible role(s) microorganisms may play in enhanced oil production. The areas of hydrocarbon and petroleum microbiology have resulted in an extensive literature over the past half century (Davis, 1967). Tremendous advances have occurred in the acquisition of definitive knowledge relating to the physiological, biochemical, and genetic properties of a wide variety of hydrocarbon-utilizing microorganisms. The application of this knowledge has yet to be realized in direct application to either EOR or oil production. A consensus of opinion exists among many microbiologists that a potential use of microorganisms and their products has application to field problems. To this end, the following discussion is directed. Microbiological processes will be discussed in context of reservoir microbiology, biopolymers, and bioengineering.

Reservoir Microbiology

The available literature concerning microbial activities within the reservoir environment is totally inadequate. Ample reports exist concerning the proliferation of sulfate-reducing bacteria and the souring of sweet crudes, plugging phenomena, and other deleterious problems (Davis, 1967). Recent studies have demonstrated rich microbial populations being obtained from cores of various reservoirs. The taxonomic classification of these microorganisms and the cataloging of their various physiological and biochemical characteristics has yet to be accomplished. The realization has surfaced that reservoirs are not sterile, biologically refractory environments. The source of the indigenous microflora is further to be determined with possible point-source contamination from surface origins or, alternatively, of subterranean origin. It has been further recognized that in situ alteration of oil is occurring as a result of some level and rate of microbial metabolic activity. Preliminary studies have indicated that the indigenous microflora can alter oil in situ through conversion of alkanes to wax esters varying in chain length from 60 to 100 carbon atoms. Other microbially induced modifications of oil are yet to be determined. Immediate concern arises as to whether producing wells decrease in productive capacity over time through microbial processes that tend to significantly alter the viscosity characteristics of the oil. If so, definitive answers to such processes would allow for meaningful treatment regimes on new wells that would control microbial activity and enhance long-term productivity.

The alternative possibility for consideration is stimulation of the indigenous microflora toward greater metabolic activity by the addition of appropriate nutrients that would allow for various hydrocarbon-oxidizing microorganisms to attack specific fractions of oil and, in turn, produce chemicals effective in increased oil production (see Bioproducts section). Evidence has appeared that indicates utilization and release of hydrocarbons and oil by microorganisms in subsurface environments by the introduction of air and mineral nutrients (Vanloocke, R., 1979).

Stimulation of Depleted Oil Reservoirs

The stimulation of depleted reservoirs by induced microbial fermentations has been subjected to study at sporadic intervals, particularly in Eastern-bloc Communist countries. Results appear to be variable and subject to varying degrees of criticism from scientific points of view. The assessment of induced bacterial fermentations in the reservoir had not been developed to any practical extent until recently. An enterprising petroleum engineer has systematically and empirically developed fermentation processes as a technology for the release of significant amounts of new oil from depleted wells (Johnson, 1979).

The oil-release technique developed is an inexpensive, anaerobic fermentation designed to function within a reservoir. The process is usable only within certain parameters. The process consists of the injection into the reservoir of certain anaerobic, spore-forming Bacillus and Clostridium species along with an aqueous solution of fermentable carbohydrate (cattle feed molasses, milk whey). If the proper oil type and reservoir conditions are chosen, an average increase in oil production in excess of 350 percent may be expected at a cost varying between 15¢ and 50¢ per additional barrel of oil.

Field and laboratory studies have shown that the technique generates beneficial metabolic products.

1. Large amounts of CO_2 and CH_4 and some N_2 are formed. The production of these gases effects a partial repressurization of the reservoir. The produced CO_2 is partially miscible with oil, causing the oil to swell, to become less viscous, and generally easier to produce.
2. Large amounts of organic acids are formed which react with reservoir carbonates to form additional CO_2.
3. A variety of surfactants are formed which tend to reduce oil-water surface tension (alcohols, ketones).

Field experiments have demonstrated that:

1. The technique is more effective with lower-viscosity oils. In general, a candidate reservoir should have an oil gravity between 15° to 30° API and a viscosity less than 400 cp.
2. Carbonate reservoirs or reservoirs with carbonate cementation respond better to microbial treatment.
3. There is strong evidence that properly staged treatments will significantly reduce formation damage within the general area of the well bore.
4. Bacterial growth is reduced if the salt concentration in the reservoir water exceeds 100,000 ppm.
5. The optimum growth temperature of the bacterial cultures is approximately 45° to 47°C.

Present production data indicate that in oil reservoirs with crudes of 15° to 30° gravity API, additional recoveries of 20 to 35 percent of the oil in place may be expected.

Biopolymers

A new fermentation industry arose in the 1950's and early 1960's around the production of extracellular microbial water-soluble biopolymers. These biopolymers have generally fallen into the general classification of polysaccharides. Several factors precipitated the rapid development and use of microbial polysaccharides. The hydrocolloids obtained from plants and seaweed have historically experienced numerous and diverse applications in the food, textile, paint, agricultural, and petroleum industries. Increased labor costs, limited sources, adverse climatic conditions, lack of physical and chemical structural alternatives, and increased demand stimulated the search for new sources and types of polysaccharides.

Extracellular polysaccharide production is an ubiquitous characteristic of microorganisms with several of these biopolymers proving to be of commercial significance. The usefulness of these bacterial polysaccharides rests with their unique physical and chemical properties determined by the individual component sugars and linkage relationships. A broad overview of microbial polysaccharide production and application may be found in Sanford and Laskin (1977).

Biopolymers have found application in enhanced oil recovery technology, where they are used to increase the efficiency by which water conducts and displaces oil from a reservoir. The function of the polymer is to reduce the mobility of injected water, mobility being defined as the flow capacity of a rock-fluid system or the volumetric flow rate per unit area under a given pressure gradient. Mobility is expressed as effective rock permeability divided by fluid viscosity in darcies (or millidarcies) per centipoise. Polymers, therefore, reduce mobility by decreasing effective rock permeability and by increasing effective fluid viscosity.

Two basic types of polymers--microbial-produced xanthan gums and hydrolyzed polyacrylamide--comprise the majority of polymers currently under study by the petroleum industry. Polyacrylamides dominate water flooding applications, while xanthans have a minor role. Increased use of xanthan gums is expected as micellar-polymer processes are tested and applications found. The factors of importance in polymer flooding are mobility reduction, polymer retention, inaccessible pore volume, transient flow behavior, mechanical degradation, injectivity behavior, and reservoir compatibility. A critical discussion of each of these parameters is beyond the scope of this chapter but has been developed by Sandvik and Maerker (1977).

At this time only three microbial polysaccharides of commercial importance are in large-scale production--dextran, polytran, and xanthan. Xanthan is currently the most important polysaccharide for EOR. Xanthan, polyacrylamide, modified starches, and cellulose derivatives (carboxymethylcellulose) are also used as additives to oil drilling muds. Xanthan gums appear to represent the preferred polymer for lubrication of bentonitic muds due to their stability in heat, pH, cations, and resistance to shear. Early estimates for EOR by polymer flooding approached 500,000 barrels/day (B/D). In 1976 revised estimates for 1986 production of residual oil were 200,000 B/D, with a starting date for large-scale development in 1979. Table 43-2 reflects those early optimistic estimates for polymer demand for use in EOR.

TABLE 43-2. Polymer demand for enhanced oil recovery

Year	Case A*	Case B*
1979	7.3	18.3
1980	14.6	40.2
1981	32.8	79.0
1982	51.1	131.5
1983	76.7	175.2
1984	110.0	215.2
1985	120.5	233.6
1986	138.5	284.2

*Thousands of lbs/day.

NOTES: 1. Polymer demand assumed to be 50 percent polysaccharide, 50 percent polyacrylamide.
2. Case A--200,000 bbl/day by 1986.
3. Case B--500,000 bbl/day by 1986.

Source: Gulf Universities Research Consortium, March 1976.

This exercise predicts markets of between 250 and 140,000 lbs of polymer demand by 1986. In 1979, the application of xanthan gum to EOR has not begun to reach those early projections for 1979 let alone expansion development projections of 1986. Industry remains very much in the experimental/research mode, testing the application of xanthan for EOR and resolving numerous disadvantages associated with currently available polymers (Table 43-3).

TABLE 43-3. Use of polyacrylamide and polysaccharide for EOR

Characteristics	Polyacrylamide	Polysaccharide	Ideal
Viscosity thickening	10-15 cp at 500 ppm	10-15 cp at 500 ppm	20 cp at 100 ppm
Salinity maximum of fluid	1,500-200 ppm	10,000 ppm	15,000 ppm
Maximum reservoir temperature	175-200°F	200-225°F	250°F
Divalent ion maximum	200 ppm	5,000 ppm	5,000 ppm
Permeability oil plugging	critical >50 md	50-100 md	>100 md
Cost $/lb	1.30 ± 0.30	2.25 ± 0.30	1.25 ± 2.00

The physical properties considered as ideal are not matched by available chemicals. Other disadvantages have become apparent as the result of field studies in complex reservoir environments.

The limited availability of diverse microbial polysaccharides for testing and evaluation in EOR has become apparent. This fact is further overshadowed by the presence of xanthan and its potential and perhaps not real potential for EOR. Table 43-4 lists a few selected conventional hydrocolloids of microbial origin and some that are not so conventional.

The extracellular biopolymers produced by *Candida albicans* and *Acinetobacter* species are atypical macromolecules. The fatty acid-polysaccharide complex has both hydrophilic and hydrophobic properties. The lipoprotein forms clear microemulsions with hydrophobic compounds. Interestingly, both macromolecules partition hydrocarbons and oil as pseudosoluble microemulsions as well as displace oil from a sandstone matrix in excess of 90 percent displacement efficiency. Further, these two biopolymers are produced only by hydrocarbon-utilizing microorganisms under conditions of active hydrocarbon metabolism. The suggestion that these biopolymers are candidates for EOR is not implied but rather that hydrocarbon-utilizing microorganisms appear to mediate the transport of hydrophobic substrates through the biosynthesis of highly species-specific biopolymers. The screening and analysis of the hundreds of

TABLE 43-4. Selected hydrocolloids of microbial origin

Organism	Biopolymer	Reference
Pseudomonas species	Polysaccharide (glucose, galactose)	Deavin et al., 1977
Azotobacter vinelandii	Alginic acid (D-mannuronic, L-guluronic acid)	Deavin et al., 1977
Xanthomonas campestris	Xanthan (mannose-6-acetate, glucuronic acid, mannose, pyruvate)	Moorhouse et al., 1977
Methylomonas mucosa	Polysaccharide	Tam and Finn, 1977
Erwina tahitica	Zanflo (uronic acid, glucose, galactose, fucose)	Kang et al., 1977
Azotobacter indicus	PS-7 (uronic acid, glucose, rhamnose)	Kang and McNeely, 1977
Alcaligenes faecalis	Curdlan (β-1,3-glucose polymer)	Harada, 1977
Leuconostoc mesenteroides	Dextran (α-D-1,6-glucose polymer)	Jeanes, 1977
?	Polytran (scleroglucan)	Jeanes, 1977
Aureobasidium pullulans	Pullulan (1,6-maltose polymer)	Jeanes, 1977
Rhinocladiella mansonii	N-acetyl-glucosamine-N-acetyl-glucosamin uranic acid	Burton et al., 1976
Candida tropicalis	Fatty acid-polysaccharide complex	Kapeli and Fiechter, 1976, 1978
Acinetobacter species HO1-N	Lipoprotein	Kappeli and Finnerty, 1979

microorganisms capable of growth on various and diverse hydrocarbons and/or hydrophobic substrates may well allow for the identification of a new class of biopolymers with more favorable characteristics and potential for application to EOR. A growing body of knowledge in hydrocarbon microbiology strongly merits further research and development of this facet of biopolymer production. Preliminary screening of some 40 cultures of hydrocarbon-oxidizing bacteria has yielded in excess of 6 macromolecules that affect over 90 percent displacement efficiency of crude oil from an inert matrix (W. R. Finnerty, unpublished studies).

Biosurfactants and Bioemulsifiers

The application of surfactants to EOR has long represented an active development area in oil production. The consideration and application of biosurfactants and/or bioemulsifiers have been generally ignored or overlooked, primarily for the lack of a definitive and cohesive literature on the subject. A recent review discusses the broad generalities of bioemulsifiers in context of sources, types, and application (Zajic and Panchal, 1976). The production of surface active compounds by microorganisms has been recognized for years. A system characterization of these compounds has been slow to emerge. However, recent studies in many laboratories have documented the production of a variety of surfactant compounds produced by microorganisms. Table 43-5 lists those surfactant types of molecules that have been chemically characterized from hydrocarbon-utilizing microorganisms.

An emerging pattern is developing that microorganisms produce a wide variety of biosurfactants. The development and application of these biologically produced molecules may indicate that they represent valuable and potentially useful compounds in addressing a myriad of recovery and production problems in the oil industry. Although biosurfactants are not currently used or considered by the industry, potential applications are suggested as aids for micellar flooding, foaming techniques, or rock-wetting agents.

Bioengineering

The application of specifically tailored microbial processes through physiological and genetic means may produce useful changes in:

1. oil, through:
    a. reduction of interfacial tension and viscosity
    b. conversion or degradation of components
    c. alteration to useful secondary metabolites or gas
    d. reduction of surface tension by biosurfactants

TABLE 43-5. Surfactants of microbial origin

Organism	Surfactant	Reference
Corynebacterium hydrocarboclastus	Protein-lipid-carbohydrate complex	Zajic et al., 1977
Corynebacterium lepus	Corynomycolic acid	Cooper et al., 1979
Arthrobacter RAG-1	Fatty acid-polysaccharide complex	Zuckerberg et al., 1979
Candida petrophilum	Proteo-lipid	Iguchi et al., 1969
Bacillus subtilis	Surfactin (proteo-lipid)	Arima et al., 1968
Arthrobacter paraffineus	Lipopolysaccharide	Suzuki et al., 1969
Acinetobacter species	Diglyceride, fatty acid	Makula et al., 1974
Acinetobacter species	Lipoprotein	Kappeli and Finnerty, 1979, 1979
Candida tropicalis	Fatty acid-polysaccharide complex	Kappeli and Fiechter, 1976, 1978

e. in situ formation of CO_2, H_2, CH_4, and other metabolic by products

2. the aqueous phase, through:
   a. increase of viscosity with biopolymers
   b. biological treatment of low-quality water used for reservoir injection
3. reservoir rock, through:
   a. alteration of rock permeability
   b. in situ leaching of rock
4. oil shales, through:
   a. surface-pile leaching
   b. enzymatic release of oil from solid surfaces
   c. alterations of kerogen or kerogen-like materials.

An example of the application of bioengineering principles can be cited in the context of bacterial desulfurization. Specific, biocatalytic systems have been developed that reduce the organic-sulfur content of high-sulfur oil by 98 percent (W. R. Finnerty, unpublished results). Three bacterial systems are available that convert dibenzothiophenes and substituted dibenzothiophenes to water-soluble sulfur containing organic compounds. Two of the bacterial systems are mixed cultures and the third, a pure culture. All of these systems reduce the organic sulfur content by 98 percent within 48 hours as a strictly biocatalytic phenomenon. The organisms do not grow at the expense of other soil components, such as alkanes, cycloparaffins, or simple aromatic compounds. The reaction is highly specific and selective for a particular class of compounds--namely polyaromatic ring structures with sulfur as the covalently bonded heteroatom. The mixed-culture systems are mutually synergistic and do not function independently in the desulfurization reaction. The pure culture is an example of a gene-engineered bacterium. A plasmid has been incorporated into a bacterial strain which lacked the ability to degrade dibenzothiophene. The stabilization of this plasmid changed the recipient bacterium to a dibenzothiophene-oxidizing organism.

Similar cultures have been isolated which convert polyaromatic ring systems with nitrogen as the covalently bonded heteroatom to water-soluble products.

Applications of bioengineering concepts to EOR and oil production and processing are contingent upon acquisition and development of base-line data and definitive information. It is imperative to clearly define and delineate the objectives to effectively utilize the extremely powerful techniques of bioengineering.

The future interface between petroleum engineering and the application of biological processes appears complementary in terms of contributions that the biologist can offer to the industry. The development of stronger cooperative ventures in addressing the complex problems of EOR and oil production are considered of significance in obtaining goals of energy independence and self-sufficiency.

REFERENCES

Arima, K., Kakinima, A., Tamura, G. 1968. Biochem. Biophys. Res. Comm. 21: 488-491.

Burton, K. A., Cadmus, M. C., Lagoda, A. A., Sandford, P. A., Watson, P. R. 1977. Biotechnology and Bioengineering 18:1669-1677.

Cooper, D. G., Zajic, J. E., Gracey, D. E. F. 1979. J. of Bacteriology 137: 795-801.

Davis, J. B. Petroleum Microbiology. Elsevier Publishing Co., 640 p.

Deavin, L., Jarman, T. R., Lawson, C. J. Righelato, R. C., Slocombe, S. 1977. American Chemical Society Symposium 45:14-26.

Harada, T. 1977. American Chemical Society Symposium 45:265-283.

Iguchi, I., Takeda, I., Ohsana, M. 1969. Agric. Biol. Chem. 33:1657-1663.

Jeanes, A. 1977. American Chemical Society Symposium 45:284-298.

Johnson, A. C. 238 E. Ellis Drive, Tempe, Arizona 85282.

Kang, K. S., and McNeely, W. N. 1977. American Chemical Society Symposium 45:220-230.

Kang, K. S., Veeder, G. T., Richey, D. D. 1977. American Chemical Society Symposium 45:211-219.

Kappeli, O., and Fiechter, A. 1976. Bioengineering and Biotechnology 18: 967-971.

Kappeli, O., and Fiechter, A. 1978. J. Bacteriol. 133:952-957.

Kappeli, O., and Finnerty, W. R. 1979. Bioeng. Biotech. in press.

Kappeli, O., and Finnerty, W. R. 1979. J. Bacteriol. in press.

Makula, R. A., Lockwood, P. J., Finnerty, W. R. 1974. J. Bacteriol. 121: 250-258.

Moorhouse, R., Walkinshaw, M. D., Arnott, S. 1977. American Chemical Society Symposium 45:90-102.

Sandford, P. A., and Laskin, A. 1977. American Chemical Symposium 45, Extra-cellular Microbial Polysaccharides, American Chemical Society, Washington, D.C.

Sandvik, E. J., and Maerker, J. M. 1977. American Chemical Society Symposium 45:242-264.

Suzuki, T., Tanaka, K., Natoubara, I., Kinoshita, S. 1969. Agr. Biol. Chem. 33:1619-1621.

Tam, K. T., and Finn, R. K. 1977. American Chemical Society Symposium 45: 58-80.

Vauloocke, R. 1979. Environmental Science and Technology 13:346-348.

Zajic, J. E., Guignard, H., Gerson, D. F. 1977. Bioengineering and Biotechnology 19:1285-1301.

Zajic, J. E., and Panchal, C. J. 1976. Critical Reviews in Microbiology 5: 39-66.

Zuckerberg, A., Diver, A., Peeri, Z., Gutnick, D. L., Rosenberg, E. 1979. Applied and Environmental Microbiology 37:414-420.

CHAPTER 44

ENERGY PRODUCTION BY ELECTRICAL INDUCTION HEATING

Sidney T. Fisher[1]

INTRODUCTION

Electrical induction heating is proposed for the following in situ production processes:

1. High-calorific-value gas and coal tar from coal.
2. Petroleum from heavy oil.
3. Bitumen from oil sand, convertible at the surface to petroleum.
4. Shale oil from oil shale, convertible at the surface to petroleum.

A fifth process proposed is the surface production by induction heating of metallurgical coke from mined coking coal.

In the case of in situ coal, the volatiles--high-calorific-value gas and coal tar--are produced by induction heating of the deposit. Then the residual coke, which is hot and permeable, is processed in situ by existing methods to produce low-calorific-value gas. This is led to the surface and burned there to generate electricity in the conventional manner. Figure 44-1 shows the process schematically. For oil-shale, oil-sand, and heavy-oil deposits the in situ induction heating results in volatilization of the hydrocarbons contained in them. These volatiles are led to the surface and further processed there by existing methods to produce petroleum. This is shown schematically in Fig. 44-2. Shafts and tunnels are drilled from the surface to form a series of adjacent loops, spaced from 25 to 50 m, surrounding part of the fossil-fuel deposit. The central shafts are drilled through the overburden and fuel layer, and into the basement rock below. The upper axial and transverse tunnels are in the overburden immediately above the fuel layer, the lower ones in the basement rock immediately below. A conductor is placed in each of these passages to form a complete loop. A large alternating current is passed through a group of loops, and this sets up electric and magnetic fields in the deposit. These fields heat the deposit by inducing displacement and eddy currents in it. This arrangement is shown by Fig. 44-3.

---

[1]F. F. Fisher's & Sons, Limited, Montreal, P.Q.

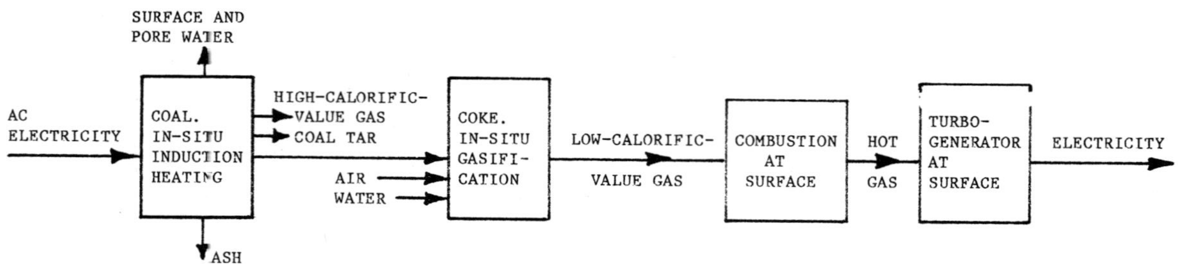

Figure 44-1. Block schematic of proposed method of utilizing coal deposits by induction heating

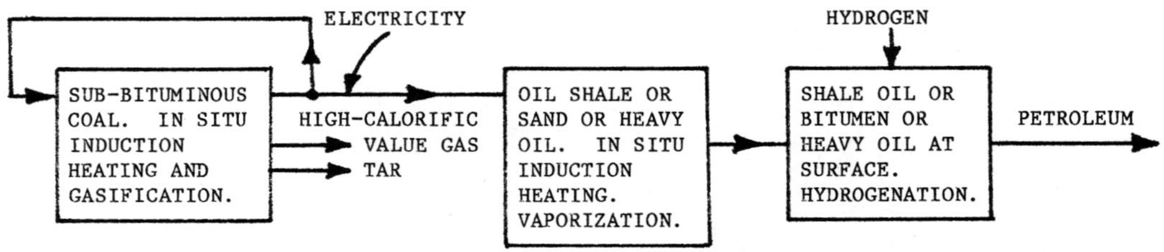

Figure 44-2. Synergetic processing by induction heating in situ

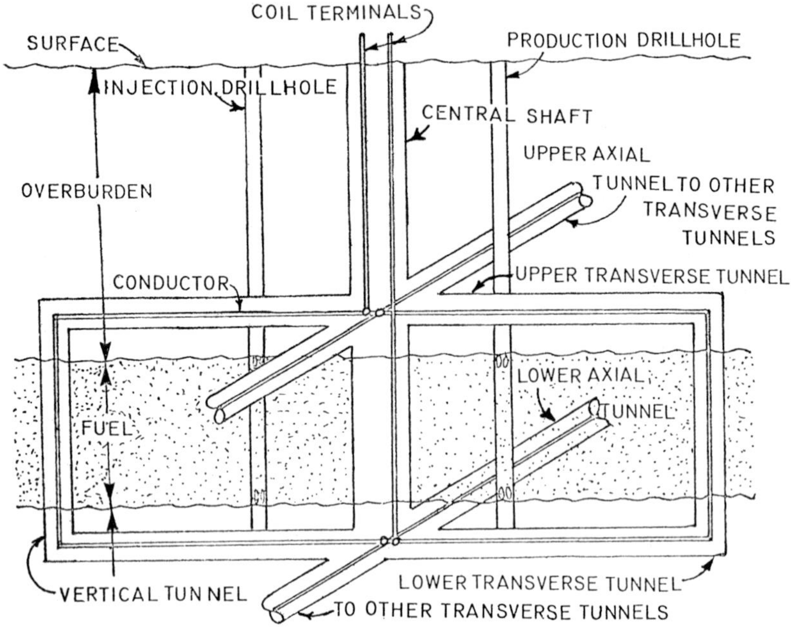

Figure 44-3. Oblique cut-away view of an in situ induction heating installation

Heating takes place at all frequencies due to both displacement and eddy currents. The heating effect due to displacement currents (Campbell, 1976) is:

$$p = \varepsilon'' E^2 \omega \, W/m^3$$

Where: ε'' is the imaginary part of the complex dielectric constant
ω is 2π times the frequency
E is the field intensity.

The heating effect due to eddy currents (DuBow, 1977) is:

$$p = \gamma E^2 \; W/m^3$$

Where: γ is the conductivity
E is the field intensity.

Both effects are included in the term induction heating. The field set up in conducting or semi-conducting material by a conductor loop carrying an alternating current induces both eddy and displacement currents in the enclosed material. As the fuels are heated, volatilization begins and continues over a range of temperatures.

THE PROCESS

In the cases of oil shale, oil sand, and heavy oil, where organic material forms a proportion of the order of 10 wt percent of the total formation, most of the organic content can be volatilized and brought to the surface, to be converted to petroleum, gas, and chemical feedstocks there.

In the case of coal, the volatilizable proportion of the formation, including water, may be of the order of 60 wt percent of the formation, inorganic material 5 wt percent, and nonvolatilizable coke, mainly carbon, 35 wt percent. This coke is hot, particulate, highly resistive to eddy currents, and permeable at the conclusion of the induction-heating process. It is then largely converted to low-calorific-value gas by interaction at high temperature with steam. The gas is led to the surface and burned there to generate electricity. Part of this electricity, about one-quarter, is returned to the adjacent loop in the coal deposit, to continue the induction-heating process, and the remaining three-quarters is transmitted to an oil-producing site, to heat by induction a deposit of oil sand, oil shale, or heavy oil.

When all the volatiles have been produced, the power source is disconnected and reconnected to an adjacent group of loops. Thus both electrical input and output of products are essentially constant and continuous. This procedure results in continuing long-term construction and operation, with production starting after the first group of loops has been constructed, and continuing until after construction of the last group. In the case of coal the underground-gasification process and generation of electricity commence when the induction heating of each group of loops is completed, so that these operations also are continuous throughout the lifetime of the coal deposit.

The world resources of coal are great in relation to other solid fossil fuels, so that procedures special to coal may be employed. One is to carry the coal installations through the induction-heating stage, producing high-calorific-value gas and coal tar, and enough electricity to continue the coal process, leaving most of the hot, permeable residual coke underground, to be utilized at some future time for the production of low-calorific-value gas and generation of electricity. The volatiles produced pay all costs of the heating operation, and the reserve of hot coke thus involves no financial charges.

A second alternative is to gasify all the coke and then generate electricity for purposes other than the induction heating of oil-producing deposits. One purpose is the production at the surface by induction heating of mined coking coal, to produce coke for metallurgical industries. In this case further high-calorific-value gas and coal tar are produced.

ADVANTAGES

This process fully utilizes most resources of coal, oil sand, oil shale, and heavy oil throughout the world, to produce petroleum, high-calorific-value gas, coal tar--mainly used as chemical feedstocks--electricity, and metallurgical coke. The following advantages over existing and other proposed methods of fossil-fuel exploitation are present:

1. The percentage of total reserves that can be processed is high.
2. The yield from the fuel processed is high.
3. The cost of the products is low.
4. The development and construction time is probably short enough that no serious energy-supply gap need be experienced.
5. The lifetime of world solid fossil-fuel deposits is measured in centuries, even at accelerated rates of use.
6. Human and environmental damage and pollution are small.
7. Since all five energy commodities can be produced in any desired mix, serious disruption to industrial and domestic users, with consequent cost and delay, will not occur.

GENERAL DISCUSSION

There are two mechanisms of heating by electrical induction, one by means of displacement currents induced from an electric field, in the process sometimes called RF or dielectric heating, and the other by means of eddy currents induced from a magnetic field, often called simply induction heating and until now applied mainly to metals.

To maximize the rate of displacement-current heating, the frequency and the imaginary component of the complex dielectric constant should be high. Dry coal, for example, has a complex dielectric constant of about 5, and high resistivity, 10^{10} ohm cm or more.

Water has a complex dielectric constant of 81; if water is injected into a solid-fuel deposit, and is saturated with metallic salts, the conductivity of the deposit becomes very high. If the water is also loaded with a suspension of particles of one of the ferrites (whose dielectric constants are as high as 10^5 or more), then the imaginary component of the complex dielectric constant also becomes high. With cold fuel and at a high frequency, the rate of displacement-current heating is therefore high. With heated fuel and at a low frequency, the rate of eddy-current heating is also high, with considerable protection in the latter case against the formation of a conducting coke shell (since the coals and other fuels we are concerned with form nonagglomerating coke), and with very wide spacing of the coil conductors possible.

The heating of oil sand only to the temperature at which the bitumen becomes mobile, and its explusion in the liquid state, is one of the possible options. It is probably not the preferred one, because of the lower yield of in situ processes of large dimensions, where the products are liquid and not gaseous. The world cannot afford to skim the cream off the pail and throw away the milk. Many deposits occur in areas where water is scarce, and bitumen drive by a surfactant aqueous solution not only requires a large amount of water, but the water used becomes heavily contaminated with surfactants, silt, and heat, and must be returned to the environment. The amount of water required to modify the electrical properties of the fuel in the induction heating process is much less and is returned to the surface in a pure state.

INJECTION OF FLUID

The world's first problem, however, before oil sand, oil shale, and heavy oil are attacked, is to deal with coal. A review of the potential of the injection of a high-conductance solution into coal deposits, such as a saturated solution of metallic salts in water, indicates that an induction-heating operation in coal would depend entirely on eddy-current heating operation. The solubility of some common salts, in grams of salt per gram of water, at 100°C, is high:

Calcium chloride	1,576
Magnesium chloride	723
Potassium chloride	566
Sodium chloride	392

The conductance of a 10 wt percent of sodium chloride is 6×10^4 times greater than that of pure water at 20°C. Since deposits of sub-bituminous and lignite coal are ordinarily saturated with water, with a wt percent content of 20 or more, and the coal is a network of interlinking fissures and passages containing water, a large amount of a high-concentration sodium-chloride aqueous solution introduced at a few points in the deposit at high pressure will quickly result in lower-concentration solution throughout the coal mass, raising its conductivity by several orders of magnitude. The solubility of sodium chloride, for example, is 36.6 percent at 50°C and 35.7 percent at 0°C, so that temperature is relatively unimportant.

The conducting solution is injected into the fuel layer, not into the overburden or the impermeable basement rock, so that the increase in conductivity takes place only in the fuel layer, and the heating effect is concentrated in the desired area.

The probable result of the injection of a conducting solution into a coal layer is that displacement-current heating can be dispensed with, even at low temperatures, and that low-frequency eddy currents become the main heating mechanism. The low conductivity may even permit the use of 50 or 60 Hz current. When all the water and other volatiles are expelled, leaving only particulate coke, mixed with inorganic material, the resistivity will be so high that the induction-heating process will stop.

In the cases of oil sand, heavy oil, and oil shale it may be necessary to initiate the operation with displacement-current heating, in order to provide a permeable region around the conductors into which the saline solution can be injected. Following this, the conductivity of the whole mass can be expected to rise quickly to a value at which eddy-current heating at a low frequency becomes feasible. One result of the injection of a high-conductance solution into the fuel mass is that the coil efficiency increases, since the copper losses assume less importance.

The critical temperature of water is 374°C, a safe margin above the point at which pyrolysis of the solid fossil fuels begins.[2] This corresponds to a vapor pressure of 218 bars, the formation pressure encountered at depths of

---

[2]The following physical data have been used in computations: heat of vaporization of water 2242 J/g; thermal conversion factor 3.6 kJ/kWh; assumed formation temperature 20°C; and heat of vaporization of hydrocarbon compounds 837 J/G.

900 to 1,000 m. Vapor pressure of 100 bars occurs at 315°C, still above the temperature of the onset of pyrolysis, and this is the formation pressure at about 450 m depth. These values are considerably influenced by the concentration of the saline solution. The electrical conductance of the solution is many times greater at high temperatures than the value at 20°C. These pressures also affect the conductor cooling-system design.

The conductance of a sodium-chloride solution increases rapidly with concentration. Although the solubility of sodium chloride in water increases only from 35.7 percent at 0°C to 39.2 percent at 100°C, this rate of increase nevertheless results in significantly higher values at temperatures above 300°C.

RESULTS

An energy input of 347 kWh is required per metric ton of coal to raise it in temperature from 20°C to 600°C and to vaporize the water and hydrocarbons. To consume coal at the rate of 1,000 t/h, an input power, including losses, of about 400 MW is required.

In oil sand a maximum temperature of about 300°C is required, and the heat input to achieve maximum volatilization is about 123 kWh/t. Oil sand occurs in beds averaging 150 m in thickness, often reaching much greater values, so that the length of tunnel and conductor required for a given power input is low.

In oil shale the maximum temperature required is 550°C, and the heat input to obtain complete volatilization is 153 kWh/t. Oil shale occurs in very thick beds, averaging several hundred metres in thickness.

Heavy oil occurs in formations and has characteristics similar to those of oil sand, with the added advantage that the product is more readily converted to marketable crude than is the bitumen derived from oil sand, or the shale oil obtained from oil shale.

There is a distinguishing characteristic of coal: the total organic content is about 50 wt percent; in the other three formations it is a much lower proportion of the formation weight, averaging 10 or 11 wt percent. The energy content of a given mass of deposit is therefore much less in these three fuels. Coal occurs in deposits of from a few metres to several hundred metres thickness.

LOSSES

Due to the injection of a saline solution into the fuel deposit, the electrical losses from an induction-heating installation are not a controlling factor.

The thermal losses to the enclosing rocks from the fuel seam are low. At the boundary between the fuel and the rock a layer of inorganic ash will likely form, of low thermal conductivity, and the greater part of the thermal gradient will occur across this ash layer. Heat conducted horizontally to the unprocessed fuel is not lost because it is retrieved during later operations. The presence of surface and pore water in the fuel will increase the energy input by the amount necessary to heat and vaporize it. In the overall process it will be beneficial, since it provides a greater distribution network for the saline solution.

Inorganic materials in the deposits absorb energy effectively only so long as they contain water, and have little effect beyond increasing to a minor degree the total amount of electrical input. This fact will permit the working of otherwise useless layered, faulted, and discontinuous fuel deposits.

Underground gasification and surface coking of coal, and underground combustion of oil shale, oil sand, and heavy oil by present methods require high temperatures, since the heated layer is propagated mainly by thermal conduction, which in these materials is low; the hydrocarbon vapors released at the heated face then flow back over the much hotter residue and are seriously degraded. This effect does not occur with induction heating, since heat is generated within the material, and the vapors are at the same temperature as the processed material they flow back over.

Some product loss may occur due to diffusion of the volatiles generated by induction heating and retained under pressure in the formation, through the overburden, the unprocessed fuel, and the basement rocks. Condensation of the tarry volatiles will tend to seal off such escape routes.

ECONOMIC CONCLUSIONS

Substantially complete exploitation is feasible of the solid fossil fuels in situ. The computed results, based on the ideal case of 100 percent utilization and yield, show the following ratios of useful thermal output to electrical input power:

Coal, coke utilized	20.6
Coal, coke not utilized	11.0
Oil sand, coke not utilized	9.3
Oil shale, coke not utilized	6.6
Heavy oil, coke not utilized	9.3

An attempt has been made to gain an idea of the economic results of the technique, based on a coal operation which would consume 1,000 t/h for 40 years, with all capital costs amortized in 40 years, the unamortized costs carrying 10 percent interest. No costs are included for leases or royalties. The results of these computations follow.

The total capital cost for the coal installation in a 50-m seam to produce gas, coal tar, and electricity is about $1,834 million, of which $1,300 million is incurred initially and $534 million is incurred over the 40-year period. This results in a total annual cost of $215 million. In the case of large, thick coal deposits considerably lower costs than those computed for a 50-m thick deposit can be expected.

For the other fuels, the values are:

Cost ($ million)	Oil sand	Oil shale	Heavy oil
Total capital	1,003	996	1,005
Annual operating	160	159	161

A brief study has been made of the revenues that may be possible, using selling prices for the commodities that are current or will be exceeded before long. These revenues retire all costs in 7 or 8 years, and thereafter result in the accumulations of large surpluses.

SUMMARY

The favorable economic estimates, the probable scientific and technological feasibility of the techniques proposed, the large fuel reserves which might be processed to give long-term supplies throughout the world of all the energy commodities, and the likely reductions in human hazards and ecological damage over other sources, argue that these suggested methods now deserve serious consideration, and that the necessary technology should be developed.

The period required for this work is of the order of 10 years. While eventually large development costs will be incurred, the work can be planned in a sequence of steps, each of relatively low cost, leading in the end to full-scale in situ processing of each of the solid fossil fuels.

The benefits of the new techniques over present methods may be large: all the energy commodities--electricity, oil, coke, gas, and chemical feedstocks--will be produced abundantly and at low cost. In addition the technique offers the possibility of utilizing most deposits of coal, oil sand, oil shale, and heavy oil, whether layered, low-grade, diffused, deep, wet, fractured, or otherwise disabled for conventional exploitation. It offers the possibility of reducing surface disturbance, atmospheric pollution, and interference with surface drainage and subsurface aquifers, and may markedly reduce thermal pollution.

REFERENCES

Berkowitz, N., and Speight, J. C., 1975, The oil sands of Alberta: Fuel, v. 54, no. 3, p. 138-149.

Camp, F. W., 1975, The tar sands of Alberta: Denver, Colo., Cameron Engineers, 77 p.

Campbell, J. H., 1976, Pyrolysis of sub-bituminous coal as it relates to in situ gasification; Parts 1 and 2: Lawrence Livermore Laboratory Paper UCRL 52035.

DuBow, J., and others, 1977, Electrical and thermal transport properties of Green River oil shale heated in nitrogen: ASME Symposium on Thermophysical Properties, 7th, New York, 1977, Proceedings, p. 396-403.

DuBow, J., and others, 1978, Thermophysical properties of oil shales: Colorado State University Department of Electrical Engineering, Review Article.

Papic, M. M., and others, 1977, Hat Creek coal utilization: Canadian Mining and Metallurgical Bulletin, v. 70, no. 787, p. 99-105.

SECTION VI. UNCONVENTIONAL NATURAL GAS

CHAPTER 45

UNDERGROUND COAL GASIFICATION: CURRENT STATUS

Edward L. Burwell[1]
Douglas R. Stephens[2]
and
Charles F. Brandenburg[3]

INTRODUCTION

A major objective of the U.S. energy program is the development of environmentally acceptable ways to use coal. Thus, development of processes to produce clean fuels from coal is a high priority task. Underground coal gasification (UCG) is one of the most promising of these processes. It offers four major potential advantages as a source of synthetic fuel: (1) pipeline quality gas at costs competitive with or lower than other synfuels; (2) domestically produced transportation fuels at costs competitive in the existing market; (3) use of as much as 1.2 trillion tons of coal (triple the present coal reserve) not economical to strip or deep mine; and (4) possibly benign environmentally. The purpose of this paper is to briefly review the state of the art in UCG, with particular emphasis on the U.S. Department of Energy (DOE) program.

Underground coal gasification is a method whereby coal in place is partially oxidized with air or oxygen and steam to produce a low- or medium-Btu gas. An array of boreholes completed in the coal seam provide for injection of reactants and production of the product gas. The coal-bed permeability must be increased because the natural permeability is too low for gasification. The bed is prepared by reverse combustion, directional drilling, or other techniques to form a linkage path near the bottom of the seam between the array of boreholes. This is particularly important and effective for low-rank coal (subbituminous and lignite), which shrinks and disaggregates upon heating. Thus coal immediately above the gasification zone falls apart and, in effect,

---

[1]Office of Deputy Assistant Secretary for Oil & Gas Technology, U.S. Department of Energy.

[2]Project Director, Lawrence Livermore Laboratory.

[3]Assistant Director, Laramie Energy Technology Center.

creates an underground packed bed of coal. If the link is at the bottom of the coal seam, the gasification proceeds from bottom to top, utilizing all the coal. The existence of a long, hot, linkage channel ensures the product gas is properly reduced and has a high heating value that remains fairly uniform with time.

Underground coal gasification has existed as a concept since 1868 and field tests have been conducted since 1912. The Soviets have executed a field program for over 40 years and have operated semicommercial UCG plants for over 20 years. Figure 45-1 shows a side view of a gasification zone. The linkage path has been formed at the bottom of the coal seam. As gasification proceeds from right to left, coal falls into the gasified cavity and creates a highly reactive rubble zone. As the system expands, the gasification zone gradually encompasses the full thickness of the coal seam and moves in a broad front towards the outlet well.

Figure 45-2 is a plan view of a field development, as applied in the USSR near Moscow, which allows controlled extraction of the entire coal resource. The dotted lines show the location of the underground linkage channels formed in the coal by reverse combustion in preparation for gasification. The production phase of gasification is carried out by forward gasification in the channels. A combination of convective and dispersive effects in a thermally evolving system permits lateral movement of oxygen a considerable distance from the injection wells, and accounts for the experimentally observed wide sweep in low-rank coal. Commercially attractive well spacings are believed to be longer than the 75 ft shown in this figure; spacings of 100-150 ft would be more appropriate in the U.S.

With the sharp increase of the cost of oil and gas in the 1970's, underground coal gasification testing was initiated by the U.S. Atomic Energy Commission (AEC) and U.S. Bureau of Mines (USBM) in 1972, and this grew into the current DOE program. A number of major milestones have been accomplished, including the following (scf - standard cubic feet; SCFD - standard cubic feet per day):

- Air injection producing low-Btu gas (165 Btu/scf) at 8.5 million SCFD. This experiment was conducted by the Laramie Energy Technology Center and was in operation continuously for 55 days.

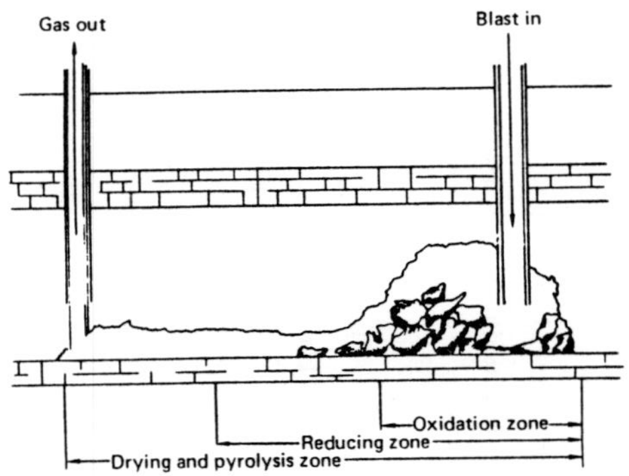

Figure 45-1. Conceptual view of a channel during gasification

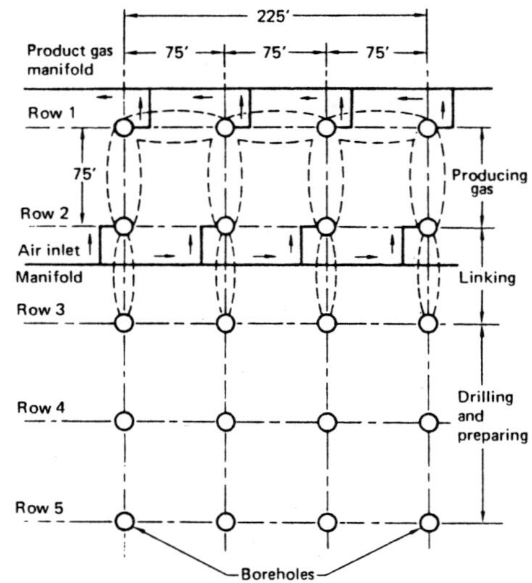

Figure 45-2. Plan of a shaftless generator Moscow region plan

- Oxygen/steam injection producing medium Btu gas (210 Btu/scf) at 4.0 million SCFD. This test was conducted for DOE by the Lawrence Livermore Laboratory and also ran for a period of 55 days, resulting in the gasification of over 4,000 tons of coal.
- Air injection into a 6 ft thick, 900 ft deep, swelling bituminous coal, resulting in production of 120 Btu/scf gas at a rate of 4.0 million SCFD. This test was conducted by the Morgantown Energy Technology Center.
- Initiation of a gasification zone in a steeply dipping coal seam (64°), with production of 180 Btu/scf gas at a rate of over 3.0 million SCFD. This test is being conducted by the Gulf R&D Company as a cost-shared project with the DOE.

In the private sector, Basic Resources, Inc., a subsidiary of Texas Utilities Services, Inc., has licensed the Soviet technology and is developing underground gasification of Texas lignites near Fairfield, Texas. Basic Resources has completed two tests and plans to build two or more 200 megawatts electric (MWe) power plants using UCG in the 1980's. The Atlantic Richfield Company has executed a successful field gasification test in a 100 ft thick subbituminous coal seam near Reno Junction, Wyoming. The Lawrence Livermore Laboratory (LLL) supported this effort with a high-frequency electromagnetic diagnosis of the burn-front propagation. Texas A&M University is carrying out UCG tests in lignite under sponsorship of a consortium of private companies.

International UCG activities include a field test carried out by the Alberta Research Council near Edmonton, Alberta, with further tests planned. The Federal Republic of Germany and Belgium, with support from the Common Market, are initiating field work for underground gasification of deep bituminous coal in Belgium. This test, which will run for about 6 years at an estimated cost of around $35 million, is being designed to determine whether UCG can technically and economically be carried out in thin, bituminous coal seams at depths of 800 to 1000 m. Variables to be studied include: reverse combustion linking, air/steam injection, effects of pressure on product quality, steam/oxygen injection, multilayer gasification, and socioeconomic, economic, and environmentally related issues. A multiwell pattern is planned for the test (Ledent, 1979).

RESOURCES

The end uses of UCG depend upon resources, markets, and cost. It is estimated that 1.2 trillion tons of coal are available for commercial UCG processes in the United States. This huge resource would more than triple U.S. proven reserves and is equivalent to more than 3,000 quads (10^{15} Btu) after process efficiencies and inaccessibility of some of the coal is assumed (Weiber, 1978).

Coal seams suitable for UCG are found at depths of 300 to 2,000 ft. Most of the new coal mines utilize strip-mineable coal at depths of less than 200 ft. Therefore, UCG is, in general, not competing with mining but is an alternate use of the resource.

Western coal is predominantly deep, thick, subbituminous, or lignitic, and thus is very desirable for UCG. These coals are highly reactive and tend to shrink and fall apart when heated. The midwestern and eastern coals tend to be thin, bituminous, and swell upon heating, which decreases the permeability and thus presents a greater challenge for UCG.

ENVIRONMENTAL ISSUES AND CONTROL

Environmental issues with UCG include air quality, subsidence, and water quality. The air quality issue is similar to aboveground coal processing with respect to fugitive emissions from the plant and can be minimized by good engineering practice. The underground process itself should have no significant impact upon air quality.

Efficient coal extraction will cause surface subsidence. Regions of surface cultural activity, such as towns, roads, and pipelines, should be avoided. This is a problem with any type of efficient coal extraction, including strip and underground mining.

Surface subsidence can be minimized by wide spacing of the rows of process wells. In this way pillars of unburned coal will support the overburden above the gasified zones, and surface subsidence should be small. This technique limits resource recovery but has little effect upon process economics. Eventually a decision will be necessary on this issue: whether subsidence is to be minimized or resource extraction is to be maximized.

In general, all coal beds are aquifers. Dangerous coal-derived contaminants, such as phenols, are introduced into the coal aquifer during and following the coal gasification process. Fortunately, the coal itself absorbs the contaminants and confines them near the burn zone. Data from the LLL Hoe Creek No. 1 experiment, which has been monitored for over a year, show that the contaminants produced from this test are immobilized and have decreased in concentration by a factor of 100 (Fig. 45-3). Thus, predictions made in the laboratory experimentally and by calculations have been verified in the field thus far, although more work remains to be done.

COST COMPARISONS

Independent cost estimates not funded by DOE have been made for underground coal gasification by Bechtel, Gulf, SRI International, the Resource Sciences Corporation, Amoco, Pacific Gas & Electric Company, and others which show that product gas from underground coal gasification is projected to cost some 65 percent to 75 percent as much as that produced by strip mining and aboveground coal gasification. These estimates have been confirmed by DOE-sponsored cost estimates.

Tables 45-1 and 45-2 present projected cost data of SNG and electricity from underground coal gasification compared with other methods. Table 45-1 shows that synthetic natural gas produced by underground coal gasification, delivered to the consumer, cannot compete economically with conventionally produced natural gas. However, underground coal gasification is competitive with LNG, the Northwest Alaskan pipeline, surface coal gasification, and with many of the unconventional sources of gas.

Table 45-2 shows that electric power produced by underground coal gasification is competitive with other sources. It is also attractive with respect to capital investment in comparison with nuclear- and coal-fired systems.

Finally, Table 45-3 presents estimated product costs of 24-31¢/gallon for methanol and 66-83¢/gallon for premium unleaded gasoline (via the Mobil process) produced from UCG-derived synthesis gas (Stephens, 1979). On a $/Btu basis, these costs are high, but, as is the case with synthetic natural gas and electricity, are significantly lower than those for aboveground coal conversion.

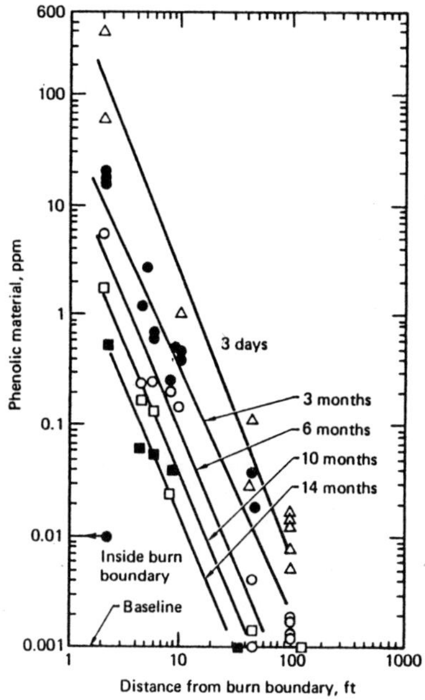

Figure 45-3. Concentrations of phenolic materials as a function of distance from the nearest burn boundary of the Hoe Creek 1 in situ coal gasification experiment. Times are measured from the end of gasification. Sampling wells are completed in the gasified coal seam and located in various directions from the gasification zone.

TABLE 45-1. Cost of pipeline gas from alternate domestic sources

Source	1978 Cost/10^6 Btu
Alaskan gas	$4.20 - 4.50
LNG	3.20 - 5.00
Devonian shale and tight sands	2.00 - 4.00
SNG from naptha	5.00 - 7.40
Geopressured methane	4.50
Surface gasification	4.00 - 4.80
UCG (MOPPS Reston Workshop)	2.70 - 3.40
UCG (MOPPS Intermediate Supply Gp.)	1.75
UCG (Gulf R&D Co.)	2.10 - 3.10

TABLE 45-2. Comparison of electricity costs from several technologies

Technology	Electricity cost cents/KWH
Direct coal fired/flue gas clean up	3.9
Direct coal fired/low-sulfur coal	3.8
Light water nuclear	3.2
Geothermal	2.5
Fluidized coal bed	
- Atmospheric	3.5
- Pressurized	3.6
Fuel cell - molten carbonate	3.5
Surface gasification	
- Commercial	3.9
- Advanced/combined	3.4
- Cycle	
Ocean thermal	4.3
MHD	3.0
UCG	2.5-3.5

TABLE 45-3. Estimated products and costs from medium-Btu synthesis gas from UCG (Stephens 1979)

Product	Cost ($1978)
Medium-Btu Gas	$1.35-2.15/10^6 Btu
Methanol	24-31¢/gal or $3.80-4.90/10^6 Btu
Gasoline	66-83¢/gal or $4.80-6.50/10^6 Btu

As a result of the estimated favorable economics, large markets, and enormous resource, the projected energy production from UCG is very significant. The first commercial plants are projected to be in place by 1987, and by the year 2000 an estimated capacity of 2 TCF of SNG, and 20 gigawatts of electric power would be achieved. A significant but unknown amount of gasoline or methanol production could also be realized by then.

RECENT UNDERGROUND COAL GASIFICATION RESULTS IN THE DOE PROGRAM

There are four field projects in the DOE UCG program: (1) Western Low-Btu Gas, directed by Laramie Energy Technology Center (LETC); (2) Western Medium-Btu Gas, directed by Lawrence Livermore Laboratory (LLL); (3) Eastern Coal Technology, directed by Morgantown Energy Technology Center (METC); and (4) Steeply Dipping Beds, executed by Gulf R&D Company. Sandia Laboratories is providing diagnostic instrumentation to LETC and LLL. There also is a supporting laboratory program, including participation by LLL, LETC, University of Wyoming, METC, University of Washington, West Virginia University, Oak Ridge National Laboratory, Argonne National Laboratory, Los Alamos Scientific Laboratory, University of Texas, and University of Alabama. A detailed review of the DOE Underground Coal Gasification Program is available in Sikri and Burwell (1979). The following sections present a brief review of recent field tests and those planned for the near future.

Low-Btu Gasification of Low-Rank Coal. Since 1972 the Laramie Energy Technology Center has been applying reverse combustion as the linking technique in a series of successful field tests at Hanna, Wyoming. The performance parameters of these tests are summarized in Table 45-4.

TABLE 45-4. Western low-Btu gas project test series

Test	Gas quality (MJ/StdM)	Production (StdM/day)	Duration (days)	Efficiency (percent)	Tonnes of coal gasified
Hanna I	4.96	42,880	100	-	3,830
Hanna II-1	5.98	72,380	38	83	1,140
Hanna II-2	6.89	227,800	25	89	2,290
Hanna II-3	5.43	321,600	38	76	3,810
Hanna III	5.12	288,000	38	76	2,590

In the conduct of these field tests 11 of 12 linkages were completed successfully. Important process performance data from Hanna II Phase 2 are shown in Fig. 45-4; a constant and high heating value of product gas was obtained. This series of tests not only demonstrated process feasibility but delineated the important design and operational factors necessary for successful gas production.

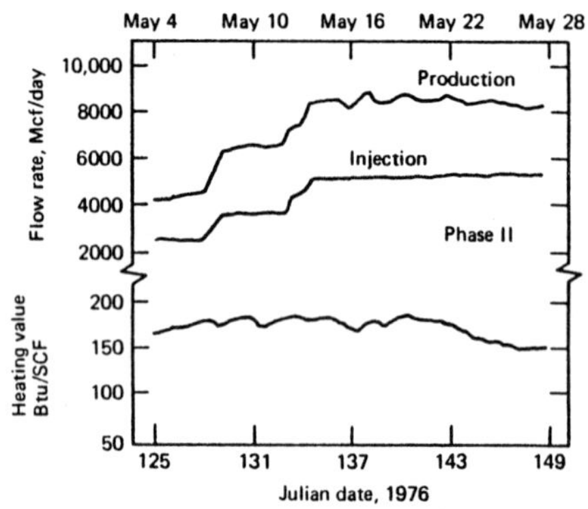

Figure 45-4. Production rates and heating values of gas from LETC experiment Hanna II, Phase 2

The Sandia Laboratories is providing valuable diagnostic information predominantly to LETC and recently to LLL. They have successfully developed both downhole (thermocouples, pressure and gas sampling, subsidence, and acoustic) and remote (electrical) techniques to analyze the propagation of the burn front to aid in the understanding and control of the underground process (Hommert, 1978). It is expected that commercial operators would prefer less costly surface diagnostic techniques to downhole instrumentation. Sandia has also been active in site-characterization for LETC.

Medium-Btu Gasification of Low-Rank Coal. Since 1972 the Lawrence Livermore Laboratory has been studying permeability enhancement, linking techniques, and steam/oxygen gasification. They have completed two underground coal gasification tests at the Hoe Creek site near Gillette, Wyoming (Table 45-5). Explosive fracturing was used to link Hoe Creek No. 1 while reverse combustion was used to link Hoe Creek No. 2.

The first 5 days of forward combustion during Hoe Creek No. 2 produced good quality gas, averaging 125 kJ/mol (140 Btu/scf). However, an override situation developed which caused combustion to occur across the top of the seam with a corresponding sharp decline in the gas quality. Switching injection to a line completed to the bottom of the coal seam again produced good quality gas until the end of the experiment.

During forward gasification 2,090 tonnes (2,300 tons) of coal were consumed producing gas which averaged above 96 kJ/gmol (108 Btu/scf), during air injection. A 2-day steam/oxygen burn produced medium-Btu gas between 220-265 kJ/gmol (250-300 Btu/scf) with no operational or safety problems. Figure 45-5 shows the product gas heating value during the oxygen phase and the results with air immediately prior to and after the oxygen phase.

TABLE 45-5. Western medium-Btu gas project test series

Test	Gas quality (MJ/StdM)	Production rate (StdM/day)	Duration (days)	Thermal efficiency (percent)	Tonnes of coal gasified
Hoe Creek 1	4.33	455,600	11	73	118
Hoe Creek 2	4.17	884,400	58	68	2,090
(Oxygen Test)	10.3		(2)		
Hoe Creek 3	8.3	1,072,000	55	70	4,000

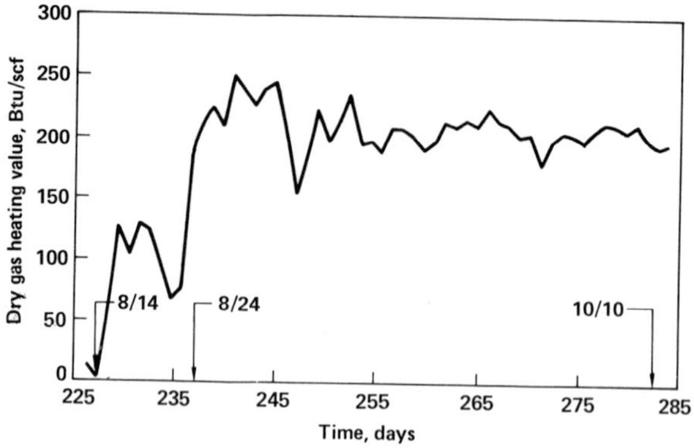

Figure 45-5. Product gas heating value during Hoe Creek No. 3

Preparations were made during 1978-1979 for the Hoe Creek No. 3 test, designed to use a known, reliable channel between the process wells. During July 1978 a directional hole was drilled from the surface, using a special rig and a Dyna-drill, producing a 750 ft long curved hole which was completed in the bottom half of the 180 ft deep coal seam for a total distance of 200 ft. Three process wells, with 100 ft spacing, were completed to the horizontal hole to constitute the gasification channel. This project is cosponsored by the Gas Research Institute. This test was ignited on August 14, 1979, and the directionally drilled hole was successfully enlarged by reverse burning to about 6 inches by August 17. An 8-day air burn from August 17-24 gasified 220 tons of coal and produced gas with a heating value of 110 Btu/scf. From August 24 to October 10 steam and oxygen were injected, resulting in the production of about 4.0 million SCFD of 210 Btu/scf gas. Gas quality was extremely stable (Fig. 45-5) and no operational problems were encountered.

Eastern Coal Technology. The Morgantown Energy Technology Center is conducting a major effort to develop a viable process for the recovery of swelling bituminous coal. Early work involved laboratory simulation studies, modeling support, and a low-priority field effort. Model development support has resulted in models to predict overburden subsidence and subsurface cavity growth from fluid flow and reaction kinetics, and to monitor subsurface cavity growth using high-frequency electromagnetic techniques. One deviated well, which opened a 500 ft long hole through and parallel to a thin coal seam, was completed, thus pioneering the use of directional drilling in UCG.

METC, in August 1979 initiated Pricetown 1, a mini-test which applied reverse combustion linkage experience in low-rank coal to the swelling bituminous coal. The results of the laboratory and modeling efforts were used in the design and fielding of the test to predict air injection flows and pressures, sweep width, temperature gradients, and product gas composition.

The test produced extremely encouraging results in that (1) a link was established, (2) a gasification zone was generated and moved through the coal bed, (3) permeability problems that were encountered appeared manageable with proper operation, and (4) a 120 Btu/scf gas was generated at a rate of about 4.0 million SCFD.

Steeply Dipping Bed (SDB) Project. DOE signed a contract with Gulf Research & Development Company in September 1977, to conduct underground coal gasification in a steeply dipping coal seam over a 5-year period. The project builds on SDB gasification technology that has been successfully tested in the Soviet Union (Kreinen, 1974).

A site was selected at North Knobs, near Rawlins, Wyoming, in a subbituminous coal seam dipping 64° and averaging 20 ft in thickness. The site has been characterized geologically and hydrologically and an environmental analysis prepared. The engineering design has been completed on the test facility for Burn #1, which consists of two process wells linked by slant drilling. This test is currently producing about 3.0 million SCFD of 180 Btu/scf gas with air injection. After about 24 days of air operation it was planned to switch to oxygen/steam injection for a similar period to allow a valid comparison of operation and production for each method.

Specific project objectives are as follows: develop technology for gasification of steeply dipping coal beds, verify the technical feasibility, environmental impacts, and preliminary economic feasibility of the technology, and provide cost estimates and a preliminary design for a pilot-scale SDB project.

ISSUES

Although the results from UCG have been encouraging, a number of key issues must be resolved to move the technology into the commercial sector. The main issues are demonstration of reliable link at bottom of coal seam, control of water influx, minimization of gas losses, subsidence control, process control with good gas quality, environmental control, technology transfer to industry, and conduct of larger multimodule tests to demonstrate control of interacting modules.

In all UCG projects the intent is to perform linear (two-to-three process well) experiments such as Hanna IV, Hoe Creek No. 3, Pricetown I, and SDB #1, in order to demonstrate reliable linking, show that water influx can be minimized while also minimizing gas losses, obtain good gas quality and good process control, which is vital for favorable economics, and demonstrate acceptable environmental impact. With success at this level, it is then necessary to scale up to sweep tests to demonstrate at a significant scale that all the above criteria are met, including acceptable subsidence control.

A typical schedule for the low-rank coal UCG projects is shown in Fig. 45-6. A deep site (coal seam at 500-1,000 ft) of potential commercial interest would be selected and characterized. A three-process well, steam/oxygen test, designated test No. 1 would be designed and tested at 100 tons per day (TPD), 5×10^6 SCFD of medium-Btu gas level. Test No. 2 would be a sweep experiment involving up to three simultaneous burns in a six-to-nine well pattern. The test would be carried out on a sufficiently large scale (300 TPD, 15×10^6 SCFD) to verify technical, environmental, and economic predictions. We anticipate executing these tests as a joint effort of DOE laboratories and industrial partners. Following the successful execution of these tests, the industrial partner would carry out the ensuing pilot-scale test with support, as appropriate, from DOE laboratories.

CONCLUSIONS

1. UCG can recover the energy in unmineable coal seams to ease demand for imported oil and natural gas.
2. The coal reserve for UCG is vast, widely distributed, and could ultimately supply at least 3,000 quads as SNG, electricity, or clean liquids.

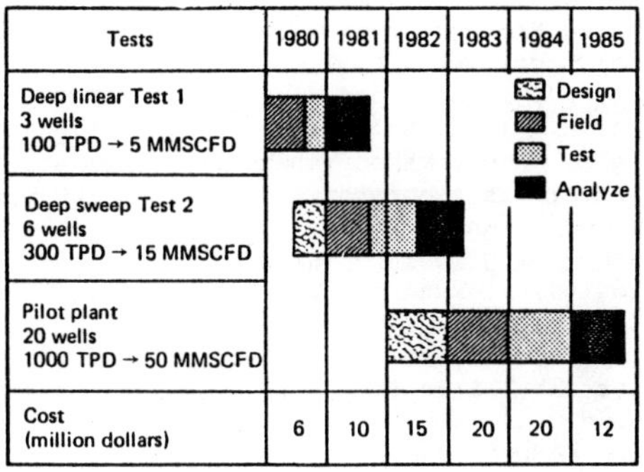

Figure 45-6. Underground coal gasification project schedule

3. The priority targets for UCG are conversion of medium-Btu gas to transportation fuels, and SNG and low-Btu gas to electrical power.
4. Tests to date confirm economic and environmental advantages.
5. Critical technical and environmental issues are being resolved.
6. The UCG processes are undergoing scale-up.
7. High industry interest and involvement in the DOE program assures rapid technology transfer and could lead to commercial UCG process by late 1980's.

REFERENCES

Hommert, P. J. and Northrop, D. A., 1978, Sandia Laboratories Project Review--Instrumentation and Process Control Development for In Situ Coal Gasification: 4th Annual Underground Coal Conversion Symposium, Steamboat Springs, Colorado, July 17-20, 1978, Proceedings, Rept. SAND78-0941.

Kreinin, E. and Revva, M., 1966, Underground Gasification of Coal: Podzemnaya Gazifikatsiya Uglei, Kemerovskoe knizhnoe izdatel'stvo; English translation, Lawrence Livermore Laboratory, UCRL-Trans-10810, December 1974.

Ledent, P. and Beckervordersandforth, P., 1979, Joint Belgian-German U.C.G. - Field Test in Deep-Lying Coal Deposits: Fifth Annual Underground Coal Conversion Symposium, Alexandria, VA, June 18-21, 1979, Proceedings, p. 119-126.

Mead, S. W., Ganow, H. C., and Wang, F. T., 1978, Groundwater and Subsidence Investigations of the LLL In Situ Coal Gasification Experiments: 4th Annual Underground Coal Conversion Symposium, D. A. Northrop and W. L. Noll, Eds., Rept. SAND78-0941.

Sikri, A. P. and Burwell, E. L., 1979, The Department of Energy Underground Coal Gasification Program: AAAS Meeting, Houston, Texas, January 3-8, 1979.

Stephens, D. R., 1979, LLL In Situ Coal Gasification Project, Quarterly Progress Report, January-March 1979, UCRL-50026-79-1.

Wieber, P. R. and Sikri, A. P., 1978, The Development of In Situ Processes for Energy and Fuels from Coal: Mining Engineering, v. 30, p. 557-565.

CHAPTER 46

NATURAL GAS FROM COALBEDS

Troyt B. York

INTRODUCTION AND BACKGROUND

This paper argues in favor of the Rural Gas Initiative, a vast expansion of the use of domestic natural gas for purposes of small business, community, and rural development. Further, it takes the position that natural gas associated with coalbeds should be specifically developed for these purposes.

In the United States, despite the waste of vast quantities of natural gas from underground coal mines, no serious effort was made to capture and use the resource until the U.S. Bureau of Mines demonstrated the feasibility of producing pipeline quality gas in commercial quantities in 1972.[1] More recently, the U.S. Department of Energy has taken an active role in accelerating the commercial development of natural gas from coalbeds. As a result, an organized and intensified industrial involvement is bringing large-scale natural gas extraction closer to reality.

THE TECHNOLOGY

The gases (Table 46-1) contained in coalbeds are predominantly methane with lesser amounts of ethane, propane, butane, and pentane, any of which may be present. Generally, the gases are accompanied by small amounts of carbon dioxide, oxygen, nitrogen, hydrogen, and helium. The gases are believed to be by-products of the coalification process but some could occur by migration from other formations.

THE COAL RESOURCE BASE

The United States is endowed with vast reserves of coal (Fig. 46-1; Table 46-2). Methane gas is always present in amounts varying from a trace to

---

[1] Fields, H. H. et al., Degasification of Virgin Coalbed Through a Large Bore Hole, U.S. Bureau of Mines (1973).

TABLE 46-1. Composition and heating value of coalbed gas and natural gas
(percent)

Source	CH_4	CH*	H_2	Inerts**	O_2	Btu/scf
Pocahontas No. 3	96.87	1.40	0.01	2.09	0.17	1,059
Pittsburgh	90.75	0.29	--	8.84	.20	973
Kittanning	97.32	.01	--	2.44	.24	1,039
Lower Hartshorne	99.22	.01	--	0.66	.10	1,058
Mary Lee	96.05	.01	--	3.45	.15	1,024
Natural Gas***	94.40	4.90	--	.40	--	1,068

* Other hydrocarbons

** N_2. CO_2 and He

*** Moore, B. J., R. D. Miller, and R. D. Shrewbury. "Analysis of Natural Gases of the United States." USBM IC 8302, 1966.

750 ft^3/ton, depending primarily upon rank, depth, pressure, temperature, absorptive capacity, and the moisture content of the coal.[2]

Only a small fraction of the natural gas resources in coalbeds has been estimated. A prior 300 TCF projection by the U.S. Bureau of Mines has been increased recently by the U.S.G.S. to 766 TCF.

The gas reserves can be calculated for all the subsurface coalbeds of the United States by determining the thickness and areal extent of the beds, and the gas content of the cored coal. If gas content cannot be determined directly, then the indirect method developed by the U.S. Bureau of Mines can be used by estimating rank and reservoir pressure (Fig. 46-2). Table 46-3 portrays the total volume of natural gas in coalbeds, assuming various gas concentrations (in ft^3 per ton) and thickness of coal (in ft per acre and ft per mi^2). If the coal is permeable or can be made permeable, 90 percent recovery is a reasonable goal if sufficient time for drainage is available.[3]

---

[2]Kim, A. G.; Estimating Methane Content in Bituminous Coal Beds from Absorption Data, U.S. Bureau of Mines (1977) p. 2-12.

[3]Duel, M.; Natural Gas From Coalbeds, National Academy of Science (1976) p. 196.

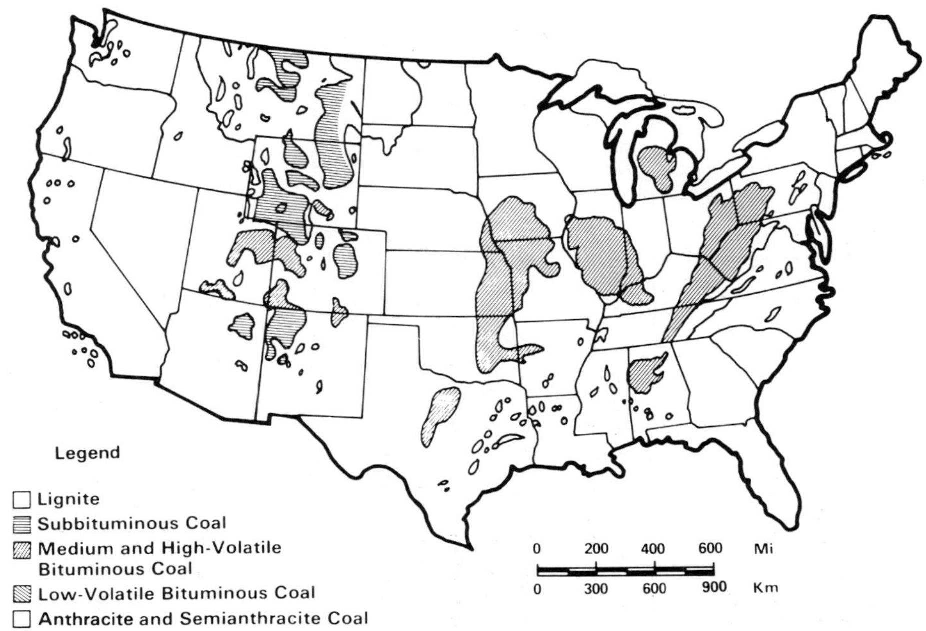

Figure 46-1. Where U.S. coal deposits are

TABLE 46-2. U.S. coal reserves and resources

	Coal billion tons	Coalbed gas trillion ft^3
Coal reserves		
Surface mining	137	
Underground mining	297	59
Identified coal resources		
0-2,000 ft	1,297	259
Hypothetical coal resources		
0-3,000 ft	1,849	370
3,000-6,000 ft	388	78
Total	3,968	766

Source: Averitt, P. "Coal Resources of the United States, January 1, 1974," U.S. Geol. Survey Bull. 1412, 1975.

Forty percent of the bituminous coal in the lower-48 States is considered unmineable by current methods. To get the most of those limited energy resources, "uneconomical" coals could be considered as reservoirs of high-Btu gas recoverable with conventional drilling and well stimulation technology. The supply picture will improve substantially as additional information becomes available on the thickness and continuity of deeper formations and the thick sequences of alternating marine and nonmarine sediments adjacent to the coal basins.[4]

PRODUCTION OF NATURAL GAS FROM COALBEDS

Demonstration projects conducted by the U.S. Bureau of Mines and Department of Energy can be considered evidence of the technical and economic feasibility of producing gas from coalbeds. Proven technology is at hand to produce gas from permeable coalbeds by any of several proved or prospective techniques.

[4] Kim; op. cit., p. 11. And Averitt, P., Coal Resources of the United States, U.S. Geological Survey (1974) p. 131.

Ann G. Kim[1]

The gas-adsorption capacity of coal depends upon pressure, temperature, and rank. Since pressure and temperature are functions of depth, the gas content of most coals can be estimated from rank and depth or calculated from the general adsorption equation. In some cases, factors such as high moisture content, low pressure gradient, and anomalous temperatures should be considered in evaluating the accuracy of the gas-content estimate. The estimates of coalbed gas content derived from adsorption data provide reasonable preliminary figures and can be developed using readily available data.

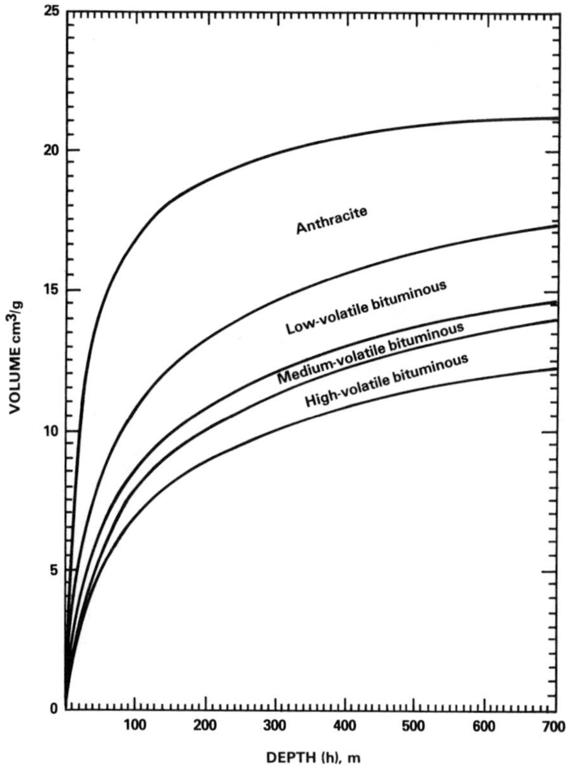

Estimated methane content with depth and rank.

[1]Chemist, Pittsburgh Mining and Safety Research Center, Bureau of Mines, Pittsburgh, Pa.

Figure 46-2. Estimating methane content in coals

TABLE 46-3. Gas from coal beds

Total Volume of Gas in Coal Beds, Assuming Various Gas Concentration (in Cubic Feet per Ton) and Thicknesses of Coal (in Feet per Acre and Feet per Square Mile)

Gas Content in Cubic Feet Per Ton of Coal	Volume → Acre-foot	1 ft/sq mi	3 ft/sq mi	6 ft/sq mi	50 ft/sq mi
Coal, tons →	1,800	1.15×10^6	3.45×10^6	6.9×10^6	57.5×10^5
50	90×10^3	58×10^6	173×10^6	345×10^6	2.9×10^9
75	135×10^3	87×10^6	255×10^6	515×10^6	4.3×10^9
100	180×10^3	115×10^6	345×10^6	690×10^6	5.8×10^9
200	360×10^3	230×10^6	690×10^6	1.4×10^9	11.6×10^9
300	540×10^3	345×10^6	1×10^9	2×10^9	—
400	720×10^3	460×10^6	1.4×10^9	2.7×10^9	—
500	900×10^3	575×10^6	1.7×10^9	3.4×10^9	—
600	1.1×10^6	690×10^6	2×10^9	4.1×10^9	—
650	1.1×10^6	748×10^6	2.2×10^9	4.5×10^9	—

Source: Devl, U.S. Bureau of Mines, 1974

Vertical Boreholes Drilled from the Surface

Methane may be recovered from coal by drilling small diameter vertical boreholes from the surface to the coalbed. Casing is set and cemented in the borehole. Subsequently, fractures in the coalbed are enlarged and extended beyond the borehole by hydraulic pressure and injection of foam or gelled sandbearing water forced through perforations in the well casing. Sand grains in the gelled fluid act as propping agents which prevent the enlarged fractures from sealing when the injection pressure is released (Fig. 46-3).

Small diameter vertical boreholes have drained methane from coalbeds with varying degrees of success. Best results are obtained in coalbeds which are deep, blocky, and moderately gassy. To date, the DOE and the Bureau of Mines have cooperative data on 39 wells in five coalbeds and eight states.[5]

Hydraulic or foam stimulation has been used to increase the gas production from 23 of the 39 wells. Prestimulation gas production rates have ranged from 500 to 10,000 ft^3 per day. After stimulation, production rates of over 100,000 ft^3 per day have been obtained.[6] In boreholes stimulated under Bureau of Mines contracts and subsequently mined through, no damage to roof or floor was observed. In all cases, stimulation served to widen and extend the fractures occurring naturally within the coal, and stimulation fractures were contained largely within the coalbed.[7]

A point should be made that virtually all of the vertical boreholes drilled have been in association with the degasification of active coal mines. Consequently, they reflect gas production from relatively shallow coalbeds. As explained earlier, gas production would be expected to be substantially higher if produced from deeper formations.

Production of natural gas from coalbeds need not be associated with coal mining operations. Conventional natural gas wells drilled into permeable coalbeds have produced commercial quantities of gas for years. Equitable Gas

---

[5] Duel, M. and Kim, A. G., Methane Drainage--An Update, The Mining Congress Journal, vol. 64, number 7, (July 1978) p. 38.

[6] Ibid., p. 39-41.

[7] Sullivan, A. M., Methane Labeled as Gas Resource, Coal Age (April 1979) p. 88.

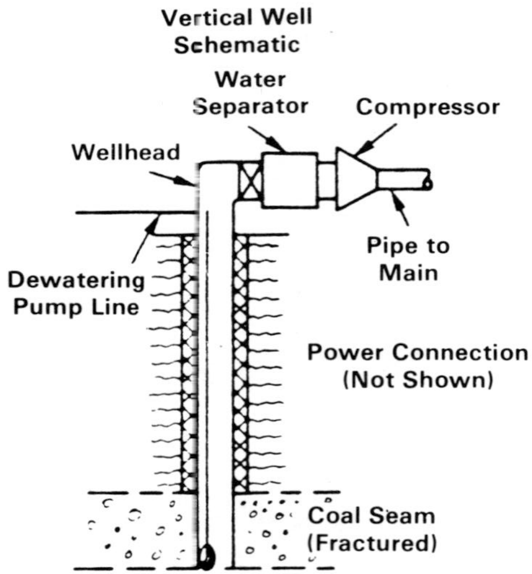

Cost Component	Rate	Cost (800 Foot Seam Depth) $000 1976 Low	High
Wellbore			
Drilling	$10-20/ft	8.0	16.0
Pipe	$4-5/ft	3.2	4.0
Grouting	$5/ft	4.0	4.0
Stimulation	$5,000-10,000	5.0	10.0
Dewatering Pump	$150-250/HP	.3	.5
Land	$5,000/Acre	3.2	3.7
		23.7	38.2

Operation and Maintenance	
1 Man/40 Wells	$500/Year
Power	$250
Total	$750/Year

Figure 46-3. Stimulated vertical well cost projections completed well, gas injected into commercial pipeline

Company, for example, farmed out several tracts in Wetzel County, West Virginia, to an independent producer who completed an initial drilling program in 1949. A total of 23 wells were drilled, the cumulative production through 1974 was 1.8 billion ft^3 of natural gas.[8]

Horizontal Boreholes from Shaft Bottoms

Another method of producing methane from coalbeds is by drilling horizontal holes from the bottom of a shaft sunk to a coalbed (Fig. 46-4). In tests conducted to date, the rate of gas flow per linear foot of 3-inch diameter horizontal hole drilled in a permeable coalbed ranges from 100 to 450 ft^3 per day. The gas produced in this way is collected in a manifold and transported to the surface by a pipeline. The possible rate of production from a shaft is estimated conservatively at 1 million ft^3 of gas per day.

Horizontal wells are being used to drain methane at the multipurpose borehole and the Honey Run shaft of the Federal No. 2 mine of Eastern Associated Coal Corporation in West Virginia. Almost 2 billion ft^3 of gas have been drained from Pittsburgh coalbed at this location. Currently, gas is being drained from these two shafts at the rate of 1.2 million ft^3 per day. About 625,000 ft^3 per day is being sold commercially.[9]

Horizontal Boreholes in Active Mines

Methane can be produced from the active working sections of a developing mine that has not been advanced to the property limits. The process consists of drilling small-diameter horizontal boreholes into the virgin coal as the working face is advanced.

Holes are drilled in outside entries at a slight angle ahead of an active face. The drained gas is conducted to the surface by pipeline. The technique is being used in the Pittsburgh and Sunnyside coalbeds, as indicated below:

---

[8] Tilton, J. G., Gas from Coal Deposits, National Academy of Sciences (1976) p. 214.

[9] Duel and Kim, op. cit., p. 41.

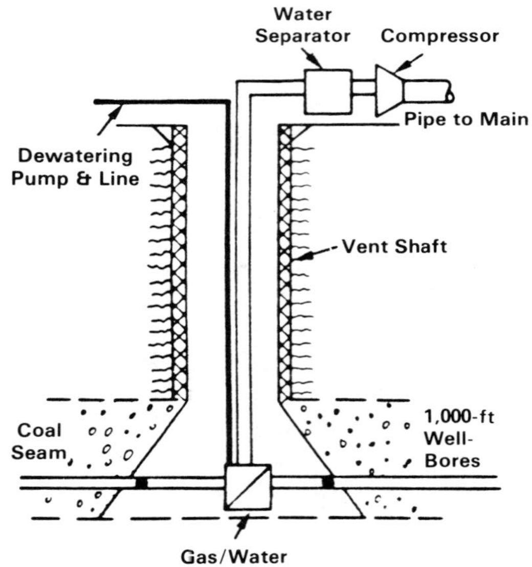

Vent Shaft and
Horizontal Well Schematic

Cost Component	Rate	Cost (800-Foot Seam Depth) $000 1976 Low	High
Vent Shaft	$180-260/Foot	184.0	208.0
Bore Holes			
Drilling & Case	$6/Foot	36.0	36.0
Collection	$2,000	2.0	2.0
Dewatering Pump	$150-250/HP	1.5	1.5
Land	$5,000/Acre	10.0	15.0
Total		193.5	262.5

Note: Cost Based Upon Constructing Vent Shaft 2-3 Years Ahead of Schedule, 5% Inflation, 50% Tax Rate, 10-Year Life for Vent Shaft, and 15% ROI.

Figure 46-4. Vent shaft with horizontal wellbores, completed system cost

Coalbed and location	Hole number	Hole length, ft.	Production rate, Mcfd	Drainage rate cfd/ft.
Pittsburgh	BM-1	1,066	180	169
Washington, PA	BM-2	2,500	330	132
Sunnyside	KS-1	450	100	222
Carbon, Utah	KS-2	450	100	222

In 8 months, 46 million ft^3 of gas has been drained through two relatively short holes in the Sunnyside coalbed. Similar drainage projects using horizontal holes are now planned for coal mines in Alabama, West Virginia, Pennsylvania, and Oklahoma.[10]

SLANT HOLE METHOD UNDER DEVELOPMENT

Another technique to drain methane from coal is currently being developed by the Bureau of Mines. The directional slant hole is drilled from the surface at an angle to intercept the coalbed horizontally (Fig. 46-5). When fully developed, the directional slant hole may be used where conventional horizontal or vertical holes are not feasible and at sites that are unsatisfactory for other types of methane drainage.

Of the methods described above, vertical boreholes, shaft, and horizontal boreholes are currently being utilized. Demonstrations show that all of these techniques are effective ways of recovering commercial quantities of natural gas from coalbeds. The slant hole technique is in the initial stages of technology development and probably will not be in regular use for several years.

DEVELOPMENT CONSTRAINTS

The problems that must be resolved before widespread development of coalbed methane can be accomplished can be described as technical, legal, marketing, and political.

---

[10] Ibid., p. 40.

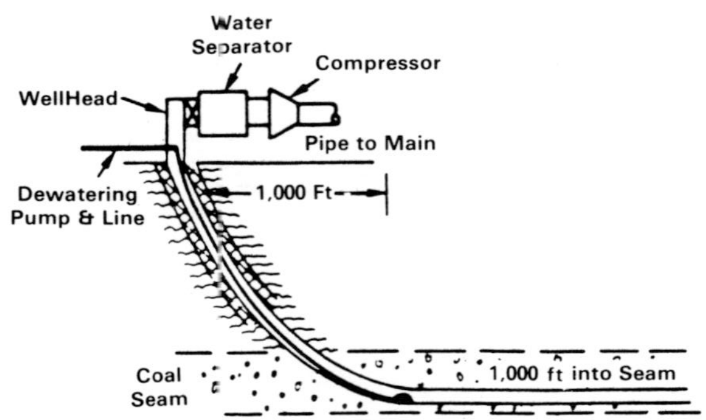

Cost Component	Rate	Cost (800-Foot Seam Depth) $000 — 1976 Dollars Low	High
Wellbore	$50-100/Foot[1]	65.0	130.0
Pipe	$4-5/Foot	2.0	2.5
Grouting	$5/Foot	2.5	2.5
Dewatering Pump	$150-250/ HP	.8	1.2
Land[2]	$5,000/Acre	3.5	5.0
		73.8	141.2

1. Projected Cost for Early Wells. Cost May Be Reduced by a Factor of 2 or More as the Technology is Developed. Longer Inseam Lengths May Also Be Achieved as the Technology Matures.
2. Land Cost Includes Row to Well Site and Well Site Area.
3. Maintenance Cost About $400/Year. Well Cleaning Cost Estimated at $5,000 After 1 Year Operation.

Figure 46-5. Horizontal wellbore-slant hole well, completed well cost; gas injected in commercial pipeline

TECHNICAL PROBLEMS

Production of methane from coal has not advanced as quickly as desirable, in part, because the techniques and the technology demonstrated in research efforts have not been extensively applied to commercial operations. This is especially true for the slanthole drilling techniques.

No satisfactory method for continuous surveying of slantholes during drilling has been developed. There is a need for a system that would provide near-continuous drill orientation while drilling is in progress. This would eliminate the need to interrupt drilling to make measurements for determining borehole azimuth and inclination.

A companion technology to the continuous surveying system ~~would~~ be an improved drill-bit guidance ~~system~~ to operate effectively in the hard sediments associated with coal.

The low permeability of coalbeds is the main impediment to producing methane at higher commercial rates-of-flow. Coalbeds have directional permeability and can be drained most efficiently if horizontal boreholes intersect the face cleats perpendicularly. Blocky coals, generally found at depths of 800 ft to 1,200 ft have prominent cleats. The friable, medium- to low-volatile bituminous coals generally have complex cleat systems, thus directional permeability is difficult to measure, and these coals require unique stimulation methods to increase their permeability.

These are all problems of technology refinement and should not be considered "show stoppers." The Bureau of Mines has already demonstrated that commercial quantities of pipeline-quality gas can be drained from bituminous coalbeds for less than \$0.50 per 1,000 ft^3, using unconventional methods.[11] The technology will be improved as we gain more experience from production operations.

LEGAL ISSUES

Some coal companies have questioned the ownership of gas found in coal horizons and have claimed that, if they own the coal, they own the gas contained in that coal. Such an opinion is entirely at odds with the oil and

[11] Ibid., p. 38.

gas industry's normal position and with several legal opinions which rule that natural gas belongs to the holder of the gas rights.

A court case currently under way in Green County, Pennsylvania, will test the validity of these divergent views of ownership. Kimlock Development, a consortium headed by Mary Cunningham, is suing for the right to extract methane from property on which the U.S. Steel Corporation holds coal rights. Cunningham contends that the methane is no different from any other natural gas and belongs to whoever has leased oil and gas drilling rights. U.S. Steel contends that methane is an essential and integral part of the coal and as such belongs to the owner of the coal. The company also claims that drilling and fracturing of its coal seam, which is necessary to release the gas, is dangerous and constitutes trespass. The case is in the final stages of resolution.[12]

Three possibilities appear available to resolve the ownership issue between private parties: litigation, legislation, and cooperation. Litigation offers the least benefits. Court cases would not necessarily set a precedent because each controversy would be settled on a specific site basis. Although legislation is a possibility, cooperation would probably prove to be more beneficial. Several cooperative projects have been negotiated and production is underway.[13]

There is a fourth major possibility that is only now being explored by the Department of Energy: the public sector as a major owner, producer, and distributor of natural gas from coalbeds. Gas districts as special units of local government could bring the power of eminent domain to bear on the issue in which case the issue could be resolved more directly.

MARKETING ISSUES

The major reasons why the coal and gas industries have not moved more aggressively into the commercial development of natural gas from coalbeds hinges on three points. First, coalbed methane is widely dispersed regionally and is produced at relatively low volume, low pressure, and over a long period of time. Second, the long term required to drain gas from tight coal formations

---

[12] Cunningham, M., Personal Interview, (August 1979).

[13] Sullivan, op. cit., p. 39.

is viewed as incompatible with the coal mining schedule. Finally, methane produced from coal is a superior fuel and is substantially less costly per million Btus than coal. Therefore, the coal market, which is depressed by lack of demand, cannot afford to have large quantities of coalbed methane in the market as a competitor. The coal industry has generally chosen to vent the gas in the name of mine safety.

The specific arguments advanced by industry against the development of coalbed methane for the national market are the most compelling reasons why municipalities, small businesses, and rural cooperatives would find ownership of the same gas resources institutionally attractive for local markets. Namely, their use of natural gas does not require large, high-pressure volumes. Long-term predictable delivery is an asset. In a competition between the use of natural gas and coal in a community setting, environmental considerations will dictate the use of gas.

POLITICAL ISSUES

A principal constraint on the municipal development and distribution of natural gas from coalbeds is embodied in two basic questions representing divergent political points of view:

> Will the sources of energy in the United States be privately owned and controlled; highly centralized; and, developed by means of capital-intensive high technology?
>
> or
>
> Will the sources of energy be treated as a utility vested with the public interest; consumer-owned and decentralized; and, developed by appropriate technologies to utilize local resources to meet local energy needs?

In an effort to solve rural energy problems, it is counterproductive to consider these questions in the either/or context. Rather, it is more appropriate to look upon the questions as a range of options, the choice of which is dictated by the availability of energy resources and local socioeconomic conditions. An even more useful perception is to consider the questions in terms of the opportunities for novel partnership programs between the public and private interests in the community.

A STRATEGY OF IMPLEMENTATION

The Congress and the President have provided the authority and the mandate to achieve the Rural Gas Initiative. Through its recent enactment of the National Gas Policy Act and the Fuels Use Act, the basic pricing and market incentives are provided to accelerate the development of gas from coalbeds.

Speaking in Des Moines, Iowa, in May 1979, President Carter made it clear that he perceives rural energy independence as a national objective.

> Rural areas offer many opportunities for . . . rural energy development. They represent a return to the idea that government exists not to stifle but to facilitate local initiatives.
>
> And these initiatives are not outmoded in modern America, they are leading us into the future. They show us the way to the security and self-sufficiency that we almost lost.[14]

In the same address the President announced a grant from the Department of Energy to the American Public Gas Association to demonstrate the feasibility of utilizing natural gas from coalbeds for purposes of small business, community, and rural development. In addition, he reserved $300 million in loan and loan guarantee resources to accelerate development of natural gas from coalbeds during fiscal years 1979 through 1981.

OBJECTIVES OF THE DEMONSTRATION PROGRAM

There are essentially two reasons why the Department of Energy has entered into an agreement with the APGA to conduct the national demonstration project. The first is to accelerate development of gas from coalbeds by sharing in the nonrecoverable pioneering costs. The second is that the public interest is served by the project by providing a hedge against unexpected events: namely the protection of the U.S. food and fiber system against uncontrollable forces in the international energy market.

MARKET DEVELOPMENT

There are over 6,500 communities of 10,000 population or under in the coal provinces of the United States. Twenty-four million Americans reside and work in these small towns. Assuming an average energy requirement

---

[14] President Carter, Rural Energy Initiatives, President Address in Des Moines, Iowa (May 1979), p. 1.

equivalent to 490 ft^3 per day per person, farms and small communities need about 4.3 trillion ft^3 per year to carry on activities at the present rate of consumption.

The DOE planning objectives are 3 to 5 trillion ft^3 of gas per year from coal and other unconventional sources by 1990 at a marginal cost of under \$4.00/thousand ft^3 (Fig. 46-6).

This production objective is feasible if four conditions are met.

1. Consumers must be willing to substitute natural gas for petroleum-based fuels.
2. The rate of market penetration must be scheduled in a way that does not cause economic disruptions.
3. Drilling capability, production equipment, and manpower must be available for an aggressive development program.
4. The existing transportation and distribution infrastructure must be utilized. Additional capacity can be justified only with the establishment of proven resources of new gas.

Natural gas from coalbeds, produced specifically for small business, community, and rural development may offer the largest supply and cheapest option to reduce oil imports between now and the turn of the century.[15]

DEVELOPMENT ASSISTANCE

Estimates are that a national program to develop 3 to 5 trillion ft^3 of gas from coalbeds would cost \$9.5 to \$16.5 billion over the next 10 years. As explained above, natural gas from coalbeds is not a viable business proposition for those producers seeking large volume wells for short payout in the interstate gas market. Secondly, municipalities have a tradition as gas distributors but not as gas producers. They will proceed with development cautiously and only if there is little risk of losing revenues from local taxes.

These circumstances require the Federal government to provide development assistance in the early stages of the national development program. The basic forms of assistance needed are policy, technical, and financial.

---

[15] Linden, H., "The Gas Option," Testimony Before the House Subcommittee on Energy Aid Power, U.S. House of Representatives (June 1979), p. 22.

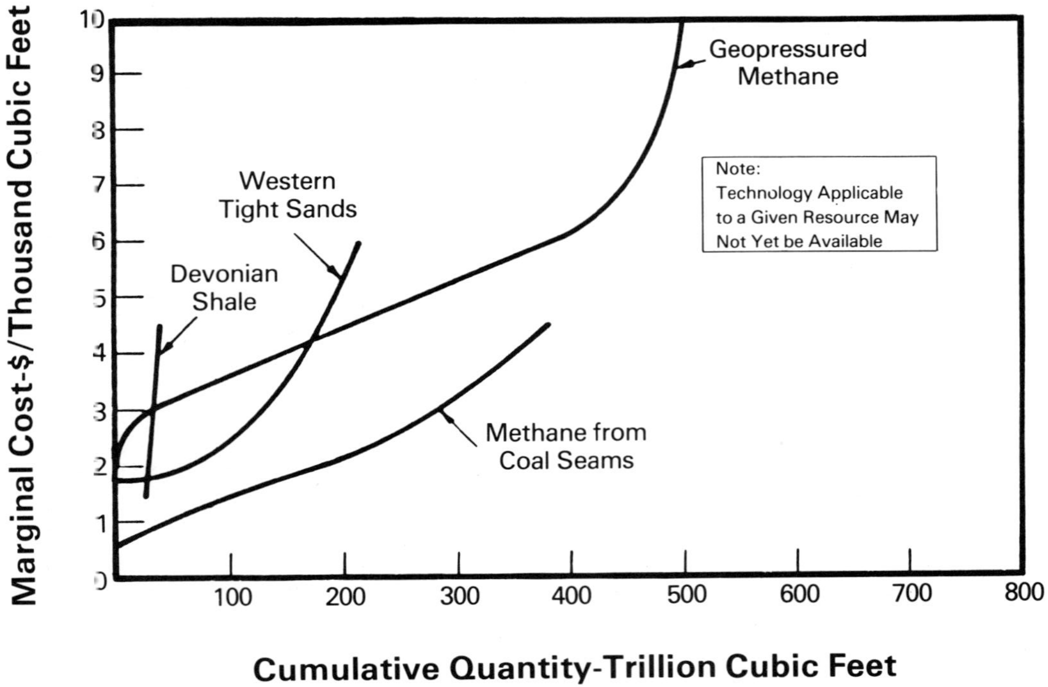

Figure 46-6. Unconventional gas resources

Policy. The national interests will best be served by an energy policy that promotes the interests of rural and small community consumers through the provision of locally available energy supplies at the lowest reasonable cost. The intent of such a policy must be to foster the good health of the nation's farms, small businesses, and municipal and consumer cooperatives involved in the production, transportation, and marketing of food, fiber, and energy. It appears that the only way to achieve such a policy is to use the resources of the Federal government to assure competition among the parties engaged in supplying energy. Only with their own source of fuel supply will the rural and small community interests be in a position to negotiate on energy prices in the marketplace.

Technical Assistance. The Federal system of government provides the opportunity for intergovernmental participation with private industry to carry out resource development programs. The normal arrangement is for the national government to conduct research, development, and demonstration programs and to make the results widely known to information users.

The state governments provide comprehensive planning, program development, and regulatory support. In addition the state governments are in a unique position to enter into cooperative agreements with the private sector for joint ventures.

Municipal governments provide the initiative in formulating project proposals and following through on project development and management. At the policy, program, and project levels a partnership of governments should be forged to provide the technical and managerial support essential to the development of local gas resources. The APGA demonstration project will define these relationships clearly.

Financial Assistance. The rationale for Federal involvement in financing local sources of gas from coalbeds is the development of a domestic natural gas reserve for unexpected contingencies.

Financial assistance should be available in a variety of forms for a variety of eligible applicants. Assistance should be available in the form of grants, contracts, cooperative agreements, and loan guarantees. Help should be available to finance any subsystem in the natural gas fuel cycle from exploration to utilization. Financial assistance should be available to any individual; private profit corporation meeting the SBA criteria of a small business; nonprofit corporation; agency of municipal, county, or regional government; consumer-owned cooperative; or Indian tribe. A qualification for financial assistance should be that the project be physically situated in the area in which the gas is to be produced and used. The DOE/APGA demonstration will assess the financial assistance requirements for municipal gas development and recommend the form and levels of funding that should be authorized to achieve the Rural Gas Initiative.

SUMMARY AND CONCLUSIONS

The Rural Gas Initiative should be developed as an element of national energy policy intended to protect the U.S. food and fiber system from uncontrollable disruptions in the international energy market. The policy should encourage the vast acceleration of local production and utilization of low-pressure sources of natural gas for purposes of small business, community, and rural development. Natural gas from coalbeds is an attractive alternative fuel supply that should be developed and committed for these purposes.

The scientific basis for the recovery of pipeline-quality gas from coalbeds has been demonstrated to be technically, economically, and environmentally feasible. All that now remains to be done is further evaluation of the gas content of the nation's coal resources and a program of accelerated development. Refinements in production technology will be accomplished as development proceeds.

Estimates indicate that enough gas from coalbeds could be produced by the turn of the century to supply the natural gas requirements for the U.S. population living in rural areas and in communities of 10,000 population and less. To achieve production objective of 3-5 trillion ft^3 of gas per year will cost between $9.5 and $16.5 billion and will require the Federal government to participate in a strong program of policy, technical, and financial assistance.

The most compelling reasons for an aggressive program of coalbed methane development focus on factors such as the use of existing technology; a large resource base; high success rate; low cost of production; nationwide distribution of resources and proximity to markets; favorable environmental circumstances; and, immediate availability. In the larger policy sense, such a program may be the least cost alternative for reducing our national dependence on foreign sources of petroleum in the near term.

REFERENCES

Averitt, P., Coal Resources of the United States, U.S. Geological Survey (1974).

Carter, President of U.S., Rural Energy Initiatives, The White House (May 1979).

Cunningham, M., Personal Interview: Kimlock Development, Inc. vs. U.S. Steel Corporation (August 1979).

Duel, M., Natural Gas from Coalbeds, National Academy of Sciences (1976).

Duel, M., and Kim, A. G., Methane Drainage: An Update, The Mining Congress Journal (July 1978).

Fields, H. H. et al., Degasification of Virgin Coalbed Through a Large Borehole, U.S. Bureau of Mines (1973).

Kim, A. G., Estimating Methane Content in Bituminous Coal Beds from Absorption Data, U.S. Bureau of Mines (1977).

Linden, H., The Gas Option, Testimony: U.S. House of Representatives (June 1979).

Sullivan, A. M., Methane Labeled as Gas Resource, Coal Age (April 1979).

Tilton, J. G., Gas from Coal Deposits, National Academy of Sciences (1976).

CHAPTER 47

PEAT GASIFICATION FOR DIESEL GENERATING STATIONS

André Chamberland[1]

AIM OF THE PROJECT

Electricity in Québec's remote areas is produced by diesel generating sets and its cost therefore depends largely on the cost of fuel oil. The aim of this project, undertaken at the request of the Government of Québec's energy department, was to find a means of replacing this fuel with peat, with minimum changes in the present generating method.

The technology of power generation from large peat-fired boilers (40 to 600 MW) has already reached an advanced stage of development in countries such as the USSR, Finland, and Ireland. There is no reason why the same technology could not be applied to smaller plants (of only a few megawatts) as in unconnected networks, but it involves a basic revision of the power generating methods currently employed in these areas. As will be pointed out in the course of this chapter, however, gasified-peat plants do not demand such far-reaching modifications.

Our intention here is to review the main stages of a preliminary study on peat gasification for the production of electricity in remote areas: selection of the site, peat extraction method, and power generation method. A means of assessing the economic aspects of dual-fuel (diesel-gasified peat) power plants is then proposed.

STATE OF THE ART

Gasification of organic matter, particularly wood, gained widespread popularity in Europe during World War II. This technology was abandoned, however, as large quantities of cheap oil became available and found widespread

---

[1]Institut de recherche de l'Hydro-Québec.

[2]Prepared in cooperation with: Poulin, Thériault, Limitée, and Professors E. Chornet and B. Coupal, Chemical Engineering Department, Université de Sherbrooke.

applications. Then, following the oil crisis in 1973, which sent oil costs spiralling, gasification again came to the fore. Many countries have devoted a great effort to research and development on different types of gasifiers, especially those designed to run on wood, lignite, or similar organic matter. However, our studies failed to reveal a system developed specifically for the gasification of peat. Yet in view of the physical and chemical similarities between peat and wood, it seems logical that the same type of gasifier could be used successfully for either of these fuels and we therefore limited our choice of gasifier to the wood and lignite models commercially available or those whose reliability has been demonstrated on a prototype.

The exploitation of peat bogs for fuel production is not new. Ireland, Finland, and the USSR have been taking advantage of such natural resources for over 25 years. The two production methods most widely used in these countries for extracting and cutting the fuel are the sod-peat and milled-peat methods. Other harvesting techniques were included in our analysis but, for the purpose of this study, we concentrated on proven methods.

SELECTION OF AN APPROPRIATE SITE

The site chosen for the pilot project had to fulfill the requirements of our mandate, namely to assure a continuous supply of electricity to one of Hydro-Québec's unconnected networks; thus the project met real needs.

From studies of local maps and an analysis of aerial photographs of possible sites, we were able to eliminate most of the areas envisaged because the quality of the peat was unsuitable or for other technical reasons. Two sites were eventually retained: Port-Menier (Anticosti Island) and Natashquan on the north shore of the St. Lawrence River.

The ground exploration, which covered a 10-km radius around the two towns, confirmed that many peat bogs exist in both areas and that the supplies of peat correspond to an energy potential much higher than local requirements. We made a detailed analysis of a few of these bogs and found that, on the whole, the peat in the Natashquan area is more fibrous, in other words less decomposed, than that at Port-Menier.

The gasification process to be used at the pilot plant calls for peat in the form of briquettes or sods, which means that the peat must be in a relatively advanced state of decomposition. The peat from Port-Menier thus seems more suitable and has therefore been selected as the site for setting up the dual-fuel power station.

PREPARATION OF THE PEAT BOG

The peat bog we selected for the project is dome-shaped with a definite slope towards the periphery providing for natural drainage. Nevertheless, it is heavily waterlogged and needs draining. Main ditches must therefore be dug first, then trees cut, stumps removed, and the surface leveled, making sure that it slopes slightly towards the secondary drainage ditches.

The peat bog covers approximately 180 ha and has a perimeter of 5 km. Based on the principle of one main ditch every 400 m, we calculate that four main ditches of 1.5 km and two of 0.8 km would be required for drainage, forming a total of almost 8 km, whereas the secondary ditches, possibly as many as 50, would total 25 km in length. Because of the natural drainage capacity, the peat extraction could probably start 2 years after the ditches are dug.

PEAT EXTRACTION AND PROCESSING METHOD

As mentioned earlier, for gasification purposes, the peat must first be extracted and cut to a suitable size. A machine already exists which combines these two functions and forms firm, dense sods as it moves over the bog. The sods are about 80 mm in diameter when they are first cut but shrink by 10 to 40 mm once they have dried out on the bog surface; their length can vary. Relatively straightforward peat machines have been designed and developed in Finland.

NATURAL DRYING

Sod-peat harvesting calls for good natural drying conditions. When first extracted, the sods contain about 88-90 percent moisture. Natural drying on the bog surface removes a large percentage of the water and, according to the prevailing weather conditions, the moisture content eventually drops to about 35 to 45 percent over a period of time, depending on a number of factors, such as wind speed, humidity of the air, air temperature, type of peat (degree of humification), extrusion pressure, sod dimensions, and natural drainage capacity of the bog.

The first three factors are closely related to the climatic conditions of the site. The total number of days without rain is significant, although not as much so as the number of consecutive days without precipitation.

Based on climatic data for the area selected for the project and on Finland's experience, such conditions should allow 30 to 40 days' harvest a year. During the remaining months, the peat will be stockpiled at the power station.

POWER STATION EQUIPMENT: DIESEL GENERATING SETS AND GASIFIERS

General Description of the Power Plant

Plans for the power plant call for a dual-fuel engine designed to operate on diesel oil and peat gas, a gasifier, and a scrubber to remove undesired components from the gas before it enters the engine. This equipment is described in more detail below.

Selection of a Suitable Type of Engine

According to the original plans, the engine would operate partly or even fully on the gases produced by a peat gasifier. If operating on lean gas only, a fairly large backup supply of gas would be required to meet output variations and, in particular, to assure protection in the case of gasifier breakdown. In fact, it would take two gasifiers for each generating set in order to guarantee a constant supply.

Lean gas from gasifiers cannot be used to supply a diesel engine because it does not allow compression ignition. Until recently, spark-ignition engines therefore had to be installed, but modern technology now allows lean-gas engines using 10 percent conventional diesel oil and 90 percent lean gas to be employed. In this case, the engine starts on pure diesel, which gives the gasifier time to reach its optimum level. Shortly after start-up, as soon as a suitable quality of gas is obtained, this replaces the diesel fuel. There is obviously a great advantage in the dual-fuel engine being able to develop its full capacity at any time, during firing of the gasifier, for instance, or during breakdown or delays caused by its automatic controls. The dual-fuel engine thus offers the same universality as a pure diesel engine and, if connected to a turbocompressor or supercharger, can achieve an efficiency of around 37 percent. On the average, a dual-fuel engine operating on lean gas uses 10 percent of the gas oil it needs when operating on diesel alone.

For these reasons and to ensure operation of the generating set if the gasifier breaks down, we recommend a supercharged dual-fuel engine operating on diesel and lean gas. This decision should result in considerable savings, since there will be no need to store gas and, instead of two gasifiers per generating set, only one of suitable capacity will be required.

GASIFICATION PROCESS

Types of Gasifiers

Gasification processes involve both pyrolysis and combustion reactions, the latter supplying--in the same reactor--the thermal energy required for the pyrolysis reactions. Pyrolysis is defined as the thermal decomposition of organic matter in an atmosphere with a low oxygen content. The products formed comprise noncondensable gases (at ambient temperature), liquid hydrocarbons (and tars), and carbonized solids. In the case of pure pyrolysis, the heat supplied for the decomposition process is generated by a system outside the reactor and conveyed to the walls of the pyrolysis vessel by conduction or via a coolant. Since gasification processes are usually less complex and easier to operate, we have decided to adopt one of these processes for our peat conversion project.

Gasification processes fall into several categories: fluidized-bed, which involves a relatively sophisticated technology, and fixed-bed, which is easier to use. A fixed-bed process was therefore selected for our purpose.

Fixed-bed Gasifiers

Fixed-bed gasifiers comprise the countercurrent type, in which the gases formed are exhausted at the top of the bed with the solids flowing downwards, and the downdraft type where gases and solids both flow downwards.

The first type, although somewhat simpler in design, produces gas that contains oils and tars which, at the gasifier outlet when the gases are very hot, are entrained in the form of aerosols. It is difficult to completely remove these products from the gas, which means that, if used to supply an engine, frequent cleaning is necessary; this results, of course, in higher maintenance costs.

The more complex downdraft gasifiers have the advantage of producing relatively tar-free gases, although they have a higher char carryover than the countercurrent type. This may be explained by the fact that the gases flow over unburnt ash before leaving the reactor and thus entrain particles and crack the tars. The ash entrained by the gas is easy to filter out and the gas is then suitable for feeding engines such as the one selected for our project.

Based on our study of commercially available fixed-bed and downdraft gasifiers, we recommend the type of gasifier made by Duvant in France, or by the German firm Imbert. Both these companies offer a full range of equipment for converting biomass into electricity.

POWER STATION INSTALLATIONS

Figure 47-1 is a schematic of a power station similar to the pilot project. It comprises a light structure about 10 m high housing the gasifier and scrubber, and a standard building for the generating set and the electrical part of the plant. An "igloo" type building is provided for storing several weeks' supply of peat while the remainder of the sods will be stockpiled under tarpaulins elsewhere on the site.

The construction program for this power station calls for an initial 750 kW module in 1979-1980 followed by others according to needs. The program also covers the purchase of regular diesel generating sets to provide backup and peaking capacity for the power station.

PRELIMINARY ECONOMIC EVALUATION OF POWER GENERATION FROM GASIFIED PEAT

Generation Costs

Different analyses have shown that the price of peat varies with the production volume and year of production. Based on this information and on a study of other economic factors, we were able to establish that for the period 1980-1994 electricity at Port-Menier would cost 220 mills/kWh if produced by standard diesel generating sets compared to 223 mills/kWh if produced by a dual-fuel system. The results of this comparison are presented in Fig. 47-3. It may be seen that at first the dual-fuel option costs are higher than those of the pure diesel option but equal out between 1987 and 1990 and, from then on, are markedly lower.

Fuel costs for the first and second option show opposite tendencies. Oil prices continue to increase with time whereas the price of peat tends to decrease. This may be explained by the larger quantities of peat used as

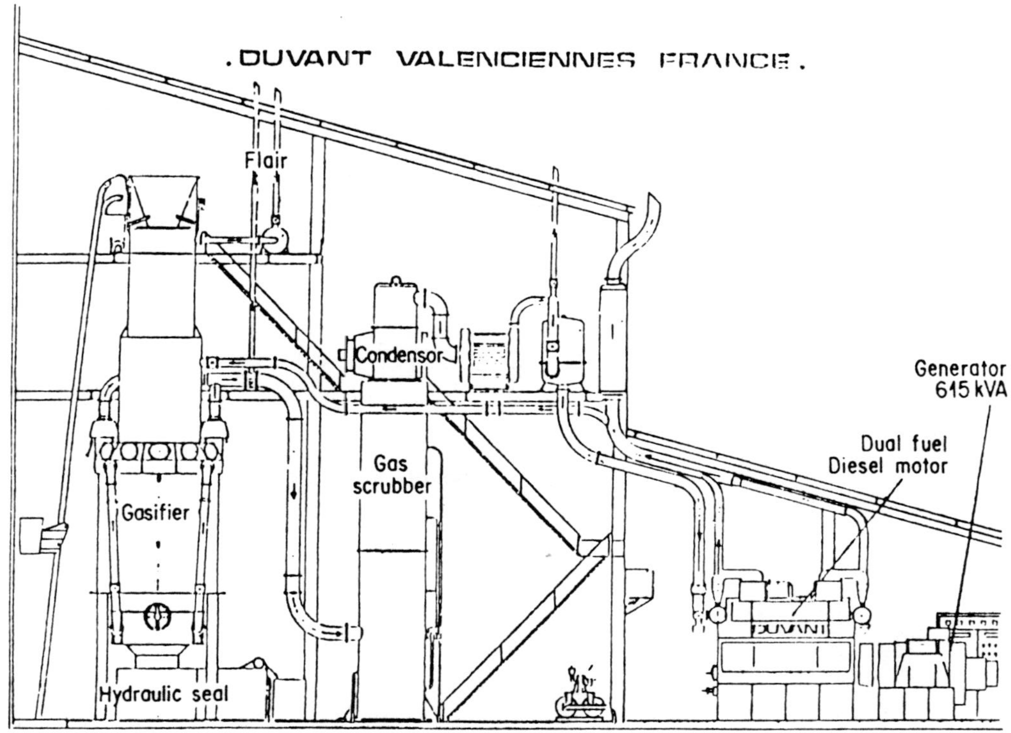

Figure 47-1. Dual-fuel power plant

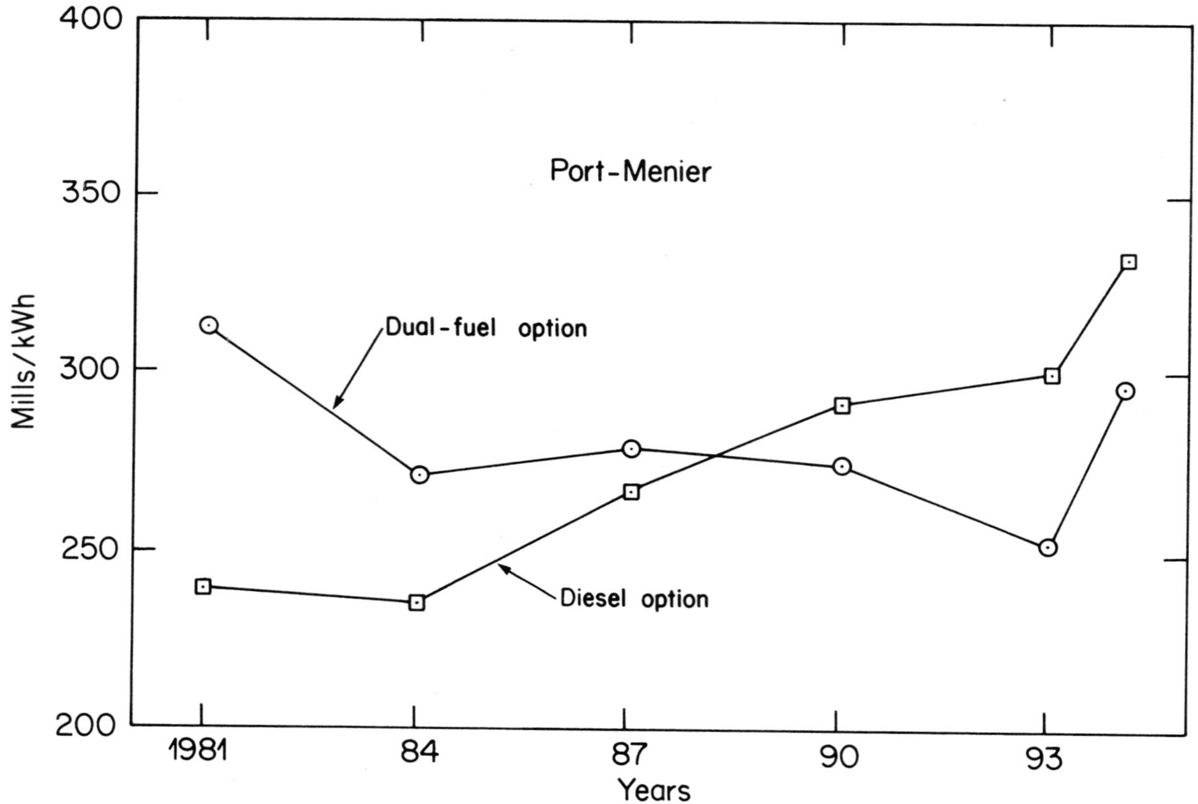

Figure 47-2. Total unit cost (in 1979 dollars)

production expands. In the diesel option, the fuel cost represents 28 percent of the total cost at the beginning of the period studied and rises to 54 percent by the end of the period. In the dual-fuel option, the opposite trend is observed: the fuel cost as a percentage of the total cost drops from 38 percent to 32 percent. On the other hand, operating and maintenance costs are higher with the dual-fuel option, representing 40 percent of the total cost, whereas with the diesel option they do not exceed 30 percent. This is due mainly to the additional manpower required for operating the gasifiers.

DISCUSSION

Our study has shown that the equipment required for small dual-fuel (diesel + gasified organic matter) generating stations already exists or can be adapted for peat. It also revealed that, from the economic point of view, this type of plant seems very promising for use in remote areas. In fact, they will doubtless prove even more profitable than predicted because oil prices are likely to increase at a rate far in excess of the reference figure used in our study. It has also been shown that the greater the peat consumption, the lower the cost of the power generated. The social and economic repercussions of setting up a gasified-peat generating station in remote areas have not yet been analyzed but cannot fail to be positive. A large part of the fuel cost, in fact, will be distributed in the form of remuneration for those working the peatlands near the power station. This type of dual-fuel plant requires additional personnel and will therefore create more jobs than a conventional power station.

The dual-fuel power station described here will enable us to familiarize ourselves with the related technology. This small-scale project offers the possibility of developing the techniques of producing peat for fuel and, also, of estimating the real costs of this technology in the Québec context.

In conclusion, I would like to point out that gasification is the first step in the conversion of biomass to synthetic gas, methanol, and other petrochemicals essential to our present way of life. The gasification of peat is a particularly interesting technology to develop since it offers applications to the gasification of wood and other forms of biomass, of which abundant supplies are found in Canada.

CHAPTER 48

METHANE IN GEOPRESSURED SUBSURFACE WATER

Dr. Claude R. Hocott[1]

It is likely that all of the active sedimentary basins of the world are prone at depth to have geopressured formations. Whether or not natural gas is ubiquitous in these formations is not known, although some evidence of its presence is noted in most marine sedimentary environments. Certainly significant quantities of methane are dissolved in the subsurface waters of those provinces known through oil and gas exploration to be petroliferous. A great many large natural gas (as well as oil) reservoirs have been developed in geopressured zones in various parts of the world, and frequently these fields have strong natural water drive from surrounding geopressured aquifers. Based on the experience with hydropressured formations, it is estimated that waters adjacent to natural gas accumulations are saturated with methane at formation pressure; however, the lateral extent of this condition is not known.

The only area of the world known to be thoroughly tested for natural gas dissolved in subsurface waters is that of the hydropressured formations of the coastal region of the Gulf of Mexico basin. Here, the subsurface waters, by the time they become saline, contain varying concentrations of hydrocarbon gases, chiefly methane, in solution. It is this region that shall serve as a basis for the discussion presented in this chapter.

GEOLOGY

The Cenozoic geopressured-geothermal zones of the Gulf of Mexico sedimentary basin underlie a large portion of the coastline from northern Mexico to the north Florida embayment. They cover a strip which extends landward up to 100 miles and seaward certainly to the continental slope. The sediments have a maximum thickness of about 15,000 m and are composed of normal alternating sand-shale sequences, with the shaliness usually increasing with depth.

---

[1]Gulf Universities Research Consortium.

Geopressured zones have been noted as shallow at 1,000 m, but are not usually encountered above 2,500 m. These formations are nearly always geopressured below 5,000 m. Geopressured systems are generally believed to have developed under rapid compaction as great loads of successive sediments are dumped basinward. Further, the geopressured sands are presumed to have been vertically isolated early by alternating layers of low-permeability shales and laterally by growth faults. The major faults, commonly down-to-the-basin, together with concurrent transverse faults, have resulted in the creation of closed rectangular sedimentary blocks containing connate waters which have not had sufficient time to diffuse to the normal hydrostatic level. Hence, we have an unstable, geodynamic situation which must be considered in any evaluation of resource potential.

Any estimate of the quantity of natural gas in place in these aquifers is subject to a great deal of speculation, and the amount that is potentially recoverable, even more so. Confusion has risen because various reports fail to distinguish these two estimates, leading to false hopes and wishful thinking on the part of many people, some in high places of influence. Nevertheless, there does seem to be a resource potential of sufficient magnitude to justify an extensive evaluation program.

PRESSURE

The subsurface pressure in the upper sediments of the coastal region approximates 0.465 lbs per in^2 per ft of depth. This is generally referred to as the normal hydrostatic pressure based on fluid continuity through a column of the normally saline water. When the pressure encountered exceeds the "normal gradient," this "abnormal pressure" is that now commonly called geopressure, since the fluids contained in these isolated sedimentary zones must bear part of the overburden load of the geologic column. Pressure gradients in the deeper geopressured zones frequently approach lithostatic pressure of 1 lb/in^2/ft of depth.

The onset of geopressure with depth is quite varied. It may come on gradually, or, especially when encountered at deep levels, quite precipitously. It was first noted in oil and gas drilling by the increased density of drilling fluid required to prevent formation fluids from entering the well bore. In the early days, if abnormal high pressures were tapped suddenly, well blowouts were common. Today, techniques for advance indications of approaching geopressure have usually enabled operators to prepare accordingly.

The varied nature of geopressured gradients is illustrated on Fig. 48-1 (1). Case A shows that the gradient increased very slowly for a considerable depth and then abruptly rose to a high level. Case B shows a greater rate of increase but still a gradually increasing effective gradient with depth. In neither of these cases is there hydrostatic equilibrium between the successive permeable zones in the sediments. If opened throughout the section to the same well bore, subsurface flow would occur from deeper to shallower strata. Case C is an example of abrupt onset of geopressure, to a gradient of about 0.8 lbs per in^2 per ft, then increasing at a hydrostatic gradient as depth is increased. In this case fluid continuity is indicated throughout the geopressured depth penetrated.

The nature of the continuity of the geopressured fluids, of course, greatly influences the potential producibility of the water-bearing strata. Cases A and B would have very low possibilities, with Case C deserving greater evaluation for probable deliverability and block size. To date, no data are available on the dimension of the self-contained, continuous fluid-producible blocks. Many geologists believe that they are of relatively small size, thereby accounting for their abnormal pressure. Others feel that they cover broad areas in the form of sheet sands. Limited observations tend to favor the first conclusion.

DISSOLVED GAS

Any consideration of the dissolved gases in the geopressured-geothermal waters of the coastal belt of the Gulf of Mexico should look first at the dissolved gases in the hydropressured regions of the target geological formations. Whereas there have been few samples taken from the geopressured zones, the hydropressured horizons have been examined extensively.

It has been found that at depths below a few thousand ft the water (usually brine) that fills the pore space of subsurface formations in the area of the Gulf of Mexico Basin bordering the coast generally contains appreciable quantities of dissolved hydrocarbon gas plus occasional minor amounts of CO_2. The chief hydrocarbon constituent is usually methane; however, measurable proportions of ethane, propane, and butane are frequently present. A typical gas consists of:

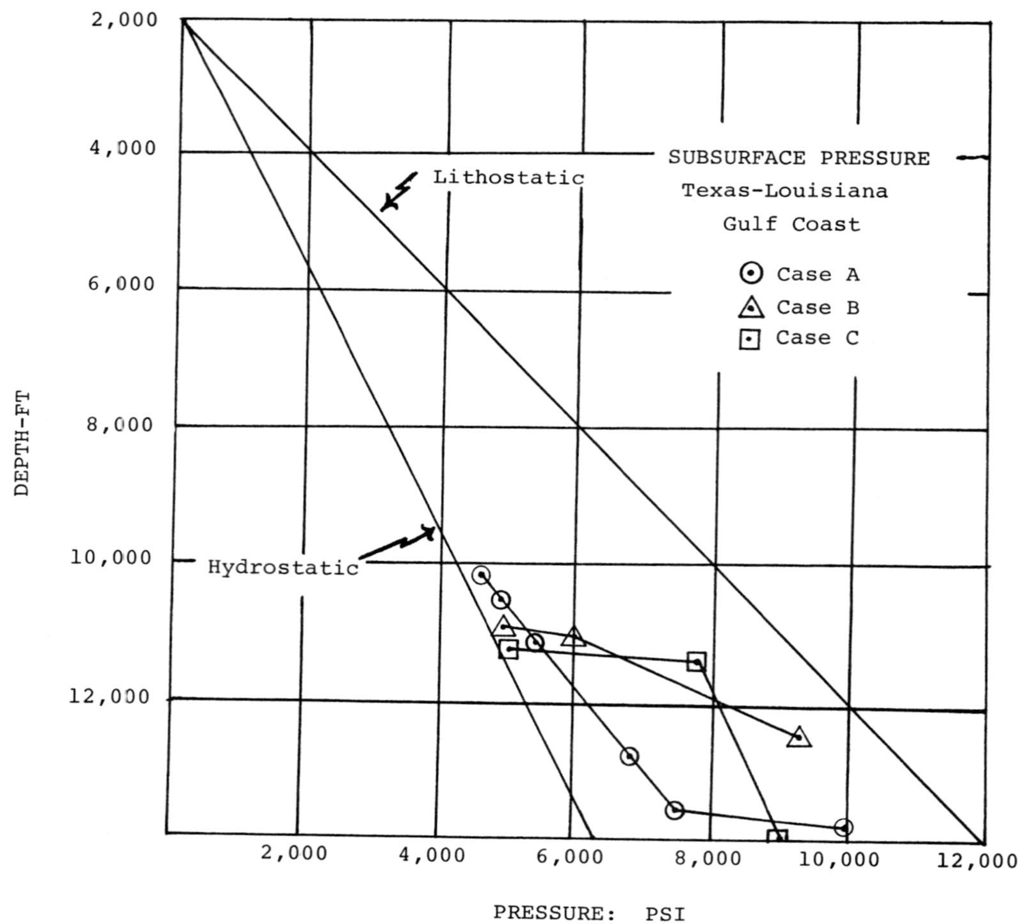

Figure 48-1. Subsurface pressure, Texas-Louisiana Gulf Coast

CH_4	90.1%	
CO_2	7.8%	(CO_2 may vary from 0-10%)
C_2H_6	1.8%	
C_3H_8	0.2%	
$C_4H_{10}+$	0.1%	
	100.0%	

The total quantity of gas dissolved in the water of subsurface formations undoubtedly exceeds the known proved gas accumulations heretofore discovered in commercial deposits.

In order to gain some insight into the nature of this intriguing potential future source of natural gas, a brief look at some representative data from the hydropressured waters referred to earlier is required. Figure 48-2 shows the quantity of gas dissolved in the subsurface waters of the Woodbine sand in East Texas, as a function of depth (2). Although there is considerable scatter, the quantity of dissolved gas generally increases with depth. It should be noted that both local and regional exceptions have been observed. Figure 48-3 presents the escaping pressure (or bubble-point pressure) corresponding to these data. In each case the escaping pressure is below the hydrostatic pressure for the given depth (2). In other words, all water samples were undersaturated with gas at formation conditions of temperature and pressure. This is by far the most frequent situation in over 1,000 subsurface samples taken from over 800 wells from south Florida to the Rio Grande.

Further insight into the distribution of gases in subsurface waters in a single formation is illustrated in Fig. 48-4, which shows the bubble-point pressure for water in a natural gas field in the Lower Tuscaloosa formation of southern Mississippi (2). The shallowest sample was taken below only a 9 ft shale break separating the water from contact with a free gas reservoir; yet, it is undersaturated with gas. As samples were taken at successively lower depths in this sand-shale sequence, the escaping pressure and the dissolved gas content continued to decrease. This is most disconcerting for the potential of deeper geopressured zones.

Two recent tests of geopressure waters from below 15,000 ft had dissolved gas contents of 20 to 24 ft^3 per barrel of water. Although escaping pressures have not yet been measured, the quantities of gas suggest that these waters, too, are undersaturated with methane. In any event these concentrations are probably too low to be considered as current resources for domestic gas supply.

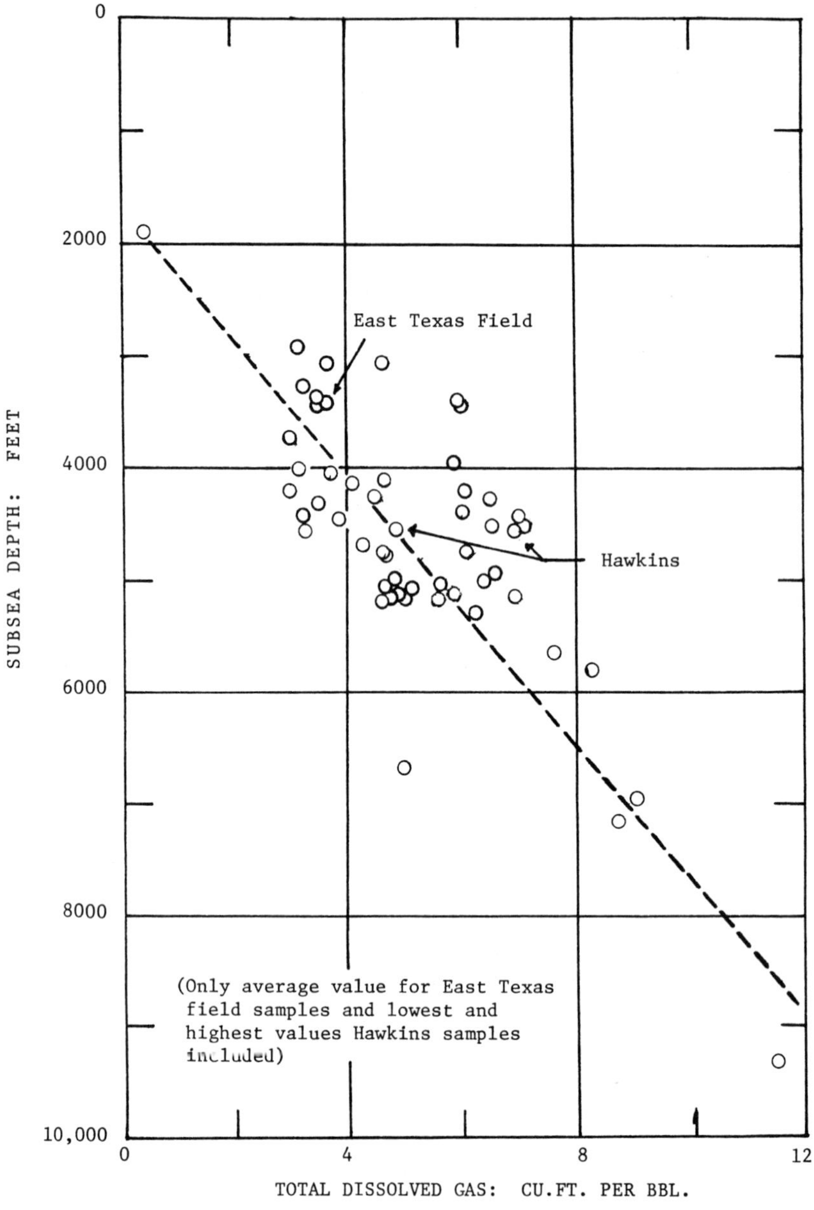

Figure 48-2. Dissolved gas content of Woodbine water, East Texas area

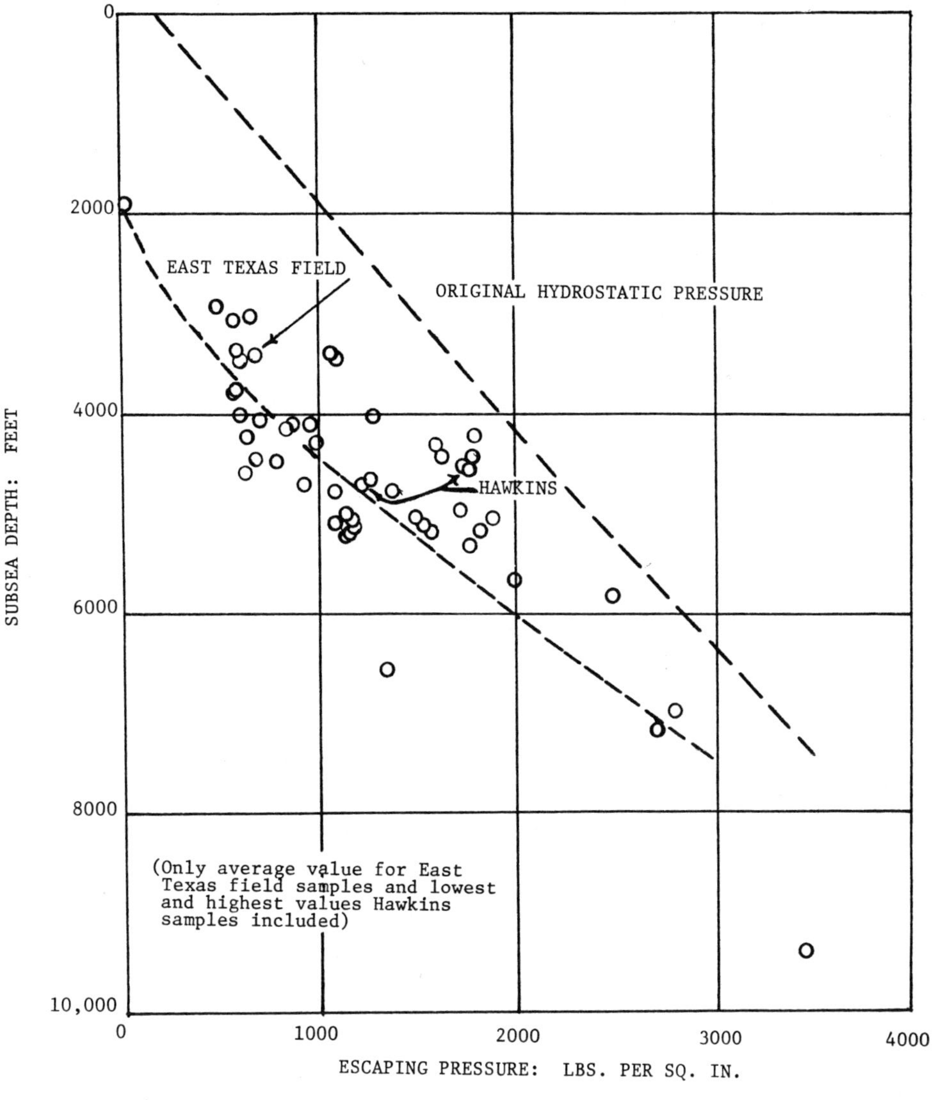

Figure 48-3. Escaping pressure of Woodbine water, East Texas area

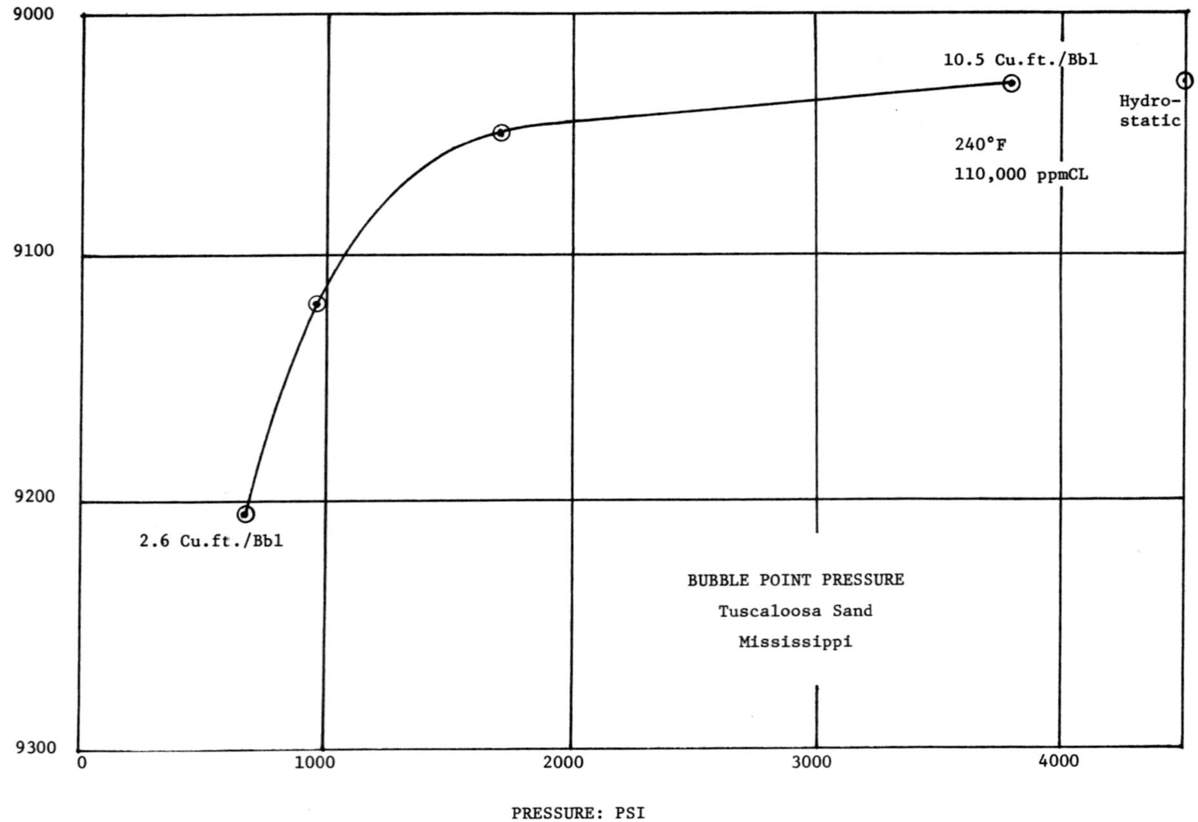

Figure 48-4. Bubble-point pressure, Tuscaloosa Sand, Mississippi

PRODUCTION PERFORMANCE

In attempting to arrive at some criteria for judging potential producibility and total production from geopressured zones, a number of production calculations were made on the University of Texas Reservoir Simulator. Parameters of the formation and reservoir fluids were selected so as to approximate those which might be encountered in the deeper Gulf Coast geopressured zones. Table 48-1 presents the assumed parameters and the results of two typical runs, to give some idea of the range of performance that might be expected (3).

The calculated flowing production lives for the two runs ranged from approximately 5 years for a 125 ft thickness to approximately 23 years for the 500 ft thickness. Recovery efficiency was about 15 percent of the brine originally in place.

One of the critical factors, of course, in a reservoir simulator such as this is the effective composite compressibility of both the reservoir rock and the bulk compaction of the rock mass. The factor used was an empirical value derived from the behavior of a deep aquifer associated with a producing field.

No attempt has been made to estimate economics for such a case due to inflation and the rapid escalation of both capital and operating costs. The above physical performance will provide the basic information for economic calculations.

TABLE 48-1. Simulated geopressured reservoir performance generation enhancement

Area	4 sq. miles
Depth	17,000 feet
Initial reservoir pressure	15,000 PSI
Final reservoir pressure	7,500 PSI
Porosity	20 percent
Permeability	50 millidarcies
Thickness - Run 1	125 feet
- Run 2	500 feet
Bulk volume modulus compressibility	4 microsip
Reservoir temperature	325° F.
Total dissolved solids	20,000 ppm
Dissolved gas content (initial)	50 ft^3/bbl
(final)	29 ft^3/bbl
Production rate	40,000 Bbls/day

REFERENCES

1. Cannon, G. E., and Craze, R. C., 1938, Excessive Pressures and Pressure Variations with Depth of Petroleum Reservoirs in the Gulf Coast Region of Texas and Louisiana: Petroleum Development and Technology, v. 125, p. 31.
2. Buckley, S. E., Hocott, C. R., and Taggart, M. S., Jr., 1958, Distribution of Dissolved Hydrocarbons in Subsurface Waters; <u>in</u> Habitat of Oil: AAPG, p. 850.
3. Malott, J. P., Olsen, N. H., Purser, P. E., and Zimmerman, F. W., 1977, Potential Producibility and Recovery of Natural Gas from Geopressured Aquifers of the Cenozoic Sediments of the Gulf Coast Basin: GURC, Report no. 161.

CHAPTER 49

HYDROPRESSURED NATURAL GAS RESERVOIRS

Professor S. S. Marsden[1]

INTRODUCTION

In our quest for natural gas other than that associated with crude oil or that found as free gas, we are considering many diverse sources. That in coalbeds has been known and exploited in some countries for many years. Potential new sources, such as the Devonian shales of the eastern United States and the geopressured brines of the Gulf Coast area of the United States, are well known and under active investigation.

While teaching in Japan many years ago, I became aware of another natural gas resource which was being produced in significant amounts. This is a natural gas, mostly methane, dissolved in brine. Unlike deep geopressured reservoirs, these are relatively shallow, the gas-containing brine existing under hydrostatic pressure. A couple of decades ago, these reservoirs were the main source of natural gas in Japan; even with present restrictions on production, they are still a major source.

If this kind of natural gas was found only in Japan, there would be little point in discussing it in an international meeting. But there are reports in the literature of its existence and exploitation in Italy (Reeves, 1963), in the USSR, in China (Motojima, 1979; Kuo, 1979), and in the Philippines (Villalva, 1976). There are also reports of its existence without exploitation in the United States (Buckley et al., 1958). A more thorough study of the literature than we have been able to carry out to date may well demonstrate the existence and even possibly the exploitation on a small scale of this interesting resource elsewhere.

The purpose of this presentation is to give a general description of the characteristics of the Japanese reservoirs and the gas they produce. For more detailed information, the reader is referred to the extensive literature in

[1]Department of Petroleum Engineering, Stanford University, Stanford, California.

the Japanese language and to papers written by the author in collaboration with Japanese colleagues from the University of Tokyo, (the late) Professor Kōzō Kawai (Marsden and Kawai, 1965) and Professor S. Tanaka (Marsden, 1979).

RESERVOIR CHARACTERISTICS

In Japan, this gas dissolved in brine is usually found in relatively shallow reservoirs at depths from 500 to 1,000 m. Some is produced from as deep as 2,000 m, and small quantities are sometimes produced for local consumption by farmers and village people from wells drilled to the depths of ordinary water wells.

Because both the hydrostatic pressure and the gas solubility increase with depth, the producing gas-water ratio (GWR) does also. But the cost of drilling and completing the wells also increases with depth; the combination of these several factors together with geological conditions in Japan have established this usual depth range mentioned above. It may well be different under conditions in other countries.

Petroleum and natural gas are generally found in geologic structures such as anticlines, fault traps, or stratigraphic traps. Because of the density differences between the two fluids, on the one hand, and underground waters, on the other, the oil and gas accumulate and are held in hydrostatic equilibrium in the reservoir rocks until found by the drill bit. In contrast to this, the gas dissolved in brine is found in reservoir rocks in synclines or monoclines, types of structures practically always ignored in petroleum and natural gas exploration. The upper or shallower portions of the reservoir rocks in these synclines contain meteoric water which serves to confine the gas-bearing brine. The slightly greater density of the brine relative to the confining water again establishes hydrostatic equilibrium. In a few cases, gentle anticlines and faults exist, but these are the exception rather than the rule.

Although this kind of gas is found on all of the four large islands of Japan, as well as some of the smaller ones, the major deposits are in two areas of Honshu. One is the Chiba Prefecture, just east of Tokyo, and the other, the Niigata Prefecture, on the western coast of northern Honshu. In the former, the reservoir rocks are poorly consolidated sandstones and siltstones having permeabilities in the range of 70 to 200 md (69 to $197 \times 10^{-4}\ \mu m^2$). In the latter, they are poorly consolidated sands and gravels having higher permeabilities, generally from less than 1 to 50 darcies (1 to $70.7\ \mu m^2$). In all cases, the porosities range from 20 percent to 30 percent. Because cores are difficult and expensive to take, the porosity and permeability data are not extensive.

While the most productive reservoirs are of Pliocene and Pleistocene age, others vary from Recent to Miocene. The larger ones are found in what were marine or lagoonal sedimentary basins having wide areal extent and thick deposits, but some smaller ones are also found in what were inland lacustrine sediments. The Pliocene and Pleistocene reservoirs have intergranular porosity and high permeabilities, which give them acceptable productivity despite the relatively low solubility of the gas in the brine. The Miocene reservoirs, however, must have both intergranular and fracture porosity to make them productive.

The gas-in-brine-bearing reservoir rocks are known by experience to exist over wide areas, and so it is not really necessary to seek specific geological structures for drilling. If enough geological or geophysical information exists to indicate the presence of a syncline, then drilling is done along its axis. Experience has shown that such wells produce at a higher GWR than do those drilled on the flanks.

Two of the Pliocene reservoirs of Chiba Prefecture have maximum thicknesses of 270 and 370 m. They are interbedded every meter or so with thin mudstone layers which also contain gas-bearing brine and contribute to the production.

THE GAS AND BRINE

The composition of the gas varies from one field to the next. Most contain at least 90 percent methane, with usually no more than 5 percent carbon dioxide, but sometimes as much as 10 percent. Nitrogen is often present up to several percent, but where the brine has been diluted with meteoric water, there can be up to 40 percent. Hydrocarbons other than methane are essentially absent. Hydrogen sulfide is found in only a few ppm (cm^3/m^3) at most. Thus, the gas is a good feedstock for those petrochemical processes utilizing catalysts poisoned by hydrogen sulfide. Simple combustion of the gas for domestic, commercial, or industrial uses results in minimal air pollution from oxides of sulfur.

Because of the lack of other hydrocarbons, this gas is believed to be genetically different from the natural gas found associated with crude oil or as ordinary free natural gas. Extensive isotope ratio and other geochemical studies in Japan have shown that the methane and carbon dioxide probably derive from bacterial decomposition of some of the organic material in the sediments and the nitrogen and occasional traces of argon are of atmospheric origin.

The composition of many of the brines is tabulated elsewhere in the literature, as is that of many of the natural gases (Marsden and Kawai, 1965). It differs from many other subsurface brines in having iodide ion (I^-) concentrations of as much as 80 mg/ℓ or more. After separation of the natural gas at the surface of the ground, this brine is sent to plants for recovery of the iodine, either by ion exchange or by an air-blowing process. Most of the iodine in international commerce comes from these Japanese brines.

A simple correlation between high chloride ion (Cl^-) or high ammonium ion (NH_4^+) concentration, on the one hand, and good gas production potential, on the other, has been observed in some cases. The gas-bearing brine is a colorless fossil sea water and the meteoric water which confines it is an oxygen-rich fresh water. Between the two is a zone of brownish water low in (Cl^-).

DRILLING AND PRODUCTION PRACTICES

The rocks drilled are usually soft, so that rotary drilling with inexpensive fish-tail bits is appropriate. Since the brines are corrosive and no really high pressures are encountered, PVC (polyvinylchloride) tubing and casing are used. Gas lift with tubing flow has been used, but casing flow has been found to be more economical. In some cases, larger diameter holes have been drilled and submersible pumps employed. Screens of 80-mesh (sometimes preceded by 30-mesh) have eliminated sand problems and although gravel packing has been tried, it is not really necessary. In Chiba Prefecture, multizone completions with screens several hundred m in length are used while in Niigata Prefecture, where the productive formations can be recognized on electric logs, single-zone completions of 20 to 40 m are used.

The initial producing GWR is usually what we would expect from the solubility of methane in the brine at the reservoir temperature and pressure. Most of the fields have strong water drives created by the meteoric water upstructure from the brine. If the reservoir transmissability is impeded by faults or other structural features, production may be by solution-gas drive or by a combination of the two methods. As is the case with solution gas-drive oil reservoirs, the producing GWR increases with time and can go through a maximum just as is the case with GOR (gas-oil ratio) for depletion-type oil reservoirs.

It is now recognized by the Japanese that there are two distinct types of reservoirs producing this kind of natural gas. For one, which is known as the "ordinary-type," the GWR is what one would calculate from the gas solubility at reservoir temperature and hydrostatic pressure. These are found where the permeabilities are high in Niigata Prefecture and in most of Japan, excluding the eastern part of Chiba Prefecture. In the latter, where the permeabilities are low, we have Mobara-type reservoirs with GWRs starting at about 3 m^3/m^3 and going as high as 20 m^3/m^3. Sometimes the wells flow spontaneously and without artificial gas lift. One can see that the Mobara-type have economic advantages over the ordinary-type both because artificial lifting is not needed and because less brine must be handled for a given gas production.

The pressure decline with time for the Mobara-type reservoir is greater than that for the ordinary-type. The gas comes not only from the sandstone beds, but also the interbedded mudstone layers by flowing into and through the sandstone.

Subsidence and Brine Reinjection

Production of underground fluids sometimes leads to subsidence of the surface of the land (Marsden and Davis, 1967). This has been observed in many parts of the world when oil and fresh water are produced, and it also is the case with the Japanese brines. This subsidence can be arrested and often stopped completely by simultaneously injecting one fluid while producing another, thereby maintaining underground fluid pressures.

For many years, the natural gas and then the iodine were stripped from the brine produced from these fields, after which the brine was discarded into the ocean or nearby rivers. Slow, and sometimes rapid, subsidence became commonplace in some productive areas. At that time both the cost of brine reinjection and the engineering problems encountered dictated this disposal practice.

In 1970, both the government-instigated regulations and gas-company field tests were directed to eliminating urban subsidence. Brine production was decreased and eventually stopped entirely in eastern Tokyo as well as in and near the city of Niigata. All produced brine had to be reinjected. Topographic surveys were made regularly by both the government and the companies. In essence, subsidence has been stopped by proper reinjection of the produced brine in all areas where it might be damaging.

Production and Reserves

Almost four-fifths of Japan's natural gas production now comes from reservoirs having free gas. Of the remainder, 10 times as much comes from gas dissolved in brine as from gas dissolved in oil. Most (82 percent) of the gas dissolved in brine comes from Niigata Prefecture, while most of the remainder (16 percent) comes from Chiba Prefecture.

The Geological Survey of Japan has surveyed the reserves of this kind of gas for the entire country. Fields have been found from Hokkaido through Honshu to Kyushu, and even in Okinawa. Because porosities of the reservoir rocks are usually not known with any great accuracy, porosities of 25 percent and 30 percent are usually used to calculate gas in place in the reservoirs. In 1975, the estimates made on this basis were 739 and 887 billion $(10^9)m^3$ (26.1 and 31.3 trillion (10^{12})CF) of gas.

REFERENCES

Buckley, S. E. et al., 1958, Distribution of dissolved hydrocarbons in subsurface brines, in Habitat of Oil: AAPG, p. 850.

Kuo Shang-ping, Lanzhou Inst. of Glacialogy & Crypedology, Academia Sinica, PRC: personal communication.

Marsden, S. S. and Kawai, K., 1965, Suiyjoseiten'nengasu--A special type of Japanese natural gas resource: Am. Assoc. Pet. Geol. Bull., v. 49, no. 3, p. 286.

Marsden, S. S. and Davis, S. N., 1967, Geological subsidence: Sci. Am., v. 216, no. 6, p. 93.

Marsden, S. S., 1979, Natural gas dissolved in brine--A major energy resource of Japan: SPE of AIME, 54th Annual Fall Meeting, Las Vegas, Nevada, Sept. 23-26, 1979, Paper SPE 8355.

Motojima, K., 1979, Some considerations on the petroleum geology in China: J. Jap. Assoc. Pet. Tech., v. 44, no. 2, p. 42.

Reeves, F., 1953, Italian oil & gas resources, Am. Assoc. Pet. Geol. Bull., v. 37, p. 601.

Villalva, E. B., 1976, Utilization of low pressure natural gas occurrences in the Philippines: The Comvol Letter, v. 8, nos. 5 & 6, p. 1.

CHAPTER 50

GAS IN SUBSURFACE WATERS

Lawrence C. Bonham[1]

INTRODUCTION

In this time of widespread concern over energy resources, considerable publicity has been given to huge estimates of unconventional natural gas which conceivably could be produced with improved technology and increased prices. Most of this gas is not proved to exist, so the estimates of gas in place, although highly speculative, are not impossible. Even if the gas actually exists, we cannot now predict that significant amounts can be produced economically. Thus the speculative estimates are not of proven reserves but mainly of undiscovered resources most of which may not be producible under any foreseeable conditions. Nevertheless, any comprehensive survey of energy resources must include these unconventional gases just as a survey of the world's metal resources should include the gold, silver, lead, copper, zinc, and other metals contained in seawater, even though the metals are not recoverable by today's technology and economics.

This chapter is concerned with one type of unconventional natural gas, that is, gas dissolved in waters beneath the surface of the earth. Dissolved gases can conveniently be divided into two groups: (1) those in which the fluid is at or near hydrostatic pressure and (2) those in which the fluid pressure is considerably above hydrostatic; i.e., the geopressured fluids.

GAS DISSOLVED IN HYDROSTATIC WATER

Japan

In two areas in Japan (Fig. 50-1), natural gas occurs dissolved in brines in aquifers, mostly at depths of 500 to 1,000 m (Marsden and Kawai, 1965; Marsden, 1979). In the Niigata area, gas has been produced from shallow

[1]Chevron Oil Field Research Company, La Habra, California.

Figure 50-1. Areas of dissolved gas production in Japan. After Marsden, 1979

water wells since the last century. Following World War II, additional wells were drilled into the gas-bearing brines and production increased to about 700,000 m^3/d ($25x10^6$ ft^3/d) in 1959.

The second area is in the southern Kantō plain of Chiba Prefecture. It underlies part of Tokyo and extends eastward about 50 km to the coast. First production was in 1939, but many additional wells were drilled as deep as 2,000 m following World War II. Maximum production was about 564,000 m^3/d ($20x10^6$ ft^3/d) in 1970. Total reserves for the two areas were estimated in 1962 at about $96x10^9$ m^3 ($3.4x10^{12}$ ft^3).

Development of dissolved gas production in Japan came about not only because of the postwar fuel shortage, but also because the associated brines contain a high iodide ion concentration, commonly more than 80,000 ppm I^-. The iodine, which is removed from the brines produced with the gas, meets Japan's domestic needs and provides most of the iodine in international trade.

A detrimental effect of gas and brine production from the Japanese deposits has been land subsidence. By 1958, subsidence in the city of Niigata led to pumping restrictions and development of a program of brine reinjection (Okumura, 1969). Subsidence in the Chiba area resulted in regulations adopted in 1970 to prohibit production in the area of maximum subsidence in eastern Tokyo, require all produced brine to be reinjected, and establish minimum well spacing.

Italy

Reeves (1953) described the Polisine gas-producing area in the eastern Po River Plain of Italy (Fig. 50-2). A 2,000 km^2 area of the Po Delta between the Adige and Po rivers east of Rovigo yields gas and brine wherever wells are drilled. First production was in 1934; by 1953 production amounted to 800,000 m^3/d ($28x10^6$ ft^3/d). Most of the wells were less than 500 m deep and produced about as much brine as gas. Ultimate yield in the producing areas was estimated at 2,300,000 m^3/km^2.

USSR

Zor'kin and Stadnik (1975), in discussing the genesis of petroleum hydrocarbons, state that gigantic volumes of hydrocarbons are dissolved in the

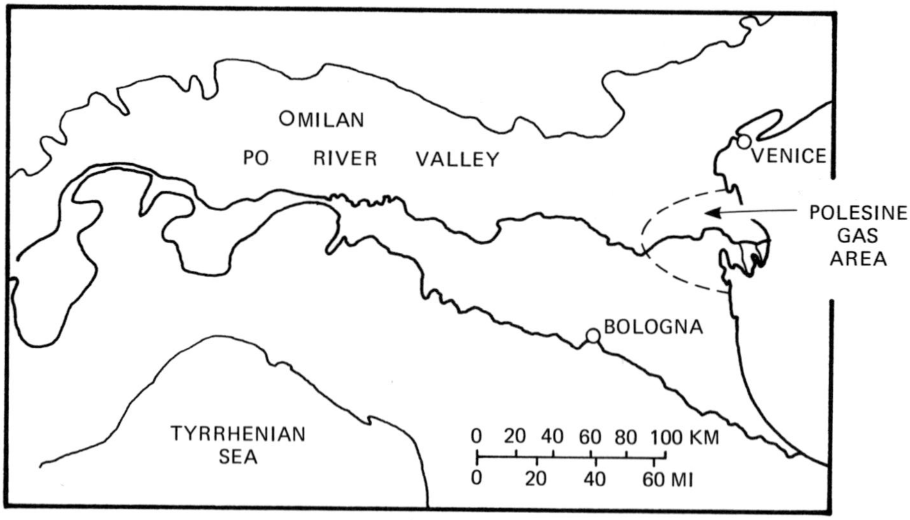

Figure 50-2. Area of dissolved gas production in Italy. After Reeves, 1953

formation waters of oil- and gas-bearing basins. In fact, the dissolved gas resources exceed the known commercial accumulations, an observation also made by Buckley et al. (1958). Zor'kin and Stadnik point out that geologically young basins are distinguished by high gas saturation of the entire geologic section. They attribute this to active hydrocarbon generation processes during which the hydrocarbons are redistributed between the gas-generating beds and the formation waters. Table 50-1 summarizes the estimates by Zor'kin and Stadnik of dissolved gas resources in several oil- and gas-bearing basins of the USSR and compares the dissolved gas volumes to the volumes in known and probable conventional accumulations.

California

Gas dissolved in subsurface water is not a newly recognized resource in California (Fig. 50-3). Many references cite examples of natural gas bubbling from springs and shallow water wells (Jenkins, 1943). For example, California's first natural gas production was from an artesian water well drilled in

TABLE 50-1. Gas in selected petroliferous basins of the USSR

Area	Gas resources, $m^3 x 10^{12}$		
	Conventional deposits		
	Proven reserves	Predicted-probable reserves	Dissolved in water
Ural-Volga	1.4	5.3	80
Timan-Pechora	0.2	3.1	300
Dnieper-Donets	0.7	2.1	75
Caspian	0.2	1.9	980
West Siberian	2.7	14.8	1,000

Source: From Zor'kin and Stadnik, 1975.

Stockton in 1858 (Stalder, 1943). Many water wells drilled later in that area produced dissolved gas which was used for domestic and manufacturing purposes. The greatest yearly production was 300 million ft^3 in 1910. The city gas supply in Stockton and Sacramento was augmented for many years by gas produced from water wells (Symons, 1935). Farther south, in Kings County, gas in water wells along the west side of the Tulare Lake bed was cited by Corwin (1960) as the reason for the drilling that discovered the Dudley Ridge gas field.

Oklahoma

In northwestern Oklahoma, near the town of Woodward, Amoco operates a plant for extraction of iodine from brines (Coffer, 1977). The brines are produced from a buried river channel, which is about 3 mi wide and 50 mi long. Production is from a depth of 2,100 to 2,300 m (7,000 to 7,500 ft). It was found necessary to add equipment to the iodine extraction plant to handle the natural gas produced with the brine. Gas production ranges from 3 to 264 m^3 of gas per m^3 of brine (16 to 1,500 ft^3 per barrel). Not all of this gas is in solution; some is produced by coning down of gas from overlying, separate gas-phase zones. Difficulties are encountered when the brine wells are shut in for short periods of time. Pressure reduction from brine production causes exsolution of gas bubbles in the sands. If the pore space contains about 10 percent free gas, the permeability is reduced to about 40 percent of the gas-free permeability.

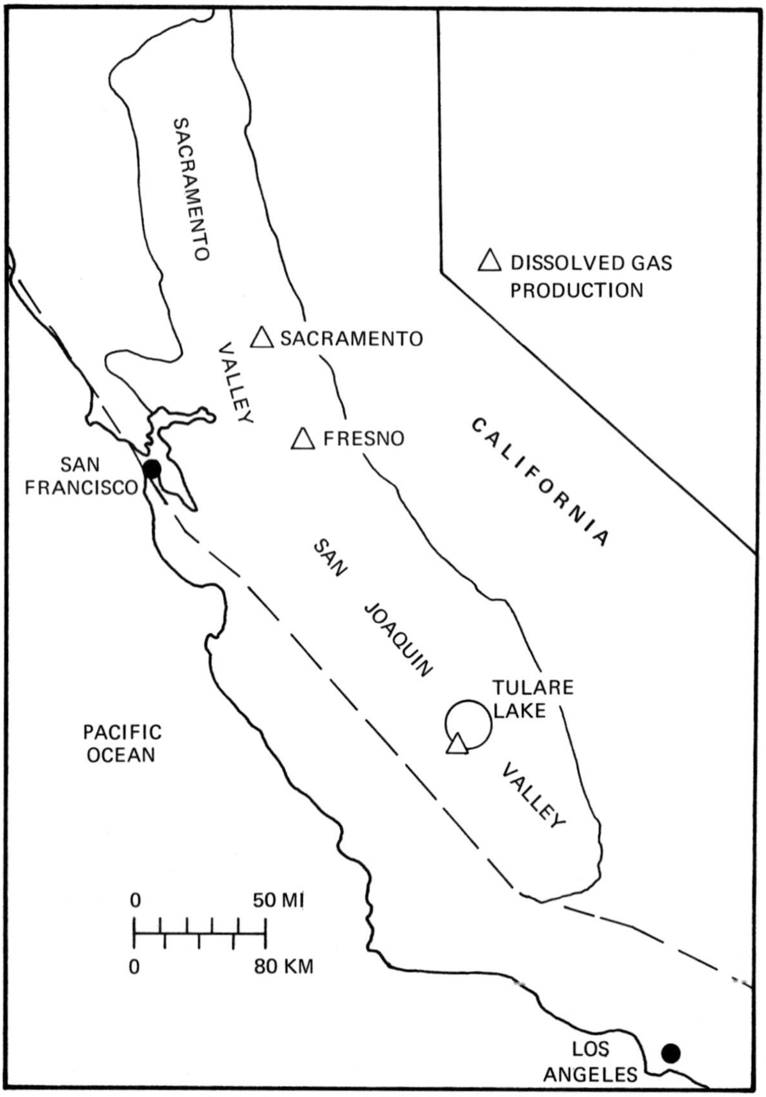

Figure 50-3. Areas of dissolved gas production in California

Texas

Data are scarce on the actual gas content of subsurface waters in other parts of the United States. The work of Buckley et al. (1958) is often cited as evidence that gas dissolved in subsurface water is not uncommon. From about 1945 to 1955, they conducted an extensive investigation of dissolved hydrocarbons in several hundred wells, mostly in Texas and Mississippi. Samples were obtained during conventional drill-stem tests of water-bearing horizons. Water samples were collected with an evacuated subsurface sampling device which was lowered into the drill pipe on a wire line. The sealed sampling device was transported to a laboratory, where analyses were made for the amount and type of gases dissolved in the water. Although not stated, it is assumed that all samples were obtained from aquifers at or near hydrostatic pressure.

Buckley et al. found that many subsurface formations in the area of the United States bordering the Gulf of Mexico contain appreciable quantities of dissolved natural gas, mostly methane. The quantities amount to as much as 2.5 m^3/gas/m^3 water (14 ft^3/bbl).

Figure 50-4 shows the total dissolved gas in a number of Woodbine water samples from East Texas. These waters range in salinity from 60,000 to 140,000 ppm total dissolved solids. Curves are shown for the solubility of methane in pure water and in 100,000 ppm salt water. Most of the samples are close to their saturation volumes in salt water even at these shallow depths under hydrostatic pressure. Buckley et al. (1958) also state (p. 867) that Frio water was found to be saturated or nearly saturated with gas in almost every well sampled. This included 32 drill-stem tests of 25 wells in the upper Gulf Coast area of Texas.

ESTIMATES OF GEOPRESSURED GAS IN THE UNITED STATES

During the past 4 years, public attention has been drawn to high temperature formation waters in the Gulf Coast area (Fig. 50-5) as a potential energy resource of gas dissolved in high pressure. Waters in the deep Tertiary aquifers were first considered as a low temperature, hot water geothermal heat source. A study by the Dow Chemical Company, however, indicated that electricity generation from the heat in these waters was not economically feasible even with allowances for recovery of dissolved methane (Wilson et al., 1974).

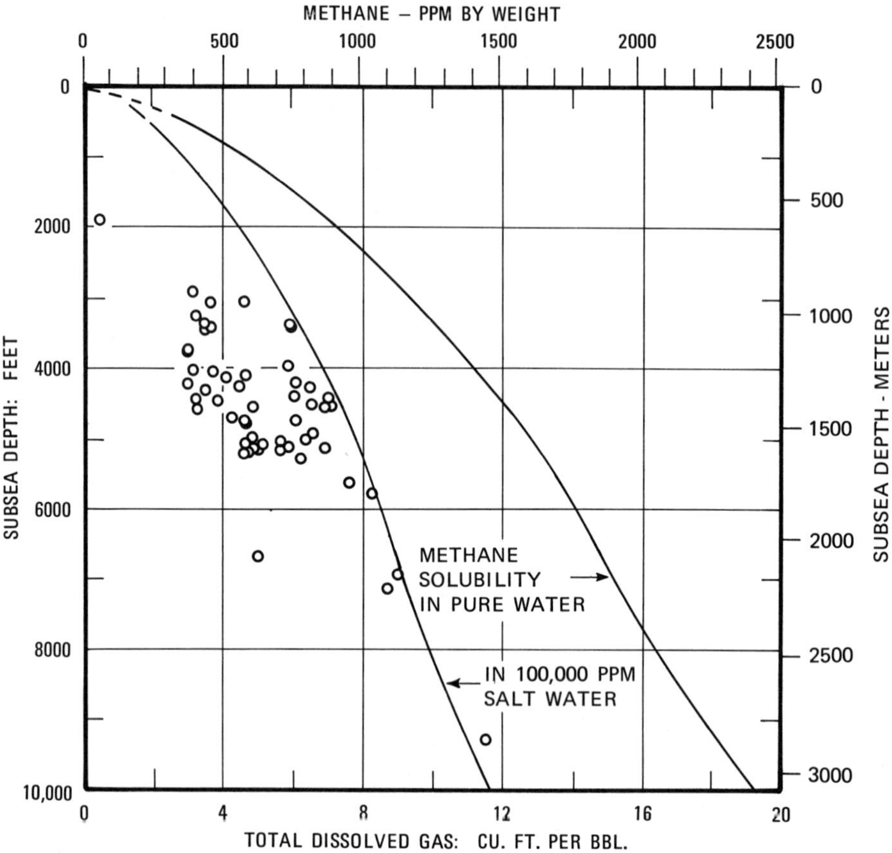

Figure 50-4. Dissolved gas in Woodbine water, East Texas. Data from Buckley et al., 1958

An article in Science (Maugh, 1976) provided three conflicting estimates of the magnitude of methane that could be dissolved in Gulf Coast geopressured waters (Table 50-2).

TABLE 50-2. Estimates of U.S. natural gas resources

	Amount		Source of information
	($m^3 x10^{12}$)	($ft^3 x10^{12}$)	
Gas in Gulf Coast geopressured zones			
Geopressured sands	84	3,000	Hise, 1976
Onshore geopressured zone	669	23,600	Papadopulos et al., 1975
Geopressured sands	1,372	49,000	Jones, 1976
Conventional U.S. gas resources			
Cumulative production	13	480	Miller, 1975
Proved reserves	7	237	Miller, 1975
Annual consumption	0.56	20	Ryan, 1979

As a means of understanding and evaluating the various estimates, model studies were made of the Gulf Coast geopressured dissolved gas resource. Factors considered in the model follow.

Area

The model represents the volume of sediments, Cretaceous to Recent, for the onshore and offshore Gulf of Mexico area from Florida to Mexico (Fig. 50-5).

Sediment Volumes

The sediment volumes used in the model are consistent with published data for Cenozoic and Mesozoic sediments of the Central Gulf Coast (Murray, 1952) and the Western Gulf Coastal Plain (Colle et al., 1952).

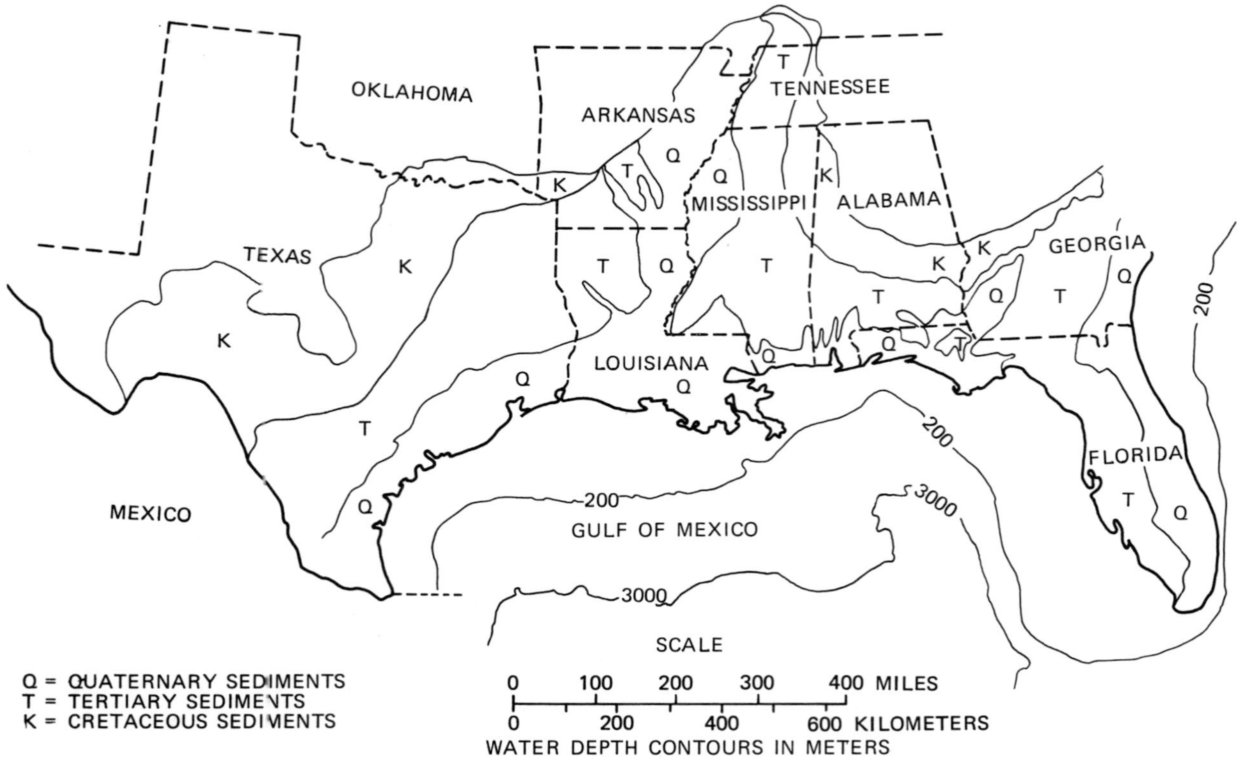

Figure 50-5. Geologic map of Gulf Coast area, United States

Geometry of the Model

The amount of gas dissolved in pore waters is dependent upon the sediment porosity, temperature, and pressure. Each of these factors is related to depth. It was necessary, therefore, to specify shapes and dimensions of prisms to which the sediment areas and volumes could be allocated. Figure 50-6 is the simplified geometric form used for modeling the Central Gulf Coast Area. For the gas volume calculations, each prism was divided into 1,000 ft depth slices so that appropriate porosity, temperature, and pressure factors could be applied.

Sand Porosities

The data of Atwater and Miller (1965) were used to assign porosity values to sands. Their depth-porosity curve for more than 17,000 sand samples extrapolates to 40 percent porosity at zero depth and zero porosity at about 32,000 ft. The rate of decrease is about 1.265 porosity units per thousand ft of depth (Fig. 50-7).

Geothermal Gradient

The assumed geothermal gradients are typical of those found in the Gulf Coast area: 2.73°C/100 m (1.5°F/100 ft) of depth to a temperature of 100°C (212°F) at about 2,885 m (9,467 ft) and 3.645°C/100 m (2.0°F/100 ft) below this depth (Fig. 50-7).

Fluid Pressures

Fluid pressures (Fig. 50-7) were assumed to be hydrostatic to a depth of 2,885 m (9,467 ft) with a pressure gradient of 107.4 $g/cm^2/m$ (.465 psi/ft). The fluid pressure gradient in the geopressured zone was assumed to be 161.8 $g/cm^2/m$ (0.7 psi/ft).

Lithology

The nature of the sediments filling the Gulf Coast Basin is well known from the thousands of boreholes drilled for petroleum exploration and production. In these model studies the sediments were assumed to be 50 percent sand and 50 percent shale to a depth of 10,000 ft. Below that depth, the sediments

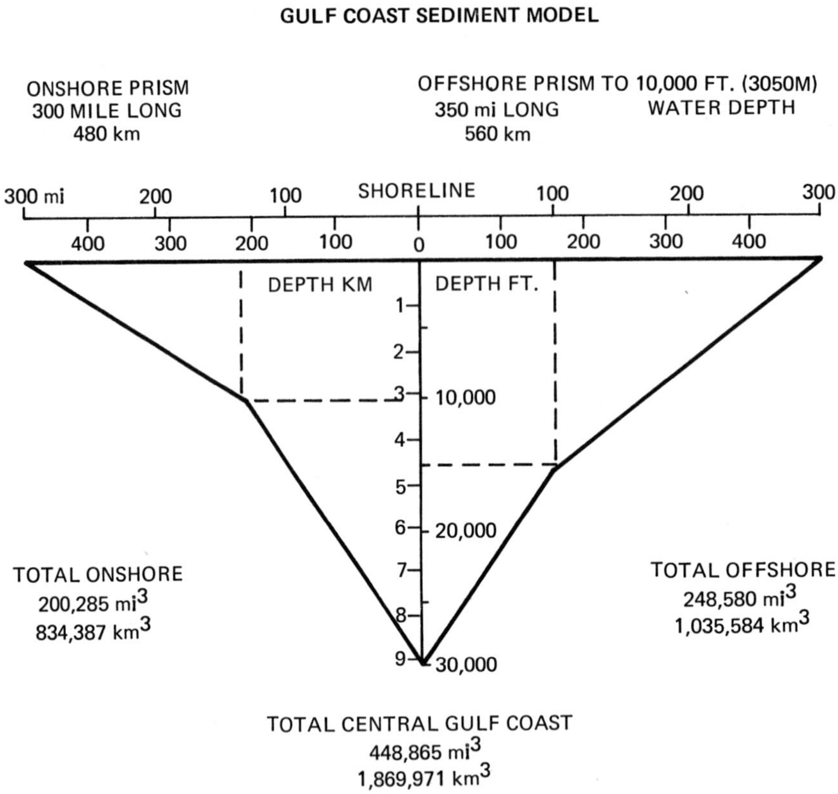

Figure 50-6. Shape and dimensions of Central Gulf Coast sediment model

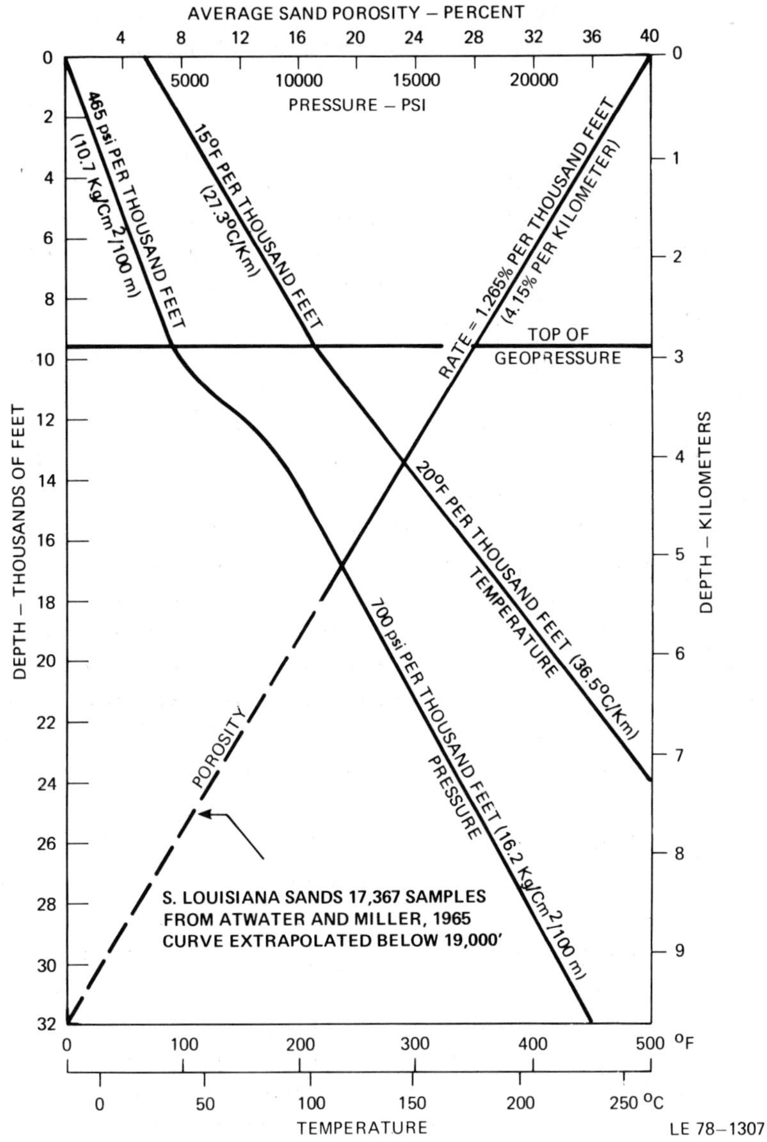

Figure 50-7. Porosity, temperature, and pressure gradients used for model calculations

were assumed to be 10 percent sand and 90 percent shale to a depth of 20,000 ft. Sediments below 20,000 ft were assumed to be all shale.

Methane Solubility

The most widely used data on the solubility of methane in water are those of Culberson and McKetta (1951), but they cover a range from 25° to 170°C (77° to 340°F) and up to 700 kg/cm^2 (10,000 psi) only. Sultanov et al. (1972) have extended the range of temperature to 360°C (680°F) and pressure to 1,100 kg/cm^2 (15,645 psi). Bonham (1978) combined these two sets of data into a single set of curves which were used for the model studies (Fig. 50-8).

Several investigators have established a decrease in methane solubility in saline waters as compared to fresh waters. Figure 50-9 (Knapp and Isokrari, 1976) is an example of available data. Gulf Coast formation waters commonly increase in salinity with depth to a maximum of over 100,000 ppm total solids a short distance above the geopressured zone.

Waters recovered from geopressured zones commonly have lower salinities than the shallower waters. Bebout (1976) predicted that deep fluids from geopressured zones would have salinities less than 20,000 ppm, perhaps as low as 5,000 ppm. In this study, methane solubilities in pure water were used for calculations in the geopressured zone where the salinities are anticipated to be low. Calculations of methane volume in the normally pressured zone were reduced 30 percent from fresh water solubilities on the assumption that the average water salinity would be about 80,000 ppm in this zone (see Fig. 50-9).

Results of the Model Calculations

Results of the model study are illustrated in Fig. 50-10, which shows the dissolved methane capacity of sands in 1,000 ft depth segments of the Gulf Coast Basin. The calculations for each 1,000 ft slice are based on conditions at the midpoint depth.

The data in Fig. 50-10 indicate that huge volumes of methane could be contained in the formation waters of sands in the Gulf Coast area. The calculated capacity of the geopressured sands, according to this model, is 240×10^{12} m^3 ($8{,}621 \times 10^{12}$ ft^3), which is more than 12 times the cumulative U.S. production plus proved reserves. The calculated capacity of the normally

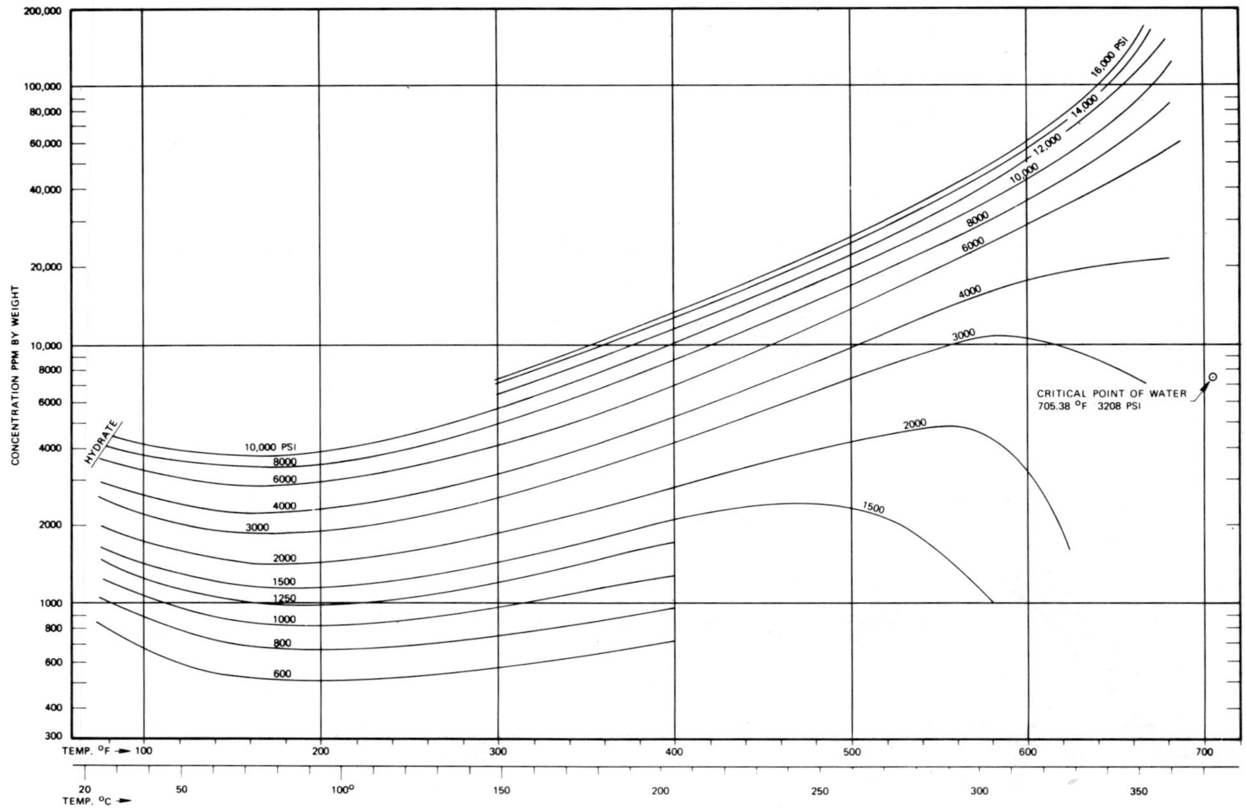

Figure 50-8. Solubility of methane in water. From Bonham, 1978

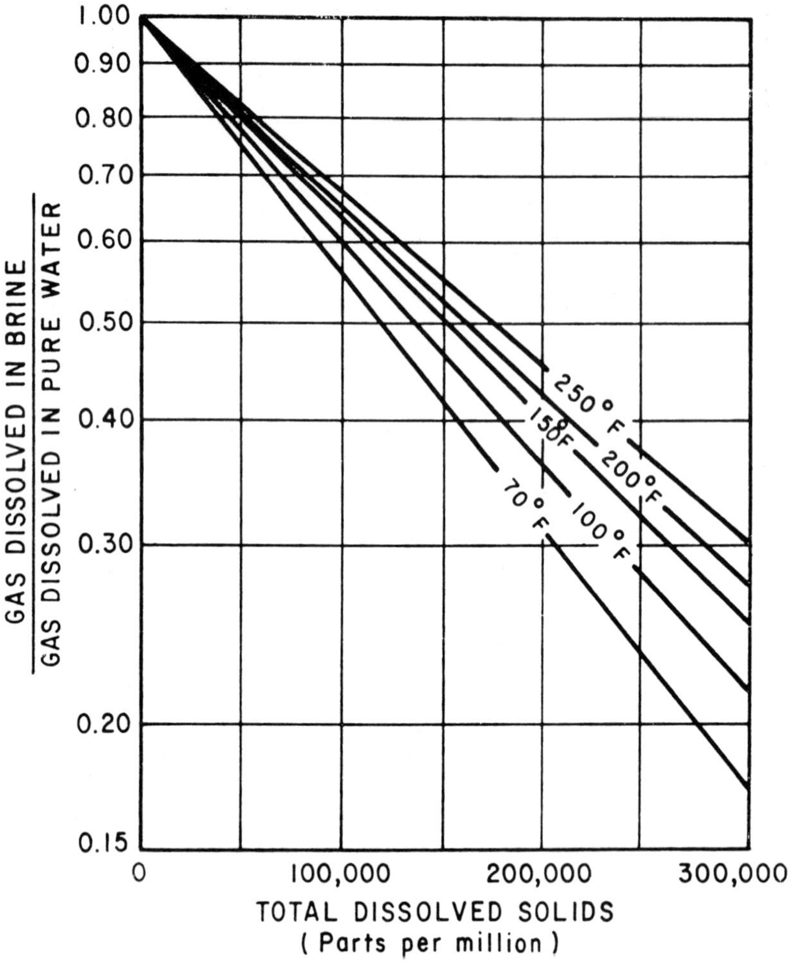

Figure 50-9. Effect of salinity on the solubility of gas in water. From Knapp and Isokrari, 1976

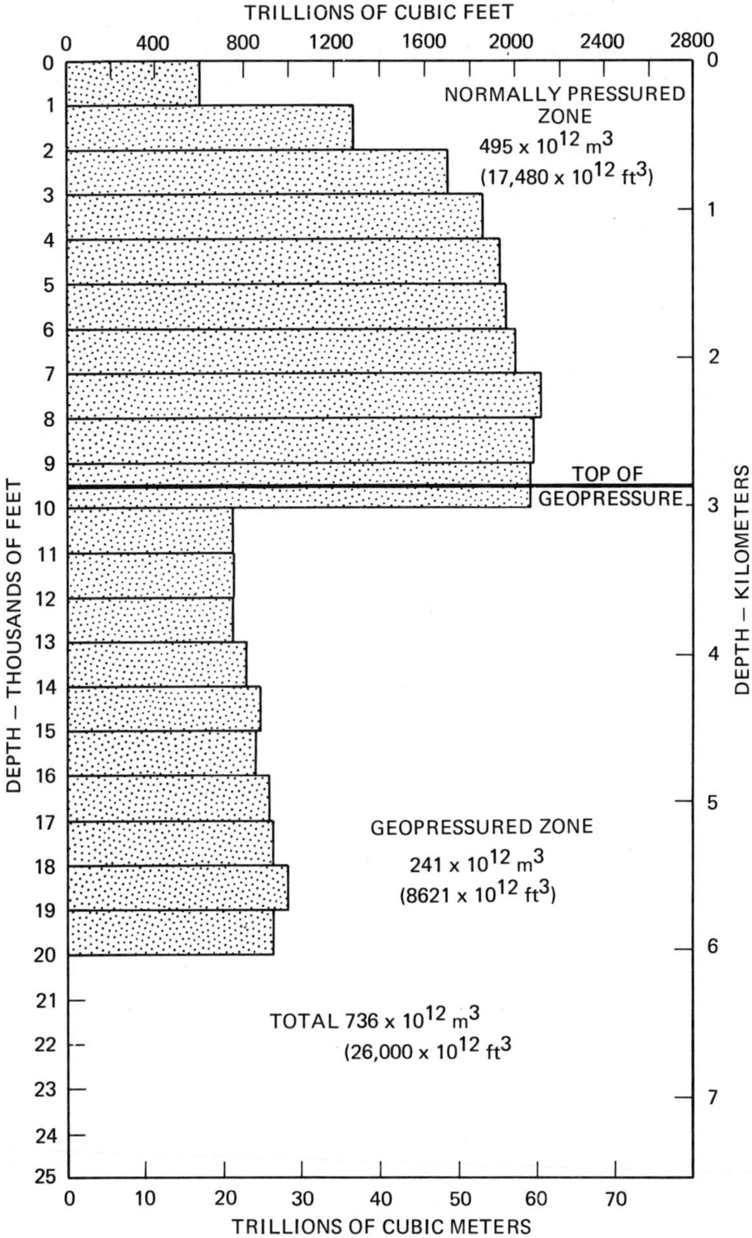

Figure 50-10. Results of model studies: methane capacity of pore waters in Gulf Coast sands

pressured sands is even greater, 495×10^{12} m^3 ($17,670 \times 10^{12}$ ft^3). This is due to the high sand percentages and high porosities in the normally pressured section.

Other Recent Estimates

Harbaugh (1979) summarizes studies of dissolved gas capacities of South Louisiana onshore and offshore sand bodies. He reports data from over 1,700 wells organized into 1,000 ft depth intervals from 2,000 to 19,000 ft beneath the surface. From this was calculated a potential resource base of 91×10^{12} m^3 ($3,264 \times 10^{12}$ ft^3).

Adams (1979) describes estimates of up to $.25 \times 10^{12}$ m^3 (9×10^{12} ft^3) of dissolved methane in the Tuscaloosa Formation of Lower Gulfian age in South Louisiana. His calculations are based on estimates of 40 mi^3 of geopressured sands in the South Louisiana Tuscaloosa trend, with temperatures ranging from 110° to 200°C (231° to 391°F) and pressures over 1,100 kg/cm^2 (16,000 psi).

Discussion

Estimates of the magnitude of the dissolved gas resource in the Gulf of Mexico area of the United States vary widely, depending on assumptions of reservoir volumes, porosity, temperature, pressure, salinity, methane solubility, and gas saturation. An even more difficult task is to estimate the volume of natural gas that could be economically recovered from the dissolved gas resource.

One study anticipates that up to 85 percent of the gas in solution could be recovered from waters produced from geopressured aquifers, but only 2 to 5 percent of the reservoir water could be produced in 30 years before the reservoir drive mechanism was exhausted. This would mean that less than 5 percent of the gas in place could be recovered (Kuuskraa et al., 1978, p. 2-6).

A recent study (Doscher et al., 1979) concludes that significant quantities of geopressured gas will not be produced in the United States at prices less than \$35/100 m^3 (\$10/MCF). This does not include costs that would be incurred if the water produced with the gas had to be reinjected. Reinjection costs alone would exceed \$14/100 m^3 (\$4/MCF).

The Task Force on Nonconventional Natural Gas Resources in the United States (Harbaugh, 1979) recently concluded that "the present lack of basic technical information related to the amount of recoverability of methane dissolved in water precludes the derivation of a meaningful estimate of the amount recoverable at any particular well head price."

At present, active programs are under way to evaluate the potential geopressured dissolved gas resource in the Gulf Coast area. The U.S. Department of Energy program is aimed at confirmation of selected optimum reservoirs by an extensive drilling and testing program and through reservoir model studies. Other programs are in progress in various industry and academic groups, but definitive answers to important technological questions cannot be expected much before the mid-1980's.

CONCLUSIONS

Available information suggests that the formation waters in many petroleum-bearing basins of the world may contain large quantities of dissolved natural gas. The deep, geopressured dissolved gas has received much attention recently, but several years of research and development must ensue before its economic potential can be evaluated. One thing is sure, very little geopressured gas will be economic at today's well head prices in the U.S. of less than $2.50.

Several writers have speculated that dissolved geopressured gas would be economic at significantly higher prices and this incentive, perhaps, would result in some production from this unconventional resource. At the same time, higher prices would result in production from many conventional reservoirs that cannot be produced economically at today's prices.

With respect to shallower gas dissolved in normally pressured waters, significant resources undoubtedly exist in many parts of the world. As pointed out by Marsden and Kawai (1965), these reservoirs may not compete with conventional gas sources, but the dissolved gas could be a useful energy source in countries with critically short fuel resources.

REFERENCES

Adams, J. B. (1979), Tuscaloosa geothermal-geopressure resource of South Louisiana: (abstract) Lafayette Geological Society Bulletin, Sept. 19, 1979, p. 3.

Atwater, G. I. and Miller, E. E. (1965), The effect of decrease in porosity with depth on future development of oil and gas reserves in South Louisiana: Unpublished manuscript of paper presented at Am. Assoc. Petrol. Geol. Annual Meeting in New Orleans.

Bebout, D. B. (1976), Resource assessment: Proceedings - Second Geopressured Geothermal Energy Conference: v. II, Univ. of Texas at Austin.

Bonham, L. C. (1978), Solubility of methane in water at elevated temperatures and pressures: Am. Assoc. Petrol. Geol. Bull., v. 62, p. 2478-2481.

Brill, J. P. and Biggs, H. D. (1975), Two phase flow in pipe: Univ. Tulsa, Oklahoma.

Buckley, S. E., Hocott, C. R., and Taggart, M. S., Jr. (1958), Distribution of dissolved hydrocarbons in subsurface waters in Habitat of Oil, Am. Assoc. Petrol. Geol., Tulsa, Okla., p. 850-882.

Coffer, H. F. (1977), Minutes of ERDO-Industry Workshop on geopressured geothermal resource development: Dallas, Texas, Sept. 28-29.

Colle, J., Cook, W. F., Denham, R. L., Ferguson, H. C., McGruit, J. H., Reedy, F., and Weaver, P. (1952), Volume of Mesozoic and Cenozoic sediments in Western Gulf Coastal Plain of United States: Bull. Geol. Soc. Am., v. 63, p. 1193-1200.

Corwin, C. H. (1960), Dudley Ridge gas field: Calif. Dept. Nat. Res., Div. Oil and Gas, Summary of Operation, California Oil Fields: v. 46, no. 1, p. 5.

Culberson, O. L. and McKetta, J. J. (1951), The solubility of methane in water at pressures to 10,000 psia: AIME Petrol. Trans., v. 192, p. 223-226.

Doscher, T. M., Osborne, R. H., Wilson, T., and Rhee, S. (1979), The technology and economics of methane production from geopressured aquifers: Soc. Petr. Eng. Preprint SPE 7827, 54th Annual Fall Tech. Conf., Las Vegas, Nevada.

Harbaugh, J. W. (1979), Nonconventional natural gas resources, sub Task Force 1 - Gas dissolved in water: Report to Federal Energy Regulatory Commission, U.S. Dept. of Energy report DOE/FERC-0029.

Hise, B. R. (1976), Natural gas from geopressured aquifers in Natural Gas from Unconventional Geologic Sources, National Academy of Sciences, NTIS report FE-2271-1, p. 41-63.

Jenkins, O. P., ed. (1943), Geologic formations and economic development of the oil and gas fields of California: Calif. Dept. Nat. Res., Div. Mines, Bull. 111, p. 604-606.

Jones, P. H. (1976), Natural gas resources of the geopressured zones in the Northern Gulf of Mexico Basin in Natural Gas from Unconventional Geologic Sources, National Academy of Sciences, NTIS report FE-2271-1, p. 17-33.

Knapp, R. M. and Isokrari, O. F. (1976), Aspects of numerical simulation of future performance of geopressured geothermal reservoirs: Proceedings - Second Geopressured Geothermal Energy Conference, v. III, p. 103, Univ. of Texas at Austin.

Kuuskraa, V. A., Brashear, J. P., Doscher, T. M., and Elkins, L. E. (1978), Enhanced recovery of unconventional gas: Report HCP/T2705-02 prepared for U.S. Dept. of Energy by Lewin and Associates, Inc., Washington, D.C., v. II.

Marsden, S. S. (1979), Natural gas dissolved in brine - a major energy resource of Japan: Soc. Petr. Eng., Preprint SPE 8355, Annual Fall Tech. Conf., Las Vegas, Nevada.

Marsden, S. S. and Kawai, K. (1965), "Suiyōsei-ten'nengasu," a special type of Japanese natural gas deposit: Bull. Am. Assoc. Petrol. Geol., v. 49, no. 3, p. 286-295.

Maugh, T. H. (1976), Natural gas: United States has it if the price is right: Science, v. 191, p. 546-550.

Miller, B. M., Thomsen, H. L., Dalton, G. L., Conry, A. B., Hendricks, T. A., Lennartz, F. F., Powers, R. E., Sable, E. G., and Varnes, K. L. (1975), Geological estimates of undiscovered recoverable oil and gas resources in the United States: U.S. Geological Survey Circular 725.

Murray, G. E. (1952), Volume of Mesozoic and Cenozoic sediments in Central Gulf Coastal Plain of United States: Bull. Geol. Soc. Am., v. 63, p. 1177-1192.

Okumura, T. (1969), Analysis of land subsidence in Niigata <u>in</u> Land Subsidence: Proc. Tokyo Symposium, v. 1, p. 130-143.

Papadopulos, S. S., Wallace, R. H., Jr., Wisselman, J. B., and Taylor, R. E. (1975), Assessment of onshore geopressured-geothermal resources in the Northern Gulf of Mexico Basin <u>in</u> Assessment of Geothermal Resources of the United States - 1975: U.S. Geol. Survey Circular 726, p. 125-146.

Ryan, S. B., ed. (1979), Monthly energy review: U.S. Dept. of Energy, DOE/EIA-0035/8(79).

Stalder, W. (1943), History of exploration and development of gas and oil in Northern California <u>in</u> Geologic Formations and Economic Development of the Oil and Gas Fields of California: Calif. Dept. Nat. Res., Div. Mines, Bulletin 118, p. 75.

Sultanov, R. G., Skripka, V. G., and Namiot, A. Yu. (1972), Solubility of methane in water at high temperatures and pressures: Gasovia Promyshlennost: v. 17, no. 5, p. 6-7.

Symons, H. H. (1935), California mineral production: Calif. Dept. of Nat. Res., Div. Mines, Bulletin 111, p. 208-221.

Wilson, J. S., Shepherd, B. P., and Kaufman, S. (1974), An analysis of the potential use of geothermal energy for power generation along the Texas Gulf Coast: Report prepared by Dow Chemical, U.S.A. for the Governor's Energy Advisory Council, State of Texas.

Zor'kin, L. M. and Stadnik, E. V. (1975), Unique features of gas saturation of formation waters of oil and gas basins in relation to the genesis of hydrocarbons and formation of their accumulations: Izv. Vyssh. Uchebn. Zaved., Geol. Razved. no. 6, p. 85-99.

CHAPTER 51

CHEMICAL AND BACTERIAL TREATMENT OF DEVONIAN SHALE FOR GAS RECOVERY AND PRODUCTION: CHARACTERIZATION AND MATRIX MODIFICATION

Teh Fu Yen[1]

INTRODUCTION

Numerous organic-rich shale deposits from widely different geologic ages occur throughout the United States. These reserves of black shale[2] and related minerals represent an important potential source of gas and oil as well as valuable metals. Notable is the black shale of Paleozoic Age of the Appalachian, Illinois, and Michigan basins. Through extraction and subsequent reaction of the organic components of black shale, one can expect to obtain high- and low-Btu gases. Also, trapped within the thick layers of shale are enormous amounts of methane. But unless a method is found to interconnect the tightly trapped, isolated pockets of gas, it would be economically infeasible at the present time to exploit these deposits.

It is the purpose of this chapter to introduce some novel approaches to exploit the shale, by the use of chemicals to cause fractures and the use of bacteria to cause erosion of the shale. Initial experiments yielded encouraging results. Potentially, therefore, recovery and production of gas from black shale is feasible.

Devonian Shale

In the Appalachian basin, during Devonian time, microscopic plant and shell life thrived in the shallow water. Dark shales of the prodelta complex

---

[1]University of Southern California.

[2]The term black shale, as used here, is not restricted to a single lithologic type of carbonaceous claystone, but includes many dark-colored, fine-grained rocks. Some are siltstone (Chattanooga shale), others are siliceous mudstone (Vinini Formation, Utah). Still others are marlstone, or in between claystone and coal (Pennsylvanian shale in western Indiana). The terms used here will exclude all the lacustrine and fresh-water formations.

were preserved for 110,000 mi^2. The thick laminated sediments, which are included in the black shales of the Upper Devonian Sonyea group of New York as well as in the Upper Cretaceous Pierre shale of Western Interior, reflect the anoxic environment of shelf muds. Distinct biofacies are observed in the organic-rich biogenic sediments (1), but the organic matter has not been carefully studied. In general, due to metamorphism, it is depleted in hydrogen. Recent studies show the kerogen from the black shales of the eastern U.S. to be definitely more aromatic than that of the western Green River oil shale. The more aromatic nature of the eastern kerogens may be of value for in situ recovery, such as in direct gasification from the site.

Gas Potential

One of the more significant features is the tight trapping of gas in the black shale formations. Within the thick layers of shale are enormous amounts of methane produced in the oxygen-deficient environment. One such reservoir alone, in the Appalachian basin, is estimated to contain 1.6 trillion ft^3 of gas (4). The total gas trapped in Devonian shale is estimated to amount to as much as 500 quadrillion ft^3.

Natural gas is thought to have originated from source materials and processes similar to those responsible for petroleum formation--the decomposition of marine organisms in the absence of oxygen. Sources of naturally occurring methane could come from (1) thermal or microbial degradation of organic matter in sediments, (2) maturation of coal beds, and (3) thermal alteration of reservoir oil or asphalt (5). Methane occurs in reservoirs under geological conditions similar to those of petroleum deposits: (1) an adequate reservoir rock of appropriate porosity and permeability, (2) a relatively impervious cap rock, and (3) a structural trap to confine the gas. For the black shale deposits it is safe to state that the associated gas cannot be recovered with ease unless the permeability and porosity of the black shales are increased, to permit the gas to escape through faults, fissures, internal channels, and natural fractures.

At the present time, there is no feasible process for artificially releasing the entrapped natural gas. Technology for deep-well hydraulic-pressure fracturing by injection of a fluid under high pressure is still in the developmental stages. Thus, there exists an urgent need for the development of alternative methods for energy recovery of the trapped gas or even the associated minerals.

Furthermore, as a long-term goal, feasibility studies on underground gasification of some selected sites of black shale should be explored. The apparently tremendous quantity of gas which could be recovered far exceeds the estimated natural gas, both onshore (712-1414 trillion ft^3) and offshore (418-834 trillion ft^3), in undiscovered conventional reservoirs.

CHARACTERIZATION OF DEVONIAN SHALE

Sample of Black Shale and Analysis

Two cored samples (No. 11740 and 12041), obtained from the Morgantown Energy Research Center, from a depth of 3,200-3,800 ft near Cottageville, Jackson County, West Virginia, were used in the investigation (Fig. 51-1). The samples represent the middle section of the brown shale (Marcellus shale) of the Devonian age. From the elemental analysis it seems that the amount of organic carbon content is proportional to that of the pyrite components (Fig. 51-2). Chemical composition studies indicate nitrogen, in general, is extremely low. Besides aluminum and iron, the magnesium, potassium, and titanium contents are significant. The more abundant trace elements are barium and manganese. The clay minerals, illite and chlorite, and pyrite are prominent in the mineralogical analysis.

Isolation of Bitumen and General Properties

Ground shale samples (-200 mesh) were used for both the Soxhlet extraction (continuous mode at refluxing temperature of solvent) and hot soaking (stationary mode at 85°C) to achieve the highest efficiency. The solvent systems, in general, were benzene, benzene-methanol, acetonitrile, ethylene diamine, and 1-methylnaphthalene. In order to make a comparison, they were applied over an arbitrarily chosen period of 10 hrs. From the data it is evident that increase in polarity of the solvent systems facilitates the extraction (1.4 to 3.4 percent). The system that used benzene-methanol as an azeotrope composition appears to be the best one.

The H/C atomic ratio is a good indicator of the extent of saturated carbon content and aromatic carbon content of the given fractions. The range of

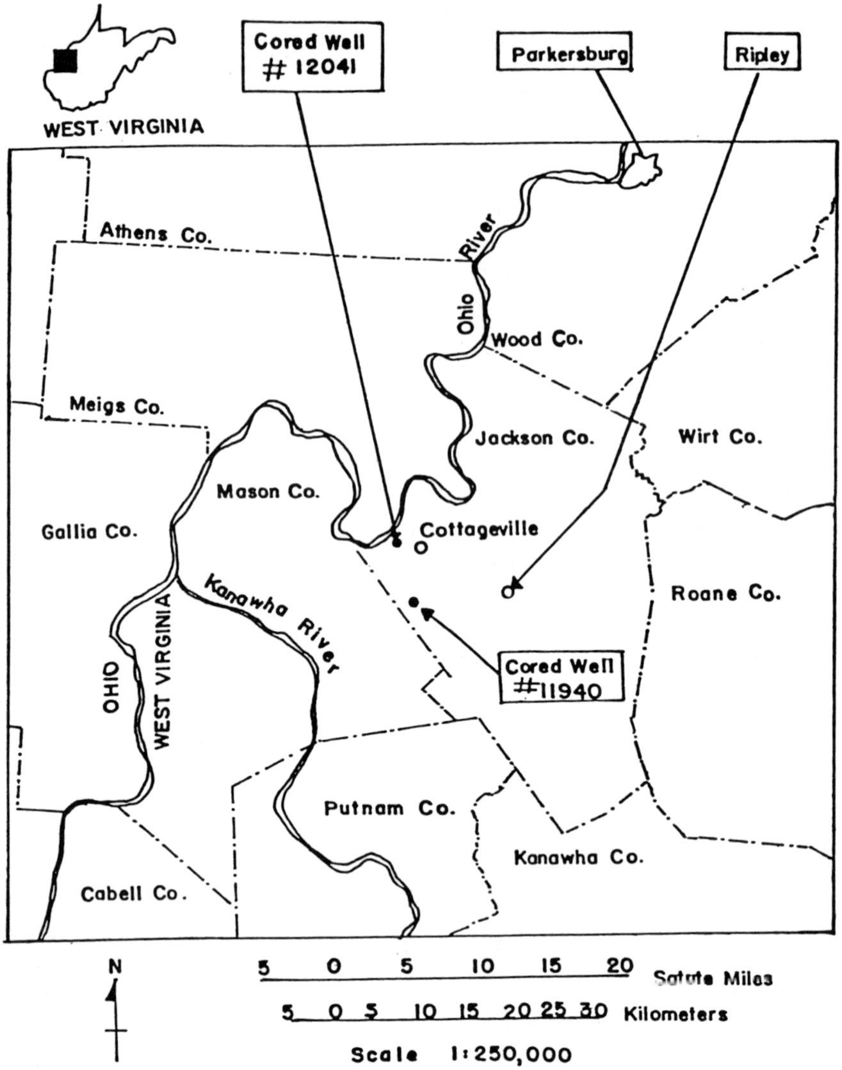

Figure 51-1. Location of cored wells

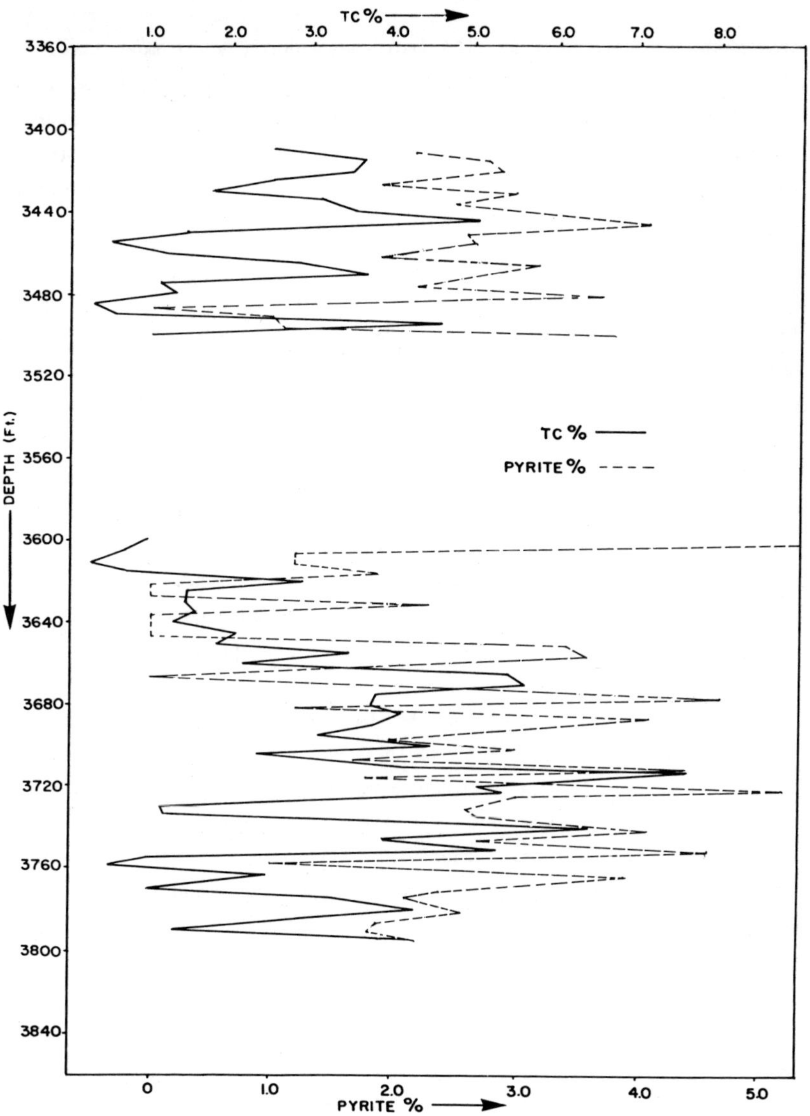

Figure 51-2. Weight percent of total organic and pyrite in Cottageville Devonian shale

Devonian shale bitumens in 1.52-1.56 indicates that they are most likely to have small amounts of aromatic molecules. The Green River oil shale bitumen has an H/C value of 1.68, which indicates that it has an even more saturated carbon content. Molecular weights of the Devonian bitumens were 378 (No. 8), 395 (No. 43), and 760 (No. 25) by vapor pressure osmometry. They probably contain molecules ranging from 100 to 900.

Carbon Preference Index

Gas chromatography has been used to characterize Devonian shale bitumens. The example shown in Fig. 51-3 is derived from No. 8 of Core No. 11940. The column used was 3 percent Dexsil 300 on chromosorb Q. n-Alkanes are found to be predominant (6).

In this manner, the carbon preference index (CPI) has been calculated for correlation with maturation. Usually, maturation causes a more random distribution of n-alkanes, and CPI is expressed as the ratio of the amount of odd carbon-numbered n-alkanes to that of even carbon-numbered ones. Three samples (#8, #43, and #55-56, Core #11940) fall between the geological ages of Chattanooga shale (Devonian, 300×10^6 years) and Appalachian shale (Ordovician, 400×10^6 years).

The distribution of n-alkanes in #43 bitumen extracted by different benzene-methanol solvent systems is shown in Fig. 51-4. Effects of the solvent systems on the distribution of dominant components in bitumen are investigated by using the solubility parameter. The solubility parameter, δ, is defined as the square root of the potential energy per millimeter of material. The effective solubility parameter of the mixture of components 1 and 2 with ν_1 and ν_2 volume fractions can be estimated from the following formula:

$$\delta_{mixture} = \nu_1 \delta_1 + \nu_2 \delta_2$$

In the benzene-methanol systems, the effective solubility parameter is increased by increasing the amount of methanol. Usually the higher the solubility parameter is, the larger the carbon number in dominant components that is obtained.

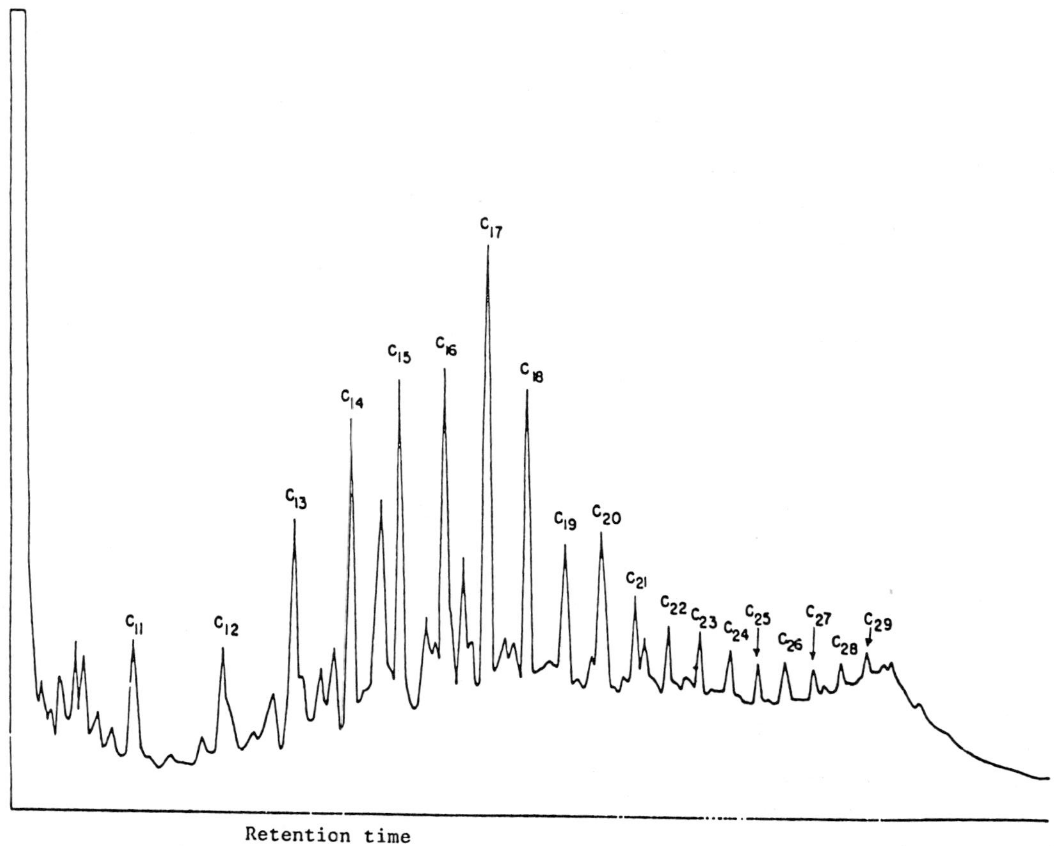

Figure 51-3. Normal alkane distribution in the bitumen of Devonian shale No. 8 from core 1194C

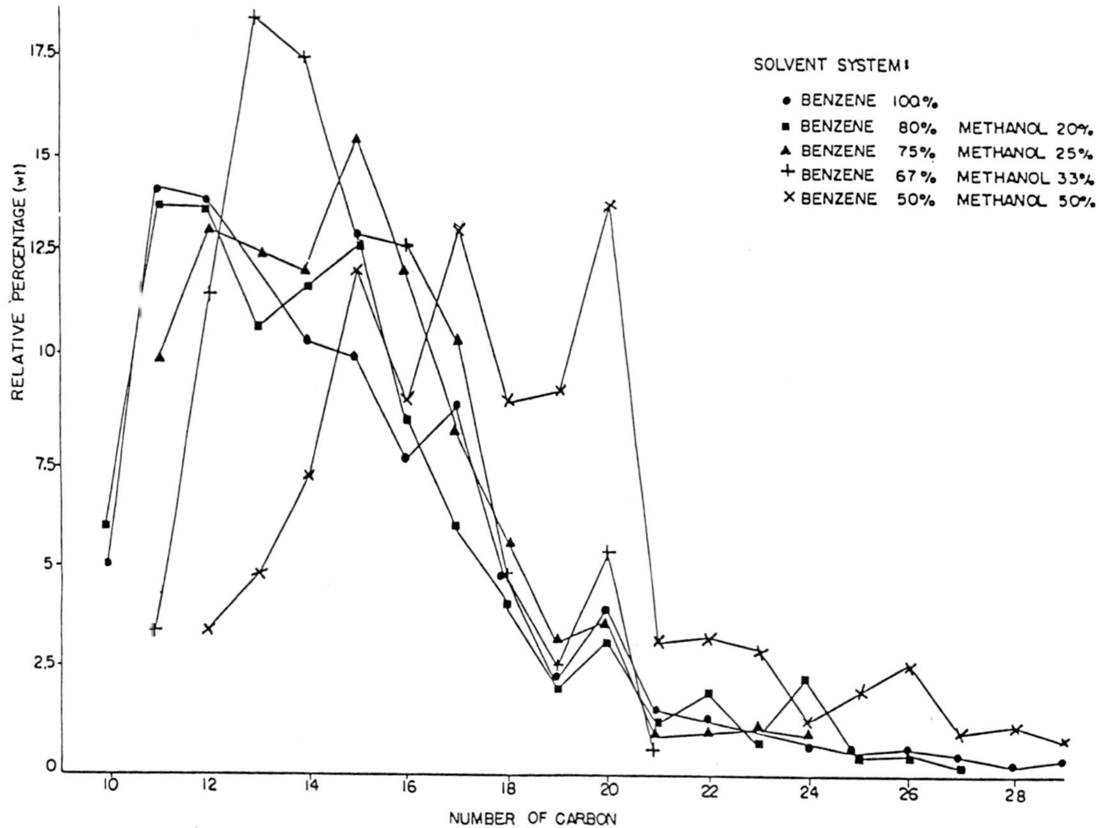

Figure 51-4. Effect of solvent system on distribution of dominant hydrocarbons

Fractionation of Bitumens

The benzene-extracted bitumen was precipitated with excess n-hexane to recover the asphaltene. The deasphaltened bitumen was loaded onto a silica gel column and a neutral oil fraction was first recovered. Subsequently, benzene-n-hexane, as well as tetrahydrofuran (THF), was used as an eluate to obtain the polar materials. The product distribution was as follows: neutral oil (58 percent), benzene-hexane (50:50) cut (12 percent), THF cut (18 percent), asphaltene (2 percent). The H/C ratio decreases from 1.69 to 1.36 and then to 1.32.

Asphaltene from Bitumens

This is the first case in which asphaltene has been separated from a Devonian shale. The asphaltene is a dark brown, powdery material, soluble in benzene and other aromatic solvents. NMR studies indicate that the asphaltene fraction is the most aromatic. This is supported by the evidence of C-H out-of-the-plane bending vibrations in the infrared spectra. The three bending vibrations at 880, 780, and 740 cm^{-1} are similar to those of petroleum-derived asphaltene. In contrast to the Green River bitumen, asphaltene is quite aromatic. The 731 cm^{-1} polymethylene band in the infrared spectrum of the Green River bitumen asphaltene is unique. The long-chain methylene structure in Green River kerogen is known. The aromaticity value of Devonian shale bitumen asphaltene as determined by the x-ray method is 0.49, in contrast to the low value of that from Green River bitumen asphaltene, which is 0.23 (7).

The geochemical significance of the asphaltenes is essential. The Green River bitumen asphaltene is less aromatic; Devonian shale bitumen is more aromatic. Thus, the nature of the asphaltene may be a useful indicator for maturation studies.

Chemical Structure of Bitumens

The straight-chain hydrocarbons were removed from the neutral oil fraction by urea adduction. The paraffin-free neutral oil was analyzed by gas chromatography-mass spectrometry (GC-MS). This fraction contains largely 2- and 3-ring condensed aromatics substituted with mono-, di-, tri-, and tetra- C_1

to C_3 alkyl substituents with the highly substituted ones as major components. Nuclear magnetic resonance spectra have been obtained for all fractions, and the neutral oil fraction definitely contains aromatic systems. The highly-condensed ring systems in Devonian shale bitumens are unique in that there is nil aromatic structure in the bitumen fractions of Green River oil shale (8).

Isolation of Kerogen Concentrates

Kerogen concentrates were prepared under an acid-leaching method which was modified from the method used for Green River oil shale previously (9). The removal of particular types of mineral in Devonian shales is possible. A summary of the results of kerogen concentrates is that about 9 percent (by weight) of the kerogen concentrate was obtained from #43 (approximately 3,714 ft) sample, 8 percent from #8 (approximately 3,445 ft) sample, and 0.5 percent from #25 (approximately 3,625 ft) sample.

The Cottageville raw shale contains 1-3 percent (by weight) of pyrite. Pyrite is extremely difficult to separate from the shale organic material. We currently have three methods for removal of pyrite from kerogen concentrate. These are (1) lithium aluminum hydride reduction, (2) electrolytic treatment, and (3) the microbiological method.

Aromaticity Evaluation of Kerogen Concentrate

The X-ray diffraction method developed by us, based on mesomorphic substance, was used to study the Devonian shale kerogen concentrate. The low-ash kerogen sample #43 (Core #11940) was subjected to X-ray diffraction studies (10). The aromaticity value was computed to be $f_a = 0.37$. It should be noted that the kerogen concentrate from Green River oil shale is $f_a = 0.15$. For this value both the X-ray method and the solid state nuclear magnetic resonance method yield identical results. It should also be recalled that the Appalachian black shale has an aromaticity value of $f_a = 0.50$.

Oxidative Degradation of Kerogen Concentrate

Oxidative degradation of the kerogen concentrate has also been studied. The method is similar to the one applied to Green River oil shale kerogen (9), where smaller quantities of aqueous alkaline permanganate were used for the stepwise, controlled oxidation of the kerogen molecule. The same method was applied previously to lignite (11) with success. During the course of the degradative studies, the larger molecules appeared to be released in the intermediate steps (Fig. 51-5).

A careful study of a middle cut of the oxidized kerogen revealed that the major components are C_7 to C_{16} dicarboxylic acids as well as C_9 to C_{19} mono-carboxylic acids of monobenz system. A close study of the minor peaks revealed the presence of over 100 different compounds.

MATRIX MODIFICATION OF DEVONIAN SHALE

Chemical Fracture of Devonian Shale by Ammonia

Core samples of Devonian shale from Richland County, Ohio, at 349 and 367 ft, as well as Chautauqua County, New York, at 2,658 and 2,095 ft were placed in a sealed vessel in contact of ammonia vapor. Visible fractures parallel to the bedding plane in a minimum time of 5-30 min were observed. A few of these fractures observed under the microscope did have connections. The cracks are about 0.1 mm apart as seen in thin section over polarized microscope. Both ammonia vapor as well as liquid can cause fractures. However, the shale of New York did not fracture, perhaps due to the low concentration of organic content (ca. 4-9 percent), in contrast to the other shales (11-15 percent). It is possible that ammonia specifically attacks the bonding between inorganic and organic matter. Therefore a higher organic content, which means a higher percentage of organic-inorganic bondings among the interlayer bondings in the shale, will render the shale more susceptible to the attack by ammonia. Dissolution of organic matter by ammonia is unlikely because shale kerogen isolated was found to be insoluble in liquid ammonia at room temperature (12).

Bioleaching of Devonian Shale

During preliminary bioleaching, finely pulverized shale was used in the laboratory. The mechanism of pyrite removal was expressed as follows:

$$FeS_2 + H_2O + 3\tfrac{1}{2}O_2 \longrightarrow Fe^{2+} + 2SO_4^{2-} + 2H^+$$

$$Fe^{2+} + \tfrac{1}{2}O_2 + H^+ \longrightarrow Fe^{3+} + \tfrac{1}{2}H_2O$$

The resulting ferric ion can further be used for desulfurization

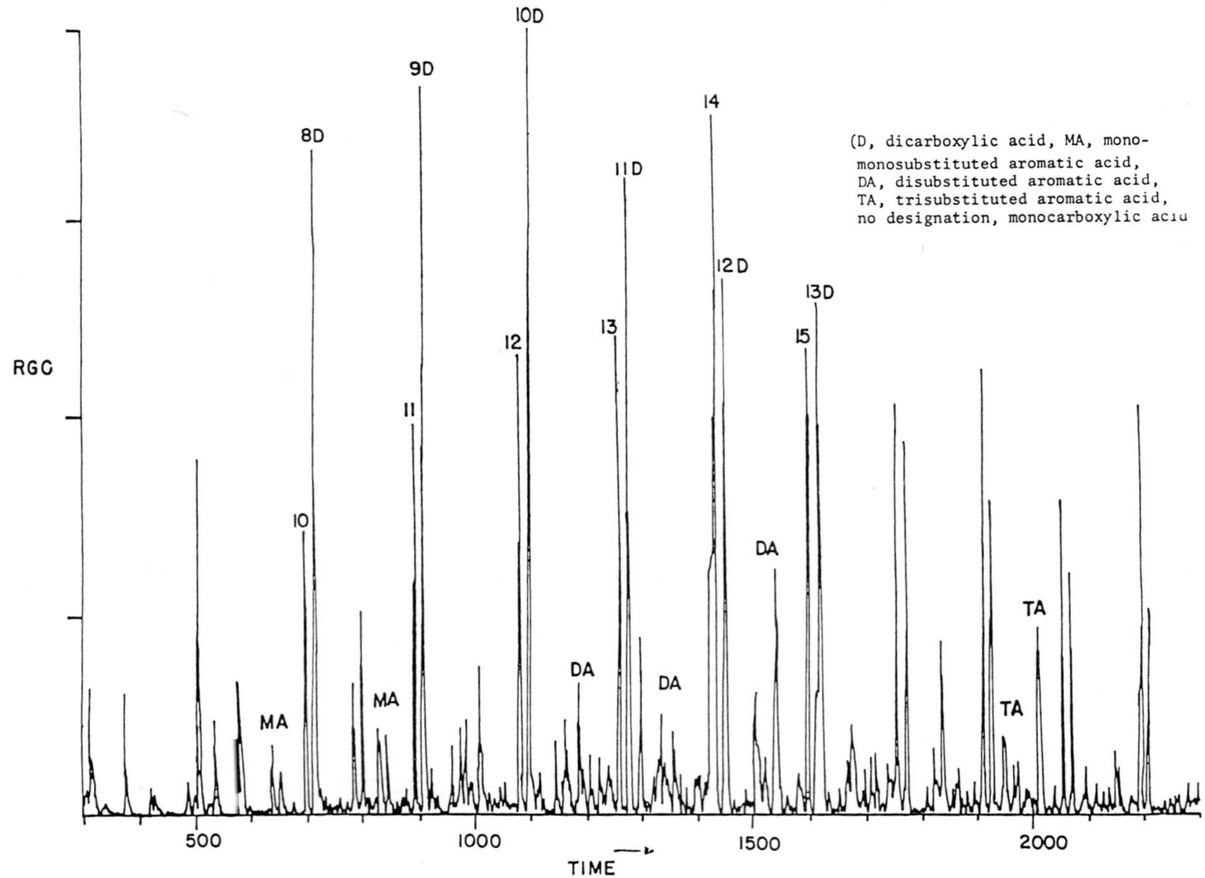

Figure 51-5. GC-MS of an oxidized Devonian shale kerogen

(D, dicarboxylic acid, MA, mono-
monosubstituted aromatic acid,
DA, disubstituted aromatic acid,
TA, trisubstituted aromatic acid,
no designation, monocarboxylic acid

$$2Fe^{3+} + FeS_2 \longrightarrow 3Fe^{2+} + 2S$$

and

$$S + H_2O + 4\ (O) \longrightarrow SO_4^{2-} + 2H^+$$

A variety of strains of bacteria was explored for the pyrite removal (Table 51-1). The course of the reaction in a 10-14 day period was closely followed for pH change as well as dissolved oxygen (DO) variaions. The consumption of oxygen to form sulfate can be monitored.

The rate of pyrite removal was followed by an infrared method by using the 420 cm^{-1} wavelength band in the far infrared region. The gradual decrease of this absorption frequency is illustrated in Fig. 51-6. Complete removal of pyrite could be found when the sample was upgraded to kerogen concentrate to remove more minerals and subjected to X-ray powder pattern studies (13).

TABLE 51-1. Microbiological leaching of Devonian shale

Culture	Medium	Pyrite Removal, %
Ferrobacillus ferrooxidans (ATC2370 via W.N. Leathen)	9K	50
F. ferrooxidans + Thiobacillus thiooxidans	9K + Wakesman	60
Thiobacillus ferrooxidans (TCC19859 via P. C. Trussell)	No. 64	70
J. V. Beck's mixed culture	No. 64	70

Microbial Oxidation

Fatty acids present no major problems, for oxidative degradation has been indicated by our work on the biodegradations of the oil shale retort water. Aromatic ring systems and heterocyclic compounds are generally known to be refractoric under microbial oxidations. For this reason a blend of six model compounds of pyridine, phenol, xylene, caproic acid, thiophene, and quinoline was subjected to Pseudomonas.

In this manner, Pseudomonas fluorescens, ATCC strain number 11172 and 12842, and Pseudomonas putida, ATCC strain number 12633, were used to degrade

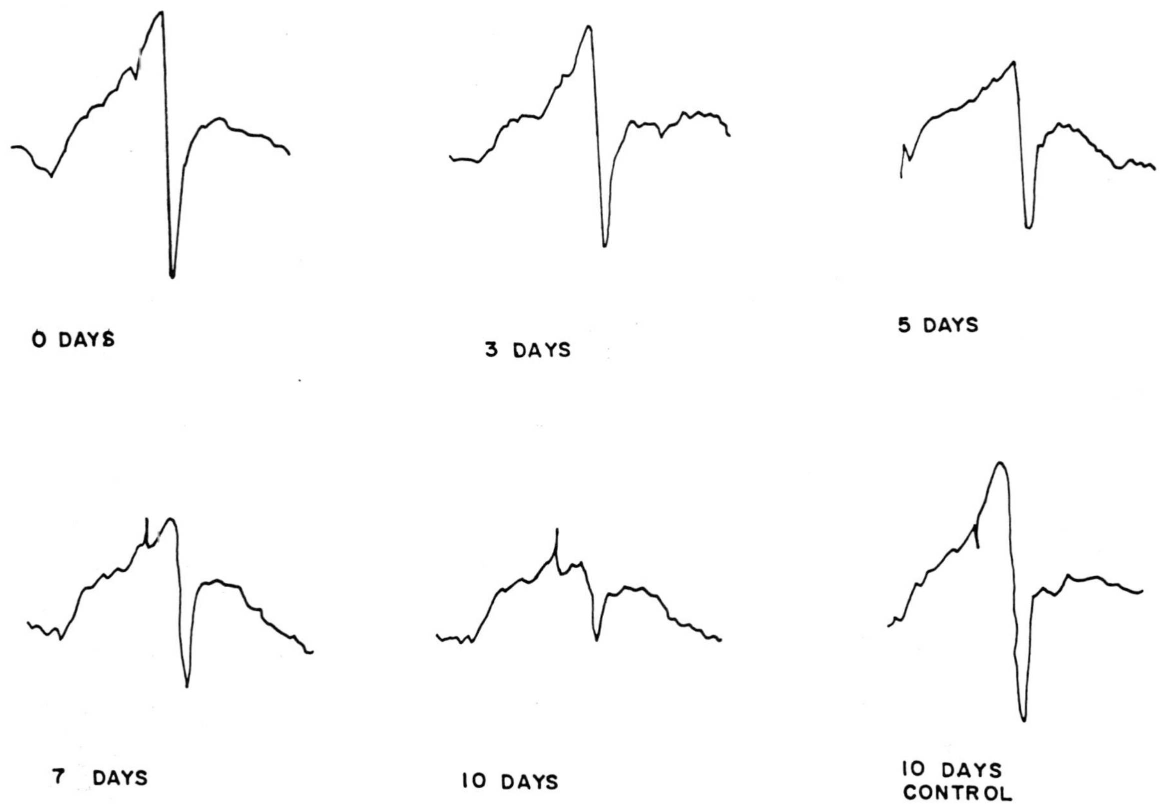

Figure 51-6. IR spectra - T. ferrooxidans leaching of 420 cm^{-1}

the model compounds. Experiments were followed daily by measuring both total organic carbon (TOC) as well as ultraviolet absorption. The latter technique was based on UV absorption at 300-350 nm (quinoline), at 250-270 nm (phenol-xylene), and at 200-250 nm (pyridine-thiophene). Nearly 77 percent of TOC was observed when a mixed strain was used for individual strains; the amount of reduction was 50 to 70 percent (Fig. 51-7). Similar results were ascertained by UV absorption (Table 51-2).

TABLE 51-2. Biodegradation of model components with Pseudomonas

		Percent of reduction*	
Strain	Quinoline	Phenol/Xylene	Thiophene/Pyridine
12633	55	64	41
11172	70	79	56
12842	75	78	62
mixed	82	88	74

* Based on percent reduction of UV peak height.

We also demonstrated the physical as well as the chemical in situ oxidation techniques for the Devonian shale. The chemical methods include ozone, hypochlorite, and hydrogen peroxides. Physical oxidations will entail the electrolytic oxidation process.

Preliminary Fermentation Studies

Anaerobic digestion sludge is still the best seed to promote the fermentation of simple carbon sources to produce methane. Digesters have been built and model compound studies have been made. Anaerobic sludge fermentation has been applied to glucose and succinic, malonic, and benzoic acids. The amounts of gas produced from these carbon sources are summarized in Table 51-3. The purpose of these experiments is to illustrate that the oxidized products of Devonian shale, which consisted of both benzoic acids and alkane dicarboxylic acids, can be suitable substrates for the fermentation discussed. It has been demonstrated that, even under very mild oxidation conditions, Devonian shale will produce lower homologs of n-alkanes (Fig. 51-8). These are the trapped

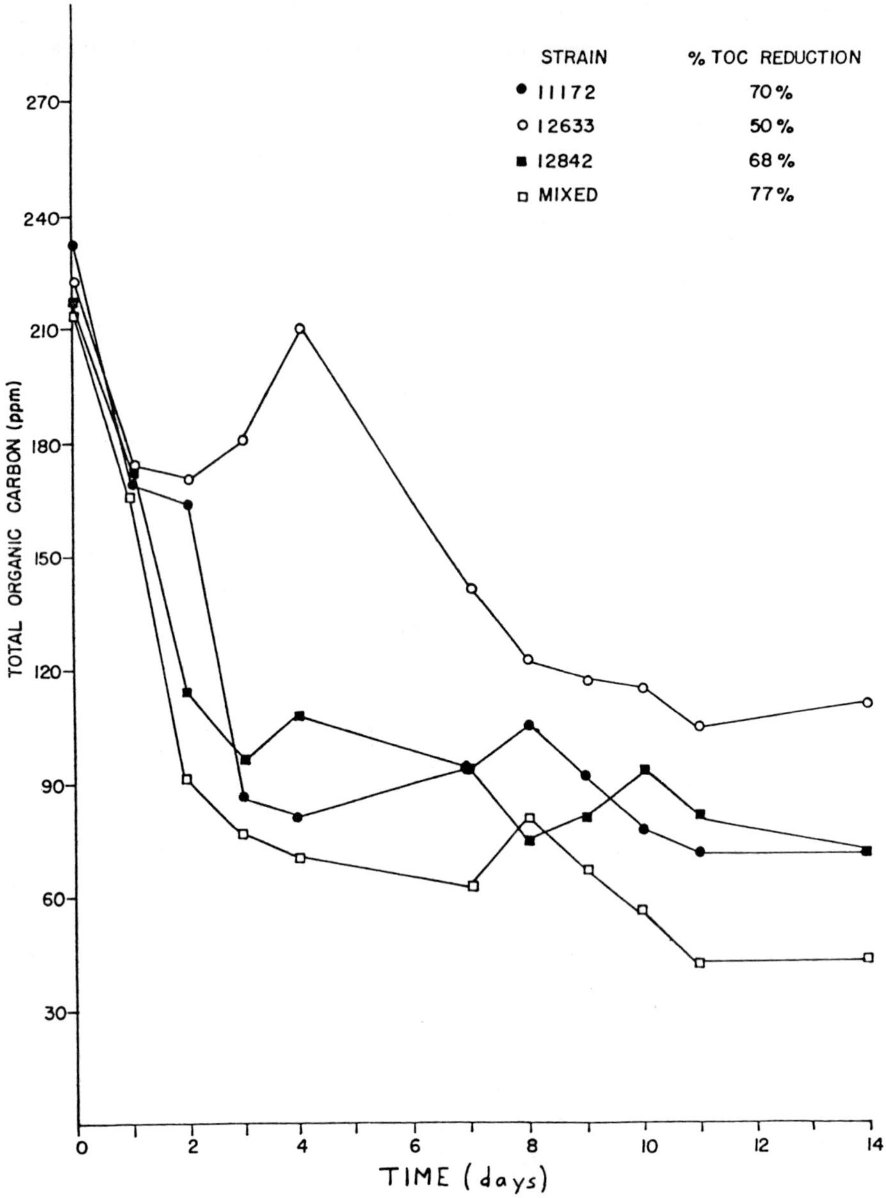

Figure 51-7. Pseudomonas growth on model compound

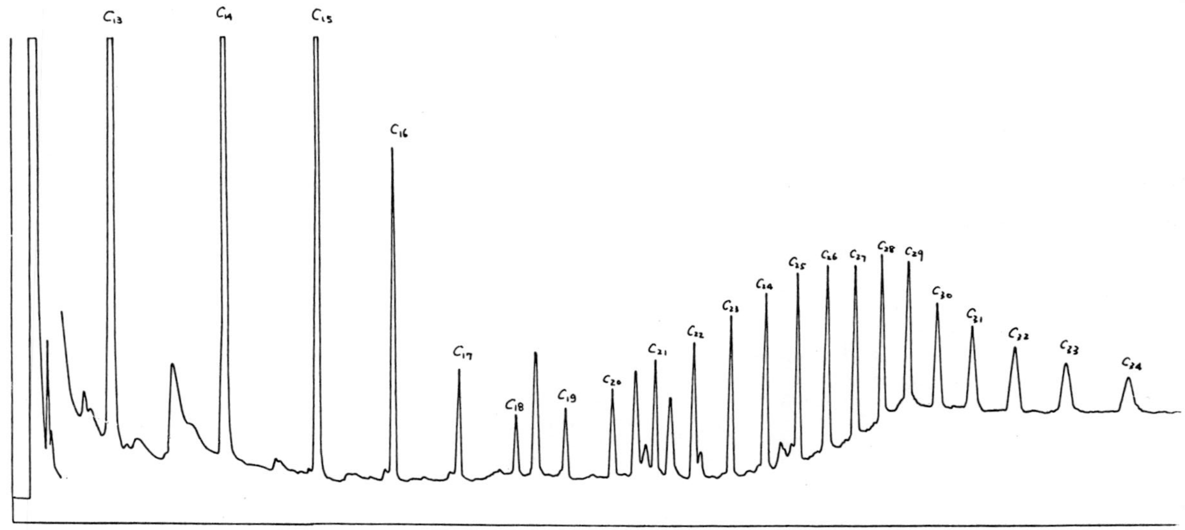

Figure 51-8. n-Alkane fraction isolated from Appalachian black shale after electro-oxidation (unpublished work of Wen and Yen). The first distribution centered at C_{13}, is from kerogen oxidation; the second distribution C_{20} - C_{34} is from bitumen

TABLE 51-3. Anaerobic sludge fermentation of 10 days

Substrate	Total gas (ml)	C input (M mole)	CH_4 output (M mole)	CH_4/CO_2
Sludge control	259	13.2	1.63	2.85
Malonic acid	203	21.9	1.32	3.04
Glucose	255	15.6	2.20	1.01
Succinic acid	280	20.0	2.11	3.46
Benzoic acid	237	13.3	2.77	2.71

species in the crosslink network. The network also can be ruptured and thus many carboxylic acids species can be formed. The recovery of gas then depends on the fermentation of the acids and the released hydrocarbons.

ACKNOWLEDGEMENT

Support of the work through U.S. DOE, EY-76-S-03-0113, EY-77-G-21-8104, EW-78-G-21-8349, and 79 EY-10017•000, U.S. EPA, R-806167-01-0 and GRI, 5010-362-0036 are appreciated.

REFERENCES

1. Byers, C. W., II, 1973, Biogenic structures of black shale paleoenvironments: Yale University, Ph.D. dissertation.
2. Young, D. K., and Yen, T. F., 1977, The nature of straight-chains aliphatic structure in Green River kerogen: Geochim. Cosmochim. Acta, v. 41, p. 1411-1417.
3. Yen, T. F., 1976, Structural aspects of organic components in oil shales; in T. F. Yen, and G. V. Chilingar, eds., Developments in petroleum science, v. 5, Oil Shale: Elsevier, p. 129-147.
4. Columbia Gas System, 1972, Operating, financial, and statistical data.
5. Hunt, J. M., 1979, Petroleum geochemistry and geology: Freeman, 595 p.
6. Yen, T. F., and others, 1978, Structural characterization of bitumen and kerogen from Devonian shale: TIC Gas-Shale Symposium, 1st, Proceedings, p. 572-588.
7. Yen, T. F., Wen, C. S., Kwan, J. T., and Chow, E., 1978, The nature of shale-oil derived asphaltene; in O. P. Strausz, ed., Oil sand and oil shale: Verlag Chemie, p. 309-324.
8. Yen, T. F., Tang, J. I. S., Young, D. K., and Chow, E., 1978, Second Eastern Gas Shale Symposium, Preprints, v. 1: METC, p. 330-338.
9. Young, D. K., Shih, S., and Yen, T. F., 1976, Mild oxidation of bioleached oil shale; in T. F. Yen, ed., Science and technology of oil shale: Ann Arbor Science, p. 65-72.
10. Kwan, J. T., and Yen, T. F., 1976, Aromaticity determination of coal, oil shale and their derivatives by X-ray diffraction: ACS, Div. Fuel Chemistry, Preprint, v. 21, no. 7, p. 67-76.
11. Young, D. K., and Yen, T. F., 1976, Oxidation of lignite into water-soluble organic acids: Energy Sources, v. 3, no. 1, p. 46-53.
12. Chan, M. K., and Yen, T. F., 1979, ACS, Div. Geochem., Paper 8, Washington, D.C.
13. Wen, C. S., Kwan, J. T., and Yen, T. F., 1976, Exploratory experiments on pyrite removal from oil shale by an electrolytic process: Fuel, v. 55, p. 75-78.

CHAPTER 52

GAS FROM SHALES RICH IN ORGANIC DETRITUS

Wallace de Witt, Jr.[1]

ABSTRACT

Dark brown to black gas-producing shales of Devonian age in the Appalachian basin contain as much as 20 percent by volume of a mixture of marine and terrestrial kerogens. The shales underlie more than 150,000 mi^2 of the basin and their volume is about 12,600 mi^3. They are generally well laminated, and individual gas-shale units have great lateral continuity. In contrast, the geochemistry of the shale shows marked vertical inhomogeneity. Because of their large organic component, the gas shales are excellent source beds and have the capability of generating about 3×10^{15} standard ft^3 (STF) of gas. Most of the gas is adsorbed on organic matter and clay in the shale, and less than 0.3 percent has been produced by existing extraction techniques. Gas is produced mainly from fractures in the gas shales, and can only be produced commercially in areas where extensive open natural fracture systems abound. To date, the gas shales have produced slightly more than 3×10^{12} standard ft^3 (TSCF) of gas, mainly in the Big Sandy gas field of eastern Kentucky. More than 90 percent of the Big Sandy's gas wells required stimulation by massive charges of explosive or, more recently, by hydraulic fracturing.

Resource estimates of recoverable gas in the Devonian gas shales of the Appalachian basin, which are based upon a complex of geologic, engineering, and economic factors, range from 8-285 TSCF. A detailed investigation now in progress should yield a better estimate of the recoverable resources.

THE DEVONIAN GAS SHALES

Gas-producing or potentially gas-producing black, dark brown, or dark gray shales of Devonian age, rich in organic detritus, and herein referred to as gas shales, underlie large parts of the Appalachian, Illinois, and Michigan basins of the eastern and central United States (Duncan and Swanson, 1965, p. 11). Similar gas shales of Devonian age are found west of the Mississippi

---

[1]U.S. Geological Survey, Reston, Virginia 22092.

River, but their extent and thickness are not as well known as their eastern equivalents. At present studies to characterize the gas shales, to determine their hydrocarbon potential, and to devise methods to enhance recovery of the contained hydrocarbons are being conducted by the U.S. Department of Energy and its cooperators in all three eastern basins.

My report deals with the Appalachian basin, the area of the Appalachian Mountains from south-central New York south to central Alabama, because the U.S. Geological Survey has the responsibility to characterize the Devonian gas shales and associated rocks in this basin as part of a study being made for the U.S. Department of Energy. Studies now in progress in the Illinois and Michigan basins by other geological groups show that the dark Devonian gas shales in the two basins have properties similar to the gas shales of the Appalachian basin. Many of the properties mentioned in this report are characteristic of gas shales of different geologic ages and in many different geographic localities.

The dark bituminous gas shales of Devonian age are composed of 30 to 50 percent clay minerals, 15 to 25 percent silt-sized detrital quartz particles, and 4 to 20 percent kerogen--degraded organic detritus. Illite is the most abundant clay mineral present; kaolinite occurs locally; and montmorillonite is present sparingly in a few layers of volcanic ash. A percent or two of silt-sized feldspar grains and carbonate minerals are commonly present. Ubiquitous pyrite locally makes up as much as 10 percent of the gas shale by weight. It forms small euhedral crystals, framboids, irregular masses, or discrete nodules. Limestone and dolomite also occur in nodules and concretions, ranging in diameter from 1 in (2.5 cm) to more than 8 ft (2.4 m).

The gas shales are characteristically well laminated, having dark layers of clay and organic detritus intercalated with lighter layers made up largely of silt-sized detrital quartz grains and pyrite crystals. The laminae are generally only a few millimeters thick but have great lateral extent. They are not readily seen in unweathered rock. Weathering accentuates the laminae. The shaly character of the gas shales is apparent in outcrop, where the shale disintegrates into irregular discs and sharp-edged chips about 0.2-1 in (5-25 mm) thick. Commonly the chips are coated by a film of dark reddish-brown iron oxide, which was produced by the weathering and oxidation of pyrite in the shale.

The dark gas shales are more resistant to erosion than the associated gray shale and mudrock of the Devonian shale sequence. They tend to crop out in cliffs and steep-walled canyons at many places in the western part of the Appalachian basin. Commonly, the cliffed outcrops show a ribbed and furrowed surface parallel to bedding, resulting from subtle differences in resistance to erosion by the black and dark brown components in the gas shales. The black shale is more resistant and weathers in relief between shallow reentrants cut in the softer dark brown shales. A rib and furrow couplet, which may range in thickness from 1 to 4 ft (0.3-1.2 m), appears to have been deposited as a depositional cycle. Commonly, many cycles can be observed in a thick sequence of gas shales. The ribbing of the shales clearly illustrates the great lateral homogeneity of the Devonian gas-shale sequence.

Extent and Volume

The Devonian gas shales underlie more than 150,000 mi^2 (389,000 km^2) in the Appalachian basin, including about 125,000 mi^2 (324,000 km^2) in the Appalachian Plateaus and about 25,000 mi^2 (65,000 km^2) to the east in the western part of the folded and faulted Valley and Ridge province. The total volume of the gas shales is more than 12,600 mi^3 (52,520 km^3), and they weigh about 1.5×10^{14} tons (1.37×10^{14} metric tons) (de Witt, Perry, and Wallace, 1975). The gas shales range in thickness from less than an inch (25 mm) in southern Tennessee and Alabama to more than 1,500 ft (457 m) in north-central Pennsylvania. They were deposited in a shallow epicontinental sea west of the rapidly expanding Catskill Delta. Consequently, the gas-shale sequence is thinnest and contains the greatest amount of organic detritus in the western part of the Appalachian basin adjacent to the Cincinnati arch. The sequence is thickest close to the delta where the shales interfinger with sequences of distal turbidites from the Catskill deltaic clastic rocks. From central Tennessee north to southern Ohio along the east side of the Cincinnati arch, the Devonian shale sequence--the rocks below the Lower Mississippian Berea Sandstone and above the Middle Devonian Columbus Limestone or older beds--ranges in thickness from 25-250 ft (7.6-76 m) and is composed of more than 80 percent dark gray, dark brown, or black shale. In contrast in eastern Pennsylvania, where the gas shales intertongue with coarser-grained rocks of the Catskill deltaic sequence, the stratigraphic equivalents of the gas shale sequence of the western part of the basin are more than 7,500 ft (2,285 m) thick, and less than 1 percent of these rocks are black shale.

Commonly, individual beds of black shale are of great lateral extent, which aids greatly in establishing a regional stratigraphic framework and in delineating with a high degree of confidence areas that contain thick sequences of black shale. In the Devonian gas-shale sequence of western and central New York, individual beds of black shale less than 6 in (0.15 m) thick have been traced fron Canandaigua Lake westward 85 mi (137 km) to Lake Erie (de Witt and Colton, 1978). Our stratigraphers have been able to trace the black Pipe Creek Shale Member of the Java Formation by its characteristic gamma-ray and compensated density-log signatures in the subsurface from the member's type section in southwestern New York southward across western Pennslyvania, western West Virginia into the Greendale syncline in the Valley and Ridge of southwestern Virginia, a distance of more than 500 mi (805 km). Throughout this distance, the Pipe Creek black shale is rarely more than 25 ft (7.6 m) thick. These data illustrate the great lateral extent of individual beds, members, or formations within the Devonian gas-shale sequence.

Gas Shale as Source Beds

The most important property related to gas productivity of the Devonian shales is the organic content of the rock. Geochemical and paleobotanical analyses show that the gas shales contain kerogens (organic detritus) of both marine and terrestrial origin. Proportions of the two types of kerogen are not constant throughout the basin. Terrestrial kerogens consist mainly of macerated plant matter derived largely from plants growing on the Catskill delta complex or in highlands east of the delta. Streams swept plant debris into the Devonian sea at the edge of the delta. Marine currents, especially turbidity flows, entrained the water-logged detritus and spread it widely across the sea floor under the quiet waters in the western part of the basin. Large quantities of marine algae proliferated in the Devonian sea. Accumulations of marine kerogens from the algae are characterized by the presence of the dark reddish-brown discoidal spore, Tasmanites. Locally the spores occur in sufficient abundance to impart a deep reddish-brown color to the gas shale. Well drillers examining cable-tool well cuttings saw a resemblance of the chips to ground cinnamon or to coffee grounds; hence the derivation of the local driller's names "cinnamon" or "coffee" shales for the dark brown and black shales in the gas-shale sequence. In general, decomposition of marine kerogens during thermal maturation produces petroleum, whereas decomposition of terrestrial kerogens yields mainly methane during thermal maturation. Kerogens in the shallow western part of the Appalachian basin are thermally immature and yield mostly biogenic or very early thermo-maturated gas. The gases are isotopically light and the ratio of C^{13} to C^{12}, $-\delta^{13}C$ is between -90 and -55 per mil (Claypool, Threlkeld, and Bostick, 1978, p. 56). Gas in this range of isotopic values indicates the early pre-oil stage of thermal maturation. In the central part of the basin, from southeastern Kentucky to western New York, a broad area of gas shales yields gas having a considerable component of heavier hydrocarbon molecules in addition to methane. The isotopic value, δC^{13}, of these wet gases is in the range of -55 to -50 per mil and is characteristic of thermal-maturation temperatures that break down marine kerogens to yield heavier hydrocarbons as well as methane. East of the area of maximum wet-gas generation near the center of the basin is a broad area in which the gas shales yield only dry methane having carbon isotope ratios, $\delta 13^{C}$, in the range of -50 to -35 per mil. These dry gases were generated largely from terrestrial kerogens during the late-mature post-oil generation stage of thermal maturation. Thus, within the Appalachian basin, the Devonian gas shales show a full spectrum of gas generation by thermal maturation from low-temperature (less than 60°C) biogenic gas to high-temperature (more than 200°C) late thermogenic gas. These data clearly show that the Devonian gas shales are excellent source beds.

Claypool (Claypool, Threlkeld, and Bostick, 1978, p. 59) applied a conversion factor for thermal maturation to the original gas-generating capacity of the gas shales on the basis of his organic-analysis data and concluded that a very preliminary estimate of the amount of gas generated in the gas shales of the Appalachian basin would be about 3.1×10^{15} standard ft^3 (SCF). Unfortunately, for the gas industry, most of the gas generated in the gas shales is adsorbed on the organic matter, in the shales (Schettler, 1978, p. 371), and relatively little gas can be extracted by existing methods of drilling and well stimulation. Production of 3 trillion ft^3 (TCF) in the Big Sandy gas field of eastern Kentucky, the principal producing area for Devonian shale gas in the basin plus an estimate of about 1 TCF for all other shale gas produced in the basin is equivalent to approximately 0.3 percent of Claypool's estimated 1.5×10^{15} SCF gas generated in the shale-gas sequence.

Reservoir Rocks

Reservoir rocks and impermeable strata in addition to source beds are needed for the accumulation of commercially exploitable amounts of hydrocarbons. Although gas has been produced for domestic and semicommercial use from the Devonian gas shales along the south shore of Lake Erie since 1820 and gas has been produced commercially from the gas shales of Kentucky's Big Sandy gas field since the early 1920's, few data are available concerning reservoirs in the gas shales. After much debate, during a period of about 40 years, geologists generally agreed that open fractures and joints are the main components of the reservoir system in the Big Sandy gas field of southeastern Kentucky (Hunter and Young, 1953, p. 285). Locally, as in the Boyd County gas field at Ashland, Ky., stringers and thin beds of siltstone and very fine-grained sandstone are secondary reservoirs in the gas shale (Browning, 1932). In the Rome trough segment of the Eastern Interior aulacogen, which extends from central Kentucky to north-central Pennsylvania and includes most of the Big Sandy gas field (Harris, 1978, p. 72), repeated movement along the trough during the late Paleozoic fractured the Devonian gas shales and produced a complex network of fractures that acted as gas reservoirs as well as channels for gas moving from the shale matrix into the well bore.

Doscher in the Lewin Report (Lewin and Associates, 1978) emphasized that commercially exploitable volumes of gas could not be extracted from the Devonian shales unless an extensive natural fracture porosity existed. His observation appears to be correct, and one task of the exploration geologist is to locate and delineate areas of extensive fracture systems in the gas-shale sequence. Fortunately, large areas of the Appalachian basin appear to contain such systems. These areas include:

1. The Rome trough of the Eastern Interior aulacogen
2. The area of thin-skinned faulting in the Appalachian Plateaus
3. The area of salt tectonics in the Appalachian Plateaus
4. The area of glacial rebound on the south shore of Lake Erie.

The Rome trough contains widespread zones of fracture porosity induced by several episodes of warping during the late Paleozoic. The major producing fields from the Devonian shale-gas sequence lie mainly within the area of the Rome trough.

Structure of the eastern half of the Appalachian Plateaus from southern New York to Alabama is controlled by thin-skinned tectonics. Thrust faults ramping from Cambrian or Ordovician decollement levels into the Upper Silurian or Lower Devonian rocks duplicate beds and form anticlinal folds containing extensively fractured Devonian gas shales. Splay folds from decollements in the Silurian sequence also produce zones of fractures in the overlying gas shales. Commonly, good shows of gas are encountered in the gas-shale sequence in wells drilled to test deeper buried rocks in this area.

In the south-central New York and adjacent Pennsylvania complex thrust faulting in the Upper Silurian salt beds is superimposed upon the deeper thin-skinned thrusting. Anticlines in the Devonian shales above duplicated and deformed salt beds commonly contain sequences of fractured gas shales. The old low-pressure gas field in the black Devonian Marcellus shale at Watkins at the south end of Seneca Lake was probably related to local faulting and fracturing produced by movement in the underlying salt beds.

Although the data are not conclusive, the area of low-pressure gas in the Devonian shales along the south shore of Lake Erie was loaded by as much as a mile (1.6 km) of ice during the Pleistocene and is now in a state of rebound and gentle uplift. Movement has opened the regional joint system to a depth of 1,000 to 1,500 ft (305-457 m) and permitted movement of gas from the gas shales into adjacent sequences of fine-grained sandstone and siltstone. Locally, the joints are open to the surface, and gas escapes in seeps in the shallow water of Lake Erie or in its tributary streams.

In many places in the central part of the Appalachian basin from northwest Pennsylvania south to western Virginia, sequences of turbidite siltstone and fine-grained sandstone are intercalated with or are adjacent to the gas shales. Where naturally fractured, the sequences of brittle reservoir rocks and shaly source beds commonly contain concentrations of gas.

Production and Well Stimulation

Natural gas from the Devonian gas shales bubbling up through the shallows of Canadaway Creek at Fredonia in southwest New York led to the drilling of the first wells specifically for gas in the United States and to the founding of the natural gas industry. The first well was drilled in 1820 using the crude tools invented by the Appalachian basin salt-well drillers (Ley, 1935, p. 950). The well greatly enhanced the volume of gas obtained from the seep in Canadaway Creek, and the gas was used locally for both fuel and illumination. The well, which was eventually deepened to 1,000 ft (305 m), yielded

3 million SCF per year at few pounds above atmospheric pressure. Apparently the reservoir leaked at other places in the creek, and the well produced from underpressured fracture porosity. The well was characteristically long lived, as most shale-gas wells are, and produced about 195 million SCF in its 65-year life. Its annual yield was about 3 million SCF when the well was plugged and abandoned in 1885 (Ley, 1935, p. 951).

Drilling for low-pressure gas in Devonian shale spread west along the south shore of Lake Erie to Erie, Pa., before 1860 and to Cleveland, Ohio, about 1880. Most shale-gas wells were shallow, commonly less than 1,500 ft (457 m) deep; were not stimulated with explosives; and produced small to moderate volumes of gas in the range of 5 to 100 thousand SCF per day at low pressures. Many wells were abandoned for convenience rather than because the supply of gas was exhausted. The gas was used locally for fuel and illumination because the extensive network of the modern gas-transmission system did not then exist.

In the early 1920's, about 100 years after the drilling of the first Devonian shale-gas well at Fredonia, N.Y., gas was produced commercially in the Big Sandy gas field of southeastern Kentucky. Wells drilled there to test the oil potential of the older Devonian limestones occasionally released as much as 1 million SCF of gas per day from the dark Devonian shales. Encouraged by the unexpected flow of gas but puzzled because the gas did not occur in the same stratigraphic position in adjacent wells, gas companies began an extensive drilling campaign for shale gas in southeastern Kentucky. Well drillers soon made the astounding discovery that wells that were dry holes when drilled--wells that did not contain detectable amounts of gas--could be induced to yield gas in commercial quantities when stimulated by large amounts of explosive. Field procedures in the Big Sandy were to drill a 7- or 8-in (18-20 cm) hole through the 600 to 1,000 ft (180-305 m) thick Devonian shale sequence; fill the hole with 3 to 5 tons of gelatinated nitroglycerine; and detonate the explosive by means of a time bomb below a gravel or water tamp. The massive shots commonly increased well yields about 10 to 1,000 thousand SCF per day but produced prolonged clean-cut jobs because of extensive rubble formed by fracturing and crushing the shaly rock around the well bore. According to Ray (1976, p. 103), before the introduction of hydrafrac well stimulation more than 90 percent of the wells in the Big Sandy area were shot with explosives and after shooting, about 11 percent remained nonproductive.

Characteristically the Big Sandy shale-gas wells were long lived and produced 35 to 70 years before reaching the limit of economic productivity. An average well produced about 400 to 450 million SCF before abandonment (Avilla, 1976, p. 83; Ray, 1976, p. 103), although individual wells have produced as much as 7,000 million SCF during a 35-year period.

Production records and shot records of wells that did not yield gas without stimulation indicate that shooting opened pathways from the well bore to the extensive natural fracture system. The size, extent, and area of the fracture system largely governed the well's productivity. Wells in which shooting did not break into the fracture system remained dry holes.

When gamma-ray well logs and well stimulation by hydraulic fracturing were introduced in the Appalachian basin in the 1960's, several major gas companies began experimenting with hydrafrac stimulation rather than using the time-honored massive-shooting technique. Gamma-ray, density, and temperature logs permitted closer delineation of gas-productive zones in the shale sequence and the selection of zones for hydraulic stimulation. Treatments using water as the fracture fluid were successful, resulting in little apparent swelling of the shale. The yield per well was increased as much as 70 percent when compared with similar wells stimulated by shooting (Ray, 1976, p. 107-110).

Early in their experiments, one company attempted a fracture treatment using diesel oil as the frac fluid. The experiment was an unqualified disaster. The diesel oil combines with some of the organic matter in the black shale to form a sticky viscid emulsion that blocks the fractures in the shale and positively refuses to leave the fractures and reenter the well bore. The experiment was quickly terminated, and to my knowledge, nobody else has attempted a second "diesel" frac.

Encouraged by the increased yield of Devonian shale-gas wells by hydraulic fracture stimulation, the U.S. Department of Energy's Morgantown Energy Technology Center is conducting a series of investigations of the relative merits of different fracturing methods using water, foam, liquid carbon dioxide, or nitrogen as the fracturing fluid. Because water has an adverse effect on some components of the gas shales that may induce blockage in the fractures adjacent to the well bore during stimulation, the Department of Energy's studies are intended to determine which of the various frac treatments will yield optimum recovery rates for Devonian shale gas. Also under study is the effectiveness of Magna Frac, the use of much larger volumes of fracturing fluid and propping agent than is employed in a standard frac treatment. Other stimulation techniques being studied include the use of liquid explosive as the fracturing fluid and the drilling of deliberately slanted or inclined holes designed to intersect a greater number of fractures in the shale sequence.

The purpose of the Department of Energy's studies (Wise, 1979, p. 7) is to encourage industrial exploitation of the Devonian gas shales by demonstrating techniques to:

1. Increase the daily yield per well from 100 to 300 thousand SCF per day.
2. Increase the ultimate yield of a Devonian shale-gas well by 200 to 300 million SCF in a 30-year period.

Resource

Unquestionably the Devonian gas shales are excellent source beds throughout much of their extent in the Appalachian basin. A preliminary estimate of the amount of gas generated by thermal maturation of the Devonian gas shales in the basin is about 3.1×10^{15} SCF. If the average gas content per ft^3 of gas shale is 0.42 SCF (Wise, 1979, p. 6) and the volume of black shale is 12,600 mi^3 (52,520 km^3) the volume of free gas in the Devonian gas shales would be about 7.8×10^{14} SCF. Smith (1978, p. 78) gives estimates of gas content in the shales of the basin ranging from 2×10^{14} to 1×10^{15} SCF using various combinations of basic resource data. All these estimates indicate that a very large volume of gas is present in the source beds. Unfortunately drilling and production data indicate that very little contained shale gas can be recovered economically by existing extraction technology.

Several geologic and engineering studies of the gas shales have produced a wide variety of estimates of the volume of gas producible from the gas shales in the Appalachian basin. The wide variety of estimates indicates the degree of uncertainty surrounding many of the factors used in the evaluations. The estimates include 8 trillion SCF (Lewin and Associates, 1970), 15 to 35 trillion SCF (Office of Technology Assessment, 1977, p. 4), 149 trillion SCF (U.S. Geological Survey, 1976), and 285 trillion SCF (Orlofsky, 1975). Many geologic, engineering, and economic factors were considered and evaluated to derive these estimates. The interplay of these factors and the weight assigned to each resulted in the wide range of estimates of producible gas. A part of the present Department of Energy's study of the Eastern gas shales will be to provide an assessment of the gas potential of the Devonian gas shales.

SUMMARY

The black and brown gas shales of Devonian age of the Appalachian basin are unquestionably good source beds for gas. They have the capability to have generated several thousand trillion cubic feet of gas, much of which is adsorbed on the organic matter in the shale and cannot be recovered economically by existing well-completion and stimulation techniques unless an extensive natural fracture system is present in the shale beds. Because of low porosity and extremely low permeability of the gas shales, commonly measured in microdarcys, wells may produce erratically even when natural fractures are locally abundant.

Because the daily yield of shale-gas wells is relatively low, ranging from 10 to 500 thousand SCF, shale-gas fields will require many thousands of producing wells if this nonconventional source of gas is to have major effect on the Nation's fuel supply. However, because shale-gas wells commonly have a long productive span of 30 to 50 years, an extended drilling campaign will certainly add to the proven reserves in the Appalachian basin in an area close to major markets.

The Devonian gas shales, in common with other shales rich in organic detritus, are a slow-yield, long-lived fuel source which ought not to be ignored because of the large volume of gas present in these shales in the Appalachian, Illinois, and Michigan basins. Proximity to markets, long well life, and increasing demand for gas ought to make the Devonian gas shales good exploration targets for many years to come. At present, the challenge is to find environmentally acceptable methods for extracting shale gas relatively inexpensively.

REFERENCES

Avilla, John, 1976, Devonian shale as a source of gas, in Natural gas from unconventional geologic sources: National Academy of Sciences, Washington, D.C., p. 73-85.

Browning, I. B., 1932, Relation of structure to shale gas accumulation: Appalachian Geological Society, Devonian Shales Symposium, Charleston, W. Va., Mimeograph.

Claypool, G. E., Threlkeld, C. N., and Bostick, N. H., 1978, Natural gas occurrence related to regional thermal rank of organic matter (maturity) in Devonian Rocks of the Appalachian Basin; Second Eastern Gas Shales Symposium: U.S. Department of Energy, Morgantown Energy Technology Center, Morgantown, W. Va., Preprints, v. 1, p. 54-65.

deWitt, Wallace, Jr. and Colton, G. W., 1978, Physical stratigraphy of the Genesee Formation (Devonian) in western and central New York: U.S. Geol. Survey Prof. Paper, 1032-A, 22 p., maps.

deWitt, Wallace, Jr., Perry, W. J., Jr., and Wallace, L. G., 1975, Oil and gas data from Devonian and Silurian Rocks in the Appalachian Basin: U.S. Geol. Survey Misc. Inv. Ser. Map I-917-B. 4 sheets.

Duncan, D. C. and Swanson, V. E., 1965, Organic-rich shale of the United States and world land areas: U.S. Geol. Survey Circ. 523, 30 p.

Harris, L. D., 1978, The Eastern Interior aulacogen and its relation to Devonian shale-gas production; Second Eastern Gas Shale Symposium, U.S. Department of Energy, Morgantown Energy Technology Center, Morgantown, W. Va., Preprints, v. II, EG5-60, p. 55-72.

Hunter, C. D. and Young, D. M., 1953, Relationship of natural gas occurrence and production in eastern Kentucky (Big Sandy gas field) to joints and fractures in Devonian bituminous shale, in Symposium on fractured reservoirs: Am. Assoc. Petroleum Geologists Bull., v. 37, no. 2, p. 292-299.

Lewin and Associates, 1978, Enhanced recovery of oil and gas: Lewin and Associates, Inc., Washington, D.C., Feb. 1978, Report prepared for U.S. Dept. of Energy.

Ley, H. A., 1935, Geology of Natural Gas [edited with forward by Henry A. Ley]: Am. Assoc. Petroleum Geologists, Tulsa, Ok., 1,227 p.

Office of Technology Assessment, 1977, Status report on the gas potential from Devonian Shales of the Appalachian Basin: Office of Technology Assessment, U.S. Congress, Nov. 1977, 87 p.

Orlofsky, Seymore, 1975, The Eastern Gas Shales: Statement before the Conservation, Energy and Natural Resources Subcommittee of the Committee on Government Operation, Washington, D.C., May 5, 1975.

Ray, E. O., 1976, Devonian Shale Development on Eastern Kentucky, in Natural gas from unconventional geologic sources: National Academy of Sciences, Washington, D.C., p. 100-111.

Schettler, P. D., 1978, The implication of specific degasibility for models of gas production from shale; Second Eastern Gas Shale Symposium: U.S. Department of Energy, Morgantown Energy Technology Center, Morgantown, W. Va., Preprints, v. 1, EGSP-96, p. 370-376.

Smith, E. C., 1978, A practical approach to evaluating shale hydrocarbon potential; Second Eastern Gas Shale Symposium: U.S. Department of Energy, Morgantown Energy Technology Center, Morgantown, W. Va., Preprints v. 22, EGS-70, p. 73-87.

U.S. Geological Survey, 1976, New look at old gas deposits in Appalachia: U.S. Geol. Survey, News Release, 3/31/76, 3 p.

Wise, R. L., 1979, Status Report of the Eastern Gas Shales Project; Third Eastern Gas Shales Symposium: U.S. Department of Energy, Morgantown Energy Technology Center, Morgantown, W. Va., Proceedings, p. 3-13.

CHAPTER 53

GAS HYDRATES

Richard D. McIver[1]

INTRODUCTION

The report of the discovery of natural gas in the "frozen state" in Siberia (Makogen, 1965a), triggered widespread petroleum-industry interest in naturally occurring gas hydrates. Later reports (Anonymous, 1970a, b; Chersky and Makogen, 1970) of more than 535 trillion ft^3 of gas in this form in the Soviet Union stimulated interest even further.

Petroleum company explorationists and economists soon found that hydrates of gases had been studied in the laboratory for years before naturally occurring ones were recognized.

Although the first hydrate, that of chlorine, was observed in 1810 (Davy, 1811), the true nature of hydrates remained a mystery for more than a hundred years. In the meantime, hydrates of a large number of gases and numerous halogenated hydrocarbons were prepared in the laboratory. It was established that the equilibrium between water plus gas and the hydrate is univariant, i.e., the equilibrium pressure changes with temperature and is a fixed value at each temperature for a pure gas or for a gas of constant composition. It was also found that these compounds are well behaved, i.e., their formation and decomposition are predictable.

For many years, the ratio of water to gas in hydrates was a matter of debate. Chemical methods of measuring the proportions were somewhat inaccurate because of the difficulties of collecting pure hydrates without included unhydrated water in the crystals. Moreover, personal prejudice, that the value of n in the formula, $gas \cdot nH_2O$, must be a small whole number, led many investigators to attribute fractional results to errors in analyses. More recently, however, the nonstoichiometry of the hydrates has been demonstrated.

The study of hydrates took a practical turn in the petroleum industry after Hammerschmidt (1934) noted that plugging of natural gas pipelines was often not due to freezing of associated water at 32°F, but was due to the

[1]GeoChem Research Incorporated, Houston, Texas.

formation of hydrates of methane, ethane, propane, and isobutane at temperatures up to 60°F. The Bureau of Mines subsequently took the lead in studying many aspects of hydrocarbon hydrates. Their work culminated in the report by Deaton and Frost (1946).

Because of the unusual properties of hydrates, several interesting applications were proposed and tested extensively by industry; one to store natural gas as the hydrate, reviewed by Parent (1948); one to transport gas as hydrate rather than as LNG; and still another to use hydrates as desalination agents, i.e., hydrate the water molecules from sea water, then separate and wash the hydrate crystals (which contained only the Freon gas and pure water), and finally decompose the hydrate and release the pure water (Donatz, 1959; Koppers, 1964; Williams, 1964). These processes, though technically feasible, were found to be more expensive than competing alternate technology.

Another characteristic of these clathrates is that the host lattice would be thermodynamically unstable except for the presence of the hydrated molecules. The forces holding the hydrated molecules in place must be similar to the intermolecular forces that hold molecules in solution in water. Therefore, gas-hydrate clathrates may be considered to be solid solutions of gas molecules in a metastable lattice. The number of gas molecules that can be accommodated in the lattice will vary between some minimum (below which the hydrate would not be stabilized) and some maximum (equivalent to saturation).

While it is difficult to predict which gas molecules will form stable hydrates, it can be generalized that all molecules that fit into the clathrate cavities have the potential to stabilize the lattice unless they are quite small and inert (e.g., He or Ne) or unless they strongly interact with the water (e.g., HCl, HBr, NH_3, etc.). The latter gases form stoichiometric species which are hydrates, but not crystalline hydrates. Therefore, for the sake of this discussion, when hydrates are mentioned, the reference will be to clathrate hydrates.

X-ray studies show that gas hydrates commonly crystallize into two cubic structures, types I and II, in which the hydrated (or dissolved) gas molecules are located in voids surrounded by a latticework of water molecules held together by hydrogen bonds. In both types of hydrates the water molecules are tetrahedrally coordinated as in ordinary ice. Only type I is likely to occur in nature because it is formed from the most commonly found natural gases--methane, ethane, H_2S, and CO_2, with or without smaller amounts of propane and isobutane. Type II is formed when the gas is made up predominantly of propane or isobutane. However, these larger molecule gases are seldom, if ever, found alone in nature. Evidence is that still larger hydrocarbon molecules do not form either type I or II hydrates.

The crystal structure of the type I hydrate was elucidated by Pauling and Marsh (1952), who found that the crystal was cubic. A unit cell was found to be made up of 46 water molecules and 8 cavities of two types. Smaller cages, positioned at the corners and center of the cube, are made up of pentagonal dodecahedra each with 20 water molecules. The second type of cageformers, sharing some of the same water molecules as the dodecahedra, are tetrakaidecahedra each made up of 24 water molecules and having 2 hexagonal faces (opposite each other) and 12 pentagonal ones. These latter cages may be thought of as occupying long columns cutting each of the six faces of the cube. The cavities remaining between these columns are the dodecahedra.

The molecules of natural gas able to stabilize type I hydrates are relatively small and generally (for CH_4, C_2H_6, H_2S, and CO_2, but not C_3H_8) occupy both sizes of cavities (5.1 and 5.8 Angstroms in diameter, respectively). Propane and isobutane barely fit in the larger cavities. If every cavity in a unit structure were to be occupied by a gas molecule, the composition would be 46/8, or 5.75, water molecules for each molecule of gas. But a certain fraction can remain unoccupied when an undersaturated hydrate is in equilibrium with water (Pauling and Marsh, 1952). In fact, with a large excess of water, virtually all the free gas will enter the hydrate. The number of water molecules would probably not approach the theoretical minimum (5.75) but could be as high as 20 or 22 for the undersaturated type I structure (Trebin et al., 1966). With an excess of gas, virtually all the water would enter the hydrate, and the number would approach the minimum.

The type II hydrate contains more dodecahedra per unit cell than type I, but the larger cavities are hexakaidecahedra (16 sides), which are larger than the tetrakaidecahedra (14 sides) of type I. Type II is stabilized by propane and isobutane but can contain smaller amounts of CH_4, C_2H_6, H_2S, or CO_2. The smaller molecules can occupy either size cavity. However, if there is a preponderance of the smaller molecules, the type I hydrate will be more stable and will form preferentially.

Chemical Control of Hydrate Occurrence

Figure 53-1 illustrates the pressures and temperatures at which methane hydrates are stable, based on data presented by Katz et al. (1959). The large area on the lower right defines the pressure-temperature combinations where methane and water occur as gas and liquid, respectively. To the left of the water freezing-point line, water exists as solid ice and methane as gas. The curving diagonal boundary defines the conditions above (for pressure) and to the left of which (for temperature) methane-water mixtures form the solid hydrates. Under any combination of conditions defined by this area of the diagram, the two components combine as the hydrate until equilibrium is reached. With a large excess of water, virtually all the free gas enters the hydrate. With an excess of gas, virtually all the water forms hydrate. To the left of the water freezing-point line and above the diagonal, the water and gas molecules will also combine to form the hydrate, with the excess occurring as ice or gas. There is no known upper temperature limit for methane hydrate formation, but above 80°F, the pressure requirements for stability increase so rapidly and to such high values (>9,000 psi) that this region can be ignored in a discussion of naturally occurring hydrates.

The data in Fig. 53-1 apply to methane and fresh water mixtures. Conditions for methane hydrate formation change only slightly if the water is of sea water salinity, but the presence of just a few percent of gases other than methane changes the pressure and temperature requirements drastically. Most of the normal secondary constituents of the gas (such as ethane or propane, carbon dioxide or hydrogen sulfide but not nitrogen) make hydrate formation easier; however, for this discussion of the geological aspects of hydrates, the boundaries on Fig. 53-1 are applied unless otherwise noted in the text. In the laboratory, hydrate formation is accelerated if the system is agitated to mix the two components (Evrenos et al., 1971). The resulting hydrate is a silver-gray "slush," much like that observed in gas pipelines (Hammerschmidt, 1934). However, if the reaction is allowed to proceed slowly without agitation the hydrate is solid and clear, much like ice (Evrenos et al., 1971). The latter would probably characterize the process in nature.

At temperatures below the freezing point of water, but at pressures above the diagonal in Fig. 53-1, the water and gas molecules will also combine to form the hydrate with the excess occurring as ice or gas. There is some question of just how much ice might be converted to the hydrate if pure ice were

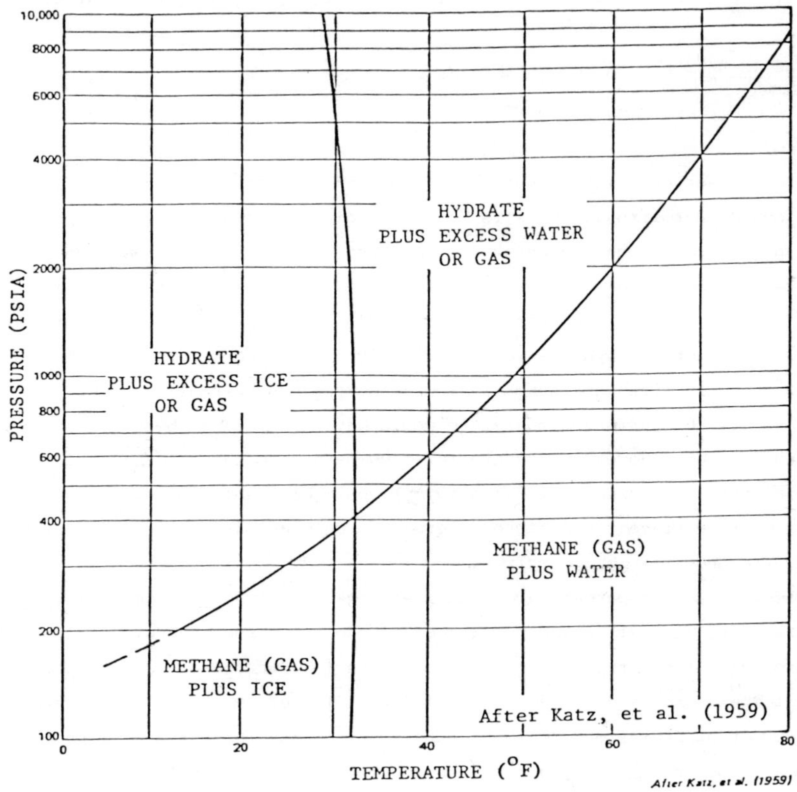

Figure 53-1. Hydrate formation in the methane-fresh water system

placed in the presence of gas under sufficient pressure to stabilize the hydrate. The reaction would proceed quickly at the surface of the ice to form a surficial layer of hydrate, but further conversion would proceed slowly because of the slow rate of diffusion of the gas through the hydrate lattice.

HYDRATES IN NATURE

On Land

Significant deposits of naturally occurring hydrates on land are limited to regions of permafrost. This is because ordinary geothermal gradients are too large for low-enough temperatures to exist at depths where high enough pressures are encountered. For example, with a methane-sea water system, the lowest pressure of hydrate stability, i.e., 430 psi, is normally encountered at 1,000 ft. Methane hydrate would form only if the temperature at this depth were 32°F. Even with an abnormally low subsurface geothermal gradient of 1.0°F/100 ft, the surface temperature would have to average about 20°F to meet this requirement. However, hydrates of a 90 percent methane-10 percent ethane mixture (approximately 0.6 gravity) could be stable in sediments in regions where surface temperatures average as much as 35°F. But, most of these regions are underlain by permafrost.

The presence of permafrost greatly enlarges the area of possible hydrate occurrence (Katz, 1971). The lower edge of the permafrost coincides with the 32°F isotherm, i.e., the melting point of water. (Because of salt effects, the actual temperature is a degree or two lower.) Therefore, where there is at least 800 ft of permafrost, the pressure requirements for methane hydrate formation are just reached along the 32°F isotherm. Where permafrost is thicker, a hydrate-favoring zone exists both beneath and within the permafrost. With 2,000 ft of permafrost and a geothermal gradient of 1.5°F/100 ft in the sediments immediately beneath it, the methane hydrate-favoring zone would be about 1,700 ft thick beneath the permafrost. For a 90 percent methane, 10 percent ethane mixture, the zone would be 2,550 ft thick.

Permafrost covers a significant portion of the earth's surface. Permafrost in the northern hemisphere generally covers most land area north of 40° north latitude. In the Soviet Union the area extends as far south as 50° north latitude. Makogen (1965b) estimates that permafrost underlies approximately 10.6 million km^2, or about half of the total land area of that country. In

Alaska, the permafrost region covers well over half the entire state (Black, 1954); and in Canada, it covers about half the land area (National Research Council of Canada, 1957). However, in the Soviet Union only about half of the permafrost area is underlain by sedimentary basins; in Alaska it is somewhat less than half, and in Canada only about a fourth.

Because the center of the permafrost itself is generally at temperatures considerably lower than 32°F, the resulting lower pressure requirements for hydrate formation make it possible that gas hydrates could be mixed or interspersed with the permafrost in many areas. Even where the pressure and temperature combinations favor it, gas must have been formed and have been concentrated in sufficient quantity for hydrates to form, i.e., in sufficient amounts to saturate the water.

In Ocean Basins

Another place where the conditions are favorable for hydrate formation is under the oceans. The combination of low bottom temperatures and high hydrostatic pressures suggests not only that the sediments immediately beneath the sediment-water interface could contain hydrates if gas were present, but also that any gas bubbles that seeped into the water might also be hydrated.

Different ocean-temperature gradients control the depths at which hydrate-favoring zones may occur and their thickness. In areas where cool water approaches the surface (i.e., a thin thermocline), e.g., with 40°F water temperature at 500 ft below sea level, the methane hydrate-favoring zone would exist only where the sediments were under 1,250 to 1,300 ft of water. With 40°F water at about 3,000 ft water depth, the sediments would have to be under 2,100 ft or more of water. For 0.6 gravity gas, the two depths would be about 550 and 1,400 ft, respectively. In the open ocean, the thermocline is generally at about 3,000 ft; but along continental margins, upwelling currents may bring the cool water up the slopes to much shallower depths (Arx, 1962).

There is only circumstantial evidence to date that gas hydrates do exist under the ocean, but it is strong evidence. First, of course, the pressure and temperature conditions are met. Second, shallow cores taken in the deep (>1,000 ft) ocean by the Deep Sea Drilling Project (DSDP) generally have large quantities of gas (McIver, 1974).

The gas is there, the conditions are right, and the hydrate is probably there, too. But the investigation that provided the first strong evidence that gas hydrates occur naturally under the deep ocean centered around the JOIDES Deep Sea Drilling Project core holes 102, 103, and 104 on Leg 11 (Lancelot and Ewing, 1972). These holes were drilled in 11,000 to 13,000 ft of water, into the Blake-Bahama Outer Ridge to determine the cause of a bottom-paralleling seismic reflector that lies about 0.6 second (750 ft) below the sea floor.

The drilling results turned up clues that suggested gas hydrates as the possible cause of the reflector. Almost every core brought up from the three coring sites on the outer ridge contained large amounts of gas. The gas was primarily methane with traces of ethane. Unfortunately, there was no effort to estimate the quantity of gas released from the core after it was recovered aboard ship, much less any way to estimate the quantity of gas in place. However, there was enough gas in these cores to cause them to expand--so much so that as much as half of the sediment was extruded from the core liners. Sometimes the expansion continued for 2 hours after the cores were recovered.

Consideration of the sedimentology, the drilling breaks, and the geophysical results gave a calculated interval velocity of about 2 km/sec (6,600 ft/sec) for the layer above reflector Y. This is a surprisingly high velocity for deep-sea muds, but it was later confirmed with independent sonobuoy shooting (Bryan, 1974). From this information and the fact that the pressure and temperature are such that the hydrates should be stable, Stoll et al. (1971) deduced that the sediments above the reflector contained methane hydrate. Subsequently, Stoll was able to form the hydrate in a 20-30 mesh Ottawa sand pack at the temperature and pressure of the deep-sea muds; he observed that sonic velocity in that core increased from 1.7 km/sec (5,600 ft/sec) before hydration to 2.5 km/sec (8,200 ft/sec) after the hydrate was allowed to form for several hours.

After the deployment of a new piston corer, Deep Sea Drilling Project scientists apparently recovered some crystalline hydrate offshore Mexico (JOIDES, 1979) at Sites 490, 491, and 492 of Leg 66.

There are still many questions about possible hydrates under the ocean. To date most of the supporting evidence is circumstantial: the sediments are generally gassy, and the physical conditions are right. Moreover, there is the bottom-paralleling reflector, which has been noted several other places (Stoll, 1972). This reflector may mark the lower edge of the hydrated layer.

HYDRATE ACCUMULATIONS

Hydrates filling reservoir rocks at shallow depths can contain more gas in a unit volume than could be held as free gas in that same space, under the same temperature and pressure conditions. This could be a consideration in calculating reserves and determining the economics of developing shallow gas reservoirs, although, of course, the problems of producing hydrated gas have not been solved. The maximum volume-enrichment decreases rapidly with depth, from about 5.5 at 1,000 ft (under permafrost), to unity at about 5,300 ft. The value would be reduced significantly if larger than theoretical amounts of water were hydrated in the lattice, which may be the case (Pauling and Marsh, 1952; Trebin et al., 1966). The enrichment factor would also vary slightly with the thickness of the permafrost, with different geothermal gradients, and with changes in gas composition.

As mentioned earlier, the Russians (Anonymous 1970a, b; Chersky and Makogen, 1970) reported 535 TCF of hydrated gas in reservoir rocks under permafrost in the Soviet Union. The permafrost occupies about 3.8×10^6 mi^2 (Black, 1954) of which about one-half is underlain by sedimentary basins. Therefore, reservoired hydrate volumes are about 280 million ft^3 per mi^2. Applying this factor, we calculate that there are about 200 TCF of hydrated gas in reservoirs under permafrost in Alaska, the Northwest Territories, and Arctic Islands. While this number seems large, it is far below the gas-generating capacity of an ordinary sediment (McIver, 1967) and constitutes only about 100,000 ppm (by volume) of gas in the total hydrate-favoring zone. This is in the upper range of values of gases in cuttings of fine-grained rock. However, the lack of confirmation of any large accumulations of hydrates from drilling programs in North America to date suggests that the 200 TCF figure may be very optimistic. There is always the possibility that several large accumulations may be encountered. We must be alert for these.

If the Russian hydrate factor is applied to the area of the hydrate-favoring zone under the oceans, the volumes of expected reservoired hydrated gas appear to be enormous.

The quantity of hydrated gas under continental margins could be large because the process of hydrate formation under the ocean is different than the process under the permafrost. In the marine situation, the gas is formed very early by the action of microbes on the sedimentary organic matter. Since this gas already is in the hydrate-favoring environment it is probably hydrated almost immediately after formation. But as this zone is buried increasingly deeper, it eventually reaches a temperature where the hydrate is no longer stable, and the hydrate decomposes. Free to migrate and diffuse, much of the gas travels upward where it again encounters the hydrate-favoring zone.

This process repeated over and over again could scavenge much of the gas formed microbially at that particular site and immobilize it in hydrate form in the first few hundred or few thousand ft of sediment. Hence, the amount of gas in hydrate form could include, as a maximum, the total quantity of gas generated by the underlying sediment wedge when it was buried less deeply, plus some upward-migrating gas from thermal diagenesis of organic matter, or, as a minimum, most of the microbially generated gas within the present hydrate-favoring zone.

If we assume that the average organic carbon content of young, unconsolidated sediment is 0.5 percent, and that only 1 percent of that organic carbon is converted to methane by microbial action, the yield would be 2.2×10^{10} ft^3 or 22 BCF of gas per cubic mile of sediment. This is probably a very conservative estimate. From the elemental composition of kerogens in suites of three marine shale units of western Canada, it was estimated (McIver, 1967) that thermal diagenesis of these rocks yielded from 30 to 70 times this much gas per unit volume, i.e., of the order of 1 TCF per mi^3.

If the factor 22 BCF/mi^3 is applied to the sediment volume of the world, we estimate that the hydrate-favoring sediments of the world would contain 110,000 TCF methane, and those sediments off the continental North America would contain about 19,000 TCF. If only 1 percent of this were accumulated in reservoir-type rocks, the figures for methane in porous rocks would be 1,100 TCF and 190 TCF for the world and North America, respectively. If the hydrate scavenged gas migrating from deeper sediments, the quantities could be even larger.

PROPERTIES OF HYDRATES

Most of the existing information on the properties of hydrocarbon has been obtained in attempts to define and alleviate problems associated with hydrate formation in gas-transmission lines (Katz et al., 1959). This information focuses on pressure-temperature-composition equilibrium phase relations, water-hydrocarbon solubilities, gas dew points, and the effects of hydrate inhibitors (e.g., methanol) on the equilibrium phase relations. Russian experience with loss of gas productivity by hydrate formation around the well bore (Kovotaev et al., 1969) and production of naturally occurring hydrates (Sheshukov et al., 1972) has also stimulated work on hydrates. However, the literature surveyed to date emphasizes hydrate densities and heats of fusion, with little information on the rate of formation of hydrates and no information on their decomposition rates porous media. Values for heats of fusion below 32°F are substantially lower than those above 32°F. If 5.75 moles of water are associated with each mole of methane in methane hydrate, the heat of fusion per pound of hydrate is about 160 Btu/lb, or 11 percent higher than the 144 Btu/lb value for water.

FUTURE PRODUCTION OF HYDRATES

Production of fully hydrated reservoirs does not appear to be an imminent possibility. Enough energy must be provided to the system to melt what is equivalent to ice filling the pores. Slugs of antifreeze or salt water will only succeed in stimulating production briefly. The hydrate simply cannot be decomposed to release the hydrocarbon gas without a massive supply of energy as in steam injection of heavy oil reservoirs.

It is more likely that the first production of gas from hydrates will be (or has been, in the Soviet Union) production of gas from beneath a hydrate accumulation with the resulting lowering of reservoir pressure that in turn causes enough hydrate along the gas/hydrate interface to decompose, thus maintaining reservoir pressure. Perhaps the process can be stimulated and accelerated by steam injection or by injection of hot water from deeper formations.

CONCLUSIONS

Recognition of naturally occurring hydrates and of widespread hydrate-favoring zones in sediments has generated considerable interest among earth scientists. Initial studies of hydrates are focused on the location and positive identification of hydrated zones, recognition and solution of engineering problems associated with them, and further elucidation of their properties, particularly the kinetics of hydrate decomposition.

Whatever the outcome of that work, it is important for geologists to recognize that hydrates may exist naturally over large regions of the globe. Petroleum scientists and engineers should be aware of the potentially large quantities and should be considering possible stimulation methods for recovering the gas from at least a portion of the large quantities that may exist. While the short-term outlook is that production of gases from hydrate accumulations will not be economic, a discovery of multitrillion ft^3 accumulations would justify large expenditures for methods to exploit them.

REFERENCES

Anonymous, 1970a, Russians claim to have found big deposits of "solid gas": Gas World, v. 171, no. 1, p. 86.

________, 1970b, USSR Oil and Gas International, v. 10, no. 2, p. 1a (Newsletter).

Arx, W. S. von, 1962, An introduction to physical oceanography: Reading, Mass., Addison-Wesley Publishing Co.

Black, R. R., 1954, Permafrost-A review: Geol. Soc. America Bull., v. 65, p. 839.

Bryan, G. M., 1974, In situ indications of gas hydrate, in Natural Gases in Marine Sediments, ed., I. R. Kaplan, Plenum Press, New York, p. 299-308.

Chersky, N. and Makogen, Y. F., 1970, Solid gas-world reserves are enormous: Oil and Gas Internat., v. 10, no. 8, p. 82.

Davy, H., 1811, Royal Soc. London Philos. Trans., v. 101, p. 155.

Deaton, W. M. and Frost, E. M., 1946, Gas hydrates and their relation to the operation of natural gas pipelines: U.S. Bur. Mines Mon. 8.

Donatz, W. E., 1959, Method and apparatus for producing purified water from aqueous saline solutions: U.S. Patent 2,904,511, assigned to Koppers Corp., Inc.

Evrenos, A. J., Heathman, J., and Ralston, J., 1971, Impermeation of porous media by forming hydrates in situ: Jour. Petroleum Technology, v. 23, p. 1959.

Hammerschmidt, E. G., 1934, Formation of gas hydrates in natural gas transmission lines: Indus. Eng. Chemistry, v. 26, p. 851.

JOIDES Journal, 1979, v. V, no. 3, p. 14-25.

Katz, D. L., 1971, Depths to which frozen gas fields (gas hydrates) may be expected: Jour. Petroleum Technology, v. 23, p. 419.

Katz, D. L., and others, 1959, Handbook of natural gas engineering: New York, McGraw Hill, p. 189-221.

Koppers Corp., Inc., 1964, The Koppers hydrate process for saline water conversion-experimental and engineering studies: U.S. Dept. Interior Office Saline Water, Research and Devel. Rept. no. 90.

Kovotaev, Y. P., Kozlov, A. L., and Smiynov, V. S., 1969, Investigation of characteristics of gas wells in the Far North in relation to possible hydrate formation in strata around the well: Gazov. Prom., v. 5, p. 12.

Lancelot, Y. and Ewing, J. I., 1972, Correlation of natural gas zonation and carbonate diagenesis in Tertiary sediments from the northwest Atlantic, in Hollister, C. D., et al., Initial reports of the Deep Sea Drilling Project: Washington, D.C., U.S. Govt. Printing Office, v. 9, p. 791-799.

Makogen, Y. F., 1965a, Hydrate formation in gas-bearing strata in permafrost regions: Gazov. Prom. Izd. Nedra, v. 5.

____________, 1965b, Particulars in calculating gas reserves for gas fields in permafrost regions: Nauch. Tech. SB. Ser. Gazov. Delo, v. 7, no. 6, p. 9.

McIver, R. D., 1967, Composition of kerogen-Clue to its role in the origin of petroleum: World Petroleum Cong., 7th, Mexico, Proc., v. 2, p. 25.

___________, 1974, Hydrocarbon gases in Deep Sea Drilling Project cores, in Natural Gases in Marine Sediments, ed., I. R. Kaplan, Plenum Press, New York, p. 63-70.

National Research Council of Canada, 1957, Permafrost-A digest of current information: Associate Committee on Soil and Soil Mechanics, Ottawa, Tech. Memo. no. 49.
Parent, J. D., 1948, The storage of natural gas as hydrate: Inst. Gas Technology, Illinois Inst. Technology, Bull. no. 1.
Pauling, L. and Marsh, R. E., 1952, The structure of chlorine hydrate: Natl. Acad. Sci. Proc., v. 38, p. 112.
Sheshukov, N. L., Beznosikov, A. F., Khramenkov, E. N., and Efremov, I. D., 1972, Gazov. Delo, v. 6, p. 8.
Stoll, R. D., Ewing, J., and Bryan, G. M., 1971, Anomalous wave velocities in sediments containing gas hydrates: Jour. Geophys. Research, v. 76, p. 2090.
Trebin, F. A., Khoroshilov, V. A., and Demchenko, A. V., 1966, Kinetics of hydrate formation of natural gas: Gazo. Prom., v. 11, no. 6, p. 10 (in Russian).
Williams, V. C., 1964, Hydrate forming saline water conversion process: U.S. Patent 3,155,610, assigned to Sweet Water Development Co.

CHAPTER 54

GAS HYDRATES: A POSSIBLE ENERGY SOURCE

V. M. Dobrynin, Yu. P. Korotajev, and D. V. Plyuschev[1]

More than a half of the present-day power requirements of mankind are met by oil and natural gas. But the already discovered hydrocarbon resources, as well as those of the promising fields, are not unlimited. According to the International Gas Union the start of the coming century will mark oil and gas peak production, with a subsequent decrease of production rates.

There are two possible ways to compensate for the natural depletion of the hydrocarbon resources: to discover new oil and gas fields in those areas of our planet which have not been so far explored by man and to develop gas hydrate deposits, the existence of which in nature was proved by Soviet scientists in the 1960's.

Gas hydrates are solid intrusion solutions. They consist of a crystal grating formed by water molecules. The space in the grating contains molecules of methane. The interaction between the two types of molecules may be classified as a relatively weak Van-der-vaal's interaction.

Natural gas hydrates, as compared to other condensed phases, possess a number of unique properties. For example, they may form at temperatures above $0^{\circ}C$. Saturated vapor pressure above hydrates is 1.5 to 2 times less than that above ice. The accumulating capacity of the crystal grating of hydrates is high, so that one volume of grating may contain as much as 200 volumes of gas.

Recent studies by Soviet scientists (Trofymuk) show that the hydrate formation zone covers a large part of North America, practically the whole of Greenland and the Antarctic continent, two-thirds of the USSR, and 90 percent of the world ocean area.

According to rough estimates, the recoverable resources of gas hydrates in terms of tons of coal equivalent comprise: 11,300 billion tons in the world ocean area and 53 billion tons on the continents.

For comparison let us note that the recoverable resources of coal and lignite comprise only 3,000 billion tons of coal equivalent, and those of oil,

---

[1]Moscow Gubkin Institute of Petrochemical and Gas Industry, USSR.

415 billion tons of coal equivalent, and of natural gas, 280 billion tons of coal equivalent.

Thus it is that in the world ocean, where the basic resources of gas hydrates are concentrated, 35 percent of them occur in the shelf zone. Gas production from the near-bottom deposits, cemented with hydrates, poses considerable difficulties. The existing designs for gas hydrate extraction from the sea bottom of the shelf involve both earth dragging and hydromonitoring. One m^3 of the rock with porosity of 0.2 percent may contain 30-36 m^3 of gas, if the pores of deposit are completely filled with hydrates. A suspension of the rock with hydrates is pumped through special pipelines into bell-type gathering tanks for gas extraction.

Attention must be paid to the possibility of utilization of the bottom deposits when, together with gas hydrate silt (a good fertilizer) and other minerals such as manganese concretions, they are lifted to the surface.

The idea put forward by U.S. specialists of hydrate transportation in underwater tanks should be given due consideration.

The Messojakha field is an example of a continental gas hydrate deposit; so far, it is the only one which is being successfully developed. The gas deposit in this field consists of two sand strata: an upper one, in which gas is partly in a hydrate state, and a lower one, where there is only free natural gas. To extract gas from the hydrates, methyl alcohol is pumped into the production well. There is a dual mechanism of hydrate destruction in a layer, due to the pressure drop below equilibrium pressure of the hydrate formation when free gas is extracted and to the action of an antihydrate inhibitor (methyl alcohol). Data analysis of the Messojakha field development shows that at the present technological level there exists a possibility of profitable gas hydrates extraction, provided the concentration of hydrates in the effective porous space of the productive layers does not exceed 10 percent, whereas 90 percent of the space is filled with free gas.

Gas extraction from reservoirs saturated with gas hydrates is also possible by heating of the reservoirs. The thermal method of gas extraction, in our opinion, will be the most profitable when high-temperature reservoir waters underlie gas hydrate deposits.

This information proves that the contribution of gas resources located in continental and world ocean hydrate formation zones to the fuel potential can be large. But as the calorific value of hydrates as a fuel is low (1,000-1,800 kcal/kg), the problem of commercial development of hydrate fields can be considered as actual only after other resources of mineral fuels are completely exhausted. Probably, the prospects of gas hydrate field development are closely associated with future needs of the chemical industry.

REFERENCE

Trofymuk, A. A., Tcherski, N. V., and Tsarev, V. P., 1979, Gas hydrates as a source of hydrocarbons: Priroda, 1979, no. 1, p. 18-27.